CELL AND MOLECULAR RESPONSES TO STRESS

Volume 2

Protein Adaptations and Signal Transduction

Cover illustration: Fig. 8.2 from Chapter 8 'Early responses to mechanical stress: From signals at the cell', by Matthias Chiquet (with permission from the author).

PROTEIN ADAPTATIONS AND SIGNAL TRANSDUCTION

Edited by

K.B. STOREY and J.M. STOREY

Institute of Biochemistry
Carleton University
Ottawa, Ontario
Canada

2001

ELSEVIER
Amsterdam – London – New York – Oxford – Paris – Shannon – Tokyo

ELSEVIER SCIENCE B.V.
Sara Burgerhartstraat 25
P.O. Box 211, 1000 AE Amsterdam, The Netherlands

First edition 2001

Library of Congress Cataloging in Publication Data
A catalog record from the Library of Congress has been applied for.

ISBN: 0-444-50759-0

∞ The paper used in this publication meets the requirements of ANSI/NISO Z39.48-1992 (Permanence of Paper).

Printed in The Netherlands.

Preface

Looking out our lab window on a cold March morning we see few signs of life. Thick snow still blankets the ground and flurries drift down from gray skies. A black squirrel bounds across the snow and up a tree and looking over to the river we can spot a few ducks paddling about in a small section of open water over the rapids. But we know with certainty that all this will change within the next few weeks. A month from now groundhogs will wake up from winter hibernation and lumber out of their burrows onto revitalized campus lawns. And we'll be out wading in forest ponds to catch wood frogs when they are massed together at breeding pools—although right now they are tucked up in sheltered sites on the forest floor and frozen solid! The re-awakening of life in the spring represents a tremendous success by living organisms in overcoming environmental stress. To survive the winter, organisms must not only endure very cold temperatures but are variously challenged by a lack of food, tissue freezing, desiccating effects of cold dry air, and oxygen limitation for those species that are locked under ice-covered waters.

This volume of *Cell and Molecular Responses to Stress* has two broad themes: an examination of selected protein adaptations that support stress tolerance and an analysis of signal transduction systems, those critical links between the perception of stress and the activation of the coordinated metabolic responses that ensure survival. Several chapters deal with adaptive responses to environmental cold temperature and highlight novel advances in mammalian hibernation, low temperature enzyme function, cold-shock and antifreeze proteins, and freezing survival. Other chapters stretch out to explore biochemical responses to diverse stresses including water stress, mechanical stress, nutrient availability, oxygen limitation and oxidative stress. The integral roles of protein kinases, transcription factors, oxygen free radicals, and oxygen-sensitive ion channels in the detection and mediation of stress responses are explored. The multiplicity of stress responses is emphasized and shows us the vast potential of cells and organisms to respond to innumerable stresses, great and small, and the regulatory principles and mechanisms that are used to allow life to adapt and endure in every environment on Earth.

We would like to extend our thanks to all of the authors who contributed chapters to this volume. Their excellent writing skills and intriguing stories make every chapter a pleasure to read.

Kenneth B. Storey
Janet M. Storey
Ottawa, Ontario, Canada

List of Contributors

T. Abee
Laboratory of Food Microbiology, Wageningen University, Bomenweg 2, 6703 HD Wageningen, The Netherlands

Salvino D'Amico
Laboratory of Biochemistry, Institute of Chemistry B6, University of Liège, B-4000 Liège, Belgium

Bradford C. Berk
Center for Cardiovascular Research, University of Rochester Medical Center, 601 Elmwood Avenue, Box 679, Rochester, New York 14642, USA
E-mail: bradford_berk@urmc.rochester.edu; Tel: 716-275-0810; Fax 716-273-1497

David Bloom
Dept. Pharmacology, Baylor College of Medicine, One Baylor Plaza, Houston, TX 77030, USA

Stephen P.J. Brooks
Nutrition Research Division, Health Products and Food Branch, Health Canada, 2203C Banting Research Centre, 1 Ross Ave., Ottawa, Ontario, Canada K1A 0L2
E-mail: Steve_Brooks@hc-sc.gc.ca; Tel: 613.941.0451; Fax: 613.941.6182

Claudia M. Celli
Dept. Pharmacology, Baylor College of Medicine, One Baylor Plaza, Houston, TX 77030, USA

Sajal Chakraborti
Department of Biochemistry and Biophysics, University of Kalyani, Kalyani 741235, West Bengal, India
E-mail: s_chakraborti@hotmail.com; Tel: 0091-33-5828220/5828750/5828477 (O); Fax: 0091-33-5828282

Tapati Chakraborti
Department of Neurosciences, Brain Institute, University of Florida, Gainesville, Florida 32610, USA

Matthias Chiquet
M.E. Müller-Institute for Biomechanics, University of Bern, Murtenstrasse 35, P.O.Box 30, CH-3010 Bern, Switzerland
E-mail: chiquet@mem.unibe.ch; Tel: 41-31-632 8684; Fax: 41-31-632 4999

Paule Claverie
Laboratory of Biochemistry, Institute of Chemistry B6, University of Liège, B-4000 Liège, Belgium

Tony Collins
Laboratory of Biochemistry, Institute of Chemistry B6, University of Liège, B-4000 Liège, Belgium

Sudip Das
Department of Biochemistry and Biophysics, University of Kalyani, Kalyani 741235, West Bengal, India

Tanya Das
Bose Institute, Animal Physiology Section, Calcutta-700054, India

Peter L. Davies
Department of Biology, Queen's University, Kingston, Ontario, Canada K7L 3N6

Saravanakumar Dhakshinamoorthy
Dept. Pharmacology, Baylor College of Medicine, One Baylor Plaza, Houston, TX 77030, USA

Bernard P. Duncker
Department of Biology, Queen's University, Kingston, Ontario, Canada K7L 3N6. *Current address*: Department of Biology, University of Waterloo, Waterloo, Ontario, Canada N2L 3G1

Pingke Fang
Renal Unit and Program in Membrane Biology, Massachusetts General Hospital and Harvard Medical School, Boston, MA 02114, USA

Georges Feller
Laboratory of Biochemistry, Institute of Chemistry B6, University of Liège, B-4000 Liège, Belgium

Martin Flück
M.E. Müller-Institute for Biomechanics, University of Bern, Murtenstrasse 35, P.O.Box 30, CH-3010 Bern, Switzerland

Daphné Georlette
Laboratory of Biochemistry, Institute of Chemistry B6, University of Liège, B-4000 Liège, Belgium

C. Gerday
Laboratory of Biochemistry, Institute of Chemistry B6, University of Liège, B-4000 Liège, Belgium
E-mail: ch.gerday@ulg.ac.be; Tel: +32 4 366 33 40; Fax: +32 4 366 33 64

Geoff Goldspink
Anatomy and Developmental Biology, Royal Free and UCL Medical School, University of London, Rowland Hill St. London NW3 2PF, UK
E-mail: g,goldspink@rfc.ucl.ac.uk; Tel: +44 (0)20 7830 2410; Fax: +44 (0)20 7830 2917

Emmanuelle Gratia
Laboratory of Biochemistry, Institute of Chemistry B6, University of Liège, B-4000 Liège, Belgium

Hasem Habelhah
Ruttenberg Cancer Center, Mount Sinai School of Medicine, 1425 Madison Ave, New York, NY 10029, USA

D. Grahame Hardie
Wellcome Trust Biocentre, School of Life Sciences, Dundee University, Dundee, DD1 5EH, Scotland, UK
E-mail: d.g.hardie@dundee.ac.uk; Tel: +44 (1382) 344253; Fax: +44 (1382) 345783

Marcelo Hermes-Lima
Oxyradical Research Group, Departamento de Biologia Celular, Universidade de Brasilia, Brasilia, DF 70910-900 Brazil
E-mail: hermes@unb.br; Tel: (+55) 61-307-2192

Klaus P. Hoeflich
Department of Medical Biophysics, Ontario Cancer Institute, University of Toronto, 610 University Avenue, Toronto, Ontario M5G 2M9, Canada

Alfred N. Van Hoek
Renal Unit and Program in Membrane Biology, Massachusetts General Hospital and Harvard Medical School, Boston, MA 02114, USA
E-mail: vanhoek@receptor.mgh.harvard.edu; Tel: (617) 724 8493; Fax:(617) 726 5669

Anne Hoyoux
Laboratory of Biochemistry, Institute of Chemistry B6, University of Liège, B-4000 Liège, Belgium

Yan Huang
Renal Unit and Program in Membrane Biology, Massachusetts General Hospital and Harvard Medical School, Boston, MA 02114, USA

Anil K. Jaiswal
Dept. Pharmacology, Baylor College of Medicine, One Baylor Plaza, Houston, TX 77030, USA
E-mail: ajaiswal@bcm.tmc.edu; Tel: 713 798-7691; Fax 713 798-3145

Zheng-Gen Jin
Center for Cardiovascular Research, University of Rochester Medical Center, 601 Elmwood Avenue, Box 679, Rochester, New York 14642, USA

Oscar P. Kuipers
Laboratory of Food Microbiology, Wageningen University, Bomenweg 2, 6703 HD Wageningen, The Netherlands

Amritlal Mandal
Department of Biochemistry and Biophysics, University of Kalyani, Kalyani 741235, West Bengal, India

Malay Mandal
Department of Biochemistry and Biophysics, University of Kalyani, Kalyani 741235, West Bengal, India

Marie-Alice Meuwis
Laboratory of Biochemistry, Institute of Chemistry B6, University of Liège, B-4000 Liège, Belgium

Alexandra C. Newton
Department of Pharmacology, University of California at San Diego, La Jolla, CA 92093-0640, USA
E-mail: anewton@ucsd.edu; Tel: (858) 534-4527; Fax (858) 534-6020

Chris Peers
Institute for Cardiovascular Research, University of Leeds, Leeds LS2 9JT, UK
E-mail: c.s.peers@leeds.ac.uk; Tel: (+113) 233 474; Fax: (+113) 233 4803

Derrick E. Rancourt
Department of Biology, Queen's University, Kingston, Ontario, Canada K7L 3N6. *Current address*: Department of Medical Biochemistry, University of Calgary, Calgary, Alberta, Canada T2N 4N1

Prasanta K. Ray
Department of Surgery, Beth Israel Hospital, A.J. Antenucci Medical Research Building, Albert Einstein College of Medicine, Room 301, 432 W. 58th Street, New York, NY 10019, USA
E-mail: prasantaray@netscape.net; Tel: (718) 430-3518; Fax: (718) 430-3099

Frank M. Rombouts
Laboratory of Food Microbiology, Wageningen University, Bomenweg 2, 6703 HD Wageningen, The Netherlands

Ze'ev Ronai
Ruttenberg Cancer Center, Mount Sinai School of Medicine, 1425 Madison Ave, New York, NY 10029, USA
E-mail: ronaiz01@doc.mssm.edu; Tel: 212 659 5571; Fax: 212 849 2425

Alexander M. Rubtsov
Department of Biochemistry, School of Biology, Lomonosov Moscow State University, 119899 Moscow, Russia
E-mail: am_rubtsov@mail.ru; Tel: +7 (095) 939-4434; Fax: +7 (095) 939-3955

Gaurisankar Sa
Bose Institute, Animal Physiology Section, Calcutta-700054, India

Kenneth B. Storey
Institute of Biochemistry, College of Natural Sciences, Carleton University, 1125 Colonel By Drive, Ottawa, Ontario, Canada K1S 5B6
E-mail: kenneth_storey@carleton.ca; Tel: (613) 520-3678; Fax: (613) 520-2569

Janet M. Storey
Institute of Biochemistry, College of Natural Sciences, Carleton University, 1125 Colonel By Drive, Ottawa, Ontario, Canada K1S 5B6
E-mail: jstorey@ccs.carleton.ca; Tel: (613) 520-3678; Fax: (613) 520-2569

Alex Toker
Department of Pharmacology, University of California at San Diego, La Jolla, CA 92093-0640, USA

Howard C. Towle
Department of Biochemistry, Molecular Biology and Biophysics, University of Minnesota, 6–155 Jackson Hall, 321 Church St. SE, Minneapolis, MN 55455, USA
E-mail: towle@mail.ahc.umn.edu; Tel: (612) 625-3662; Fax: (612) 625-5476

Michael G. Tyshenko
Department of Biology, Queen's University, Kingston, Ontario, Canada K7L 3N6

Willem M. de Vos
Wageningen Centre for Food Sciences (WCFS), Diedenweg 20, 6703 GW Wageningen, The Netherlands

Virginia K. Walker
Department of Biology, Queen's University, Kingston, Ontario, Canada K7L 3N6
E-mail: walkervk@biology.queensu.ca; Tel: 613-533-6123; Fax: 613-533-6617

Wei Wang
Dept. Pharmacology, Baylor College of Medicine, One Baylor Plaza, Houston, TX 77030, USA

James R. Woodgett
Department of Medical Biophysics, Ontario Cancer Institute, University of Toronto, 610 University Avenue, Toronto, Ontario M5G 2M9, Canada
E-mail: jwoodget@uhnres.utoronto.ca; Tel: (416) 946-2962; Fax: (416) 946-2984

Jeroen A. Wouters
Laboratory of Food Microbiology, Wageningen University, Bomenweg 2, 6703 HD Wageningen, The Netherlands
E-mail: Jeroen.Wouters@micro.fdsci.wau.nl; Tel: +31-317-484981; Fax: +31-317-484893

Shi Yu Yang
Anatomy and Developmental Biology, Royal Free and UCL Medical School, University of London, Rowland Hill St. London NW3 2PF, UK

Laurent Zecchinon
Laboratory of Biochemistry, Institute of Chemistry B6, University of Liège, B-4000 Liège, Belgium

Contents

Protein Adaptations and Signal Transduction. Edited by K.B. Storey and J.M. Storey

CHAPTER 1

Signal Transduction and Gene Expression in the Regulation of Natural Freezing Survival

Kenneth B. Storey and Janet M. Storey
Institute of Biochemistry, College of Natural Sciences, Carleton University, Ottawa, Ontario, Canada K1S 5B6

1. Strategies of winter survival in animals

Winter poses severe challenges to the survival of ectothermic organisms for exposure to subzero temperatures is frequently lethal. Most living things are killed when their tissues freeze and some are even chill intolerant and killed by simple exposure to temperatures below 0°C. For many organisms, the optimal way to survive the winter is to find ways to avoid having to deal with subzero temperatures. Behavioral options can be used including migration to a warmer climate or a retreat into thermally buffered sites such as under water or deep underground. Life cycles may also be modified so that many organisms overwinter in a relatively undifferentiated embryonic form (e.g. egg, cyst, spore, seed) that can often be engineered to contain little or no freezable water. However, many other cold-blooded species still spend the winter in terrestrial habitats that offer little or no protection from ambient temperatures. There they may need to endure exposures to temperatures that are far below the freezing point of their body fluids. In general, two strategies of cold hardiness have developed: freeze avoidance and freeze tolerance (Storey and Storey, 1988, 1989; Duman et al. 1991; Zachariassen, 1985).

Freeze avoiding animals use adaptations that maintain their body fluids in a liquid state down to temperatures that are well below the expected environmental minima for their habitat. They do so by exploiting the phenomenon of supercooling—the ability of watery solutions to chill well below their equilibrium freezing point (FP, the temperature at which an introduced ice crystal begins to grow). For example, many invertebrates living in the leaf litter and upper soil layers can stay liquid down to –10 or –15°C, temperatures that are substantially below the minima that are typically experienced in the insulated environment under the snow. Others that winter in sites that are exposed above the snowline (e.g. under the bark of trees) or in extremely cold polar or alpine environments have even more impressive abilities and can often stay liquid down to –40°C or lower. Freeze avoiding organisms use a variety of adaptations to lower both the FP and the crystallization temperature (Tc, temperature at which spontaneous freezing occurs) as well as to widen the gap between these two parameters (Zachariassen, 1985; Storey and Storey, 1989; Duman et al. 1991; Lee and Denlinger, 1991). Firstly, animals minimize their contact with potential nucleators—any molecules, particles or surfaces that could seed ice formation at temperatures at or below the FP. Gut contents are emptied to expel foreign bacteria and food particles, selected blood proteins may be deleted during the winter months, and animals may wrap themselves in waterproofing (e.g. cocoons) to avoid contact with the most potent nucleator of all, ice crystals growing in their external microhabitat. Secondly, specific antifreeze proteins are synthesized and loaded into blood or hemolymph (see Duncker et al., 2001 this volume). These adhere to any microscopic ice crystals that may form and prevent them from growing to a size that could do physical damage. By doing so, they effectively lower the FP without changing melting point (MP), resulting in a thermal hysteresis between FP and MP that is the key to the detection and analysis of these proteins by researchers. Thirdly, many organisms also accumulate extremely high levels of sugar alcohols in their body fluids. In some insects, for example, glycerol levels can rise to over 2 M and may represent ~20% of the total body mass over the winter months (Storey and Storey, 1989, 1991). Like the ethylene glycol used in the radiator of a car, polyols strongly depress both the FP and the Tc of body fluids by colligative means. Overall, these strategies form an effective way to prevent body fluids from freezing. The freeze avoidance strategy has one major downfall, however, and that is that if environmental temperature drops below the Tc or if supercooled body fluids come in contact with a powerful nucleator, freezing will occur instantaneously and freezing is lethal for these animals. Hence, the freeze avoidance strategy is a bit of a gamble but one which serves many species well, ensuring the winter survival of the overall population, although not always of all individuals.

Freeze tolerant animals have an even more amazing strategy of winter survival. They can survive for days or weeks with as much as 65% of total body water converted to ice (Storey and Storey, 1988, 1992, 1996a). For example, in wood frogs (*Rana sylvatica*) ice propagates through the lumen of blood vessels, fills the ventricles of the brain, freezes the lens and bladder water, grows in sheets between skin and skeletal muscle, and fills the abdominal cavity. Only intracellular water remains unfrozen, its liquid state defended by high concentrations of sugars or sugar alcohols. Freeze tolerance is less common than freeze avoidance, probably because it is a more complex strategy to implement but, nonetheless, it has developed in diverse groups of organisms. Hundreds of insect species and other terrestrial invertebrates are freeze tolerant as are a variety of marine invertebrates that inhabit the intertidal zone on northern seashores (Storey and Storey, 1988). Various terrestrially hibernating amphibians and reptiles also endure freezing during hibernation (Storey and Storey, 1992, 1996a). In some cases, freeze tolerance appears to be the only suitable winter hardiness strategy for species that can neither escape a subzero thermal environment nor protect themselves from contact with environmental ice. For example, terrestrially hibernating frogs, such as wood frogs, have a highly water permeable skin and must hibernate in the moist microhabit of the forest leaf litter to prevent their bodies from desiccating. When the leaf litter freezes, however, so must the frogs for it is impossible for them to resist nucleation when they come into contact with environmental ice. Various species of intertidal molluscs and barnacles living at high latitudes also tolerate freezing which can occur at every low tide when the animals are exposed to winter air temperatures. Even when their shells are closed, their tissues are still bathed in seawater which freezes at about –2°C so seeding cannot be avoided. Frozen animals show no vital signs (no movement, no breathing, no heart beat) yet within minutes after thawing, all these processes resume. Studies of the adaptations that support freezing survival have been a focus of our research for many years including extensive work with freeze tolerant frogs and insects as well as exploration of freezing survival by turtles, snakes and lizards and by marine bivalve and gastropod molluscs (Storey and Storey, 1988, 1992, 1996a, 1999).

At least three serious problems threaten freezing survival: (1) ice crystals can cause physical damage, especially because water expands on freezing so that ice can rupture delicate tissues such as capillaries, (2) freezing halts delivery of oxygen and nutrients to organs via the blood, and (3) the conversion of up to two-thirds of body water into ice has major osmotic effects on cells including extreme cell volume reduction and a large increase in cellular osmolality and ionic strength. Specific adaptations of freeze tolerant animals help to deal with each of these problems (Storey and Storey, 1988, 1996a, 1999; Lee and Costanzo, 1998; Lee et al., 1998). These include: (1) methods to trigger nucleation just below 0°C (via ice nucleating proteins or heterologous nucleators) so that ice growth can be slow and controlled as it propagates through body fluid spaces, and to minimize recrystallization, the tendency of small crystals to regroup into larger crystals over time, (2) good ischemia resistance to aid tissue viability while frozen including ATP production via fermentative reactions, metabolic rate depression, and antioxidant defenses to deal with oxyradical stress when oxygen is reintroduced upon thawing (see Hermes-Lima et al., 2001 in this volume), (3) accumulation of high levels of sugars or polyhydric alcohols that act as cryoprotectants to minimize intracellular volume reduction during extracellular freezing, and (4) synthesis of other low molecular weight cryoprotectants (e.g. trehalose, proline) that stabilize membrane bilayer structure against the compression stress of cell shrinkage.

The current review focuses on new advances in understanding the cell and molecular responses to freezing stress in animals. Particular emphasis will be placed on new studies of the role of gene expression in supporting freeze tolerance and the mechanisms of signal transduction that mediate freeze-induced responses.

2. Freeze-induced gene expression

Changes in gene expression underlie the seasonal acquisition of cold or freeze tolerance in both animal and plant systems (Storey and Storey 1999; Thomashow 1998; Warren 2000). Evidence for this has been available for many years because the levels of selected proteins with functions in cold hardiness typically rise during the autumn months. For example, the activities of glycogen phosphorylase (GP) and various other enzymes involved in polyol synthesis in insects increase in the early autumn prior to the induction of cryoprotectant synthesis by cold exposure (Joanisse and Storey, 1994a,b). Antifreeze proteins or ice nucleating proteins also appear in the blood or hemolymph of various species during the autumn (Duman et al., 1991; Davies et al., 1999). Other proteins disappear over the winter; for example, no antimicrobial peptides could be detected in skin of wood frogs upon emergence from winter hibernation but a peptide of the brevinin-1 family was induced rapidly when the animals were warmed to higher temperatures and began to eat again (Mattute et al., 2000). The above cited examples represent changes of a preparative nature that are designed to alter the metabolic make-up of the organism prior to the arrival of cold weather. Often the trigger for these preparations is decreasing daylength; for example, induction of antifreeze protein synthesis in insects is triggered by a critical photoperiod (Duman et al., 1991). In some species, however, the induction of cold hardiness adaptations is obligately linked with a particular developmental stage; such is often the case in univoltine insects. In addition, whereas preparative measures (e.g. enzyme levels, glycogen accumulation) occur prior to cold exposure, the actual synthesis of carbohydrate cryoprotectants is typically triggered either by a critical temperature (5°C exposure triggers glycerol synthesis in many insects) or by freezing itself (glucose synthesis by wood frog liver is triggered within 2 min after the skin begins to freeze) (Storey and Storey, 1988).

Until recently, the identification of cold- or freeze-induced genes relied on identifying metabolic adaptations that support cold hardiness (e.g. cryoprotectant synthesis) and then working backwards to determine which proteins/enzymes were induced or up-regulated to support this function. For example, the use of glucose as a cryoprotectant by frogs suggested that increased numbers of plasma membrane glucose transporters would be needed during the winter months in freeze tolerant species and, indeed, this was found to be the case (King et al., 1995). It became increasingly obvious, however, that this approach is limited because of its dependence on a fore-knowledge of which adaptations are important for freezing survival. Techniques were needed that allow an unbiased evaluation of cold- or freeze-induced changes in gene expression. We began with an evaluation of freeze-induced protein/gene expression. Because freezing is an ischemic stress where energy is limited (cellular ATP levels fall to about 50% of normal; Storey and Storey, 1985; 1986), it seems reasonable to assume that the frozen state should be one where energy-expensive biosynthetic reactions, such as protein synthesis, are generally minimized. Hence, examples of freeze-stimulated gene expression and protein synthesis should represent protein products that have critical functions for freezing survival and the identification of these proteins should lead to critical advances in understanding the mechanisms of freeze tolerance.

Initial studies analyzed patterns of freeze- or thaw-induced protein synthesis in wood frog organs using ^{35}S-methionine labeling techniques. Intraperitoneal injection of ^{35}S-methionine was used to evaluate protein synthesis *in vivo* under two forms of water stress, thawing after 12 h frozen at –1.4°C and dehydration/rehydration (27 or 40% of total body water lost and rehydration after 40% dehydration) (Storey et al., 1997). Wood frogs can readily withstand the evaporative loss of 40–50% of total body water which mimics one aspect of freezing (the steep reduction in cell volume that occurs when up to 65% of total body water freezes as extracellular ice) and we have previously shown that dehydration of wood frogs at 5°C stimulates the same massive liver glycogenolysis and hyperglycemia as occurs during freezing (Churchill and Storey, 1993). Changes in protein patterns during freezing or thawing were also evaluated by isolating the mRNA transcripts present in the tissues of control (5°C acclimated), frozen (24 h at –2.5°C) and thawed (24 h at 5°C after 24 h frozen) frogs and subjecting these to translation *in vitro* in the presence of ^{35}S methionine/cysteine (White and Storey, 1999). For both experimental approaches, analysis of radiolabeled protein products using isoelectrofocusing and SDS-gel electrophoresis showed both freeze- and thaw-stimulated changes in the synthesis of selected proteins. Of special interest in both studies was the strong labeling of proteins of 15–20 kDa (Storey et al., 1997; White and Storey, 1999). For example, *in vitro* translation of mRNA isolated from liver of freeze-exposed frogs showed the presence of several new translation products (proteins of 45, 33.9, 21.5, 16.4, 15.8 and 14.8 kDa) as compared with controls (Fig. 1.1) (White and Storey, 1999). However, *in vitro* translation of mRNA from liver of thawed frogs showed no new protein peaks in comparison with either control or frozen profiles and the loss of several proteins of 16–22 kDa that were present in frozen

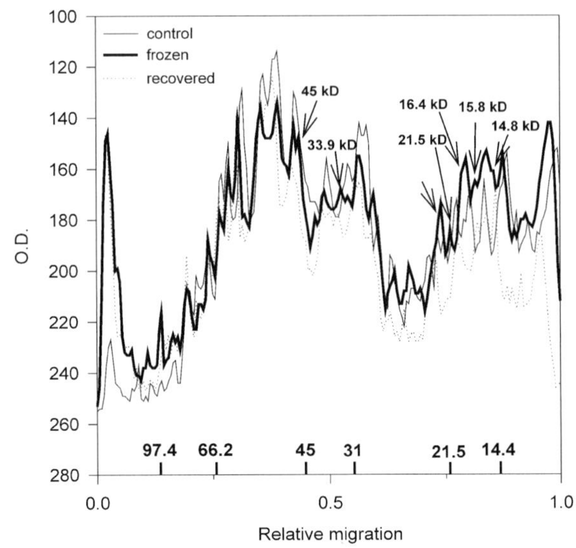

Fig. 1.1. Effect of freezing and thawing on the pattern of *in vitro* translation products produced from mRNA in wood frog liver. Total RNA was isolated from liver samples of control (5°C), frozen (24 h at –2.5°C) and thawed (24 h at 5°C after 24 h frozen) wood frogs and translated in a cell free system (wheat germ extract) followed by separation of ^{35}S-labeled proteins by SDS-PAGE. After autoradiography, densitometry scans showed the distribution of ^{35}S-labeled proteins. Peaks representing proteins that were new or enhanced in the frozen state are indicated by arrows along with their approximate molecular weights. Lines are: (thin), control; (thick), frozen; (dashed), thawed recovery. The positions of molecular weight standards are shown on the inner side of the x-axis. From White and Storey (1999).

and/or control frogs. However, neither of these methods was conducive to easy identification of the newly synthesized proteins and so we turned to techniques of cDNA library construction, differential screening, northern blotting, and DNA sequencing to isolate and identify genes, and their protein products, that are up-regulated during freezing.

2.1. Freeze-induced gene expression in wood frogs

In our first studies, a cDNA library was prepared from liver of wood frogs that were frozen for 24 h at –2.5°C. After differential screening with ^{32}P-labeled single-stranded total cDNA probes from liver of control (5°C-acclimated) vs frozen frogs, several unique freeze-responsive cDNA clones were found. DNA sequencing and Genbank searches identified two of these as the genes for the α and γ subunits of fibrinogen, a plasma protein involved in clotting that is synthesized by liver (Cai and Storey, 1997a). Both showed >70% identity of amino acid residues in the translated protein sequence with the corresponding sequences of the mammalian proteins. The gene for ADP/ATP translocase (AAT) was also freeze up-regulated (Cai et al. 1997); this protein of the inner mitochondrial membrane mediates the exchange transport of ADP and ATP. Another clone could not be identified from Genbank searches but its cDNA sequence of 457 bp had a single open reading frame that could encode a small protein of 90 amino acids with a molecular weight of ~10 kD (Cai and Storey 1997b). The deduced amino acid sequence of this novel protein, which we named FR10, showed an N-terminal region of 21 residues that contained ~80% hydrophobic residues and had a potential nuclear exporting signal (LALVVLVIAISGL). The predicted secondary structure contained long sections of α helix as well as coiling structures distributed in four narrow regions and β sheet structures in the N-terminus.

Changes in the levels of the mRNA transcripts of these four genes in wood frog liver were monitored by northern blotting over the course of a 24 h freezing exposure at –2.5°C followed by 24 h thawing at 5°C. As Fig. 1.2 shows, the genes for the two fibrinogen subunits were coordinately expressed with mRNA transcript levels of both rising by more than 3-fold after 8 h freezing and remaining at ~70% of this maximum after 24 h frozen (Cai and Storey, 1997a). When frogs were thawed, however, fibrinogen transcript levels fell and were again near control values within 24 h. The time-dependent expression of *Fr10* transcripts followed a similar pattern (Cai and Storey, 1997b) but AAT expression was different. AAT transcripts rose 4.5-fold after 8 h freezing but declined sharply with longer freezing and fell to less than control values after the 24 h thaw. AAT protein levels in liver were also monitored using immunoblotting and these followed an offset pattern with the maximum increase in protein content being ~2-fold after 24 h freezing (Cai et al., 1997).

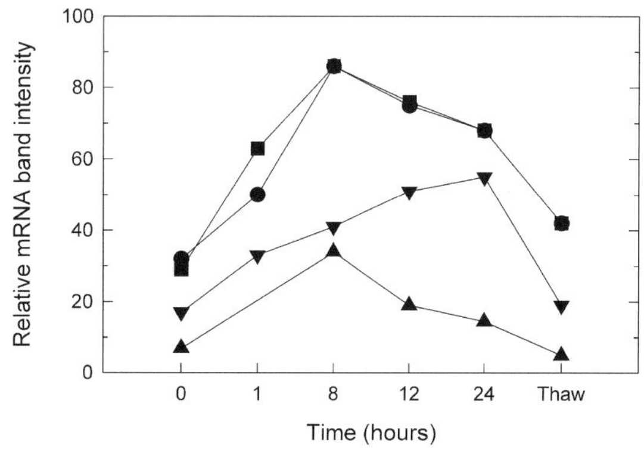

Fig. 1.2. Effect of freezing and thawing on mRNA transcript levels of four genes in wood frog liver as determined by relative band intensities on northern blots. Symbols are: (circles) fibrinogen α; (squares) fibrinogen γ; (triangles up), ATP-ADP translocase; (triangles down), FR10. Control frogs (0 h) were held at 5°C; freezing was at –2.5°C for up to 24 h, and thawed frogs were frozen for 24 h followed by 24 h back at 5°C. Compiled from Cai and Storey (1997a,b) and Cai et al. (1997).

Organ-specific patterns of gene up-regulation were also revealed. *Fr10* transcripts were found in all eight organs tested and strong up-regulation by freezing was seen in all organs except kidney and muscle (Fig. 1.3A). This suggests a near universal expression of FR10 protein in frog organs and hints at a role in freezing protection in all organs. For example, a possible role as a freeze-specific transcription factor might be proposed, accounting for the wide organ distribution of FR10, its small size and its nuclear exporting signal. This idea is currently being pursued. On the other hand, mRNA transcripts for fibrinogen α and γ subunits showed a much narrower organ distribution. Fibrinogen is viewed as liver-specific in mammals and not surprisingly, frog liver showed the highest transcript levels of all organs tested. However, low levels of fibrinogen transcripts were also found in lung, bladder and gut (Fig. 1.3 B,C) and, as in liver, transcript levels in these three organs rose significantly in 24 h frozen frogs. AAT transcripts showed another pattern. They were elevated in liver, lung and bladder during freezing, fell in kidney, heart and gut and did not change in brain and muscle; western blots revealed that AAT protein levels followed much the same pattern (Cai et al., 1997).

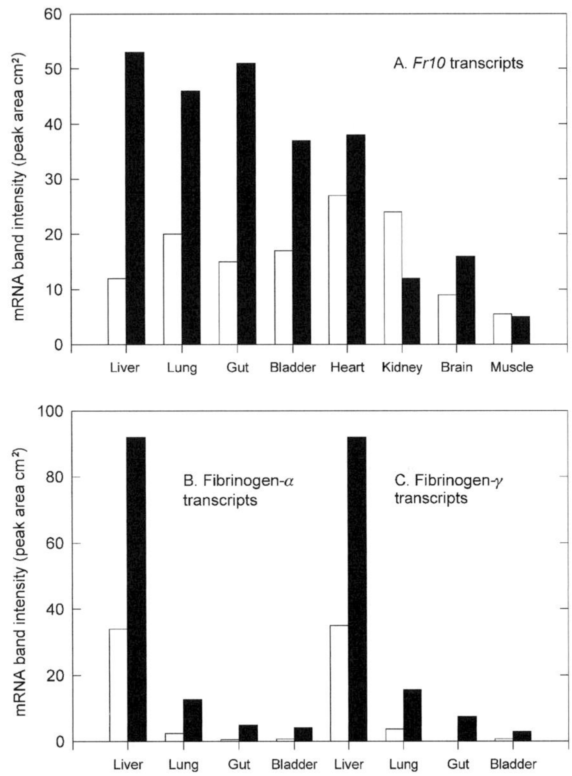

Fig. 1.3. *Fr10* (A) and fibrinogen α and γ (B) mRNA transcripts levels in organs of control (open bars) and 24 h frozen (solid bars) wood frogs. Controls were held at 5°C; frozen frogs were at –2.5°C. Total RNA was isolated from each tissue and transcript levels were analyzed via northern blots followed by autoradiography and densitometry; the 18S rRNA band was used for standardizing. Freezing had no effect on the negligible levels of fibrinogen transcripts in heart, brain, kidney or skeletal muscle. Data compiled from Cai and Storey (1997a) and Cai et al. (1997).

Freezing involves multiple stresses on cells and organs (ischemia and dehydration being two major ones) and freeze-stimulated metabolic adaptations may, in fact, be targeted to address only one of the consequences of freezing. One way to help determine the role of genes/proteins that are up-regulated during freezing is to look at their responsiveness to the various component stresses of freezing. We took this approach first when examining the control of cryoprotectant synthesis and found that the extreme hyperglycemia that is triggered by freezing is stimulated just as strongly when autumn frogs were dehydrated at 5°C (at a rate of ~1% body water lost per hour) (Churchill and Storey, 1993). Like freezing, dehydration also stimulated the rapid increase in cAMP and an activation of protein kinase A in liver that stimulated glycogenolysis whereas, by contrast, anoxia exposure (N_2 atmosphere at 5°C), had no effect on this signal transduction cascade (Holden and Storey 1997). This suggested that the activation of cryoprotectant biosynthesis responded to cell volume signals and supported the demonstrated role of glucose is minimizing cell volume reduction during freezing (Storey et al., 1992).

When the same comparison of dehydration versus anoxia effects was applied to characterize freeze-induced gene expression in wood frog liver, two types of responses were found. Fibrinogen and *Fr10* transcripts were up-regulated just as strongly by dehydration as by freezing but under anoxic conditions, their transcript levels were down-regulated and virtually undetectable after as little as 30 min (Cai and Storey, 1997a,b). However, AAT transcripts in liver responded oppositely with levels increasing strongly under anoxic conditions (1–24 h) but showing no response to dehydration or rehydration (Cai et al., 1997). Combining these data with the responses to freezing, we can suggest that both fibrinogen and FR10 may have roles in dealing with some aspect of water balance during freezing which could include functions in cell volume regulation or in the accommodation of extracellular ice. By contrast, AAT probably has a role in ischemia resistance.

A probable reason for fibrinogen up-regulation during freezing could be its known role in repairing tissue injury. Fibrinogen is an acute-phase plasma protein. It is synthesized mainly by liver and secreted into the plasma with production stimulated by stresses including infection, inflammation, and tissue injury (Huber et al., 1990). The protein has two halves, each made of three subunits (Aα, Bβ and γ) and as the final step in the coagulation cascade, thrombin cleaves near the N-termini of the Aα and Bβ chains to release the A and B fibrinopeptides and expose sites for polymerization into the fibrin mesh of a growing blood clot. Notably, although our first study retrieved clones for just the α and γ fibrinogen subunits (Cai and Storey,

1997a), in new work we have isolated clones encoding the β and γ subunits of fibrinogen when a liver cDNA library made from glucose-loaded frogs was screened (K.B. Storey, unpublished). Hence, coordinate expression of all three subunits appears likely as a response to both freezing and high glucose. A stimulation of fibrinogen biosynthesis when frogs freeze would ready the animal to deal with any internal bleeding injuries that occur upon thawing. Ice crystals can do serious physical damage to tissues, particularly as a result of ice expansion within the lumen of microcapillaries. Ice can easily rupture vessel walls so that, upon thawing, vascular integrity is destroyed. Indeed, vascular injury is one of the most widespread and devastating problems faced in cryomedicine by researchers that are trying to develop mammalian organ freezing technology (Rubinsky et al., 1987).

Freeze tolerant animals need to address this problem to minimize ice damage to their tissues. One way to do this is to limit ice growth within organ capillaries. This is done by substantially dehydrating organs during freezing and moving water out of organs to freeze in extra-organ ice masses. This is readily evident upon examination of frozen frogs. The animals have a huge mass of ice in the abdominal cavity that surrounds very shrunken organs; large sheets of ice are also sandwiched between skin and skeletal muscle layers. Quantitatively, up to 25% of organ water can be lost during freezing (Costanzo et al., 1992) and this significantly reduces the amount of extracellular ice that can form within organs. However, some physical damage by ice is still likely to occur and, indeed, hematomas are quite commonly seen in leg muscles after thawing. Thus, a freeze-induced elevation of plasma clotting capacity (involving fibrinogen synthesis and possibly other proteins of the clotting cascade) would enhance the frog's ability to deal with any internal bleeding during thawing.

The role of ADP/ATP translocase in wood frog liver during freezing is not yet clear. The enzyme catalyzes the transmembrane exchange of cytosolic ADP with mitochondrial ATP generated via oxidative phosphorylation. AAT probably has a role in dealing with freeze-induced ischemia, possibly in regulating the intramitochondrial adenine nucleotide pool size. However, what is interesting is the fact that enzymes of mitochondrial energy metabolism are appearing frequently as we search for stress up-regulated genes among freeze-tolerant and anoxia-tolerant animals.

2.2. Freeze-induced gene expression in hatchling turtles and mitochondrial gene expression

Other examples of freeze-induced gene expression come from our studies of anoxia tolerance and freeze tolerance in freshwater turtles. Adult turtles of the *Trachemys* and *Chrysemys* genera have the best developed anoxia tolerance among vertebrates and use this ability to hibernate for 3–4 months each winter on the bottom of ponds without breathing air and with only a minor ability to take up oxygen by extrapulmonary means (Ultsch, 1989). Anoxia tolerance is also put to use when diving at other seasons of the year and, intriguingly, by newly hatched juvenile turtles to aid in their winter freezing survival (Storey et al., 1992). Young turtles hatch in September but instead of exiting their terrestrial nests, they stay hidden underground for their first winter. This strategy lowers their risk of being eaten by aquatic predators but it means that they must have a way to survive when temperatures in their shallow nests drop below 0°C. For Canadian populations of the painted turtle, *C. picta*, this has meant the development of freeze tolerance (Storey et al., 1988; Churchill and Storey, 1992a), a capacity that is also rudimentary in the southern U.S. species, the red-eared slider *T. scripta elegans* (Churchill and Storey, 1992b).

Screening of a cDNA library made from heart of adult *T. scripta elegans* searched for genes that were induced or up-regulated when animals were given anoxia exposure (20 h submergence in N_2-bubbled water at 7°C) (Cai and Storey, 1996). Up-regulation was confirmed for two genes of the mitochondrial genome that encode subunits of electron transport chain proteins: subunit 5 of NADH-ubiquinone oxidoreductase and subunit 1 of cytochrome c oxidase. Transcripts of both increased by 3-4.5 fold in heart within 1 h of anoxic submergence, remained high over 20 h of anoxia and fell to control values again during aerobic recovery. Anoxic up-regulation of *Cox1* also occurred in brain, kidney and red skeletal muscle and *Nad5* transcripts were high in anoxic kidney and skeletal muscle. What was interesting is that ND5 and CO1, which are up-regulated by anoxia in adult turtles, were also up-regulated by freezing in juvenile *C. picta* (Cai and Storey, 1996). In heart, *Cox1* transcripts were 3.5-fold higher in frozen (24 h at –2°C) hatchling turtles than in 5°C-acclimated controls whereas *Nad5* transcripts were increased by about 70%. Freeze up-regulation of both transcripts also occurred in gut and kidney.

New studies are expanding the links between freeze tolerance and mitochondrially encoded genes even further. Further analysis of the wood frog liver cDNA library highlighted freeze up-regulation of subunit 4 of NADH-ubiquinone oxidoreductase whereas differential screening of a cDNA library made from wood frog brain showed freeze up-regulation of the mitochondrially encoded genes for ATPase subunits 6 and 8 of the F_0F_1ATPase complex (S. Wu and K.B. Storey, unpublished). This new data is providing us with many new avenues to explore to determine why and how the up-regulation of mitochondrially encoded genes contributes to anoxia and freezing survival.

3. Freeze tolerance, glucose metabolism and signal transduction

3.1 Unique glucose metabolism of freeze tolerant frogs

Freeze tolerant frogs have harnessed liver glycogen metabolism for a unique purpose, the synthesis of massive amounts of glucose for use as a cryoprotectant. Within 2–5 min after freezing begins on the skin surface of the frog, GP in liver has been activated, glycogenolysis is stimulated and glucose levels in liver and blood are rapidly rising. Blood and liver glucose rises from control values of ~5 mM to levels of ~40 mM within an hour and continues upwards to reach 150–300 mM in core organs fully frozen animals (Fig. 1.4) (Storey and Storey, 1986, 1988). Glucose export from liver is facilitated by seasonally high numbers of glucose transporters in liver plasma membranes that are about 8.5-fold higher in autumn- than in summer-collected frogs (King et al., 1995). Blood glucose is readily taken up by all other organs but a gradient is seen between core organs and brain where glucose is highest and the peripheral skeletal muscle and skin where glucose is lower (30–60 mM) (Fig. 1.4) (Storey and Storey, 1988). Differential organ glucose contents result because the freezing front propagates inward from the periphery and progressively cuts off blood flow (and hence glucose delivery) to tissues as ice moves inward. ^{1}H-NMR images show that heart and liver are the last organs to freeze (Rubinsky et al., 1994) and, hence, these have the highest final glucose concentrations. Interestingly, because the higher the glucose concentration, the lower the MP of cellular fluids, frogs thaw from the inside out when temperature is raised (Rubinsky et al., 1994), an effect that has the advantage of allowing the heart to recover and resume beating as soon as possible.

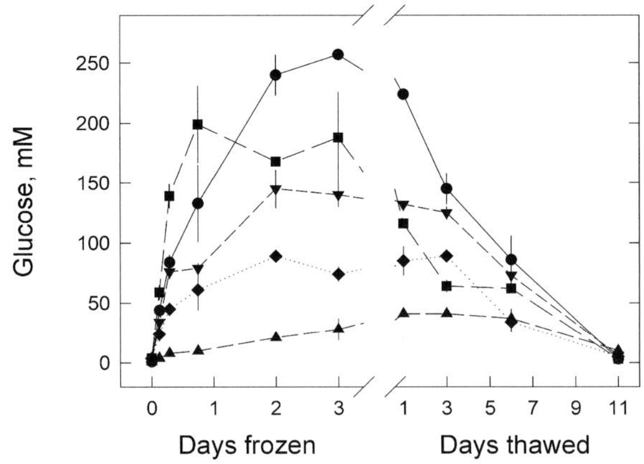

Fig. 1.4. Glucose levels in frog organs over a course of freezing at –2.5°C and thawing at 5°C. Data are means ± SEM, n=3. Symbols are: (circles), blood; (squares), liver; (triangles up), skeletal muscle; (triangles down), heart; (diamonds), kidney. Data compiled from Storey and Storey (1986).

Although glucose is readily taken up by all organs during freezing, its catabolism must be strictly limited. This is necessary in order to sustain the cryoprotectant pool and so, despite the fact that glucose is normally a very good anaerobic fuel for cells, its catabolism must be inhibited under the ischemic conditions of the frozen state. A study with wood frog erythrocytes confirms this. Figure 1.5A shows that D-[U-^{14}C] glucose was readily taken up by isolated wood frog erythrocytes at all incubation temperatures between 23 and 4°C (Brooks et al., 1999). Once taken up, ^{13}C-NMR revealed that glucose in erythrocytes was also readily catabolized at higher incubation temperatures with a linear rates of 0.91 ± 0.02 and 1.27 ± 0.02 mol/h/10^{16} cells at 12 and 17°C, respectively (Fig. 1.5B). However, when incubated at 4°C, glucose was not catabolized by wood frog red cells (Fig. 1.5B), suggesting that some form of metabolic inhibition occurs to block the catabolism of glucose in situations where a high pool size must be retained to provide cryoprotection.

The site of this blockage is likely the hexokinase reaction (glucose + ATP → glucose-6-P + ADP) the necessary phosphorylation step that converts glucose into a hexose phosphate that can enter glycolysis (or the reactions of glycogen synthesis). This must be so because glucose uptake by transporters is not blocked at low temperature nor is glycolysis itself since products of anaerobic glycolysis (lactate, alanine) accumulate over time in frozen tissues (Storey, 1987; Storey and Storey, 1986). Although it has never been proven experimentally, endogenous glycogen in organs is undoubtedly the fermentative substrate that supports anaerobic glycolysis during freezing whereas glucose use must be restricted to a cryoprotectant role by inhibiting hexokinase. Brooks et al. (1999) also noted a probable restriction on overall glycolytic flux at low temperatures. ^{13}C-NMR was used to monitor the incorporation of label from D-[2-^{13}]-glucose into glycolytic intermediates in wood frog erythrocytes incubated at 4°C. The results showed that label mixed into the hexose and hexose phosphate pools but ^{13}C-labeled fructose-1,6-bisphosphate was not found. This suggests a metabolic block at the phosphofructokinase (PFK) locus at low temperature in red blood cells, an inhibition that may help to block glucose catabolism but would also contribute to an overall metabolic rate suppression at low or freezing temperatures. The same metabolic block at the PFK locus is prominent in liver during freezing where its purpose is to inhibit glycolysis so that glycogenolysis is directed

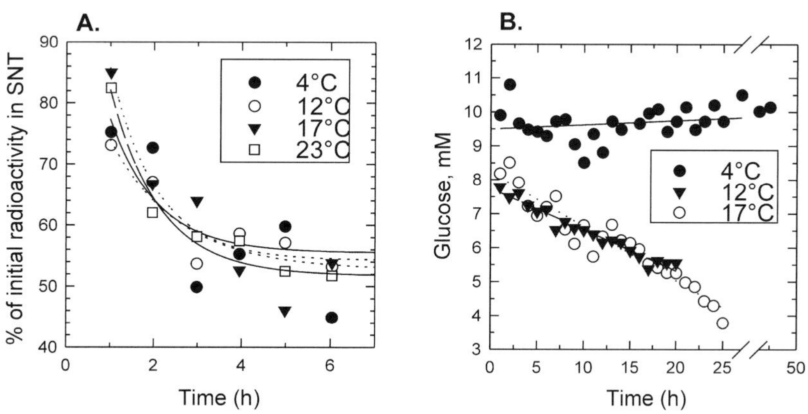

Fig. 1.5. (A) Glucose depletion from the incubation medium due to uptake by *R. sylvatica* erythrocytes. Aliquots of 130 μl erythrocytes (66% hematocrit) were incubated with 10 mM glucose containing 0.6 μCi [U-^{14}C]D-glucose at four temperatures. Samples were removed at timed intervals, centrifuged to pellet cells, and the percentage of radioactivity remaining in the supernatant (SNT) was measured. Lines show theoretical fits. (B) ^{13}C-NMR determination of glucose utilization by wood frog erythrocytes at three temperatures. Samples were matched for equal numbers of cells and then given 10 mM [2-^{13}C]D-glucose. Glucose concentration remaining at any given time was determined by comparison of the glucose peak height at 76.7 ppm with the *p*-DP standard peak height at 127.5 ppm and confirmed by comparison with known concentrations of [2-^{13}C]D-glucose measured under identical conditions. From Brooks et al. (1999).

into glucose synthesis and export (Storey, 1987). A key component of PFK inhibition in liver is the freeze-induced suppression of fructose-2,6-bisphosphate levels (a strong activator of PFK) and inhibition of the enzyme (6-phosphofructo-2-kinase) that synthesizes it (Vazquez-Illanes and Storey, 1993).

Glucose levels in vertebrates are normally strictly controlled within narrow limits (typically ~5 mM in blood) for a good reason. Sustained high glucose (10–50 mM), such as in diabetes, causes severe metabolic injuries, several of which are due to chemical effects of high glucose. Two of these are the nonspecific glycation of long-lived proteins and the pro-oxidant actions of glucose in generating reactive oxygen species (Ruderman et al., 1992; Kristal and Yup, 1992). Indeed, the pro-oxidant nature of high glucose may be the reason that wood frogs show stronger antioxidant defenses (higher activities of antioxidant enzymes and higher glutathione concentrations) than do comparably cold-acclimated leopard frogs (Joanisse and Storey, 1996; Hermes-Lima and Storey, 1996; see also Hermes-Lima et al., 2001 this volume). The potential damage that can be done by high glucose is also probably the reason that frogs do not maintain high cryoprotectant levels over the entire winter as cold-hardy insects do with the polyhydric alcohols (e.g. glycerol, sorbitol) that are their cryoprotectants. Instead, glucose production is triggered immediately when freezing begins, is sustained throughout the freeze, and then reconverted to liver glycogen reserves when the animals thaw (Fig. 1.4) (Storey and Storey, 1986).

The unique glucose metabolism of wood frogs during freezing appears to arise from adaptations of both a quantitative (e.g. higher enzyme activities) and a qualitative (e.g. changes in regulatory mechanisms) nature. Some adaptations are clearly quantitative. For example, the activity of GP in freshly isolated hepatocytes of autumn wood frogs is ~13-fold higher than in hepatocytes of leopard frogs (*R. pipiens*), a species that shares a similar range with wood frogs but is freeze intolerant and hibernates underwater (Mommsen and Storey, 1992). Similarly, the number of glucose transporters in liver plasma membranes of wood frogs was 5-fold higher than in identically acclimated (5°C) leopard frogs (King et al., 1993). Responses to hyperglycemic stimuli are also more pronounced in wood frogs than in leopard frogs. For example, in wood frogs both freezing and dehydration stimulate a rapid increase in the percentage of liver protein kinase A that is present as the active catalytic subunit (PKAc); this rose from 7–10% in controls (5°C acclimated) to 62% within 5 min after freezing began or to 31% when frogs were dehydrated by 20% (Holden and Storey, 1996, 1997). Freezing effects cannot be assessed in leopard frogs but 20% dehydration had no significant effect on the % PKAc, values being 22% in controls versus 28% in dehydrated animals (Holden and Storey, 1997). Such quantitative differences as well as 6–7 fold higher glycogen reserves in wood frog liver (Mommsen and Storey, 1992) help to make the difference between a 4-fold rise in liver glucose stimulated by the loss of 25% of body water in autumn *R. pipiens* and a 300-fold increase in liver glucose in *R. sylvatica* under the same conditions (Churchill and Storey, 1993, 1996). It should be noted, however, that the low but pronounced hyperglycemic effect of dehydration in leopard frogs led us to suggest that the cryoprotectant synthesis response to freezing by freeze tolerant frogs probably grew out of a pre-existing hyperglycemic response to dehydration that may be common to all anuran species (Churchill and Storey, 1993).

Qualitative changes in the regulation of glucose metabolism in freeze tolerant frogs are also needed to control glucose levels within a narrow range under normal (unfrozen) circumstances, just like in other vertebrates, versus to allow the development of extreme hyperglycemia during freezing. Glycogen metabolism in vertebrate liver is controlled externally by hormones (insulin, glucagon, adrenaline) and internally by a cascade of protein kinases and phosphatases that allow responses to hormonal and other stimuli. At least one part of the regulatory system must be altered to override the normal homeostatic control of glucose levels during freezing and permit the massive glycogen breakdown that allows glucose to soar during freezing. Several new studies shed light on the controls involved.

3.2. Structural modification of insulin in wood frogs

The regulation of glucose in vertebrates is normally strictly regulated by the opposing actions of two pancreatic hormones, insulin and glucagon. Insulin is quickly secreted as blood sugar levels rise so that in healthy humans, glucose rarely rises above 8 mM. Insulin stimulates the uptake and storage of glucose by organs either as glycogen or as fat via stimulation of fatty acid biosynthesis. Wood frogs, however, allow blood glucose to soar to concentrations as high 150–300 mM in core organs and blood during freezing. How this extreme hyperglycemia is permitted and how frogs avoid the negative metabolic consequences of prolonged exposure to high glucose remains to be determined but some clues have been found.

The loss of homeostatic control over glucose levels during freezing might be linked to a change in the hormonal regulation of glucose at this time. Insulin secretion from the pancreas might be inhibited during freezing or, alternatively, frog liver may become refractory to insulin stimulation at this time via a mechanism that interferes with insulin receptors on liver cell membranes. Another possibility could be a change in the structure of wood frog insulin that alters its function, particularly under subzero or freezing conditions. A recent study provides evidence for this latter possibility (Conlon et al. 1998). Table 1.1 shows the N-terminal sequences of insulin from four frog species compared with human insulin. Wood frog insulin shows some unusual features. Firstly, the wood frog hormone has a two amino acid extension (lysine-proline) on the N-terminus of the A chain. Although shared by other ranid frogs, this extension does not occur in other vertebrates. Its role remains unknown. Secondly, wood frog insulin shows some unique amino acid substitutions. The serine residue at position A23 in wood frog insulin (A21 of human) is an asparagine in all other species and the aspartic acid at B13 in wood frog insulin is glutamic acid in nearly all tetrapods. Both residues are known to play key roles in insulin function. The A21 (bonding to B22/23) helps to maintain the biologically active protein conformation and B13 is involved in binding to the insulin receptor (Markussen et al. 1988; Kristensen et al. 1997). One or both of these substitutions in wood frog insulin may impair its function. Notably, in the only other known instance of a Glu to Asp substitution at B13 (in the coypu; Bajaj et al. 1986), the change creates a low potency insulin. It is also conceivable that these substitutions may be particularly effective in disrupting insulin conformation

Table 1.1. N-terminal sequences of insulin A and B chains in wood frog, bullfrog, green frog, clawed toad and humans. From Conlon et al. (1998)

	Insulin A-chain					
Rana sylvatica	KP	GIVEQ	CCHNM	CSLYD	LENYC	S...
Rana catesbeiana	--	-----	----T	-----	-----	N...
Rana ridibunda	--	-----	----T	-----	-----	N...
Xenopus laevis		-----	---ST	--F--	--S--	N...
Human		-----	--TSI	----Q	-----	N...
	Insulin B-chain					
Rana sylvatica	FPNQH	LCGSH	LVDAL	YMVCG	DRGFF	YSPRS...
Rana catesbeiana	----Y	-----	--E--	-----	-----	-----...
Rana ridibunda	----Y	-----	--E--	-----	E----	-----...
Xenopus laevis	LV---	-----	--E--	-L---	-----	-Y-KV...
Human	-V---	-----	--E--	-L---	E----	-T-RT...

or receptor-binding ability at subzero temperatures but might be of lesser consequence to hormone action at warmer temperatures where a functional insulin would definitely be needed to regulate the disposition of glucose arising from dietary intake. The probable importance of these novel features are also underlined by the fact that whereas wood frog insulin shows unique features, wood frog glucagon does not; it is identical to glucagon from bullfrogs and has only one amino acid substitution as compared with the human hormone (Conlon et al. 1998).

3.3. Adrenergic control of freeze-induced glucose production

The signals involved in regulating the freeze-induced production of glucose as a cryoprotectant in wood frogs have been extensively studied by our lab. Synthesis is triggered by ice nucleation on the skin and signals are immediately transduced to the liver to activate glycogenolysis. The signal appears to be mediated through β-adrenergic receptors on hepatocyte plasma membranes because intraperitoneal injection of propranolol (a β-adrenergic blocker) just before freezing exposure reduced the hyperglycemic effect of freezing by ~50% whereas administration of the α-adrenergic blocker, phentolamine, had no effect on glucose accumulation (Storey and Storey, 1996b). Propranolol also reduced the strong activation of liver GP that supports freeze-induced hyperglycemia so that the activity of phosphorylase *a* rose by only 8-fold in propranolol-injected frogs during freezing as compared with the 30-fold increase seen in control animals (Storey and Storey, 1996b). The participation of β-adrenergic receptors in mediating freezing-induced cryoprotectant synthesis by wood frog liver was also confirmed by a study that quantified the numbers of α_1, α_2 and β_2 adrenergic receptors in liver plasma membranes by monitoring the binding of radiolabeled inhibitors of the three receptor types (prazosin, yohimbine, and iodopindolol, respectively) (Hemmings and Storey, 1994). This analysis demonstrated that β_2 adrenergic receptors dominated in the plasma membranes of control frogs, with levels several-fold higher than those of α_1 and α_2 receptors. β_2 receptor levels remained high (not different from controls) over the early hours of freezing (1, 12 h at –2.5°C) when rates of glucose production are highest but when frogs were fully frozen after 24 h β_2 receptor binding had decreased by 73%. After 24 h thawing at 4°C, β_2 receptor levels were even further suppressed (by 83% as compared with controls) (Hemmings and Storey, 1994). This suggests that the capacity to respond to hyperglycemic stimuli is reduced in thawed animals, a situation that would favor the clearance and restorage of cryoprotectant as hepatic glycogen. This agrees well with the rapid suppression of GP activity after thawing (Storey and Storey, 1988) and the strong resurgence of glycogen synthase, the amount of active, glucose-6-phosphate independent glycogen synthase activity rising from 0.38 U/g wet mass in liver of 24 h frozen frogs to 3.72 U/gwm after 24 h thawing (Russell and Storey, 1995).

3.4. Protein kinase A

Cyclic AMP binding to the two regulatory subunits of protein kinase A (PKA) causes the dissociation of the inactive tetramer to release the two catalytic subunits (PKAc) of the enzyme. PKAc then phosphorylates and activates glycogen phosphorylase kinase which in turn phosphorylates and activates GP. This normal pattern of activation of glycogenolysis in vertebrate liver is also stimulated by freezing in wood frogs. Within 5 min after freezing begins, the % PKAc rose from 7% in controls to over 60% (Holden and Storey 1996) and was closely followed by a rise in the percentage of GP present as the active a form and by elevated glucose output (Fig. 1.4). Although tetramer dissociation to release the catalytic subunits is the primary mode of PKA control *in vivo*, analysis of the properties of wood frog liver PKAc suggest that these could also influence enzyme function under freezing conditions (Holden and Storey 2000). Thus, the kinetic properties of purified wood frog liver PKAc were significantly affected by assay temperature with low temperature having positive effects on the enzyme (Table 1.2). Affinity for both Mg-ATP and the phosphate-accepting

Table 1.2. Kinetic parameters of the purified free catalytic subunit of PKA from *Rana sylvatica* liver.

	22°C	5°C
K_m kemptide (μM)	9.0 ± 0.1	6.4 ± 0.3[a]
K_m Mg-ATP (μM)	51.8 ± 1.0	24.8 ± 1.4[a]
I_{50} KCl (mM)	495 ± 10	720 ± 10[a]
I_{50} NaCl (mM)	562 ± 16	700 ± 12[a]

Data are means ± SEM, *n* = 3 separate preparations of purified frog liver PKAc enzyme. Kemptide (LRRASLG) is a synthetic peptide with the sequence of the phosphorylation site in pyruvate kinase.
[a]Significantly different from the corresponding value at 22°C, $P < 0.05$. From Holden and Storey (2000).

substrate, kemptide, increased at low temperature; Km Mg-ATP was 50% lower and Km kemptide 33% lower at 5°C, compared with 22°C. PKAc also showed reduced sensitivity to KCl and NaCl inhibition at low temperature with I_{50} values 45 and 25% higher at 5°C than at 22°C (Table 1.2). This lower sensitivity to salt would allow better enzyme function under the conditions of rising cytoplasmic ionic strength that occur as more and more body water freezes out in extracellular ice masses.

3.5. Protein phosphatase-1

Opposing PKA in the homeostatic control of glycogen metabolism is protein phosphatase-1 (PP-1). Under normal conditions in vertebrate liver if glucose rises above about 7–8 mM, PP-1 intervenes to halt further glycogen breakdown by dephosphorylating GPa and phosphorylase kinase. However, in the liver of freezing frogs, glucose production continues unabated to levels of 200–300 mM. Obviously, the off-switch must be missing or inactivated during freezing. To determine whether differential regulation of PP-1 was responsible for the apparent loss of regulatory control over glycogenolysis during freezing, we analyzed the properties of this phosphatase in wood frog liver. Opposite to our expectations but in line with the normal behaviour of the enzyme when glucose concentrations rise, we found that the amount of active PP-1 actually rose progressively over the first hour of freezing exposure, reaching 2.4-fold higher than control values (MacDonald and Storey, 1999). But, despite this, the normal effects of PP-1 action (GP inactivation, glycogen synthetase activation) do not occur during freezing although they are re-instated rapidly when animals thaw (Storey and Storey 1988; Russell and Storey 1995). To determine why this is so, we looked at other aspects of PP-1 regulation.

Further studies revealed that the key factor in liver PP-1 control is probably the physical location of the enzyme. Under normal conditions in vertebrate liver, PP-1 is distributed between cytosolic and glycogen-bound pools. To permit glycogen binding, the PP-1 catalytic subunit must first bind to a G subunit (glycogen binding protein) and this dimer then binds to glycogen. All three isoforms of the PP-1 catalytic subunit (α, δ, γ1) can bind to the G subunit, but the α and δ subunits are the main ones associated with glycogen *in vivo* (Alessi et al. 1993). To analyze the distribution of PP-1 between free and glycogen-bound forms in liver of control (5°C-acclimated) and frozen (12 h at –2.5°C) wood frogs we used differential centrifugation to separate cytosolic and glycogen particle fractions, followed by PP-1 isolation from each fraction using microcystin-agarose affinity chromatography, and then SDS-PAGE and western blotting using antibodies to the three PP-1 isoforms of rat liver (MacDonald and Storey, 1999). The effects of freezing on PP-1 distribution were dramatic. In control frog liver the α and δ isozymes of PP-1 were predominantly localized in the glycogen fraction with little or no crossreacting material detected in the cytosolic fraction (Fig. 1.6). However, the distribution of PP-1 isoforms changed radically with freezing. In frozen animals, virtually all of the α and δ isozyme proteins were translocated into the cytosolic fraction where they could no longer regulate GP. By contrast, the distribution of the γ1 isoform did not change and was primarily cytosolic in both control and frozen frogs.

With the translocation of PP-1 α and δ to the cytosol during freezing, GPa can then function unrestrained in frog liver during freezing and provide a continuous output of glucose for use as a cryoprotectant until glycogen is depleted. The mechanism that regulates PP-1 translocation is not

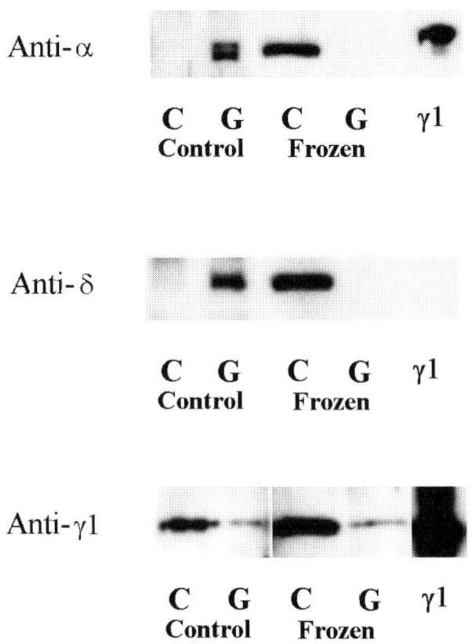

Fig. 1.6. Protein phosphatase-1 (PP-1) isozymes in liver of control and 12 h frozen wood frogs. PP-1 was partially purified from cytosolic (C) and glycogen (G) fractions of frog liver by microcystin-agarose affinity chromatography followed by SDS-PAGE and blotting onto PVDF membranes. Immunoblotting used antibodies to rat liver α, δ and γ1 PP-1. Recombinant PP-1-γ1 was present in the fifth lane as a positive control. Gels show PP-1 bands at 37-39 kDa. From MacDonald and Storey (1999).

yet known but two possibilities exist. One is a down-regulation of the G subunit during freezing. Studies have shown that G subunit levels can affect glycogen metabolism; for example, in insulin-dependent diabetes, low levels of the G subunit impair liver glycogen synthesis (Doherty et al.; 1998). Hence, freezing might stimulate a rapid decrease in the amount of the G subunit in wood frog liver which would reduce the ability of PP-1 to bind to glycogen particles and lead to the appearance of PP-1 in the cytosolic fraction of liver in frozen frogs. A second possibility is reversible control over the G subunit during freezing, perhaps via protein phosphorylation and affecting its ability to either bind PP-1 or bind glycogen. A reversible system, rather than a change in G subunit amount, provides the capacity for a rapid reversal during thawing when GP activity drops by as much as 100-fold and glycogen synthase activity soars (Storey and Storey, 1988; Russell and Storey, 1995). PKAc is known to phosphorylate the G subunit in mammalian liver which increases the rate of PP-1 inactivation by inhibitor-1 and promotes PP-1 release from glycogen (Hubbard and Cohen 1989). Thus, the sustained high PKAc activity in liver of freezing frogs (Holden and Storey, 1996) could be responsible for both GP activation and PP-1 inhibition. However, normally this system is very sensitive to rising glucose levels so a missing piece of the puzzle still remains: specifically, how is the normal sensitivity of this system to high glucose overridden during freezing yet reinstated immediately upon thawing.

3.6. PKG, PKC and MAPKs

Recent studies in our lab have also evaluated the possible role of other signal transduction pathways in mediating events of freeze tolerance. An analysis of second messenger responses to freeze/thaw showed differential responses by both cGMP and inositol 1,4,5-trisphosphate (IP_3), the second messengers of protein kinases G and C, respectively, to freezing and thawing in five wood frog organs (Holden and Storey 1996). IP_3 levels in liver were especially intriguing, rising by 70% within 2 min after freezing began but then continuing to reach a peak of 11-fold higher than controls after 24 hours of frozen. This contrasts with the pattern of cAMP changes which jump 2-fold within 2 min, are sustained over the first hour of freezing and then begin to fall. IP_3 also rose in brain to a maximum of 75% higher than controls after 8 hours frozen. The longer response time for the rise in IP_3 and the sustained high levels of this second messenger during prolonged freezing suggests a possible role for PKC in events that occur over the longer term during freezing such as ischemia resistance or cell volume regulation. Notably, IP_3 levels also rose in wood frog liver when animals were under dehydration stress; IP_3 had increased by 70% in liver of frogs that had lost 5% of total body water and peaked at 4-fold above control values in 40% dehydrated animals (Holden and Storey, 1997). Liver IP_3 levels fell again in both 24 h thawed frogs and fully rehydrated frogs. These very similar responses to freezing versus dehydration suggest that PKC may be involved in regulating cell responses to volume changes.

Mitogen-activated protein kinases (MAPKs) mediate a vast number of cellular responses including gene transcription, cytoskeletal organization, metabolite homeostasis, cell growth and apoptosis in response to many different extracellular signals (Kyriakis 1999; Hoeflich and Woodgett, 2001 this volume). Subfamilies include the extracellular signal regulated kinases (ERKs), Jun N-terminal kinases (JNKs) (also called stress-activated protein kinases) and p38. The latter is the vertebrate counterpart of the yeast Hog1 which was named for its role in the high osmolarity glycerol response. To gain an initial assessment of the roles of MAPKs in freeze tolerance, organ-specific responses of the enzymes were analyzed in wood frogs and hatchling turtles (*T. s. elegans*) (Greenway and Storey 1999, 2000). ERK activities did not change in frog organs over freeze/thaw and in turtles changed only in brain where the amount of active, phosphorylated ERK2 doubled after 30 min freezing and remained high through 4 h frozen (Greenway and Storey 1999, 2000). This limited response suggests that ERKs are not involved in transducing signals from freezing stress, which is perhaps not surprising since ERKs appear to primarily transduce signals from growth factors and mitogens.

However, both JNK and p38 responded to freeze/thaw. JNK activities did not change in wood frog organs over a 12 h freeze but fell by 40–50% in turtle liver and heart over a 4 h freeze (Greenway and Storey 1999, 2000). JNK activity showed a strong increase after 90 min thawing in both liver and kidney of wood frogs (rising ~5- and 4-fold, respectively) suggesting a role in responses activated by thawing. JNK activity was also elevated in frog heart during thawing, doubling after 4 h thawing. The p38 MAPK was the only one to show a widespread response to freezing in frogs (Greenway and Storey 2000). The amount of active, phosphorylated p38 rose by 5–7 fold in liver and kidney within 20 min post-nucleation (as measured by densitometry of immunoblots) but this was reversed by 60 min (Fig. 1.7). A role for p38 in one of the rapid, initial responses to freezing in these organs is therefore implicated. However, in heart, phospho-p38 content rose on a slower time course by about 4-fold after 1 h of freezing and 7-fold after 12 h. Heart experiences a progressive increase in work load as freezing progresses because blood viscosity increases as does peripheral resistance and hence changes in signal transduction in heart may be linked to either adjustments to heart function or implementation of freeze tolerance adaptations. Changes to the phosphorylation state of p38 also occurred during thawing and in brain, p38 was the only MAPK that was activated by thawing. A comparable analysis the effects of anoxia stress (0.5–12 h under N_2 at 5°C) or dehydration (10–40% of total body water lost) on phospho-p38 content in wood frog liver and kidney showed no changed under these stresses. This suggests that p38 might mediate metabolic responses that are unique to freezing survival.

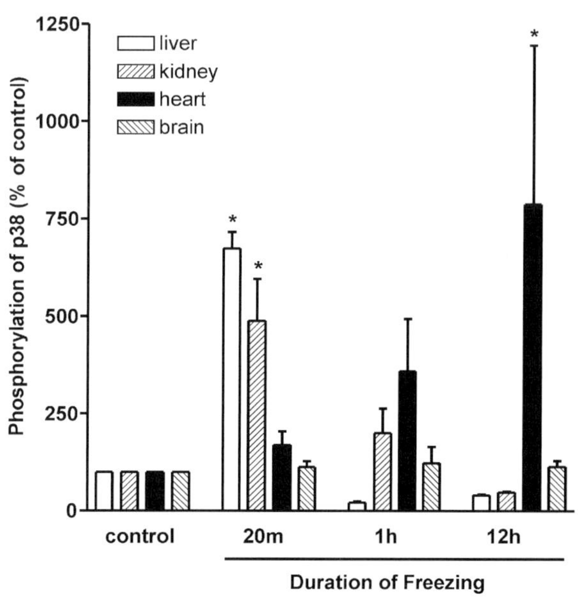

Fig. 1.7. Changes in the amount of the phosphorylated (tyr 182) form of p38 in spring wood frog tissues sampled from control frogs (5°C-acclimated) and frogs frozen for 20 min, 1 h (or 4 h frozen for brain only), or 12 h at –2.5°C. Phospho-p38 content was determined on western blots (equal amounts of protein loaded in each lane) and then blots were scanned and subjected to densitometry. Data were standardized relative to control values and are shown as means ± SEM, n = 3 except for n = 4 for control and 12 h frozen brain, and n = 2 for 1 h and 12 h frozen heart. * Significantly different from the corresponding control value, $P < 0.05$. From Greenway and Storey (2000).

4. Conclusions and future directions

Much remains to be learned about the cell and molecular responses to freezing stress and the multifaceted adaptations that allow freezing survival. Identification of genes that are up-regulated under freezing stress are leading to whole new areas of research. For example, we are finding that a common response to many forms of stress (freezing, anoxia, hibernation) is the up-regulation of genes that are encoded on the mitochondrial genome and much work remains to be done to determine why this is so. Studies with freeze tolerant insects and marine gastropods are also opening up new venues. For example, in other recent work, we have identified a metallothionein as up-regulated in response to freezing and anoxia stresses in marine snails (*Littorina littorea*) (T.E. English and K.B. Storey, unpublished results). Injury caused by reactive oxygen species (ROS) is a serious problem in mammalian systems of ischemia/reperfusion but freeze tolerant animals that undergo ischemia/reperfusion with every cycle of freeze/thaw should have adaptations that address this problem. In previous studies we have shown that freeze tolerant species have high constitutive activities of antioxidant enzymes that could minimize damage by ROS when oxygen is reintroduced during thawing (see Hermes-Lima et al., 2001 this volume). The new finding of freeze-induced up-regulation of metallothionein could also contribute to antioxidant defenses by increasing the capacity of tissues to sequester iron, which as a catalyst for Fenton reactions, plays a major role in the generation of ROS in cells. Furthermore, a general elevation of metal binding capacity in cells during freezing could be of importance to the maintenance of homeostasis in the frozen state. If two-thirds of total body water is converted to ice, then the concentrations of all ionic species in the cytoplasm will rise by about 3-fold and this could cause problems for reactions that are influenced by metal ion concentrations. Increased metal binding capacity could lower cytoplasmic metal ion concentrations back into the normal range in shrunken, freeze-concentrated cells. Hence, these recent results suggest a new potential type of freezing adaptation for exploration.

Much more also remains to be learned about how cells perceive and transmit freezing signals and coordinate both general and organ-specific responses to freezing stress. The regulation of the synthesis and distribution of cryoprotectants is now quite well understood but very little is yet known about how cells manage and regulate the very large changes in water flux, cell volume, ionic strength and osmolality that occur as a consequence of freezing. The molecular mechanisms that underlie the recovery of physiological functions (breathing, heart beat, nerve activity) after thawing are also a mystery awaiting to be explored.

Acknowledgements

Studies in the authors' laboratory have been funded by the Heart and Stroke Foundation of Ontario, the Canadian Diabetes Association, the National Institute of Health (USA) and the Natural Sciences and Engineering Research Council of Canada. We thank the many students and postdoctoral fellows who have contributed to this research in our lab including Q. Cai, S. Brooks, M. Hermes-Lima, T. Churchill, J. MacDonald, D. Joanisse, C. Holden, S. Greenway, E. Russell, and D. White.

References

Alessi, D.R., Street, A.J., Cohen, P. and Cohen, P.T. (1993). Inhibitor-2 functions like a chaperone to fold three expressed isoforms of mammalian protein phosphatase-1 into a conformation with the specificity and regulatory properties of the native enzyme. Eur. J. Biochem. 213, 1055–1066.

Bajaj, M., Blundell, T.L., Horuk, R., Pitts, J.E., Wood, S.P., Gowan, L.K., Schwabe, C., Wollmer, A., Glieman, J. and Gammeltoft, S. (1986) Coypu insulin. Primary structure, conformation and biological properties of a hystricomorph rodent insulin. Eur. J. Biochem. 238, 345–351.

Brooks, S.P.J., Dawson, B.A., Black, D.B. and Storey, K.B. (1999). Temperature regulation of glucose metabolism in red blood cells of the freeze tolerant wood frog. Cryobiology 39, 150–157.

Cai, Q. and Storey, K.B. (1996). Anoxia induced gene expression in turtle heart: up-regulation of mitochondrial genes for NADH-ubiquinone oxidoreductase subunit 5 and cytochrome C oxidase subunit 1. Eur. J. Biochem. 241, 83–92.

Cai, Q. and Storey, K.B. (1997a). Freezing-induced genes in wood frog (*Rana sylvatica*): fibrinogen upregulation by freezing and dehydration. Am. J. Physiol. 272, R1480–R1492.

Cai, Q. and Storey, K.B. (1997b). Upregulation of a novel gene by freezing exposure in the freeze-tolerant wood frog (*Rana sylvatica*). Gene 198, 305–312.

Cai, Q., Greenway, S.C. and Storey, K.B. (1997). Differential regulation of the mitochondrial ADP/ATP translocase gene in wood frogs under freezing stress. Biochim. Biophys. Acta 1343, 69–78.

Churchill, T.A. and Storey, K.B. (1992a). Natural freezing survival by painted turtles *Chrysemys picta marginata* and *C. p. bellii*. Am. J. Physiol. 262, R530–R537.

Churchill, T.A. and Storey, K.B. (1992b). Responses to freezing exposure by hatchling turtles *Trachemys scripta elegans*: factors influencing the development of freeze tolerance by reptiles. J. Exp. Biol. 167, 221–233.

Churchill, T.A. and Storey, K.B. (1993). Dehydration tolerance in wood frogs: a new perspective on development of amphibian freeze tolerance. Am. J. Physiol. 265, R1324–R1332.

Churchill, T.A. and Storey, K.B. (1996). Metabolic effects of dehydration on an aquatic frog, *Rana pipiens*. J. Exp. Biol. 198, 147–154.

Conlon, J.M., Yano, K., Chartrel, N., Vaudry, H. and Storey, K.B. (1998). Freeze tolerance in the wood frog *Rana sylvatica* is associated with unusual structural features of insulin but not glucagon. J. Mol. Endocrinol. 21, 153–159.

Costanzo, J.P., Lee, R.E. and Wright, M.R. (1992). Cooling rate influences cryoprotectant distribution and organ dehydration in freezing wood frogs. J. Exp. Zool. 261, 373–378.

Davies, P.L., Fletcher, G.L., and Hew, C.L. (1999). Freeze-resistance strategies based on antifreeze proteins. In: Environmental Stress and Gene Regulation. (Storey, K.B., ed.), pp. 61–80, BIOS Scientific Publishers, Oxford.

Doherty, M.J., Cadefau, J., Stalmans, W., Bollen, M. and Cohen, P.T. (1998). Loss of the hepatic glycogen-binding subunit (GL) of protein phosphatase 1 underlies deficient glycogen synthesis in insulin-dependent diabetic rats and in adrenalectomized starved rats. Biochem. J. 333, 253–257.

Duman, J.G., Wu, D.W., Xu, L., Tursman, D. and Olsen, T.M. (1991a). Adaptations of insects to subzero temperatures. Quart. Rev. Biol. 66, 387–410.

Duncker, B.P., Rancourt, D.E., Tyshenko, M.G., Davies, P.L. and Walker, V.K. (2001) *Drosophila* as a model organism for the transgenic expression of antifreeze proteins. In: Cell and Molecular Responses to Stress. Vol. 2, pp. 21–29. Elsevier, Amsterdam.

Greenway, S.C. and Storey, K.B. (1999). Discordant responses of mitogen-activated protein kinases to anoxia and freezing exposures in hatchling turtles. J. Comp. Physiol. 169, 521–527.

Greenway, S.C. and Storey, K.B. (2000). Activation of mitogen-activated protein kinases during natural freezing and thawing in the wood frog. Mol. Cell Biochem. 209, 29–37.

Hemmings, S.J. and Storey, K.B. (1994). Alterations in hepatic adrenergic receptor status in *Rana sylvatica* in response to freezing and thawing: implications to the freeze-induced glycemic response. Can. J. Physiol. Pharmacol. 72, 1552–1560.

Hermes-Lima, M. and Storey, K.B. (1996). Relationship between anoxia exposure and antioxidant status in the frog *Rana pipiens*. Am. J. Physiol. 271, R918–R925.

Hermes-Lima, M., Storey, J.M. and Storey, K.B. (2001). Antioxidant defenses and animal adaptation to oxygen availability during environmental stress. In: Cell and Molecular Responses to Stress. (Storey, K.B. and Storey, J.M., Eds.), Vol. 2, pp. 263–287. Elsevier, Amsterdam.

Hoeflich, K.P. and Woodgett, J.R. (2001). Mitogen-activated protein kinases and stress. In: Cell and Molecular Responses to Stress. (Storey, K.B. and Storey, J.M., Eds.), Vol. 2, pp. 175–193. Elsevier, Amsterdam.

Holden, C.P. and Storey, K.B. (1996). Signal transduction, second messenger, and protein kinase responses during freezing exposures in the wood frog. Am. J. Physiol. 271, R1205–R1211.

Holden, C.P. and Storey, K.B. (1997). Second messenger and cAMP-dependent protein kinase responses to dehydration and anoxia stresses in frogs. J. Comp. Physiol. B 167, 305–312.

Holden, C.P. and Storey, K.B. (2000). Purification and characterization of protein kinase A catalytic subunit from liver of the freeze tolerant wood frog: role in glycogenolysis during freezing. Cryobiology 40, 323–331.

Hubbard, M.J. and Cohen, P. (1989). The glycogen-binding subunit of protein phosphatase-1G from rabbit skeletal muscle. Further characterization of its structure and glycogen-binding properties. Eur. J. Biochem. 180, 457–465.

Huber, P., Laurent, M. and Dalmon, J. (1990). Human β-fibrinogen gene expression: Upstream sequences involved in its tissue specific expression and its dexamethasone and interleukin 6 stimulation. J. Biol. Chem. 265, 5695–5701.

Joanisse, D.R. and Storey, K.B. (1994a). Enzyme activity profiles in an overwintering population of freeze-tolerant larvae of the gall fly *Eurosta solidaginis*. J. Comp. Physiol. B 164, 247–255.

Joanisse, D.R. and Storey, K.B. (1994b). Enzyme activity profiles in an overwintering population of freeze-avoiding gall moth larvae, *Epiblema scudderiana*. Can. J. Zool. 72, 1079–1086.

Joanisse, D.R. and Storey, K.B. (1996). Oxidative damage and antioxidants in *Rana sylvatica,* the freeze tolerant wood frog. Am. J. Physiol. 271, R545–R553.

King, P.A., Rosholt, M.N. and Storey, K.B. (1993). Adaptations of plasma membrane glucose transport facilitate cryoprotectant distribution in freeze-tolerant frog. Am. J. Physiol. 265, R1036–R1042.

King, P.A., Rosholt, M.N. and Storey, K.B. (1995). Seasonal changes in plasma membrane glucose transporters enhance cryoprotectant distribution in the freeze tolerant wood frog. Can. J. Zool. 73, 1–9.

Kristal, B.S. and Yu, B.P. (1992). An emerging hypothesis: synergistic induction of aging by free radicals and Maillard reactions. J. Gerontol. 47, B107–B114.

Kristensen, C., Kjeldsen, T., Wiberg, F.C., Schaffer, L., Hach, M., Havelund, S., Bass, J., Steiner, D.F. and Andersen, A.S. (1997). Alanine scanning mutagenesis of insulin. J. Biol. Chem. 272, 12978–12983.

Kyriakis, J.M. (1999). Making the connection: coupling of stress-activated ERK/MAPK signaling modules to extracellular stimuli and biological responses. Biochem. Soc. Symp. 64, 29–48.

Lee, R.E. and Costanzo, J.P. (1998). Biological ice nucleation and ice distribution in cold-hardy ectothermic animals. Ann. Rev. Physiol. 60, 55–72

Lee, R.E., Costanzo, J.P. and Mugnano, J.A. (1998). Regulation of supercooling and ice nucleation in insects. Eur. J. Entomol. 93, 405–418

Lee, R.E. and Denlinger, D.L. (eds.) (1991). Insects at Low Temperature. Chapman and Hall, New York.

MacDonald, J.A. and Storey, K.B. (1999). Protein phosphatase responses during freezing and thawing in wood frogs: control of liver cryoprotectant metabolism. Cryo-Lett. 20, 297–306.

Markussen, J., Diers, I., Hougaard, P., Langkjaer, L., Norris, K., Snel, L., Sorensen, E. and Voigt, H.O. (1988). Soluble, prolonged-acting insulin derivatives. III. Degree of protraction, crystallizability and chemical stability of insulins substituted in positions A21, B13, B23, B27 and B30. Protein Engineer. 2, 157–166.

Mattute, B., Storey, K.B., Knoop, F.C. and Conlon, J.M. (2000). Induction of synthesis of an antimicrobial peptide in the skin of the freeze-tolerant frog, *Rana sylvatica*, in response to environmental stimuli. FEBS Lett. 483, 135–138.

Mommsen, T.P. and Storey, K.B. (1992). Hormonal effects on glycogen metabolism in isolated hepatocytes of a freeze tolerant frog. Gen. Comp. Endocrinol. 87, 44–53.

Rubinsky, B., Lee, C.Y., Bastacky, J. and Onik, G. (1987). The process of freezing and the mechanism of damage during hepatic cryosurgery. Cryobiology 27, 85–97.

Rubinsky, B., Wong, S.T.S., Hong, J.-S., Gilbert, J., Roos, M. and Storey, K.B. (1994). ^{1}H magnetic resonance imaging of freezing and thawing in freeze-tolerant frogs. Am. J. Physiol. 266, R1771–R1777.

Ruderman, N.B., Williamson, J.R. and Brownlee, M. (1992). Glucose and diabetic vascular disease. FASEB J. 6, 2905–2914.

Russell, E.L. and Storey, K.B. (1995). Glycogen synthetase and the control of cryoprotectant clearance after thawing in the freeze tolerant wood frog. Cryo-Lett. 16, 263–266.

Storey, J.M. and Storey, K.B. (1996b). β-Adrenergic, hormonal, and nervous influences on cryoprotectant synthesis by liver of the freeze tolerant wood frog *Rana sylvatica*. Cryobiology 33, 186–195.

Storey, K.B. (1987). Glycolysis and the regulation of cryoprotectant synthesis in liver of the freeze tolerant wood frog. J. Comp. Physiol. B 157, 373–380.

Storey, J.M. and Storey, K.B. (1985). Freezing and cellular metabolism in the gall fly larva, *Eurosta solidaginis*. J. Comp. Physiol. B 155, 333–337.

Storey, K.B. and Storey J.M. (1986). Freeze tolerant frogs: Cryoprotectants and tissue metabolism during freeze/thaw cycles. Can. J. Zool. 64, 49–56.

Storey, K.B. and Storey, J.M. (1988) Freeze tolerance in animals. Physiol. Rev. 68, 27–84.

Storey, K.B. and Storey, J.M. (1989). Freeze tolerance and freeze avoidance in ectotherms. In: Animal Adaptation to Cold (Wang, L.C.H., Ed.), pp. 51–82. Springer-Verlag, Heidelberg.

Storey, K.B. and Storey, J.M. (1991). Biochemistry of cryoprotectants. In: Insects at Low Temperature (Denlinger, D. and Lee, R.E., eds.), pp. 64–93. Chapman and Hall, New York.

Storey, K.B. and Storey, J.M. (1992). Natural freeze tolerance in ectothermic vertebrates. Ann. Rev. Physiol. 54, 619–637.

Storey, K.B. and Storey, J.M. (1996a). Natural freezing survival in animals. Ann. Rev. Ecol. Syst. 27, 365–386.

Storey, K.B. and Storey, J.M. (1999). Gene expression and cold hardiness in animals. In: Cold-adapted Organisms—Ecology, Physiology, Enzymology and Molecular Biology (Margesin, R. and Schinner, F., Eds.), pp. 385–407. Springer, Heidelberg.

Storey, K.B., Bischof, J. and Rubinsky, B. (1992). Cryomicroscopic analysis of freezing in liver of the freeze tolerant wood frog. Am. J. Physiol. 263, R185–R194.

Storey, K.B., Storey, J.M. and Churchill, T.A. (1997). De novo protein biosynthesis responses to water stresses in wood frogs: freeze–thaw and dehydration–rehydration. Cryobiology 34, 200–213.

Storey, K.B., Storey, J.M., Brooks, S.P.J., Churchill, T.A. and Brooks, R.J. (1988). Hatchling turtles survive freezing during winter hibernation. Proc. Natl. Acad. USA 85, 8350–8354.

Thomashow, M.F. (1998). Role of cold-responsive genes in plant freezing tolerance. Plant Physiol. 118– 1–7.

Ultsch, G.R. (1989). Ecology and physiology of hibernation

and overwintering among freshwater fishes, turtles and snakes. Biol. Rev. 64, 435–516.

Vazquez-Illanes, D. and Storey, K.B. 1993. 6-Phosphofructo-2-kinase and control of cryoprotectant synthesis in freeze tolerant frogs. Biochim. Biophys. Acta 1158, 29–32.

Warren, G.J., Thorlby, G.J. and Knight, M.R. (2000). The molecular biological approach to understanding freezing-tolerance in the model plant, *Arabidopsis thaliana*. In: Cell and Molecular Responses to Stress. Vol. 1, pp. 245–258. Elsevier, Amsterdam.

White, D. and Storey, K.B. (1999). Freeze-induced alterations of translatable mRNA populations in wood frog organs. Cryobiology 38, 353–362.

Zachariassen, K.E. (1985). Physiology of cold tolerance in insects. Physiol. Rev. 65, 799–832.

Protein Adaptations and Signal Transduction. Edited by K.B. Storey and J.M. Storey

CHAPTER 2

Drosophila as a Model Organism for the Transgenic Expression of Antifreeze Proteins

Bernard P. Duncker[1], Derrick E. Rancourt[2], Michael G. Tyshenko, Peter L. Davies and Virginia K. Walker*

Department of Biology, Queen's University, Kingston, Ontario, Canada K7L 3N6; Current addresses: [1]Department of Biology, University of Waterloo, Waterloo, Ontario, Canada N2L 3G1; [2]Department of Medical Biochemistry, University of Calgary, Calgary, Alberta, Canada T2N 4N1

1. Introduction

In a footnote to his 1964 study of the cryptonephridial complex in larvae of the common mealworm *Tenebrio molitor*, Ramsey reported unusual freezing behaviour for fluids from several compartments in this organism. He observed that for hemolymph, midgut fluid and perirectal space fluid, the temperature of ice crystal growth was significantly lower (≤ 10°C) than the melting temperature. This was the first report of a phenomenon, known as thermal hysteresis, that has subsequently been attributed to a group of proteins termed antifreeze proteins (AFPs) or thermal hysteresis proteins (THPs). AFPs have since been found in a wide range of organisms, both terrestrial and aquatic, exposed to the dangers of ice growth at subzero temperatures (e.g. Davies and Hew, 1990; Duman et al., 1991; Cheng, 1998; Ewart et al., 1999).

2. Properties of AFPs

Of the many known AFPs, those that are best characterized are from Arctic and Antarctic fishes. Numerous fish AFP genes and cDNAs have been cloned, including those of winter flounder (Davies et al., 1982, Gourlie et al., 1984), sea raven (Ng et al., 1986), herring (Ewart and Fletcher, 1993), smelt (Ewart et al., 1992), ocean pout (Hew et al., 1988) and wolffish (Scott et al., 1988), as well those encoding the antifreeze glycoproteins (AFGPs) of Antarctic notothenioids and Arctic cod (Chen et al., 1997a; Chen et al., 1997b). Fish AFPs have been placed into five classes based on their structure: the AFGPs, characterized by Ala-Ala-Thr tripeptide repeats with a disaccharide moiety attached to the threonyl residue, the alanine–rich type I AFPs, the globular, lectin-like type II AFPs, the globular, β-sheeted type III AFPs and the recently isolated type IV AFP, distinguished by a high glutamine content (for recent reviews see Yeh and Feeney, 1996; Davies and Sykes, 1997; Cheng, 1998; Ewart et al., 1999). Despite this remarkable structural diversity, all AFPs appear to act in a similar fashion, conferring protection to their host by interacting with and inhibiting the growth of ice crystals (DeVries, 1983).

As a result of their interactions with ice crystals, AFPs exhibit a number of properties of potential benefit to transgenic hosts. At moderate undercooling (< –0.5°C), ice grows along the basal plane (perpendicular to the c-axis), resulting in disk-like crystals. AFPs from fish are able to inhibit this growth by binding to prism faces. The bound surfaces are expressed as facets, which generally result in the formation of hexagonal bipyramids.

*Corresponding author.

AFP binding maintains these bipyramids at a relatively constant size and shape over a certain temperature range, in a concentration-dependent manner (Feeney and Yeh, 1978; Davies, 1999). This ice/AFP interaction results in a lowering of the solution freezing point in a non-colligative manner, without a significant change in its melting point (Raymond and DeVries, 1977).

A possibly related property ascribed to AFPs is the ability to inhibit the recrystallization of ice. As frozen tissue approaches its thawing temperature, the increase in kinetic energy allows water molecules to migrate from smaller to larger crystals, resulting in the formation of ice crystals, presumably big enough to cause mechanical damage to cells. This is a potential danger to freeze-tolerant organisms, which are able to cope with the formation of extracellular ice, and which undergo repeated freeze–thaw cycles. AFPs are able to prevent the formation of these large crystals by inhibiting recrystallization, thus protecting tissues from injury (Knight and Duman, 1986). This activity is exhibited at AFP concentrations far lower (<10 μg/ml) than the concentration required for effective thermal hysteresis.

It has also been claimed that AFPs protect membranes from damage at low temperatures, although this notion is somewhat controversial. Damage caused to cells under hypothermic conditions is thought to be due to a reduction in active transport that should balance the passive ion transport across cell membranes (Hochachka, 1986). One hypothesis is that AFPs are able to bind to cell membranes in such a way as to diminish passive ion transport (Rubinsky et al., 1991). Another view is that protection may be conferred by prevention of leakage from membranes during thermotropic phase transitions (Hays et al., 1996). It has been shown that fish AFP solutions are able to protect mammalian oocytes at 4°C, since they show increased survival at this temperature relative to controls incubated in phosphate-buffered saline (Rubinsky et al., 1991). Similar protective effects have been reported for whole rat livers (Lee et al., 1992) and human platelets (Tablin et al., 1996; Oliver et al., 1999). In other studies, however, AFPs were reportedly cryotoxic to both spinach thylakoid membranes (Hincha et al., 1993) and ram spermatozoa (Payne et al., 1994), and ineffective in conferring protection to rat cardiac explants (Wang et al., 1994).

Elevated gene dosage appears to be essential for the high AFP concentrations found in freeze-resistant organisms. Indeed, fish AFP genes typically exist in large families of 30 or more copies. In general, those fish which inhabit shallow waters and higher latitudes, and thus risk the greatest contact with ice, have a higher AFP gene copy number than fish found in deeper habitats (Scott et al., 1986). Why has high gene copy number evolved rather than powerful promoters to allow the expression of high AFP levels? One clue may lie in the fact that sea-level glaciation appears to be a relatively recent environmental stress (Scott et al., 1986; Davies et al., 1999). Rather than a common progenitor polypeptide amongst the various AFP producing fishes, it appears that a variety of prototype AFP genes have been duplicated and possibly modified to make better AFPs. This would account for the fact that three fishes belonging to the same family, shorthorn sculpin, sea raven and longhorn sculpin, produce type I, II and IV AFP, respectively. It is reasonable to assume then that further duplication of the modified genes was a faster response to the threat of serum freezing than the evolution of new promoter elements and/or trans-acting factors. Similar genomic response to environmental stress have been characterized in other organisms. For example, gene amplification results in the rapid increase in pesticide resistance in insects, herbicide resistance in plant cells and drug resistance in tumours (e.g. Raymond et al., 1991; Harms et al., 1992; Kuo et al., 1998).

3. *Drosophila* as a model system for fish AFP expression

The cloning of AFP-encoding DNA has offered the promise of conferring protection to animal and plant species that are normally vulnerable to the damaging effects of cold temperatures. The transgenic expression of fish AFPs has been achieved in non-AFP producing species of plants (Hightower et al., 1991; Kenward et al., 1993; Kenward et al.,

1999) and fish (Shears et al., 1991). In each case, however, the resultant levels of synthesized AFP were too low to produce significant thermal hysteresis.

Why has the transgenic expression of fish AFPs in plants and fish yielded such disappointing results? One obvious possibility is the choice of AFP to be expressed. In most of the aforementioned studies, genes encoding the type I AFP of winter flounder were used. In order to efficiently synthesize a protein with such an unusually high alanine content, winter flounder appear to have developed special adaptations such as the seasonal up-regulation of alanyl-tRNA synthesis during winter months (Pickett et al., 1983). It is not surprising then that transgenic hosts, lacking the ability for such modulation would produce suboptimal levels of AFP. Other possibilities include that fusion sites between the homologous promoters and the heterologous coding regions may result in mRNAs which are not ideal for ribosome binding and/or when translated, for subsequent protein secretion, AFP precursor processing may be incomplete, rearing conditions of the transgenics may not be optimal, and the AFP gene dosage may not be sufficiently elevated for high levels of expression.

We have tested improved AFP expression strategies by using the fruit fly, *Drosophila melanogaster* as a transgenic model system (Walker, 1989). In *D. melanogaster,* the discovery of transposable elements in wild populations has led to a reliable method of germ line transformation (Spradling, 1986). When plasmids incorporating transgenes between transposable P-element sequences are microinjected into pre-syncytial embryos, typically 7% of the embryos survive to fertile adults with stable germ line integration of the transgene (Spradling, 1986). The *Drosophila* system offers numerous additional advantages. They include the fact that the transgenic progeny of injectants can be easily identified through genetic markers that are transferred along with the desired transgene, usually resulting in a change in eye color, and that transgenic stocks can easily be established, due to the short (10 day) generation time and well characterized genetics of *Drosophila*.

3.1. Poor expression in transgenic type I AFP flies

Our initial experiments with transgenic flies involved the expression of type I AFP. A gene chimera was constructed in which the winter flounder AFP coding sequence was joined to the *D. melanogaster hsp70* promoter and used to transform *Drosophila* embryos (Rancourt et al., 1987). In resulting transformants, AFP mRNA was detected following heat shock (37°C). Immunodetection of AFP in the hemolymph of transformants demonstrated that the heterologous protein was recognized by the *Drosophila* secretory protein system. However, as with other transgenic organisms expressing type I AFP (Hightower et al., 1991; Shears et al., 1991; Kenward et al., 1993) there was insufficient accumulation of the foreign protein to result in measurable thermal hysteresis. Since transient expression from the *hsp70* promoter resulted in low levels of AFP production, a new construct was made by placing the winter flounder AFP gene under the transcriptional control of the developmentally regulated, female-specific *D. melanogaster* yolk protein 1 (*yp1*) promoter. Not only is this a powerful promoter, but the use of sex-specific regulatory DNA sequences to drive heterologous coding regions is convenient since it allows the recovery of transformants (in this case, males that do not express the AFP sequence), independently of any lethal effect of the expression of a particular foreign gene. Disappointingly, despite the presence of AFP mRNA in female transgenics, no AFP could be detected in their hemolymph (Rancourt et al., 1992). As noted above, the high alanine content of this protein, along with the frequent reiteration of the GCC codon in its coding sequence likely poses translational difficulties for the transgenic host. Although some AFP was produced when the gene was under the control of the *hsp70* promoter, this was probably due to genes with *hsp* regulatory elements being preferentially transcribed and translated during heat stress (McGarry and Lindquist, 1985). Additional properties of the type I AFP may contribute to difficulties in its expression. In flies, the proform of the AFP is not processed to the more active mature form found in flounder serum (Peters et al., 1993). Furthermore, the secondary structure of the flounder AFP is influenced by temperature. At –1°C the protein is 85% α-helical, but at 25°C, the normal *Drosophila* rearing temperature, this value drops to 47% (Ananthanarayanan and Hew, 1977). The effect of temperature on type I AFP levels in transgenic flies was highlighted by a study which demonstrated that the small amount of AFP found in the hemolymph of *hsp70*/AFP transgenic *Drosophila* persists longer if the flies are kept at a temperature (10°C) closer to that of the fish habitat (Duncker et al., 1995b). The decrease in AFP accumulation at 25°C was partially mediated by a temperature-dependent persistence in AFP mRNA, since message in transgenic flies reared at 10°C was more stable than for those reared at 25°C (Duncker et al., 1995b). In contrast, levels of α-tubulin mRNA and a heat shock protein (hsp83) mRNA were equivalent for flies at either temperature.

3.2. Antifreeze activity in flies expressing type II AFP

Type II AFP does not have the extreme bias towards one amino acid shown by type I AFPs, nor is there evidence of a temperature-sensitive conformational change. For this reason, type II-AFP-producing transgenic flies were generated by linking the coding region of a type II AFP gene from sea raven to the *Drosophila yp1* promoter, through an in-frame fusion between the sequences encoding the yolk protein secretory signal peptide and the AFP signal peptide. Analysis of transgenic females revealed that they produced both type II AFP mRNA and its corresponding protein (Duncker et al., 1996). Indeed, sufficient AFP was synthesized to produce 0.13–0.16°C of thermal hysteresis in the hemolymph (Table 2.1). Ice crystals grown during the AFP activity assay formed microscopic hexagonal bipyramids typical of those formed in the presence of type II AFP. In addition, inhibition of ice recrystallization was observed in hemolymph samples rapidly frozen to –133°C and then warmed, resulting in the formation of smaller ice crystals at –10°C, –3°C and on melting, than are seen with nontransgenic flies. This suggests that tissue damage might be less when frozen transgenic lines are thawed. Type II AFP is produced in fish as a preproprotein and, as was the case with the type I AFP, *Drosophila* fat bodies appropriately secreted this proprotein. However,

Table 2.1. A summary of antifreeze protein expression in *Drosophila* adults transformed with different plasmid constructions

Construct (promoter/gene)[1]	Conditions/Sex	Copy #	mRNA	AFP	Thermal hysteresis[2]
hsp70/Type I AFP	22°C	1–6	–	–	none
	37°C	1–6	+	+	none
yp1/Type I AFP	♂	1–2	–	–	none
	♀	1–2	+	–	none
yp1/Type II AFP	♂	2	–	–	none
	♀	2	++	++	0.13°C–0.16°C
yp1,2/Type III AFP	♂	1–2	–	–	none
	♀	1–2	++	++	0.14°C–0.20°C
yp1,2/Type III AFP/*yp1* 3′AFP	♂	1–2	–	–	none
	♀	1–2	++	++	0.12–0.35°C
yp1,2/Type III AFP/*yp1* 3′AFP	♂	4–8	–	–	none
	♀	4–8	++	++	0.29–0.43°C

[1]The fish from which the AFP genes were obtained were winter flounder (type I), sea raven (type II) and wolffish (type III). The heat shock 70 (*hsp70*), yolk protein 1 (*yp1*) or the yolk protein 1 and 2 intergenic region (*yp1,2*) promoters were used to force the expression of the AFP genes in *Drosophila*.
[2]Mean values from individual transgenic lines. The chromosomal position of the transgene can influence its transcriptional activity (position effect) and result in some variation in AFP levels.

there was again no removal of the pro sequence from the protein in the hemolymph as was found when the same protein was expressed in insect cells in culture (Duncker et al., 1995b; 1994). Thus the proteinase activity responsible for the hydrolysis to mature AFP in sea raven is absent from *Drosophila* hemolymph. This is of little consequence for freeze-protection, however, since type II proAFP has comparable activity to the mature form of the protein (Duncker et al., 1994).

3.3. Improved thermal hysteresis in flies expressing type III AFP

To avoid any uncertainties associated with proprotein processing, another gene chimera was constructed by fusing the type III AFP gene from wolffish to the *Drosophila* yolk protein promoter region. In this case, the entire 1 kb region which directs the divergent transcription of the yolk protein 1 and 2 (*yp1/2*) genes was used as the promoter DNA. Two copies of the AFP genes could then be transcribed from this single intergenic regulatory region (Rancourt et al., 1990), thus mimicking the situation in wolffish, where most of the AFP genes are organized in a divergent transcriptional orientation (Scott et al., 1988). The ligations between yolk protein and AFP coding regions were made in the sequences encoding the signal peptide hydrophobic cores. Adult females transformed with this construct, gave rise to fly lines demonstrating thermal hysteresis values ranging between 0.14 and 0.20°C (Table 1).

In an attempt to further increase antifreeze activity, improvements were made to the type III AFP expression cassette by the inclusion of additional *yp* gene sequences. When the single intron of one of the wolffish AFP genes was replaced with that of the *yp*1 gene, little change in the accumulation of the corresponding mRNA was observed. In contrast, the complete absence of an intervening sequence resulted in a 2–11 fold decrease in the AFP mRNA relative to a control gene (Duncker et al., 1997). Thus the presence of an intron *per se* appears to be important to maximize transgenic mRNA accumulation, even in an organism where introns are generally smaller and less numerous than in vertebrates. When the 3′-untranslated region of the *yp1* gene was substituted for that of one of the AFP genes in the *yp1/2*AFP construct, one of the derived transgenic lines showed the highest hemolymph thermal hysteresis value (0.35°C) obtained in a single insert stock (Rancourt et al., 1990). Ice crystals grown in the presence of hemolymph from these transgenic lines could not be distinguished from ice crystals grown in the presence of AFPs from fish serum (Fig. 2.1). Crosses involving lines containing this chimaeric gene were performed so that transgenic flies with additional copies of the AFP coding regions were generated (Table 2.1). The result was multi-insert lines with hemolymph thermal hysteresis values of up to 0.43°C, values significantly higher than that recorded in some AFP-producing fish (Duncker et al., 1999). Conventional genetic crosses clearly cannot be used to increase the transgene copy number to the 30–150 gene copies found in fish. A more promising approach would be to cross these lines to flies expressing an integrated copy of the P-element transposase, which is able to mobilize the integrated transgenes and catalyze their insertion at numerous genomic sites (Robertson et al., 1988; Meister and Grigliatti, 1993).

Despite the successes in type III AFP expression, transgenics tested for resistance to cold exposure at 0°C and at –7°C had survival rates no

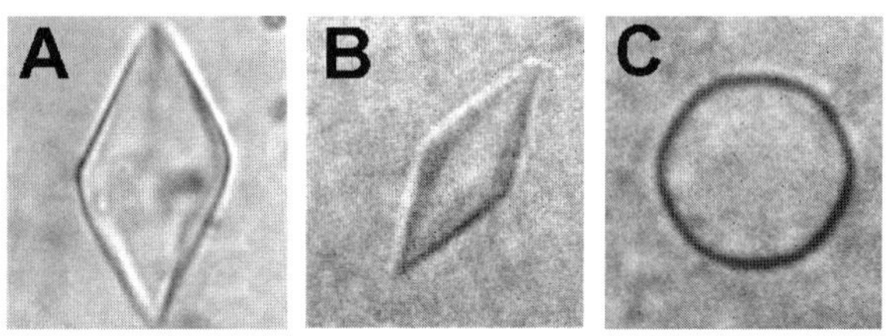

Fig. 2.1. Ice crystal morphology of *Drosophila* hemolymph transformed with wolffish, type III AFP. The ice crystals shown are representative of the ice crystals formed in the presence of (A) type III AFP purified from fish serum (B) hemolymph from female *Drosophila*, obtained from a transgenic line containing the wolffish, type III AFP coding sequences joined to a *Drosophila* yolk protein promoter (C) and hemolymph from female *Drosophila* obtained from the untransformed, host strain, showing the preferential growth of ice crystals along the a-axis, typically seen in the absence of AFP activity. Magnification was ~700-fold.

better than control flies (Duncker et al., 1995a). The flies did not freeze, even at –7°C, because they supercooled, but they did suffer mortality due to chilling injury (Chen and Walker, 1994). In retrospect, these studies could be considered as a test of the role of AFPs in membrane protection (Rubinsky et al., 1990; Rubinsky et al., 1991). It is possible that some other consequence of cold exposure had a greater influence on the insects' survival than the proposed AFP-mediated membrane protection.

4. Prospects for the transgenic expression of other AFPs

Recently, sequences encoding a variety of AFPs have been cloned from sources as diverse as plants and insects. For the most part, the known, isolated plant AFPs show low levels of thermal hysteresis activity (<0.5°C; e.g. Urratia et al., 1992; Hon et al., 1995; Worrall et al., 1998; Smallwood et al., 1999), and thus do not appear to be attractive candidates for transfer into freeze-intolerant model organisms. In contrast, AFPs or THPs from insects show hyperactivity, when compared with plant or fish AFPs (Fig. 2.2). Indeed, insects collected in midwinter have been reported to show hemolymph freezing point depression in the range of 3–6°C (Duman et al., 1991). The yellow mealworm beetle, *Tenebrio molitor*, reaches hemolymph thermal hysteresis values of 5.5°C, and at low concentrations the activity is up to 100 times that of fish AFPs (Graham et al., 1997). Rather than the characteristic bipyramid shape (Chao et al., 1995) adopted by the seed ice crystal in the presence of millimolar concentrations of any of the fish AFPs (e.g. type I, II or III), ice crystals formed in the presence of the beetle THPs are curved (Graham et al., 1997). Spruce budworm THPs direct the formation of similarly small, non-bipyramid-shaped discs (Tyshenko et al., 1997). While the 8–9 kDa THPs isolated from *Tenebrio* and the fire-colored beetle, *Dendroides canadensis*, appear to be related, sharing about 60% amino acid identity and a repetitive structure (Duman et al., 1998, Liou et al., 1999), the 9 kDa THP from the spruce budworm,

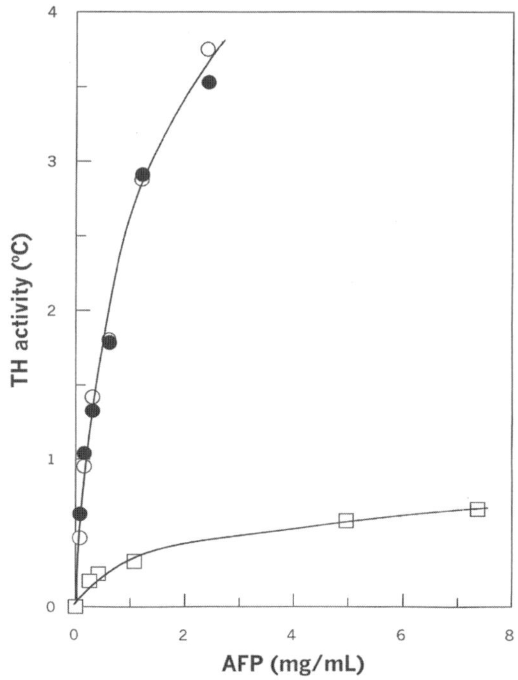

Fig. 2.2. Thermal hysteresis activity, as a function of concentration, comparing insect THPs and the fish type III AFP. Thermal hysteresis activities of purified, recombinant spruce budworm THP (open circles) and *Tenebrio* THP (closed circles) as well as type III fish AFP (open boxes) were compared at different protein concentrations, as determined by amino acid analysis.

Choristoneura fumiferama, has a distinctly different primary sequence (Tyshenko et al., 1997). Both beetle and budworm THPs, however, depend on the formation of multiple disulfide bonds for correct folding and activity, and the *Tenebrio* THPs include glycosylated isoforms (Duman et al., 1998, Gauthier et al., 1998; Liou et al., 1999). These characteristics may present a challenge for heterologous expression. In addition, the lessons learned from the fish transfer work suggest that we cannot know *a priori* which of these THPs will be the best gene sequence for transfer into *Drosophila.*

Initial constructs have used the spruce budworm cDNA joined to the *D. melanogaster actin 5c* promoter in an attempt to express the THP in the transgenic nurse cells for transport through the ring canals into the developing embryo. Since a constitutive promoter was used, it is not surprising that

THP mRNA is abundant in whole fly homogenates of all the independently isolated transgenic lines. Disappointingly, however, western blot analysis showed that the mass of the heterologous protein appears to be larger than predicted, suggesting inappropriate post-translational processing. Nevertheless, low levels of thermal hysteresis were observed in homogenates of other transgenic lines (Tyshenko, M.G., unpublished). This result suggests that by making various modifications to the constructs, including some of those used to optimize expression in the fish AFP transgenic flies, it may be possible to achieve the promise of superior low temperature protection.

5. Cautions and conclusions

The transfer of AFPs to even a generally beneficial insect such as *Drosophila melanogaster* requires that reasonable precautions should be established for biological containment. Particular care should be taken to avoid releasing transgenic insects (Hoy, 1993) bearing genes which could provide a selective advantage to their host. In our laboratory we use disabled *Drosophila* host strains; recipient strains are unable to fly due to the inclusion of the dominant mutation for flightlessness (*Ifm (3)3;* Rancourt et al., 1991; Duncker et al., 1993). Even if the transgenic flies "walked" out of the laboratory, none of their progeny from potential crosses with wild populations would be able to fly because heterozygotes also have a flightless phenotype.

The transfer and expression of any heterologous gene to an evolutionary advanced organism is not a trivial undertaking. The lessons learned from the efforts to transfer AFP genes to flies have demonstrated several important considerations, including the prudent selection of the target AFP gene, the choice of promoter, the inclusion of intronic and other regulatory sequences, copy number, post-translational processing, effective containment of the host and an appreciation of the ecology of both gene donor and recipient. It is hoped that these investigations will prove valuable for future studies on the transfer of a stress-resistant phenotype by genetic transformation.

Acknowledgements

We would like to thank our graduate and undergraduate students as well as colleagues from the laboratories of Virginia Walker and Peter Davies for their help and support of this work over the last fifteen years, including Cheng-Ping Chen, Daniel Doucet, Laurie Graham, Anne Hermans, Derek Koops, Yih-Cherng Liou, Todd MacGregor, Iain Peters, Dave Picketts, and Christine Simmons. A special debt of gratitude is owed to Sherry Gauthier. We acknowledge support from the Natural Sciences and Engineering Research Council (Canada) to V.K.W. and the Medical Research Council to P.L.D.

References

Ananthanarayanan, V.S. and Hew , C.L. (1977). Structural studies on the freezing point-depressing protein of the winter flounder *Pseudopleuronectes americanus*. Biochem. Biophys. Res. Commun. 74, 685–689.

Chao, H., C.I. DeLuca and Davies, P.L. (1995). Mixing antifreeze protein types changes ice crystal morphology without affecting antifreeze activity. FEBS Lett. 357, 183–186.

Chen, C.-P. and Walker, V.K. (1994). Cold-shock and chilling tolerance in *Drosophila*. J. Insect Physiol. 40, 661–669.

Chen, L., A.L. DeVries and Cheng, C.H.C. (1997a). Evolution of antifreeze glycoprotein from a trypsinogen gene in Antarctic notothenioid fish. Proc. Natl. Acad. Sci. USA 94, 3811–3816.

Chen, L., A.L. DeVries and Cheng, C.H.C. (1997b). Convergent evolution of antifreeze glycoproteins in Antarctic notothenioid fish and Arctic cod. Proc. Natl. Acad. Sci. USA 94, 3817–3822.

Cheng, C.H.C. (1998). Evolution of the diverse antifreeze proteins Curr. Opin. Genet. Dev. 8, 715–720.

Davies, P.L. (1999). Antifreeze Proteins. In: The Encyclopedia of Molecular Biology (Creighton, T.E., Ed.) John Wiley, New York, NY, in press.

Davies, P.L., Fletcher, G.L. and Hew, C.L. (1999). Freeze-resistance strategies based on antifreeze proteins. In: Environmental Stress and Gene Regulation. (Storey, K.B., Ed.), pp. 61–80. BIOS Scientific Publishers Ltd., Oxford.

Davies, P.L. and Hew, C.L. (1990). Biochemistry of fish antifreeze proteins. FASEB J. 4, 2460–2468.

Davies, P.L. and Sykes, B. D. (1997). Antifreeze proteins. Curr. Opin. Struct. Biol. 7, 828–834.

DeVries, A.L. (1983). Antifreeze peptides and glycopeptides in cold-water fishes. Ann. Rev. Physiol. 45, 245–260.

Duncker, B.P., Chen, C.-P., Davies, P.L. and Walker, V.K. (1995a). Antifreeze protein does not confer cold hardiness to transgenic *Drosophila melanogaster*. Cryobiology 32, 521–527.

Duncker, B. P., Davies, P.L. and Walker, V.K. (1997). Introns boost transgene expression in *Drosophila melanogaster*. Mol. Gen. Genet. 254, 291–296.

Duncker , B.P., Davies, P.L. and Walker, V.K.(1999). Increased gene dosage augments antifreeze protein levels in transgenic *Drosophila melanogaster*. Transgenic Res. 8, 45–50.

Duncker, B.P., Gauthier, S.Y. and Davies, P.L. (1994). Cystine-rich fish antifreeze is produced as an active proprotein precursor in fall armyworm cells. Biochem. Biophys. Res. Commun. 203, 1851–1857.

Duncker, B.P., Hermans, J.A., Davies, P.L. and Walker, V.K. (1993). Biological containment of transgenic flies using a flightless, white host. Drosophila Information Service 72, 133.

Duncker, B.P., Hermans, J.A, Davies, P.L. and Walker, V.K. (1996). Expression of a cystine-rich antifreeze in transgenic *Drosophila melanogaster*. Transgenic Res. 5, 49–55.

Duncker, B.P. , Koops, M.D., Walker, V.K. and Davies, P.L. (1995b). Low temperature persistence of type I antifreeze protein is mediated by cold-specific mRNA stability. FEBS Lett. 377, 185–188.

Duman, J.G., Li, N., Verleye, D., Goetz, F.W., Wu, D.W., Andorfer, C.A., Benjamin, T. and Parmelee, D.C. (1998). Molecular characterization and sequencing of antifreeze proteins from larvae of the beetle *Dendroides canadensis*. J. Comp. Physiol. B 168, 225–232.

Duman, J.G., Xu, L., Neven, L.G., Tursman, D. and Wu, D.W. (1991). Hemolymph proteins involved in insect sub-zero temperature tolerance: ice nucleator and antifreeze proteins. In: Insects at low temperatures. (Lee, R.E. and Denlinger, D.L., Eds.), pp. 94–127. Chapman and Hall, New York.

Ewart, K.V. and Fletcher, G.L. (1993). Herring antifreeze protein: primary structure and evidence for a C-type lectin evolutionary origin. Mol. Mar. Biol. Biotechnol. 2, 20–27.

Ewart, K.V., Lin, Q. and Hew, C.L. (1999). Structure, function and evolution of antifreeze proteins. Cell. Mol. Life Sci. 55, 271–283.

Ewart, K.V., Rubinsky, B. and Fletcher, G.L. (1992). Structural and functional similarity between fish antifreeze proteins and calcium-dependent lectins. Biochem. Biophys. Res. Commun. 185, 335–340.

Feeney, R.E. and Yeh, Y. (1978) Antifreeze proteins from fish bloods. Adv. Protein Chem. 31, 191–282.

Gauthier, S. Y., Kay, C.M., Sykes, B.D., Walker, V.K. and Davies, P.L. (1998). Disulfide bond mapping and structural characterization of spruce budworm antifreeze protein. Eur. J. Biochem 258, 445–453.

Gourlie, B., Lin, Y., Price, J., DeVries, A.L., Powers, D. and Huang, R.C. (1984) Winter flounder antifreeze proteins: a multigene family. J. Biol. Chem. 271, 4106–4112.

Graham L.A., Liou, Y.-C., Walker, V.K. and Davies, P.L. (1997). Hyperactive antifreeze protein from beetles. Nature 388, 727–728.

Harms, C. T., Armour, S.L., DiMaio, J.J., Middlesteadt, L.A., Murray, D., Negrotto, D.V., Thompson-Taylor, H., Weymann, K., Montoya, A.L., Shillito, R.D. and Jen, G.C. (1992). Herbicide resistance due to amplification of a mutant acetohydroxyacid synthase gene. Mol. Gen. Genet. 233, 427–435.

Hays, L.M., Feeney, R.E., Crowe, L.M., Crowe, J.H. and Oliver, A.E. (1996) Antifreeze glycoproteins inhibit leakage from liposomes during thermotropic phase transitions. Proc. Natl. Acad. Sci. USA 93, 6835–6840.

Hew, C.L., Wang, N.C., Joshi, S., Fletcher, G.L., Scott, G.K., Hayes, P.H., Buettner, B. and Davies, P.L. (1988). Multiple genes provide the basis for antifreeze protein diversity and dosage in the ocean pout, *Macrozoarces americanus*. J. Biol. Chem. 263, 12049–12055.

Hincha, D.K., DeVries, A.L. and Schmitt, J.M. (1993). Cryotoxicity of antifreeze proteins and glycoproteins to spinach thylakoid membranes-comparison with cryotoxic sugar acids. Biochim. Biophys. Acta 1146, 258–264.

Hightower, R., Baden, C., Penzes, E., Lund, P. and Dunsmuir, P (1991). Expression of antifreeze proteins in transgenic plants. Plant Mol. Biol. 17, 1013–1021.

Hochachka, P.W. (1986). Defense strategies against hypoxia and hypothermia. Science 231, 234–261.

Hon, W.-C., Griffith, M., Mlynarz, A., Kwok, Y.C. and Yang, D.S.C. (1995). Antifreeze proteins in winter rye are similar to pathogenesis-related proteins. Plant. Physiol. 109, 879–889.

Hoy, M.A. 1993. Transgenic beneficial arthropods for pest management programs: an assessment of their practicality and risks. In: Pest Management: Biological Based Technologies. (Lumsden, R.D. and Vaughn, J.L., Eds.), pp. 357–369. American Chemical Society.

Kenward, K.D., Altschuler, M., D. Hildebrand and P.L. Davies. 1993. Accumulation of type I fish antifreeze protein in transgenic tobacco is cold-specific. Plant Mol. Biol. 23, 377–385.

Kenward, K.D., Brandle, J., McPherson, J. and Davies, P.L. (1999). Type II fish antifreeze protein accumulation in transgenic tobacco does not confer frost resistance. Transgenic Res. 8: 105–117.

Knight, C. A. and Duman, J.G. (1986). Inhibition of recrystallization of ice by insect thermal hysteresis proteins: a possible cryoprotective role. Cryobiology 23, 256–262.

Kuo, M.T., Sen, S., Hittleman, W.N. and Hsu, T.C. (1998).

Chromosomal fragile sites and DNA amplification in drug-resistant cells. Biochem. Pharm. 56, 7–13.

Lee, C.Y., Rubinsky, B. and Fletcher, G.L. (1992) Hypothermic preservation of whole mammalian organs with "antifreeze" proteins. Cryo-Letters 13, 59–66.

Liou, Y-C., Thibault, P., Walker, V.K., Davies, P.L. and Graham, L.A. (1999) A complex family of highly heterogeneous and internally repetitive hyperactive antifreeze proteins from the beetle *Tenebrio molitor*. Biochemistry 38, 11415–11424.

McGarry, T.J. and Lindquist, S. (1985). The preferential translation of Drosophila hsp70 mRNA requires sequences in the untranslated leader. Cell 42, 903–911.

Meister, G.A. and Grigliatti, T.A. (1993). Rapid spread of a P element/Adh construct through experimental populations of *Drosophila melanogaster*. Genome 36, 1169–1175.

Oliver, A.E, Tablin, F., Walker, N.J. and Crowe, J.H. (1999). The internal calcium concentration of human platelets increases during chilling. Biochim. Biophys. Acta 1416, 349–360.

Payne, S.R., Oliver, J.E. and Upretti, G.C. (1994). Effect of antifreeze proteins on the motility of ram spermatozoa. Cryobiology 31, 180–184.

Peters, I.D., Rancourt, D.E., Davies, P.L. and Walker, V.K. (1993). Isolation and characterization of an antifreeze protein precursor from transgenic *Drosophila*: evidence for partial processing. Biochim.Biophys. Acta 1171, 247–254.

Pickett, M.H., White, B.N. and Davies, P.L. (1983). Evidence that translational control mechanisms operate to optimize antifreeze protein production in the winter flounder. J. Biol. Chem. 258, 14762–14765.

Ramsey, A.J. (1964) The rectal complex of the mealworm *Tenebrio molitor*. L. Philos. Trans. R. Soc. Ser. B 248, 279–314.

Rancourt, D.E., Davies, P.L. and Walker, V.K. (1992). Differential translatability of antifreeze protein mRNAs in a transgenic host. Biochim. Biophys. Acta 1129, 188–194.

Rancourt, D.E., Duncker, B., Davies, P.L and Walker, V.K. (1991). A flightless host for biological containment of P-mediated transformants. Drosophila Information Service 70, 185.

Rancourt, D.E., Peters, I.D., Walker, V.K. and Davies, P.L. (1990). Wolffish antifreeze protein from transgenic *Drosophila melanogaster*. Bio/Technology 8, 453–457.

Rancourt, D.E., Walker, V.K. and Davies, P.L. (1987). Flounder antifreeze protein synthesis under heat shock control in transgenic *Drosophila melanogaster*. Mol. Cell. Biol. 7, 2188–2195.

Raymond, M., Callaghan, A., Fort, P. and Pasteur, N. (1991). Worldwide migration of amplified insecticide resistance genes in mosquitoes. Nature 350, 151–153.

Raymond, J.A. and DeVries, A.L. (1977). Adsorption inhibition as a mechanism of freezing resistance in polar fishes. Proc. Natl. Acad. Sci. USA 86, 881–885.

Robertson, H.M., Preston, C.R., Phillis, R.W., Johnson-Schlitz, D.M., Benz, W.K. and Engels, W.R. (1988). A stable source of P element transposase in *Drosophila melanogaster*. Genetics 118, 461–470.

Rubinsky, B., Arav, A. and Fletcher, G.L. (1991). Hypothermic protection—a fundamental property of 'antifreeze' proteins. Biochem. Biophys. Res. Commun.180, 566–571.

Rubinsky, B., Arav, A., Mattioli, M. and DeVries, A.L. (1990). The effect of antifreeze glycopeptides on membrane potential changes at hypothermic temperatures. Biochem. Biophys. Res. Commun. 173, 1369–1374.

Scott, G.K., Fletcher, G.L. and Davies, P.L. (1986) Fish antifreeze proteins: recent gene evolution. Can. J.Fisheries and Aquatic Sci. 43, 1028–1034.

Scott, G.K., Hayes, P.H., Fletcher, G.L. and Davies, P.L. (1988). Wolffish antifreeze genes are primarily organized as tandem repeats that each contain two genes in inverted orientation. Mol. Cell. Biol. 8, 3670–3675.

Shears, M.A., Fletcher, G.L., Hew, C.L., Gauthier, S. and Davies, P.L. (1991). Transfer, expression and stable inheritance of antifreeze protein genes in Atlantic salmon (*Salmo salar*). Mol. Mar. Biol. Biotech. 1, 58–63.

Smallwood, M., Worrall, D., Byass, L., Elias, L., Ashford, D., Doucet, C.J., Holt, Telford, J., Lillford, P. and Bowles, D. (1999) Isolation and characterization of a novel antifreeze protein from carrot. Biochem. J. 340, 385–391.

Spradling, A.C. (1986). P-element mediated transformation. In: Drosophila: a practical approach. (Roberts, D.B., Ed.), pp. 175–197. IRL Press, Oxford.

Tablin, F., Oliver, A.E., Walker, N.J., Crowe, L.M. and Crowe, J.H. (1996) Membrane phase transition of intact human platelets: correlation with cold-induced activation. J. Cell Physiol. 168, 305–313.

Tyshenko, M. G., Doucet, D., Davies, P.L. and Walker, V.K. (1997). The antifreeze potential of the spruce budworm thermal hysteresis protein. Nature Biotech. 15, 887–890.

Urratia, M.E., Duman, J.G. and Knight, C.A. (1992). Plant thermal hysteresis proteins. Biochim. Biophys. Acta 1121, 199–206.

Wang, T., Zhu, Q., Yang, X., Layne, J.R. and DeVries, A.L. (1994). Antifreeze glycoproteins from Antarctic notothenioid fishes fail to protect the rat cardiac explant during hypothermic and freezing preservation. Cryobiology 31, 185–192.

Walker, V.K. (1989). Gene transfer in insects. Advances in Cell Culture 7, 87–124.

Worrall, D., Elias, L., Ashford, D., Smallwood, M., Sidebottom, C., Lillford, P., Telford, J., Holt, C. and Bowles, D. (1998). A carrot leucine-rich-repeat protein that inhibits ice recrystallization. Science 282, 115–117.

Yeh, Y. and Feeney, R.E. (1996) Antifreeze proteins: structures and mechanisms of function. Chem. Rev. 96, 601–617.

Protein Adaptations and Signal Transduction. Edited by K.B. Storey and J.M. Storey

CHAPTER 3

Cold-adapted Enzymes: An Unachieved Symphony

Salvino D'Amico, Paule Claverie, Tony Collins, Georges Feller, Daphné Georlette, Emmanuelle Gratia, Anne Hoyoux, Marie-Alice Meuwis, Laurent Zecchinon and Charles Gerday*
Laboratory of Biochemistry, Institute of Chemistry B6, University of Liège, B-4000 Liège, Belgium.

1. Introduction

Temperature is one of the most important environmental factors for life. Cold-adapted or psychrophilic organisms are able to thrive at temperatures around 0°C. They are either prokaryotic or eukaryotic and represent a significant portion of the living world because temperatures over a considerable portion of our planet (e.g. polar and alpine regions, deep-sea waters) are below 5°C. Evolution has allowed these adapted organisms, named psychrophiles, to survive and grow in the restrictive conditions of these cold habitats. In these environments psychrophiles display metabolic fluxes more or less comparable with those exhibited by mesophiles at moderate temperatures. The challenge for them is to counteract the reduction in chemical reaction rates due to low temperatures. The principles of enzyme cold-adaptation as understood by our current knowledge is described below.

2. The low temperature challenge

It is well known that rates of chemical reactions decrease with temperature; this is basically described by the Arrhenius equation:

$$k = Ae^{-E_a/RT} \qquad (1)$$

where k is the rate constant, A is the preexponential factor that depends on the reaction, E_a is the activation energy, R is the gas constant (8.31 kJ mol^{-1}) and T is the temperature in Kelvin. According to this equation, any decrease in temperature will induce an exponential decrease of the reaction rate. Assuming that for most biological reactions, a decrease by 10°C will divide the reaction rate by a factor 2 to 3 (Q_{10} = 2 to 3), this latter should be 16–80 times lower for a shift in temperature from 37°C to 0°C. Of course, the Arrhenius law also applies to psychrophiles but they have developed adaptation mechanisms enabling them to display, despite the low temperature of their environment, appropriate reaction rates contrary to non-adapted organisms which would not survive or grow satisfactorily in such conditions of environmental stress.

There are two possible ways to adapt to such hard conditions: synthesizing more enzymes or synthesizing cold-efficient enzymes. It can be easily understood that the first possibility is energetically expensive; therefore, the best strategy to maintain sustainable activity at low temperatures is to produce cold-adapted proteins. Another strategy, not very common, lies in the production of perfectly evolved enzymes like triose phosphate isomerase (Alvarez et al., 1998) for which the reaction rate is so fast that these enzymes are said to be diffusion controlled. In that case, the activation energy E_a is close to zero and therefore the exponential term tends to 1. As a consequence, these reactions are virtually independent of temperature.

*Corresponding author.

Table 3.1. Kinetic parameters of some psychrophilic enzymes. Temperatures of the experiments are indicated in brackets. See text for references.

Extracellular enzymes		k_{cat} (s^{-1})	k_{cat} ratio (psychrophile /mesophile)
α-amylase	Psychrophile (4°C)	490	
	Mesophile	71	6.9
Subtilisin	Psychrophile (5°C)	32	
	Mesophile	18	1.8
Xylanase	Psychrophile (4°C)	14.8	
	Mesophile	4.9	3
Intracellular enzymes		K_m (μM)	K_m ratio (mesophile/psychrophile)
Malate dehydrogenase	Psychrophile (4°C)	16	
	Mesophile	23	1.4
Phosphoglycerate kinase	Psychrophile (25°C)	210 (ATP)	
	Mesophile	370	1.8
	Psychrophile	530 (3-PGA)	
	Mesophile	590	1.1
DNA Ligase	Psychrophile (4°C)	0.165	
	Mesophile(30°C)	0.702	4.3

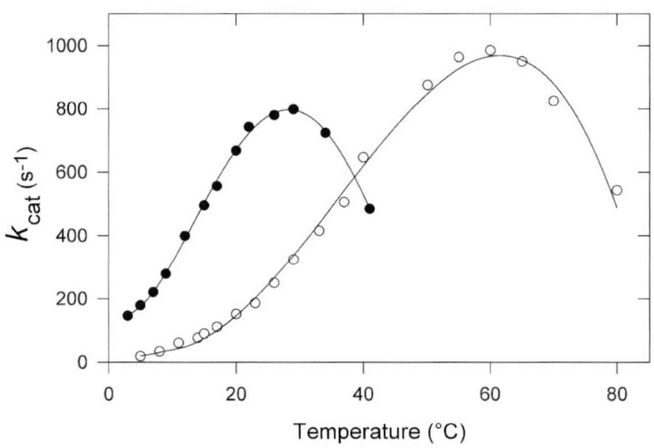

Fig. 3.1. Comparison of the temperature effect on activity of two homologous enzymes. Curves representing the thermal dependence of the specific activity of the psychrophilic α-amylase from *Pseudoalteromonas haloplanktis* (●) and of its thermostable homologue from *Bacillus amyloliquefaciens* (○).

The common strategy used by cold-adapted enzymes is to increase their catalytic efficiency k_{cat}/K_m (Feller and Gerday, 1997). For microbial extracellular enzymes that often work at saturating substrate concentrations, adaptation consists mainly of increasing k_{cat} (Feller et al., 1994; Narinx et al., 1997; Petrescu et al., 2000) (Table 3.1). The situation appears to be more complicated in the case of intracellular or extracellular enzymes facing poor substrate concentrations. In these cases, a decrease in K_m can provide a very useful higher substrate affinity (Kim et al., 1999; Bentahir et al., 2000; Georlette et al., 2000). Some enzymes can also improve both k_{cat} and K_m parameters in the course of their adaptation.

If, however, one looks to the adequacy of the adaptation achieved, it can be seen in most, if not all, cases that the specific activity displayed by psychrophilic enzymes around 0°C (Fig. 3.1) never reaches that displayed by mesophilic enzymes at 37°C. Therefore, adaptation is not complete and one can address the question: why?

3. Structural basis of adaptation to cold

3.1. Achievements

The growing interest in psychrophilic organisms and their enzymes as new tools in biotechnological processes (Russell, 1998; Gerday et al., 2000) arose from the properties of the cold-adapted enzymes which display a high specific activity at low and moderate temperatures but also a high thermosensitivity. Both properties can be useful in various biotechnological processes so the number of enzymes investigated has increased exponentially leading to the elucidation of the primary structures of numerous psychrophilic enzymes and also, in a few cases, the 3D structure (Russell, 2000). Structural models have indeed been proposed and the crystallographic structures of α-amylase (Aghajari et al., 1998a; Aghajari et al., 1998b), triose phosphate isomerase (Alvarez et al., 1998), citrate synthase (Russell et al., 1998), malate dehydrogenase (Kim et al., 1999) and Ca^{2+}, Zn^{2+} protease (Villeret et al., 1997) have also recently become available. These structures have provided new insights into understanding the molecular adaptation of cold-adapted enzymes. They reveal that only minor structural modifications are necessary to adapt a mesophilic or thermophilic homologue to cold (Fig. 3.2), as already suggested by the similarity of primary structures. For example, the α-amylase from an Antarctic bacterium shares more than 40% identity (60% similarity) with its mesophilic homologue from pig pancreas (D'Amico et al., 2000).

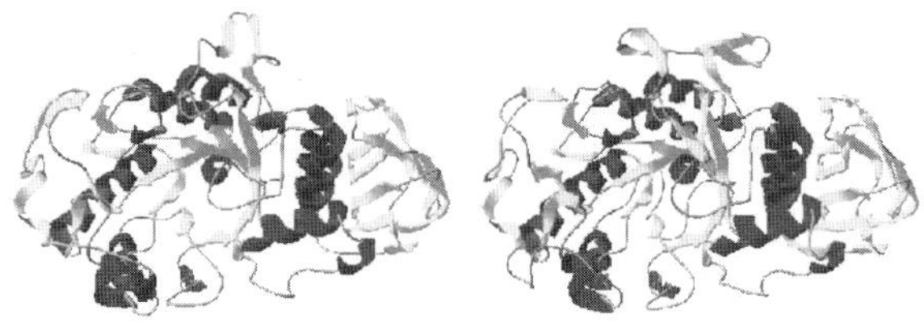

Fig. 3.2. Overall structural similarities. The psychrophilic α-amylase from the Antarctic bacteria *Pseudoalteromonas haloplanktis* is represented on the left and its homologous mesophile from the pig pancreas on the right. Picture generated by Swiss-PDBViewer (http://www.expasy.ch/spdbv/mainpage.html) with atomic coordinates from 1AQH and 1PPI, respectively (Brookhaven Protein Data Bank).

3.2. Active sites structural organization—the flexibility requirement

Investigation on active sites of psychrophilic enzymes revealed that all amino acids residues involved in the reaction mechanism are strictly conserved in comparison to their mesophilic equivalents (Holland et al., 1997; Fields and Somero, 1998). Moreover, an exhaustive comparison between complexes of a mesophilic and a psychrophilic α-amylase with an inhibitor (acarbose) revealed that all 24 residues involved in H-bonds with the pseudopolysaccharide are conserved (Qian et al., 1994; D'Amico et al., 2000). All this suggests that the molecular basis of cold adaptation has to be found elsewhere.

Low temperatures tend to improve the compacity of a protein by limiting the "breathing" of the structure corresponding to micro-unfolding processes. Therefore, at low temperatures, a mesophilic protein will intuitively lose the mobility required for its catalytic activity. The current accepted hypothesis (Fields and Somero, 1998; Zavodszky et al., 1998) suggests that psychrophilic enzymes have to increase their plasticity in order to perform catalysis at low temperatures, the increased plasticity being responsible for the generally low stability of the protein structure. This balance between flexibility and stability represents one of the crucial points in the adaptation of a protein to environmental temperature. For thermophiles, the selection pressure probably acts almost exclusively on the stability of the enzymes. For psychrophiles, if one retains the hypothesis of the need for flexibility, the main requirement is to increase the plasticity of appropriate parts of the protein to enable an easy accommodation of the substrates at low temperatures. At present, the flexibility of a protein remains a difficult parameter to measure in comparison with activity or stability since the increase in flexibility, required for catalysis, can be limited only to a small but crucial part of the protein. This is illustrated by the fact that the difference in flexibility is not often reflected by a difference in the B factor, the crystallographic temperature factor representing to a certain extent the disorder of the structure in a given region related to the square of the maximum distance of any atom from its average coordinate (Aghajari et al., 1998b). In the case of a cold-adapted citrate synthase, the B factors corresponding to the two constituting domains of the enzyme are smaller than those of the homologous citrate synthase from *Pyrococcus furiosus*. The authors tentatively explain the better catalytic efficiency of the psychrophilic enzyme by a better accessibility of the active site, possibly enhanced by an increase in the relative flexibility of one domain compared to the other. However, a few methods have been used successfully to evaluate the relative flexibility such as comparative hydrogen/deuterium exchange (Zavodszky et al., 1998), tryptophan phosphorescence (Fischer et al., 2000; Gershenson et al., 2000) and tryptophan fluorescence quenching (Chessa, 1999). The data favor the hypothesis of an inverse relation between stability and flexibility. It is worth mentioning here that homologous enzymes from different temperature environments have nearly identical conformational plasticity under their respective optimal temperature working conditions, what has been called "corresponding states" (Jaenicke, 1991).

3.3. Structural factors implicated in cold-adaptation

The refined 3D structures of psychrophilic enzymes have been quite useful in compiling an inventory of the structural factors possibly involved in cold-adaptation and also responsible for low thermal stability. In addition, some structural models are also available for cold-adapted subtilisin (Davail et al., 1994; Miyazaki et al., 2000), Antarctic fish trypsin (Genicot et al., 1996) and elastase (Aittaleb et al., 1997), lipase (Arpigny et al., 1997), β-lactamase (Feller et al., 1997b), alanine racemase (Okubo et al., 1999), xylanase (Petrescu et al., 2000), aspartate aminotransferase (Birolo et al., 2000), alkaline phosphatase (Rina et al., 2000) and phosphoglycerate kinase (Bentahir et al., 2000).

All these studies revealed that evolution made use of a specific strategy for each individual enzyme consisting of a set of structural adjustments selected from among the following possibilities:

- Decrease in:
  - Salt bridges;
  - H-bonds and aromatic interactions;
  - Ion binding constants;
  - Hydrophobic interactions;
  - Arginine content;
  - Proline residues in loops.
- Changed α-helix dipole interactions;
- Increased clustering of glycine residues;
- Insertions or deletions of loops responsible for specific properties.

Exhaustive analyses of these structural factors have been carried out previously (Arpigny et al., 1994; Feller et al., 1997a; Feller and Gerday, 1997). Only a few of these structural alterations, leading to a weakening of the stability of the molecular edifice, are used to acquire the required flexibility to be more or less adapted to the temperature of the environment. An increase in the proportion of non-polar residues exposed to the surrounding medium is also often observed causing an entropy-driven destabilizing effect (Makhatadze and Privalov, 1995). The careful analysis of the three-dimensional structures often reveals a better accessibility of the active site to the substrate. This has been clearly demonstrated from the structure of cold-active citrate synthase (Russell et al., 1998) and from the model structure of Antarctic fish elastase (Aittaleb et al., 1997) in which a loop limiting, in the mesophilic homologue, the access to the active site is deleted. This has also been shown in the Antarctic bacteria α-amylase (unpublished data) in which small deletions in loops limiting the active site induce a larger opening. The better accessibility certainly helps to reduce the energy required for substrate accommodation and/or reaction product release in the case of macromolecular substrates. It is, in this case, also a way to counteract the lower diffusion rate of molecules caused by the increased viscosity of the medium at low temperatures. However, one can question the usefulness of increasing the accessibility of the active site in the case of small substrates.

The importance of active site accessibility has been demonstrated by DNA shuffling on an indoleglycerol phosphate synthase (Merz et al., 2000). The authors selected mutants with improved catalytic activity at low temperatures; the mutated enzymes exhibit decreased affinity for both substrate and product as well as an increased catalytic turnover. The authors concluded that in the wild type enzyme the rate constant of product release is the limiting factor due to the poor flexibility of loops bordering the active site.

4. The activity–stability–flexibility trilogy

4.1. The current hypothesis

As previously seen, flexibility plays an important role in the thermal adaptation of proteins throughout the overall range of temperatures that support living organisms, from low temperatures (Fields and Somero, 1998) to high temperatures (Zavodszky et al., 1998). The relative flexibility of enzymes has been frequently suggested from structural comparisons (when available) and amino acid substitutions in homologous proteins adapted to different environments. As already mentioned, the comparison of the overall crystallographic thermal B factors often fails to reveal any significant variation between a cold-adapted enzyme and its mesophilic or thermophilic homologue. However, in the case of malate dehydrogenase from the psychrophile *Aquaspirillium arcticum* (Kim et al., 1999), it has been demonstrated that all main chain atoms and most of the side chain atoms interacting with NADH and oxaloacetate have an approximately 2-fold increase in relative B factors illustrating the importance of flexibility in the catalytic properties of the enzyme. As a matter of fact, modifications of the flexibility are often seen only in specific regions of the protein structure. This is probably due to the opposite requirements governing activity and stability. A significant increase in the overall flexibility will probably lead to enzymes of high specific activity but these may also be very unstable and at the limit of incorrect folding. To provide appropriate activity at low temperatures, an improved flexibility of crucial parts of the protein is, however, necessary. In this context, it has been proposed that the psychrophilic α-amylase from *Pseudoalteromonas haloplanktis* has reached the lowest possible stability of its native state (Feller et al., 1999). Even in this case, the specific activity of the cold-adapted enzyme at the environmental temperature (0°C) is still much lower than the specific activity of the mesophilic counterpart at 37°C, illustrating the limit of the adaptation. Millions of years of natural evolution and adaptation have partially solved the problem of the opposite requirement between activity and stability by limiting the areas of increased mobility. As a result, in naturally evolved enzymes, the changes in flexibility or plasticity are often localized in loops adjacent to active sites (for examples, see Fields and Somero, 1998; Merz et al., 2000). The reasons for this precise location are obvious: cold-adapted enzymes have to cope with the drastic decrease of environmental thermal energy. Increasing the flexibility of enzyme regions involved in catalysis probably reduces the energy barrier and so increases the catalytic activity. Moreover, as has been shown above, this could also induce a better capacity for product release and, thus, increase enzyme turnover. The consequence of such conformational alterations could be a decreased affinity of the cold-adapted enzymes for substrates. This has been observed in a limited number of cases such as triose phosphate isomerase (Alvarez et al., 1998). However, by contrast, in the case of trypsin from Antarctic fish, the catalytic efficiency is mainly improved by a decrease in the K_m value (Feller et al., 1996b). The challenge for cold adaptation is to limit the K_m increase while significantly raising the specific activity k_{cat}.

The relations between activity, stability and flexibility are therefore complex. Psychrophily requires an increased activity at low temperatures of enzymes displaying high coefficient control. This can be achieved by an increased flexibility giving rise to a decrease in the thermal stability. Does a decreased stability constitute an obligation in order to maintain activity at low temperatures? As often in life sciences, nothing is as simple as it appears.

4.2. Random mutagenesis, the perfect tool?

The first hypotheses regarding the adaptation of enzymes to low temperatures were derived from multiple amino acid sequences alignments (see for examples Somero, 1975; Zuber, 1988). A further and significant step was achieved when three dimensional structures became available, first in the form of models and then in the 1990s in the form of refined crystallographic structures, the best studied being the psychrophilic α-amylase (Aghajari et al., 1996). The alignments of primary structures revealed in psychrophiles a number of substitutions which could give rise to decreased thermostability. It was initially assumed that these substitutions were the consequence of an adaptative strategy to cold. However, when experiments with site-directed mutagenesis were attempted over the last ten years with enzymes such as lactate dehydrogenase (Zulli et al., 1991), α-amylase (Feller et al., 1996a), subtilisin (Narinx et al., 1997; Miyazaki et al., 2000), triose phosphate isomerase (Alvarez et al., 1998) or β-glucosidase (Gonzalez-Blasco et al., 2000), only a few of them confirmed the hypotheses put forward illustrating the fact that the rational approach is not the panacea leading to a direct understanding of the adaptation of a protein to a single constraint.

Very recently however, random mutagenesis techniques were applied. The principle is to induce random mutations in a gene at a controlled rate. The best known system uses error-prone PCR amplification and DNA shuffling but other systems do exist like chemical mutagenesis or use of mutator cells (Greener et al., 1997). Proteins were therefore engineered by directed evolution experiments made of mutagenesis steps, recombination and screening processes able to identify interesting mutants on the basis of an altered property (thermostability, for example). The characteristic of directed evolution is to allow the researcher to select a specific property that will be subjected to selective pressure. Directed evolution seems to be the perfect tool to understand how the activity, stability or flexibility of a protein can be modified and what is the relationship, if there is any, between these basic properties (Van den Burg et al., 1998). Until now, to our knowledge, only one psychrophilic enzyme, subtilisin S41 (Davail et al., 1994) has been submitted to such an experiment (Miyazaki et al., 2000). A variant with 7 mutations was found to exhibit higher th ermostability without compromising activity at low temperatures at least towards a synthetic substrate. In this case, an enhanced affinity for calcium was largely responsible for increased stability but not through a modification of residues directly implicated in the cation coordination. The majority of other directed evolution experiments were focused on mesophilic and thermophilic proteins and they can be divided in two categories. In the first, random mutagenesis was used to improve catalytic activity at low temperatures; this has been carried out on thermophilic indoleglycerol phosphate synthase (Merz et al., 2000), β-glucosidase (Lebbink et al., 2000) and on a mesophilic subtilisin (Taguchi et al., 1998). In the second category, selection pressure was focused on a higher thermostability, in the case of a mesophilic esterase (Giver et al., 1998; Spiller et al., 1999; Gershenson et al., 2000), subtilisin E (Zhao and Arnold, 1999), 3-isopropylmalate dehydrogenase (Akanuma et al., 1998; Akanuma et al., 1999), β-glucosidase (Gonzalez-Blasco et al., 2000) and fungal peroxidase (Cherry et al., 1999).

Some interesting points arose from these experiments. Systematically, when catalytic activity at lower temperatures was sought, a decreased stability as well as a higher (supposed) flexibility were induced, thus perfectly matching the initial working hypothesis. By contrast, when the main selection pressure was thermal stability, mutants with higher stability and unchanged or even increased catalytic activity were found in contradiction with the hypothesis that a high stability induces a low catalytic activity. As has been pointed out for the fungal peroxidase (Cherry et al., 1999), stabilized mutants selected from the first generations exhibited a lower activity in agreement with expectations; this tendency seems to be corroborated by experiments on other enzymes, e.g. an esterase (Giver et al., 1998). Several generations of mutants are, therefore, required to obtain a significant increase both in stability and activity. Indeed, 13 and 7 mutations were recorded in the most stable and active esterase and peroxidase, respectively. It is clear that maintaining a low temperature activity in thermostable mutants is a very difficult task since this requires a high selective pressure for a long period. One has also to be aware that an apparently better catalytic efficiency can, in certain cases, reflect only a better complementary for the substrate selected for the analysis.

4.3. Natural evolution vs. directed evolution

As illustrated above, it appears possible in the laboratory to increase both the activity of an enzyme at low temperatures and its stability at high temperatures. Therefore, no physical or chemical constraint seems to impair the possibility of improving both properties, within certain limits. The fact is, however, that enzymes displaying a high stability as well as a high flexibility and activity are not found in nature. The tradeoff between flexibility and stability is easy to understand and always observed without exceptions. Moreover, the correlation between these two properties has also been demonstrated by tryptophan phosphorescence experiments on laboratory-evolved thermostable esterases (Gershenson et al., 2000). As far as the relationship between activity and stability is concerned one can see that in natural environments all psychrophilic enzymes exhibit a lower stability and a higher specific activity when compared to their mesophilic counterparts provided that the appropriate substrate is used. However, random mutagenesis has demonstrated that these two properties are not necessarily incompatible and thermostable mutants can exhibit both higher stability and activity simultaneously. During natural evolution, proteins probably evolved towards a configuration displaying appropriate properties with regards to the environment through a selective pressure exerted towards the most important parameter, such as the activity at low temperatures in the case of psychrophilic enzymes. Thermal stability is not an important factor for a psychrophilic enzyme. In the laboratory, on the other hand, the parameters that were selected during the screening step were both stability and activity. This found all possible combinations between stability and activity leading to mutated enzymes displaying high stability and high activity. In nature, the selection conditions are not so drastic and the evolutionary process takes place guided by the survival requirement. As mentioned before, there is no need for a psychrophilic enzyme to be stable at high temperatures since this is never the experience in the natural environment. Therefore, we can reasonably assume that the loss of stability inherent to most psychrophilic enzymes is the result of a random genetic drift during divergent evolution from a mesophilic ancestor facilitated by the flexibility requirement. The same argument can be developed in the case of thermophilic enzymes for which the first constraint is thermal stability.

4.4. Differential scanning calorimetry

Differential scanning calorimetry (DSC) is a powerful tool to investigate the thermal unfolding of proteins and their stability. Unlike other methods, it allows direct measurement of thermodynamic parameters without extrapolation from indirect data. Indeed, the melting temperature is only one aspect of thermal stability which can also be detected by fluorescence or CD spectroscopy. DSC also allows us to calculate, in the favorable case of a reversible denaturation process, the stabilization energy which is another dimension of an enzyme's thermostability not seen by other methods. In fact, two homologous proteins can display the same melting temperature but have a different thermostability. Indeed, the stabilization energy, meaning the energy necessary to unfold the protein, can be quite different. This aspect of thermal unfolding has been discussed by Beadle et al. (1999). As an example, the stabilization energy (ΔG) curves of the psychrophilic wild-type α-amylase and of one mutant are illustrated in Fig. 3.3 (D'Amico et al., unpublished data). Despite the modest increase in melting temperature (+2.5°C), one can see that the mutant has a maximum stabilization energy higher by not less than 50% when compared to that of the wild type enzyme.

Thermograms of the psychrophilic α-amylase, its mesophile homologue, and the thermostable enzyme from *Bacillus amyloliquefaciens* are presented in Fig. 3.4. The low thermal stability of the cold-adapted protein is clearly illustrated by its low melting temperature, T_m (top of the peak

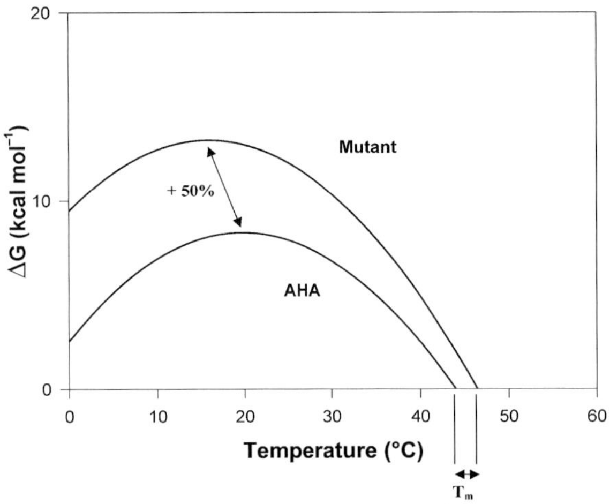

Fig. 3.3. Stabilization energy curves. The conformational stability of the psychrophilic α-amylase from *Pseudoalteromonas haloplanktis* (AHA) is compared with that of the stabilized mutant (N150D/V196F) incorporating an additional salt bridge and aromatic interaction. Note the dramatic increase of stability despite the small differences in their melting temperature (T_m).

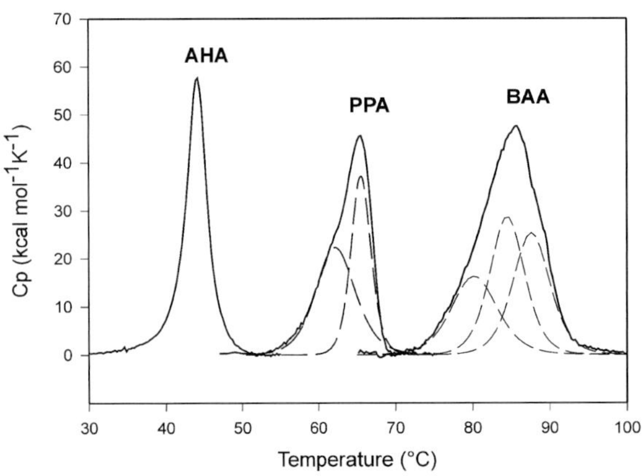

Fig. 3.4. Thermal unfolding of three α-amylases recorded by differential scanning calorimetry. From left to right: thermograms of α-amylases from the psychrophile *Pseudoalteromonas haloplanktis* (AHA), the mesophile, pig pancreas (PPA), and the thermostable enzyme from *Bacillus amyloliquefaciens* (BAA). Deconvolutions of the peaks into two or three domains, respectively, for PPA and BAA are shown in dashed lines.

representing the point of 50% denaturation) of 44°C. Unfolding enthalpy can also be calculated from the area below the curve corresponding to the heat absorbed during the unfolding process. The lower value found for the psychrophilic enzyme reflects the low amount of energy required for its denaturation. A detailed calorimetric study of the cold-adapted α-amylase can be found in Feller et al. (1999). The authors showed that this enzyme unfolds reversibly and without any stable intermediate according to a highly cooperative process. Oppositely, more stable α-amylases unfold irreversibly and the denaturation peaks exhibit distinct thermodynamic units or domains with different stability (Fig. 3.4). Therefore, it should be possible in the case of the psychrophilic enzyme to observe an overall increase in flexibility. Note that H/D exchange experiments on mesophilic and thermophilic 3-isopropylmalate dehydrogenases revealed global differences in the flexibility of the enzymes (Zavodszky et al., 1998). It is interesting that, in this case, calorimetric melting profiles also exhibit a one step cooperative denaturation.

However, from other work, it appears that the adaptation strategy can be different from that leading to the most unstable possible state of the cold-adapted global structure. The calorimetric curves for psychrophilic phosphoglycerate kinase (PGK) exhibit an unexpected profile illustrating another adaptation strategy to a cold environment (Bentahir et al., 2000). The enzyme appears to be composed of a heat-labile and a heat-stable domain (two denaturation peaks), the first being responsible for the weak thermal stability of the enzyme and the second showing a stability higher than any unit of the mesophilic counterpart. This indicates that only a part of the protein has increased its flexibility in order to increase activity at low temperatures. This fact is in concordance with what we have seen in Section 4.1 and is further emphasized by calorimetric experiments carried out on a psychrophilic chitobiase (Lonhienne, 2000). This enzyme also exhibits two domains having quite different stabilities and it has been demonstrated that the less stable one includes the active site. These results can explain how cold-adapted enzymes like the PGK and the chitobiase can increase their specific activity, by increasing the flexibility of one domain, while maintaining a reasonable substrate affinity by keeping the other domain even more stable than any domain of the mesophilic counterpart. It is worth mentioning that this unfolding behavior has also been shown and discussed in the case of a mesophilic esterase (Gershenson et al., 2000).

5. Conclusion and perspectives

Psychrophilic organisms and their enzymes have, in recent years, increasingly attracted the attention of the scientific community due to their peculiar characteristics giving a new orientation to research on protein folding and stability. Moreover, their extremophilic properties corresponding to a high specific activity at low and moderate temperatures as well as a low thermostability render cold-adapted enzymes valuable tools for biotechnological applications.

From the analysis of the available structures, it is striking that adaptation to cold can follow different strategies, each aiming to increase the catalytic efficiency at low temperatures. This can be achieved either by increasing the specific activity k_{cat} or by increasing the substrate affinity in specific cases, meaning decreasing K_m, or by improving both parameters simultaneously. In natural environments, the enhanced performance of these enzymes is always accompanied by a decrease in stability originating from a significant improvement of the flexibility of the catalytic region or of the overall structure. A comparison of the three dimensional structure of psychrophilic enzymes and their mesophilic counterparts threw some light on the possible subtle variations which could be responsible for the adaptation, but amazingly the subsequent experiments using site-directed mutagenesis to check the hypothesis have, with few exceptions, failed to clarify the problem. It is clear that in the course of natural evolution each enzyme has been the target of multiple constraints which precluded any rational approach to detect the effect of one environmental parameter, in this case, low environmental temperature.

Directed evolution, on the other hand, emerged recently as the best approach to investigate the multiple possibilities of structural modification which could give rise to a change in a specific property of an enzyme. Even in this case, however, the variety of possible strategies is such that the identified sites do not necessarily correspond to what has been achieved by nature. This means that if, for example, a mesophilic enzyme is transformed into a psychrophilic enzyme by directed evolution, the modifications to its structure will not necessarily correspond to the differences that are seen in the natural psychrophilic enzyme. This is due to the complexity of the effects caused by one single mutation and the unexpected synergy or the unexpected antagonist effects of several of them. Accumulation of mutations in clusters can lead to cooperative interactions that propagate mutational effects over a large distance. This phenomenon further complicates the predictability of the effect of a single alteration. However, directed evolution is certainly the best way to confer to a given recombinant enzyme the improved competencies required by a biotechnological process.

The quantitative evaluation of the relative flexibility of parts of the whole protein is another step necessary for understanding the relationship between stability, flexibility and activity since it is certainly a crucial factor in the thermal adaptation of enzymes. Experiments in this direction have been recently carried out using H/D exchange techniques (IR or NMR), tryptophan fluorescence quenching using acrylamide and tryptophan phosphorescence lifetime which can only be used in favorable cases. The H/D exchange technique is certainly of most general application and this, together with differential scanning microcalorimetry, gives the investigator two powerful tools to quantify the adaptation level of psychrophilic enzymes.

Acknowledgements

The authors wish to thank N. Gérardin and R. Marchand for their skilful technical assistance. The Institut Français de Recherche et de Technologie Polaire is also acknowledged for the generous support provided at the Antarctic station J.S. Dumont D'Urville. This work has been supported by the European Union Contracts: CT97-0131, CT95-

0017 and CT96-0051, the "Region Wallonne" contract no. 1928 and the Fonds National de la Recherche Scientifique: Grant 2.4523.97 to Ch. Gerday.

References

Aghajari, N., Feller, G., Gerday, C. and Haser, R. (1996). Crystallization and preliminary X-ray diffraction studies of α-amylase from the Antarctic psychrophile *Alteromonas haloplanctis* A23. Protein Sci. 5, 2128–2129.

Aghajari, N., Feller, G., Gerday, C. and Haser, R. (1998a). Crystal structures of the psychrophilic α-amylase from *Alteromonas haloplanctis* in its native form and complexed with an inhibitor. Protein Sci. 7, 564–572.

Aghajari, N., Feller, G., Gerday, C. and Haser, R. (1998b). Structures of the psychrophilic *Alteromonas haloplanctis* α-amylase give insights into cold adaptation at a molecular level. Structure 6, 1503–1516.

Aittaleb, M., Hubner, R., Lamotte-Brasseur, J. and Gerday, C. (1997). Cold adaptation parameters derived from cDNA sequencing and molecular modelling of elastase from Antarctic fish *Notothenia neglecta*. Protein Eng. 10, 475–477.

Akanuma, S., Yamagishi, A., Tanaka, N. and Oshima, T. (1998). Serial increase in the thermal stability of 3-isopropylmalate dehydrogenase from *Bacillus subtilis* by experimental evolution. Protein Sci. 7, 698–705.

Akanuma, S., Yamagishi, A., Tanaka, N. and Oshima, T. (1999). Further improvement of the thermal stability of a partially stabilized *Bacillus subtilis* 3-isopropylmalate dehydrogenase variant by random and site-directed mutagenesis. Eur. J. Biochem. 260, 499–504.

Alvarez, M., Zeelen, J.P., Mainfroid, V., Rentier-Delrue, F., Martial, J.A., Wyns, L., Wierenga, R.K. and Maes, D. (1998). Triose-phosphate isomerase (TIM) of the psychrophilic bacterium *Vibrio marinus*. Kinetic and structural properties. J. Biol. Chem. 273, 2199–2206.

Arpigny, J.L., Feller, G., Davail, S., Génicot, S., Narinx, E., Zekhnini, Z. and Gerday, C. (1994). In: Molecular adaptations of enzymes from thermophilic and psychrophilic organisms. (Gilles, R., Ed.). pp. 269–295. Adv. Comp. Envir. Physiol., Springer, Berlin.

Arpigny, J.L., Lamotte, J. and Gerday, C. (1997). Molecular adaptation to cold of an Antarctic bacterial lipase. J. Mol. Catal. B 3, 29–35.

Beadle, B.M., Baase, W.A., Wilson, D.B., Gilkes, N.R. and Shoichet, B.K. (1999). Comparing the thermodynamic stabilities of a related thermophilic and mesophilic enzyme. Biochemistry 38, 2570–2576.

Bentahir, M., Feller, G., Aittaleb, M., Lamotte-Brasseur, J., Himri, T., Chessa, J.P. and Gerday, C. (2000). Structural, kinetic, and calorimetric characterization of the cold-active phosphoglycerate kinase from the Antarctic *Pseudomonas sp.* TACII18. J. Biol. Chem. 275, 11147–11153.

Birolo, L., Tutino, M.L., Fontanella, B., Gerday, C., Mainolfi, K., Pascarella, S., Sannia, G., Vinci, F. and Marino, G. (2000). Aspartate aminotransferase from the Antarctic bacterium *Pseudoalteromonas haloplanktis* TAC 125. Cloning, expression, properties, and molecular modeling. Eur. J. Biochem. 267, 2790–2802.

Cherry, J.R., Lamsa, M.H., Schneider, P., Vind, J., Svendsen, A., Jones, A. and Pedersen, A.H. (1999). Directed evolution of a fungal peroxidase. Nat. Biotechnol. 17, 379–384.

Chessa, J.P. (1999). Molecular adaptations of a psychrophilic metalloprotease. PhD Thesis, University of Liège, Belgium.

D'Amico, S., Gerday, C. and Feller, G. (2000). Structural similarities and evolutionary relationships in chloride-dependent α-amylases. Gene 253, 95–105.

Davail, S., Feller, G., Narinx, E. and Gerday, C. (1994). Cold adaptation of proteins. Purification, characterization, and sequence of the heat-labile subtilisin from the antarctic psychrophile *Bacillus* TA41. J. Biol. Chem. 269, 17448–17453.

Feller, G., Payan, F., Theys, F., Qian, M., Haser, R. and Gerday, C. (1994). Stability and structural analysis of α-amylase from the antarctic psychrophile *Alteromonas haloplanctis* A23. Eur. J. Biochem. 222, 441–447.

Feller, G., Bussy, O., Houssier, C. and Gerday, C. (1996a). Structural and functional aspects of chloride binding to *Alteromonas haloplanctis* alpha-amylase. J. Biol. Chem. 271, 23836–23841.

Feller, G., Narinx, E., Arpigny, J-L., Aittaleb, M., Baise, E., Genicot, S. and Gerday, C. (1996b). Enzymes from psychrophilic organisms. FEMS Microbiol. Rev. 18, 189–202.

Feller, G., Arpigny, J.L., Narinx, E. and Gerday, C. (1997a). Molecular adaptations of enzymes from psychrophilic organisms. Comp. Biochem. Physiol. 118, 495–499.

Feller, G., Zekhnini, Z., Lamotte-Brasseur, J. and Gerday, C. (1997b). Enzymes from cold-adapted microorganisms. The class C β-lactamase from the Antarctic psychrophile *Psychrobacter immobilis* A5. Eur. J. Biochem. 244, 186–191.

Feller, G. and Gerday, C. (1997). Psychrophilic enzymes: molecular basis of cold adaptation. Cell Mol. Life Sci. 53, 830–841.

Feller, G., d'Amico, D. and Gerday, C. (1999). Thermodynamic stability of a cold-active α-amylase from the Antarctic bacterium *Alteromonas haloplanctis*. Biochemistry 38, 4613–4619.

Fields, P.A. and Somero, G.N. (1998). Hot spots in cold adaptation: Localized increases in conformational flexibility in lactate dehydrogenase A(4) orthologs of Antarctic notothenioid fishes. Proc. Natl. Acad. Sci. U.S.A. 95, 11476–11481.

Fischer, C.J., Schauerte, J.A., Wisser, K.C., Gafni, A. and Steel, D.G. (2000). Hydrogen exchange at the core of *Escherichia coli* alkaline phosphatase studied by room-temperature tryptophan phosphorescence. Biochemistry 39, 1455–1461.

Genicot, S., Rentier-Delrue, F., Edwards, D., VanBeeumen, J. and Gerday, C. (1996). Trypsin and trypsinogen from an Antarctic fish: molecular basis of cold adaptation. Biochim. Biophys. Acta 1298, 45–57.

Georlette, D., Jonsson, Z.O., Van Petegem, F., Chessa, J., Van Beeumen, J., Hubscher, U. and Gerday, C. (2000). A DNA ligase from the psychrophile *Pseudoalteromonas haloplanktis* gives insights into the adaptation of proteins to low temperatures. Eur. J. Biochem. 267, 3502-3512.

Gerday, C., Aittaleb, M., Bentahier, M., Chessa, J.P., Claverie, P., Collins, T., D'Amico, S., Dumont, J., Garsoux, G., Georlette, D., Hoyoux, A., Lonhienne, T., Meuwis, M.-A. and Feller, G. (2000). Cold-adapted enzymes: from fundamentals to biotechnology. Trends Biotechnol. 18, 103–107.

Gershenson, A., Schauerte, J.A., Giver, L. and Arnold, F.H. (2000). Tryptophan phosphorescence study of enzyme flexibility and unfolding in laboratory-evolved thermostable esterases. Biochemistry 39, 4658–4665.

Giver, L., Gershenson, A., Freskgard, P.O. and Arnold, F.H. (1998). Directed evolution of a thermostable esterase. Proc. Natl. Acad. Sci. U. S. A. 95, 12809–12813.

Gonzalez-Blasco, G., Sanz-Aparicio, J., Gonzalez, B., Hermoso, J.A. and Polaina, J. (2000). Directed evolution of β-glucosidase A from *Paenibacillus polymyxa* to thermal resistance. J. Biol. Chem. 275, 13708–13712.

Greener, A., Callahan, M. and Jerpseth, B. (1997). An efficient random mutagenesis technique using an *E. coli* mutator strain. Mol. Biotechnol. 7, 189–195.

Holland, L. Z., McFall-Ngai, M. and Somero, G. N. (1997). Evolution of lactate dehydrogenase-A homologs of barracuda fishes (genus *Sphyraena*) from different thermal environments: differences in kinetic properties and thermal stability are due to amino acid substitutions outside the active site. Biochemistry 36, 3207–3215.

Jaenicke, R. (1991). Protein stability and molecular adaptation to extreme conditions. Eur. J. Biochem. 202, 715–728.

Kim, S.Y., Hwang, K.Y., Kim, S.H., Sung, H.C., Han, Y.S. and Cho, Y.J. (1999). Structural basis for cold adaptation. Sequence, biochemical properties, and crystal structure of malate dehydrogenase from a psychrophile *Aquaspirillium arcticum*. J. Biol. Chem. 274, 11761–11767.

Lebbink, J.H., Kaper, T., Bron, P., van der Oost, J. and de Vos, W.M. (2000). Improving low-temperature catalysis in the hyperthermostable *Pyrococcus furiosus* β-glucosidase CelB by directed evolution. Biochemistry 39, 3656–3665.

Lonhienne, T. (2000). Cold-adaptation of chitinolytic enzymes secreted by an Antarctic marine bacterium. PhD Thesis, University of Liège, Belgium.

Makhatadze, G.I. and Privalov, P.L. (1995). Energetics of protein structure. Adv. Protein Chem. 47, 307–425.

Merz, A., Yee, M.C., Szadkowski, H., Pappenberger, G., Crameri, A., Stemmer, W.P., Yanofsky, C. and Kirschner, K. (2000). Improving the catalytic activity of a thermophilic enzyme at low temperatures. Biochemistry 39, 880–889.

Miyazaki, K., Wintrode, P.L., Grayling, R.A., Rubingh, D.N. and Arnold, F.H. (2000). Directed evolution study of temperature adaptation in a psychrophilic enzyme. J. Mol. Biol. 297, 1015–1026.

Narinx, E., Baise, E. and Gerday, C. (1997). Subtilisin from psychrophilic antarctic bacteria: characterization and site-directed mutagenesis of residues possibly involved in the adaptation to cold. Protein Eng. 10, 1271–1279.

Okubo, Y., Yokoigawa, K., Esaki, N., Soda, K. and Kawai, H. (1999). Characterization of psychrophilic alanine racemase from *Bacillus psychrosaccharolyticus*. Biochem. Biophys. Res. Commun. 256, 333–340.

Petrescu, I., Lamotte-Brasseur, J., Chessa, J.P., Ntarima, P., Claeyssens, M., Devreese, B., Marino, G. and Gerday, C. (2000). Xylanase from the psychrophilic yeast *Cryptococcus adeliae*. Extremophiles 4, 137–144.

Qian, M., Haser, R., Buisson, G., Duee, E. and Payan, F. (1994). The active center of a mammalian alpha-amylase. Structure of the complex of a pancreatic alpha-amylase with a carbohydrate inhibitor refined to 2.2-Å resolution. Biochemistry 33, 6284–6294.

Rina, M., Pozidis, C., Mavromatis, K., Tzanodaskalaki, M., Kokkinidis, M. and Bouriotis, V. (2000). Alkaline phosphatase from the Antarctic strain TAB5. Properties and psychrophilic adaptations. Eur. J. Biochem. 267, 1230–1238.

Russell, N.J. (1998). Molecular adaptations in psychrophilic bacteria: potential for biotechnological applications. Adv. Biochem. Eng. Biotechnol. 61, 1–21.

Russell, N.J. (2000). Toward a molecular understanding of cold activity of enzymes from psychrophiles. Extremophiles 4, 83–90.

Russell, R.J., Gerike, U., Danson, M.J., Hough, D.W. and Taylor, G.L. (1998). Structural adaptations of the cold-active citrate synthase from an Antarctic bacterium. Structure 6, 351–361.

Somero, G.N. (1975). Temperature as a selective factor in protein evolution: the adaptational strategy of "compromise". J. Exp. Zool. 194, 175–188.

Spiller, B., Gershenson, A., Arnold, F.H. and Stevens, R.C. (1999). A structural view of evolutionary divergence. Proc. Natl. Acad. Sci. U S A 96, 12305–12310.

Taguchi, S., Ozaki, A. and Momose, H. (1998). Engineering of a cold-adapted protease by sequential random mutagenesis and a screening system. Appl. Environ. Microbiol. 64, 492–495.

Van den Burg, B., Vriend, G., Veltman, O.R., Venema, G. and Eijsink, V.G. (1998). Engineering an enzyme to resist boiling. Proc. Natl. Acad. Sci. USA 95, 2056–2060.

Villeret, V., Chessa, J.P., Gerday, C. and Van Beeumen, J. (1997). Preliminary crystal structure determination of the alkaline protease from the Antarctic psychrophile *Pseudomonas aeruginosa*. Protein Sci. 6, 2462–2464.

Zavodszky, P., Kardos, J., Svingor and Petsko, G.A. (1998). Adjustment of conformational flexibility is a key event in the thermal adaptation of proteins. Proc. Natl. Acad. Sci. U.S.A. 95, 7406–7411.

Zhao, H. and Arnold, F.H. (1999). Directed evolution converts subtilisin E into a functional equivalent of thermitase. Protein Eng. 12, 47–53.

Zuber, H. (1988). Temperature adaptation of lactate dehydrogenase. Structural, functional and genetic aspects. Biophys. Chem. 29, 171–179.

Zulli, F., Schneiter, R., Urfer, R. and Zuber, H. (1991). Structure and function of L-lactate dehydrogenases from thermophilic and mesophilic bacteria, XI. Engineering thermostability and activity of lactate dehydrogenases from bacilli. Biol. Chem. Hoppe-Seyler 372, 363–372.

Protein Adaptations and Signal Transduction. Edited by K.B. Storey and J.M. Storey

CHAPTER 4

The Role of Cold-shock Proteins in Low-temperature Adaptation

Jeroen A. Wouters[1,2,3]*, Frank M. Rombouts[2], Oscar P. Kuipers[2,3]**, Willem M. de Vos[1,3], and T. Abee[1,2]

[1]*Wageningen Centre for Food Sciences (WCFS), 6703 GW Wageningen, The Netherlands;* [2]*Laboratory of Food Microbiology, Wageningen University, 6703 HD Wageningen, The Netherlands;* [3]*Flavour and Natural Ingredients Section, NIZO Food Research, 6710 BA Ede, The Netherlands. **Present address: Department of Genetics, Groningen Biomolecular Sciences and Biotechnology Institute (GBB), University of Groningen, 9751 NN Haren, The Netherlands*

1. Low-temperature adaptation and sensing

1.1. Low-temperature adaptation

Over the last decade, there has been great interest in the cold adaptation of food-related microorganisms. Safety and shelf life of many food products depend on cold storage. Also starter cultures, used in a variety of food fermentations, can be exposed to low temperature during storage, as well as during low-temperature fermentation and/or storage of fermented products. In combination, these applications created the need for understanding the cold-adaptation response of starter (lactic acid) bacteria, food spoilage bacteria and food pathogens (Abee and Wouters, 1999). A thorough understanding of the cold-adaptation process can be instrumental in optimizing fermentations at low temperature and may offer insight into methods to control the growth of spoilage and pathogenic microorganisms.

Research on cold adaptation of bacteria has mainly been focused on the biochemical changes in response to low-temperature exposure (Berry and Foegeding, 1997; Yamanaka et al., 1998; Graumann and Marahiel, 1998). Mechanisms that permit low-temperature growth of microorganisms include modifications in DNA supercoiling, maintenance of membrane fluidity, regulation of uptake and synthesis of compatible solutes, production of cold-shock proteins, modulation of mRNA secondary structure and, more generally, maintenance of the structural integrity of macromolecules and their assemblies, such as ribosomes (see recent reviews by Berry and Foegeding, 1997; Graumann and Marahiel, 1998; Yamanaka et al., 1998). The majority of studies on low-temperature adaptation has been performed with the mesophilic Gram-negative and Gram-positive model bacteria, *Escherichia coli* and *Bacillus subtilis*, respectively. However, currently numerous details are accumulating about the low-temperature response of other food-related bacteria (Abee and Wouters, 1999).

Research on cold adaptation has mainly focused on the synthesis of so-called cold-shock proteins (CSPs), a specific response that is shared by nearly all bacteria. These small (7 kDa) proteins are involved in gene expression, mRNA folding, transcriptional initiation and regulation and/or freeze-protection. Using primarily electrophoresis techniques other (non-7 kDa) low-temperature induced proteins have also been identified which will be referred to as cold-induced proteins (CIPs) in this review. Such proteins have been shown to be involved in a variety of cellular processes. In this

*Corresponding author.

review, novel aspects concerning the structure, function and control of CSPs and CIPs will be discussed, including a model for bacterial cold adaptation and possible mechanisms for low-temperature sensing.

1.2. Low-temperature sensing

One of the key questions to be answered is how bacteria sense low-temperature signals. The putative cellular thermosensors that have been proposed correlate to the major biochemical changes upon low-temperature exposure in bacterial cells, such as DNA topology, ribosomal structure and membrane composition (Van Bogelen and Neidhardt, 1990; Jones and Inouye, 1994; Gounot and Russell, 1999; Table 4.1). It has been well established that the sensing of heat shock interlinks with a number of two-component regulatory systems within the bacterial cell (Hoch and Silhavy, 1995). For low-temperature sensing, temperature-dependent phosphorylation and dephosphorylation of membrane and cytosol proteins have been reported for the psychrotrophic bacterium *Pseudomonas syringae*. It was suggested that both this cyclic (de)phosphorylation as well as the temperature-dependent phosphorylation of lipopolysaccharides are involved in thermal sensing (Ray et al., 1994a, 1994b). Recently, we identified a cold-induced signal transduction protein in the lactic acid bacterium *Lactococcus lactis*, which may also be involved in temperature sensing (Wouters et al., submitted).

In contrast to the situation for the heat-shock response, thus far no specific sigma factor regulating the cold-shock response has been identified. Some alternative (stress) sigma factors are also cold-induced, which suggests that these proteins are part of the cold-responsive regulon. Transcription of σ^S of *E. coli* and σ^B of the pathogen *Listeria monocytogenes* have been shown to be induced upon low-temperature exposure (Loewen et al., 1998; Becker et al., 1998). However, no evidence exists for the involvement of these sigma factors in the regulation of the CIPs characterized up till now (see below). In addition, σ^B of *B. subtilis* is not induced at low temperature as observed using *lacZ* promoter fusions and Western blotting (Becker et al., 1998). It was concluded that the control of the cold-shock stimulon is regulated in a different way than the heat-shock response (see below).

Three putative temperature sensors have been suggested to be involved in the bacterial cold-shock response (Table 4.1):

1. *Ribosomes*. A downshift in temperature causes a cold-sensitive block in the initiation of translation in *E. coli* cells, resulting in a decrease in polysomes and an increase in 70S monosomes and ribosomal subunits. VanBogelen and Neidhardt (1990) indicated that the ribosome might be the temperature sensor in bacteria, since the production of CSPs and CIPs increased upon incubation of *E. coli* cells with antibiotics specifically affecting the ribosome. This would indicate the ribosomal structure as a controlling factor for CIP production.

2. *Cytoplasmic membrane.* A decrease in temperature causes a decrease in the fluidity of the cytoplasmic membrane. In response to such a drop in fluidity, the proportion of shorter and/or unsaturated fatty acids in membrane lipids increases, allowing maintenance of optimal fluidity of the membrane. One of the most important consequences of these membrane lipid changes is the modulation of the activity of intrinsic proteins that perform functions such as ion pumping and nutrient uptake (Gounot and Russell, 1999). Temperature-induced changes in the biosynthesis of membrane fatty acids may occur by the activation or induction of enzymes (Russell, 1990). For *B. subtilis* a cold-induced membrane phospholipid desaturase (*des*) is described, which changes the saturation level of pre-existing fatty acids (Aguilar et al., 1998). Alterations in fatty acid branching, unsaturation via the anaerobic pathway or acyl chain shortening are mediated by *de novo* fatty acid synthesis, followed by integration in the membrane (Gounot and Russell, 1999). To date, only limited data have been presented describing the regulation of bacterial membrane modifying enzymes, including desaturases, and the control of their activity (Gounot and Russell, 1999). In yeast, it has been found that the disturbance of the membrane lipoprotein complexes is involved in the perception

Table 4.1. Proposed sensors for low-temperature adaptation and identified cold-induced proteins (CIPS)

Thermosensor	Cellular process	CIP	Organism	Ref.
1. Ribosome	Translation	PNPase	*E. coli, Y. enterocolitica*	1, 2
		CsdA	*E. coli*	3
		RbfA	*E. coli*	4
		IF-2	*E. coli*	1
		S6	*B. subtilis*	5
		L7/L12	*B. subtilis*	5
		Trigger factor	*E. coli*	6
2. Membrane		Des	*B. subtilis*	7
3. DNA topology	Transcription	H-NS, HU, HslA	*E. coli, B. subtilis, L. lactis*	1, 5, 8
		GyrA	*E. coli*	1
		NusA	*E. coli*	1
4. Signal Transduction		Lipopolysaccharides	*P. syringae*	9
		LlrC	*L. lactis*	8
	Amino acid metabolism	CysK	*B. subtilis*	5
		IlvC	*B. subtilis*	5
	Carbon metabolism/control	HPr	*B. subtilis, L. lactis*	5,10
		CcpA	*L. lactis*	10
		β-PGM	*L. lactis*	10
		GAPDH	*B. subtilis*	5
		TIM	*B. subtilis*	5
		AceE	*E. coli*	1
		AceF	*E. coli*	1
	Molecular chaperone	Hsc66	*E. coli*	1
		GroES	*B. subtilis*	5
	Others	PpiB	*B. subtilis*	5
		CheY	*B. subtilis*	5
		Thioredoxin	*B. subtilis*	5
		SpoVG	*B. subtilis*	5
		RisB	*B. subtilis*	5
		AtpE	*B. subtilis*	5
		PspB	*B. subtilis*	5
		RecA	*E. coli*	1
		OsmC	*L. lactis*	8

1. Jones et al., 1987; 2. Goverde et al., 1998; 3. Jones et al., 1996; 4. Jones and Inouye, 1996; 5. Graumann et al., 1996; 6. Wouters et al., submitted; 7. Hesterkamp et al., 1997; 8. Aguilar et al., 1998; 9. Ray et al., 1994A; 10. Wouters et al., 2000A.

of temperature shock and leads to transcription of genes implicated in the thermal response (Carratù et al., 1996).

3. *DNA topology*. Upon low-temperature exposure of bacteria negative DNA supercoiling of the nucleoid increases and this has important consequences for the regulation of transcription (Drlica, 1992). Modification of DNA supercoiling has been suggested as a direct effect of a change in temperature which would affect gene expression. For *E. coli* it has been shown that DNA topoisomerase activity, DNA gyrase activity and the histone-like HU protein have an important role in the process of controlling transcription at low temperature (Mizushima et al., 1997).

1.3. Production of non-7 kDa cold induced proteins

Using proteomics approaches, the rapid induction of specific sets of proteins upon cold shock is detected in a variety of bacteria. The number of CIPs may vary from approximately 12 for *L. monocytogenes*, 18 for *E. coli*, 17 for *L. lactis* to 37 for *B. subtilis* (Jones et al., 1987; Bayles et al., 1996; Graumann et al., 1996; Wouters et al., 1999). During cold-shock adaptation the synthesis of the majority of housekeeping, general stress and heat-shock proteins is blocked. The CIPs play a role in a variety of cellular processes, such as chromosome structuring, transcription, translation, general metabolism, sugar metabolism and stress response (Table 4.1). The *E. coli* CIPs include NusA (termination and anti-termination of transcription), RecA (recombination and SOS response), H-NS and GyrA (DNA supercoiling), polynucleotide phosphorylase (mRNA degradation) and *aceE* and *aceF* (pyruvate metabolism). Furthermore, the production of a set of proteins involved in the process of translation also increases at low temperature. The proteins include: initiation factor 2 (*infB*), CsdA and RbfA (a ribosome associated helicase and a ribosome binding factor A) and a trigger factor that is involved in ribosomal modification and polypeptide folding (Jones et al., 1987; Jones and Inouye, 1994; Jones and Inouye, 1996; Hesterkamp et al., 1997) (Table 4,1). In the psychrotrophic bacterium *Yersinia enterocolitica* a major role in cold adaptation was identified for *pnp*, encoding the exoribonuclease polynucleotide phosporylase (PNPase). A *pnp*-negative mutant was found to be unable to grow at 5°C and it has been suggested that PNPase is required for the correct degradation of mRNA in order to prevent trapping of the ribosome (Goverde et al., 1998; Neuhaus et al., 2000). For *B. subtilis* CIPs are described that are involved in chemotaxis (CheY), sugar uptake and global control (HPr), translation (ribosomal proteins S6 and L7/L12), protein folding (PpiB), amino acid metabolism (CysK, IlvC), glycolysis/sugar metabolism (GAPDH and TIM) (Graumann et al., 1996) and membrane composition (Des) (Aguilar et al., 1998). Similarly, for *L. lactis* CIPs have been identified that are involved in chromosome structuring (histone-like HslA-protein), stress response (OsmC), sugar metabolism and global control (β-PGM, HPr, CcpA), and signal transduction (LlrC) (Wouters et al., 2000a; Wouters et al., submitted).

In combination, these data show that in a wide range of bacteria the response to low temperature involves a variety of cellular functions (if not all) that are more or less hampered at low temperature. Increased production of specific proteins will enable bacterial cells to adapt to low-temperature conditions by which they will function in a new (sub)optimal way. Since the cold-shock stimulon seems to be quite diverse, complex and large, it is unlikely to be controlled by a single mechanism. However, there is considerable evidence for the involvement of the major CSPs in (part of) the regulation of this stimulon (see below).

1.4. Protein synthesis at low temperature and the role of ribosomes in cold adaptation

Since low-temperature exposure results, on the one hand, in a significant loss of translational activity and, on the other hand, in specific *de novo* protein synthesis, the structure and function of ribosomes seems to play a central role in the cold-adaptation process. During a cold-shock treatment the translational capacity is strongly reduced resulting in a high concentration of charged tRNA which blocks the A-site of the ribosome. This in turn would lower the (p)ppGpp concentration by the diminished synthesis of (p)ppGpp by RelA which, in turn, controls the stringent response. In *E. coli* artificially low concentrations of (p)ppGpp increase the synthesis of CIPs (VanBogelen and Neidhardt, 1990; Jones et al., 1992a; Graumann and Marahiel, 1996). In addition, upon incubation of *E. coli* and *B. subtilis* cells with chloramphenicol, a response similar to the cold-shock response develops, with the specific induction of certain CSPs and CIPs. This response was related to the inactivation of the ribosomes that are specifically blocked by chloramphenicol.

It is believed that the mRNAs of cold-induced genes are still translatable during cold shock because of the presence of *cis*-acting elements, the upstream and downstream boxes (UB and DB, respectively) that provide additional binding capacity to the ribosome. The induction of cold-shock specific ribosomal factors, such as CsdA and RbfA, leads to restoration of the ribosomal structure and the ability to form intact translation initiation complexes for translation of non cold-shock mRNAs (Jones and Inouye, 1996; Jones et al., 1996; Mitta et al., 1997). An important role in the regulation of the translation at low temperature has been attributed to CsdA, an enzyme essential for the unwinding of stable secondary mRNA structures formed at low temperature (Jones et al., 1996).

Not only the ribosomal proteins but also the structure and the number of the *rrn* operons determine the ability of a microorganism to adapt to temperature (and nutrient) changes. For *E. coli* it has been observed that the time to adapt to a temperature shift increased with decreasing numbers of intact operons (Condon et al., 1995). For the food-borne pathogen *Bacillus cereus*, differences in the structure of the *rrn* operons were observed between mesophilic and psychrotolerant species, and more specifically in the small (30S) ribosomal subunit involved in the early steps of translation initiation. Mutations are found from G and C in the mesophilic strains to A and T in the psychrotolerant strains which might result in a state of the ribosome capable of translation at low temperature, caused by the melting point reduction of the A–T bonds (Prüß et al., 1999).

2. Cold-shock proteins and their role in cold and general stress adaptation

CSPs (molecular weight of approximately 7 kDa) are observed in a wide variety of Gram-positive and Gram-negative bacteria in which they share a high degree of sequence similarity (>45%). However, CSPs have not been found in all bacteria, e.g. they are absent in the complete genomes of *Helicobacter pylori* (Tomb et al., 1997), *Campylobacter jejuni* (Hazeleger et al., 1998) and *Mycoplasma genitalium* (Graumann and Marahiel, 1996). For most bacterial species, families of CSPs consisting of two to nine members have been found. The concomitant presence of *csp* gene families probably resulted from gene duplications within the organism (Yamanaka et al., 1998). For *L. lactis*, *Y. enterocolitica* as well as for *E. coli* a clustered organization of *csp* genes on the chromosome was observed (Wouters et al., 1998; Yamanaka et al., 1998; Neuhaus et al., 1999). For *L. lactis* and *Y. enterocolita* tandem repeats of two adjacent *csp* genes, separated by an interval of only approximately 300 bp, were observed. The physiological or genetic significance of the increased *csp* gene dosage and the clustered organization of these genes remain to be elucidated.

2.1. CSPs as transcriptional activators

The most extensively studied CSPs are CspA of *E. coli* ($CspA^E$) and CspB of *B. subtilis* ($CspB^B$). The determination of their crystal structures revealed that both proteins consist of five antiparallel β-strands which together form a β-barrel structure (Schindelin et al., 1993; Schindelin et al., 1994; Newkirk et al., 1994). It was observed that $CspA^E$ contains a set of surface exposed aromatic amino acids, that is essential for RNA or DNA binding and a set of hydrophobic residues forming a hydrophobic core of the protein (Newkirk et al., 1994). The majority of CSPs have acidic isoelectric points, but several have a basic isoelectric points

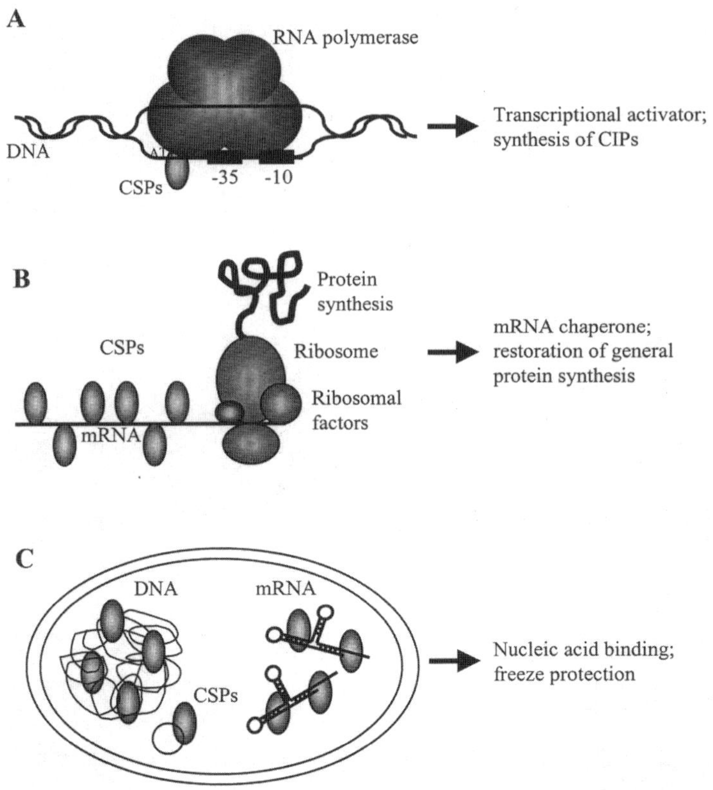

Fig. 4.1. The mode of action of CSPs as transcriptional activators, mRNA chaperones and/or freeze-protective proteins. (A) CSPs can function as transcriptional activators due to their ssDNA binding capacity and specific recognition of Y-box motifs by which they might stabilize the formed open complex during transcription initiation. (B) In response to cold shock ribosomal structure is restored by ribosomal binding factors and mRNA secondary folding is reduced by the increased number of CSPs by which translation can proceed at low temperature. (C) CSPs act as freeze-protective proteins probably by their binding to nucleic acids which are thereby protected during these freezing conditions. See text for details.

(8–10), a characteristic that might have major consequences for their RNA and DNA-binding characteristics. CSPs contain regions highly homologous to the cold-shock domain of eukaryotic DNA-binding proteins, like YB1 and FRGY2, that are known to act as transcription factors. Both $CspA^E$ and $CspB^B$ are able to bind specifically to single-stranded DNA containing the Y-box motif (ATTGG) or its complementary sequence (Newkirk et al., 1994; Graumann and Marahiel, 1994). Indeed, $CspA^E$ acts as a transcriptional activator of at least two genes encoding CIPs, GyrA (Jones et al., 1992b; Brandi et al., 1994) and H-NS (LaTeana et al., 1991), possibly by stabilization of the open complex formation by RNA polymerase (Fig. 4.1A). In addition, heterologous expression of $CspB^B$ in *E. coli* at 37°C induced a protein pattern that strongly resembled that upon cold shock, indicating that also $CspB^B$ functions of as a regulatory protein (Graumann and Marahiel, 1997). For *B. subtilis* as well as *L. lactis* changes in protein patterns were observed upon deletion of CSPs (Graumann et al., 1996; Wouters et al., submitted). Two-dimensional gel electrophoresis of cell-free extracts of *L. lactis* specifically overproducing a CSP, revealed induction of CIPs, indicating that these CSPs regulate other proteins involved in cold adaptation (Wouters et al., 2000b). These observations were verified by analysis of *L. lactis* strains carrying deletions in their *csp* genes showing a reduction in the expression of the same CIPs (Wouters et al., submitted). Strikingly, it was observed that the different lactococcal CSPs each regulate the expression of different

CIPs, which might provide a rational explanation for the existence of a CSP family (Wouters et al., 2000a; Wouters et al., submitted). Recently, Bae et al. (2000) showed that CspA, CspC and CspE of *E. coli* act as transcriptional anti-terminators. This would offer an alternative mechanism for the regulation of several CIPs, including NusA, IF2, RbfA and PNPase, by CSPs.

2.2. CSPs as RNA chaperones

Since $CspA^E$ and $CspB^B$ both posses highly conserved RNA-binding motifs, i.e. RNP-1 and RNP-2, they are also considered to be RNA-binding proteins (Jones and Inouye, 1994; Schindelin et al., 1993). It has been shown that $CspA^E$ and $CspB^B$ bind to mRNA with a broad sequence specificity (Graumann et al., 1997; Jiang et al., 1997). Recently, a more sequence specific RNA/single stranded DNA binding capacity has been observed for *E. coli* CspB, CspC and CspE (Phadtare and Inouye, 1999). The RNP motifs are located on one sheet of both $CspA^E$ and $CspB^B$ and contain several highly solvent exposed Phe residues, which are involved in DNA/RNA binding (Schröder et al., 1995). It is proposed that CSPs act as RNA chaperones, thereby minimising the increased secondary folding of nascent mRNA at low temperature. By this action, they facilitate the initiation of translation for which RNA should be in a linear form (Fig. 4.1B). The binding of CSPs to RNA is only moderately strong and it is assumed that the ribosome can detach the CSPs from the linear, nascent mRNA molecules (Graumann et al., 1997; Jiang et al., 1997).

2.3. CSPs and freeze-protection

Single deletions in the genes encoding $CspA^E$ or $CspB^B$ did not reveal a distinct phenotype in relation to growth at low temperature (Bae et al., 1997; Willimsky et al., 1992). However, multiple deletions of the *csp* genes of *B. subtilis* revealed a lethal phenotype upon deletion of all three counterparts and severe growth inhibition whenever two *csp* genes were deleted (Graumann et al., 1997). In contrast, for *L. lactis* a triple deletion of *cspABE* did not affect growth characteristics (Wouters et al., submitted), which was explained by the increased expression of remaining *csp* genes. The loss of one or two *csp* genes is compensated by an increase in the production of the remaining CSP(s) for *E. coli*, *B. subtilis* as well as *L. lactis* (Bae et al., 1997; Graumann et al., 1997; Wouters et al., submitted).

Notably, differences in freeze survival were noted in a single *cspB* deleted mutant of *B. subtilis*. The *cspB*-deleted cells showed a lower freeze survival than wild-type cells (Willimsky et al., 1992). Similarly, for *L. lactis* a freeze-sensitive phenotype was observed upon deletion of *cspA, cspB* and *cspE*. Maximal freeze protection upon exposure to low temperature could still be obtained for the triple mutant strain but the freeze-protective response was significantly delayed (Wouters et al., submitted). Strikingly, upon overproduction of CspB, CspD or CspE increased freezing survival by *L. lactis* could also be obtained (Wouters et al., 1999a; Wouters et al., 2000b). Since CSPs are recognized as nucleic acid binding proteins, they might protect RNA and DNA during the freezing process and, hence, increase the survival of the bacterial cells (Fig. 4.1C).

2.4. Role of CSPs in general stress response

Evidence has accumulated that not all members of the CSP family are cold-inducible and thus are not necessarily involved in the cold-shock response. Several counterparts of the *csp* gene families are not induced at low temperature and several cold-induced *csp* genes are also induced by other stress conditions (Mayr et al., 1996; Wouters et al., 1998; Yamanaka et al., 1998; Table 4.2). $CspD^E$ of *E. coli* and $CspB^B$ and $CspC^B$ of *B. subtilis* are induced during stationary phase conditions (Yamanaka and Inouye, 1997; Graumann and Marahiel, 1999). Moreover, two non cold-induced CSPs of *E. coli*, $CspC^E$ and $CspE^E$, have been implicated in chromosomal condensation and/or cell division (Yamanaka et al., 1994; Yamanaka et al., 1998). Recently, Hanna and Liu (1998) showed that $CspE^E$ interacts with nascent RNA in transcription complexes, indicating a role for this protein in the

Table 4.2. *csp* Genes that are not exclusively induced by cold shock

csp Gene	Organism	Cold	Stress induction	Cellular process	Reference*
cspA	*E. coli*	+	ultra high pressure		1
			dilution stationary phase		2
cspC	*E. coli*	–	unknown	Chromosomal condensation/cell division	3
cspD	*E. coli*	–	stationary phase		4
cspE	*E. coli*	–	dilution stationary phase	Chromosomal condensation/cell division	3
cspH	*E. coli*	–	unknown		5
cspB	*B. subtilis*	+	stationary phase		6
cspC	*B. subtilis*	+	stationary phase		6
cspE	*L. lactis*	–	constitutive		7
cspC	*B. cereus*	–	unknown		8
cspD	*B. cereus*	–	unknown		8
cspE	*B. cereus*	–	unknown		8

*1. Welch et al., 1993; 2. Brandi et al., 1999; 3. Yamanaka et al., 1994; 4. Yamanaka and Inouye, 1997; 5. Yamanaka et al., 1998; 6. Graumann and Marahiel, 1999; 7. Wouters et al., 1998; 8. Mayr et al., 1996.

transcription process. The disruption of *cspE*^{E}, a gene that is transiently induced during the growth lag after dilution of stationary phase cells, resulted in a longer lag period after dilution (Bae et al., 1999). However, the exact role of CspE^E in this phenomenon has not been elucidated yet.

CspA^E, has also been shown to be induced upon exposure of cells to ultra high-pressure treatments. This phenomenon might be related to non-functioning ribosomes, since it has been reported that these are also disrupted during high pressure conditions (Welch et al., 1993). Brandi et al. (1999) reported that expression of *cspA*^{E} is high upon dilution of a stationary phase culture and the *cspA*^{E} mRNA level decreases with increasing cell density. The extent of the cold-shock induction of *cspA*^{E} is inversely proportional to the pre-existing level of CspA^E. Furthermore, it has been reported that the expression of *cspA*^{E} under non-stress conditions is regulated by the antagonistic effects of the DNA-binding proteins Fis and H-NS on transcription, variation of *cspA*^{E} mRNA stability and, possibly, autoregulation (Brandi et al., 1999).

It is unclear whether the cold-induced CSPs have a role solely in cold adaptation or whether they also have functions under different stress conditions, e.g. under conditions of decreased growth rates. Since the increased production of CSPs is seen under a variety of conditions, it is likely that the roles of CSPs are not restricted to cold adaptation.

3. Regulatory elements involved in CSP synthesis

The regulation of the production of CSPs is controlled at several levels. The synthesis of CspA^E has been characterized in most detail and it was established that regulation of its expression after cold shock takes place at the level of transcription, at the level of translation, and at the level of mRNA and protein stability, involving several characteristic genetic elements. Thus, similar to sigma factors, CSPs are examples of stress proteins that are regulated at various levels (Tanabe et al., 1992; Jiang et al., 1993; Brandi et al., 1996; Goldenberg et al., 1996; Mitta et al., 1997; Fig. 4.2).

3.1. Transcriptional regulation and mRNA stability

An AT-rich sequence (UP-element) upstream of the –35 region of the *cspA*^{E} promoter enhances *cspA*^{E} transcription at low temperature (Mitta et al., 1997; Goldenberg et al., 1997). For CspA^E, CspB^B

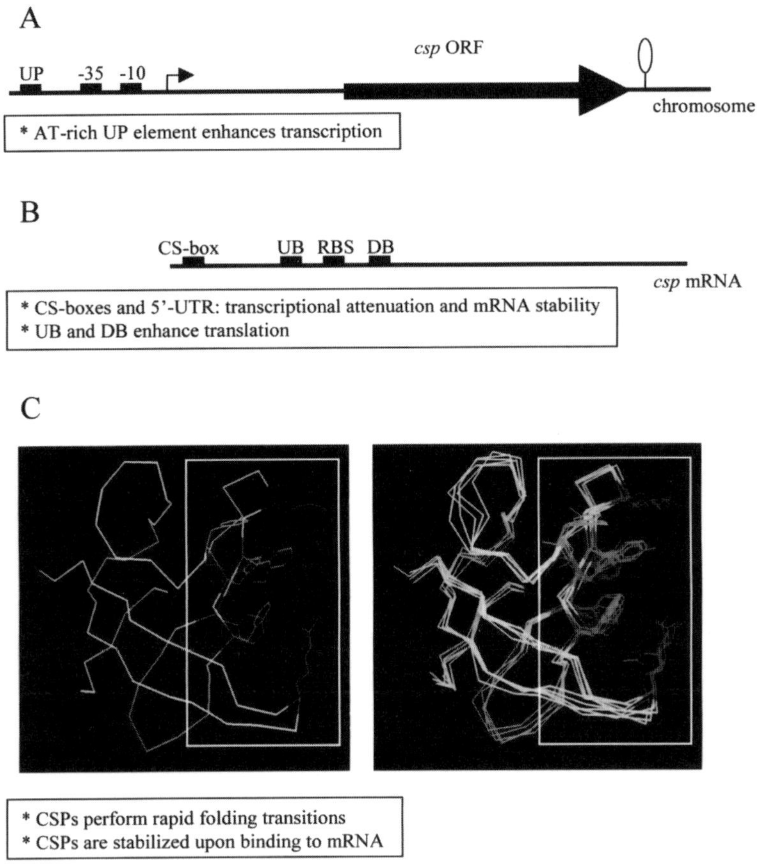

Fig. 4.2. Schematic representation of the regulatory elements involved in the expression of *csp* genes with *cspA* of *E. coli* as a model. (A) Regulatory elements at the transcriptional level. UP indicates the AT-rich UP-element, –35 and –10 indicate the respective promoter regions, the start of transcription is indicated by an arrow, the large arrow indicates the *csp* open reading frame, the hairpin indicates the terminator region. (B) Regulation at the level of translation and mRNA stability. CS-box indicates the cold-shock box, RBS indicates the ribosome binding site, UB and DB indicate the upstream and downstream box, respectively. (C) Modeling of the three-dimensional folding of CspA of *L. lactis* (left panel) and an overlay of the modeled 3D-folding structures of CspB, CspC, CspD and CspE of *L. lactis* and CspA of *E. coli* (right panel). The white lines indicate the backbone structure of the respective CSPs. Important residues for RNA and DNA binding are indicated grayish and boxed (basic K7, K13, H29, R56 or aromatic: W8, F15, F17, F27, F30). The modeling was based on the 3D-crystal structures of CspA^E and CspB^B (Schindelin et al., 1993; Schindelin et al., 1994) using Quanta/Charmm (Molecular Simulations Inc., San Diego, Ca., USA). See text for details.

as well as for CspB of *L. lactis* the analysis of promoter–reporter fusions showed increased promoter activity upon low-temperature incubation (Tanabe et al., 1992; Willimsky et al., 1992; Chapot-Chartier et al., 1997). However, the high *csp* mRNA levels after cold shock may also result from increased stability of the transcripts, whereas e.g. *cspA*^{E} mRNA is highly unstable at 37°C. This increase in stability is dependent on the unusually long, untranslated 5′-mRNA leader region (5′-UTR) of *cspA*^{E}, which is rich in secondary structure (Jiang et al., 1997). The mRNA stabilization of *cspA*^{E} upon cold shock appeared to be transient and is lost once cells have adapted to low temperature (Goldenberg et al., 1996). The *B. subtilis* CSPs have high affinity to bind to the first 25 bases of their 5′-UTRs, named cold-shock box (CS-box; Graumann et al., 1997). In addition, it was found that CspA^E negatively regulates its gene expression through a similar CS-box on its 5′-UTR

(Jiang et al., 1996; Bae et al., 1997). In this way, CSPs could down-regulate translation of their messengers to limit their own cellular concentrations. This might be an important regulatory mechanism since artificial overproduction of $CspB^B$ in *B. subtilis* had a growth inhibitory effect (Graumann and Marahiel, 1997). For *L. lactis* a highly different 5′-UTR and CS-box were found preceding the non-cold induced *cspE* gene in comparison to these regions of four cold-induced *csp* genes (*cspA* to *cspD*; Wouters et al., 1998). It was speculated that CspE destabilizes the mRNA of the cold-induced genes at high temperature by which no translation can occur. Indeed, upon disruption of CspE in *L. lactis* an increased synthesis of CspC and CspD was observed, indicating a central role for CspE in repression of the synthesis of these CSPs at normal growth temperature (Wouters et al., submitted). For *Y. enterocolitica* the degradation of *csp* mRNA is essential for the adaptation to low-temperatures. PNPase is involved in this process as was shown by using a *pnp*-negative mutant which was not able to adapt to cold-shock conditions (Neuhaus et al., 2000).

3.2. Translational regulation involving cis-acting elements

For *E. coli*, the mRNAs encoding CSPs and several CIPs are better translated than general mRNAs during cold-shock conditions, a period during which protein synthesis is blocked as a result of ribosomal malfunctioning (Mitta et al., 1997). The mRNAs of CSPs and CIPs are probably still translatable due to *cis*-acting elements that are complementary to specific regions of the 16S rRNA 3′-end, as is shown for *cspA* of and *cspB* of *E. coli* (Mitta et al., 1997). A so-called downstream box element (DB), located 12 nucleotides downstream of the initiation codon is complementary to a sequence proximal to the ribosome binding site-decoding region in 16S rRNA (Mitta et al., 1997; Etchegaray and Inouye, 1999a). This DB accounts for additional binding to the anti-downstream box (anti-DB) of 16S rRNA, thereby enhancing the formation of initiation complexes (Mitta et al., 1997). Similarly, an upstream box element (UB) has been reported for the four cold-induced *csp* genes of *E. coli* which is located 14 nucleotides upstream of the Shine–Dalgarno sequence (Yamanaka et al., 1999). The enhanced translation may also contribute to stabilization of the transcripts by protecting them from degradation. It should be noted that the actual roles of the DB and UB remain unclear and that the biochemical evidence for the DB-anti DB and the UB-anti UB interactions is still in debate (Etchegaray and Inouye, 1999b; Bläsi et al., 1999).

3.3. Protein stability

Another level by which the cellular CSP concentration is controlled is the protein stability. CSPs perform highly rapid folding and unfolding transitions and they are marginally stable in solution. Recently, it was shown that the stability of the CSPs of *B. subtilis* was significantly increased by binding to a nucleic acid ligand (Schindler et al., 1999). It has been stated that acidic CSPs have an ideal structure to act as RNA chaperones since they possess a positively charged RNA-binding epitope that is backed by a negatively-charged surface that would prevent approach of RNA by charge repulsion (Graumann and Marahiel, 1998). The role for several residues in the functioning and the stability of CSPs was shown by site-directed mutagenesis (Schröder et al., 1995; Hillier et al., 1998; Schindler et al., 1998; Wouters et al., 2000b). Mutation of Phe residues in the RNP motifs of both $CspB^B$ and $CspA^E$ not only resulted in a reduction of the nucleic acid binding capacity but also in a decreased protein stability (Hillier et al., 1998; Schindler et al., 1998). CspC of *B. subtilis* contains an Ala residue at position 58 whereas CspB and CspD of *B. subtilis* contain a Pro residue at this position and it was found that $CspB^B$ and $CspD^B$ were far more stable than $CspC^B$. This can be explained by the high stability of the protein as a consequence of the reduced entropy (Schindler et al., 1999). Similarly, a CspA mutant of *L. lactis* containing a R58P substitution was significantly stabilized. In addition, the decrease in iso-electric point may also contribute to the increased CspA-mutant stability (Wouters et al., 2000b).

4. Perspectives

The increasing number of studies of the cold-shock response in a variety of organisms allows a comparison of their responses. The regulation of synthesis and the functioning of the 7-kDa CSPs upon cold shock have been elucidated in the greatest detail, but a large number of questions have remained unanswered. The reasons for the existence of CSP families, of which the members show highly similar primary and predicted three-dimensional structures, is still unclear. Upon deletion of the genes encoding CSPs compensatory effects of the remaining counterparts are noted, which points to the presence of a tightly controlled expression network and to a (partly) replaceable mode of action of the CSP family members. For their action as RNA chaperones the need for a combined action and dimerization of CSPs has been reported (Schindelin et al., 1993; Mayr et al., 1996; Graumann et al., 1997). CSPs function as transcriptional regulators by which they regulate several CIPs and consequently fulfil a central role in cold adaptation. The increasing number of complete genome sequences, the development of the micro-array technology and the further development of proteomics analysis will undoubtedly significantly contribute to the unraveling of CSP functioning, and in particular to the exploration of their role in global regulatory phenomena.

Research on cold adaptation might yield direct applications with respect to food preservation methods and fermentation technology, for example, in the survival after freezing of starter bacteria and in the development of methods to control the growth of bacteria that challenge the safety of refrigerated and other processed foods.

Acknowledgements

The authors would like to thank Bernadette Renckens and Roland Siezen for their expert help with the modeling of the lactococcal CSPs.

References

Abee, T. and Wouters, J.A. (1999). Microbial stress response in minimal processing. Int. J. Food Microbiol. 50, 65–91.

Aguilar, P.S., Cronan, J.E. jr. and DeMendoza, D. (1998). A *Bacillus subtilis* gene induced by cold shock encodes a membrane phospholipid desaturase. J. Bacteriol. 180, 2194–2200.

Bae, W., Jones, P.G. and Inouye, M. (1997). CspA, the major cold shock protein of *Escherichia coli*, negatively regulates its own gene expression. J. Bacteriol. 176, 7081–7088.

Bae, W., Phadtare, S., Severinov, K. and Inouye, M. (1999). Characterization of *Escherichia coli cspE*, whose product negatively regulates transcription of *cspA*, the gene for the major cold-shock protein. Mol. Microbiol. 31, 1429–1441.

Bae, W., Xia, B., Inouye, M. and Severinov, K. (2000). *Escherichia coli* CspA-family RNA chaperones are transcription terminators. Proc. Natl. Acad. Sci. USA 97, 7784–7789.

Bayles, D.O., Annous, B.A. and Wilkinson, B.J. (1996). Cold stress proteins induced in *Listeria monocytogenes* in response to temperature downshock and growth at low temperatures. Appl. Environ. Microbiol. 62, 1116–1119.

Becker, L.A., Cetin, M.S., Hutkins, R.W. and Benson, A.K. (1998). Identification of the gene encoding the alternative sigma factor σ^B from *Listeria monocytogenes* and its role in osmotolerance. J. Bacteriol. 180, 4547–4554.

Berry, E.D. and Foegeding, P.M. (1997). Cold temperature adaptation and growth of microorganisms. J. Food Prot. 60, 1583–1594.

Bläsi, U., O'Connor, M., Squires, L. and Dahlberg, A.E. (1999). Misled by sequence complementarity: does the DB-anti-DB interaction withstand scientific scrutiny? Mol. Microbiol. 33, 438–441 (MicroCorrespondence).

Brandi, A., Pietroni, P., Gualerzi, C.O. and Pon, C.L. (1996). Post-transcriptional regulation of CspA expression in *Escherichia coli*. Mol. Microbiol. 19, 231–240.

Brandi, A., Pon, C.L. and Gualerzi, C.O. (1994). Interaction of the main cold shock protein CS7.4 (CspA) of *Escherichia coli* with the promoter region of *hns*. Biochimie 74, 1090–1098.

Brandi, A., Spurio, R., Gualerzi, C.O. and Pon, C.L. (1999). Massive expression of the *Escherichia coli* major cold-shock protein CspA under non-stress conditions. EMBO J. 18, 1653–1659.

Carratù, L., Franceschelli, S., Pardini, C.L., Kobayashi, G.S., Horvath, I., Vigh, L. and Maresca, B. (1996). Membrane lipid perturbation modifies the set point of the temperature of heat shock response in yeast. Proc. Natl. Acad. Sci. USA, 93 3870–3875.

Chapot-Chartier, M.-P., Schouler, C., Lepeuple, A.-S., Gripon, J.-C. and Chopin, M.-C. (1997). Characterization of *cspB*, a cold-shock inducible gene from *Lactococcus lactis*, and evidence for a family of genes homologous to the *Escherichia coli cspA* major cold shock gene. J. Bacteriol. 179, 5589–5593.

Condon, C., Liveris, D., Squires, C., Schwartz, I. and Squires, C.L. (1995). rRNA operon multiplicity in *Escherichia coli* and the physiological implications of *rrn* inactivation. J. Bacteriol. 177, 4152–4156.

Drlica, K. (1992). Control of bacterial DNA supercoiling. Mol. Microbiol. 6, 425–433.

Etchegaray, J.-P. and Inouye, M. (1999a). A sequence downstream of the initiation codon is essential for cold shock induction of *cspB* of *Escherichia coli*. J. Bacteriol. 181, 5852–5854.

Etchegaray, J.-P. and Inouye, M. (1999b). DB or not DB in translation? Mol. Microbiol. 33, 438–441 (MicroCorrespondence).

Goldenberg, D., Azar, I. and Oppenheim, A.B. (1996). Differential mRNA stability of the *cspA* gene in the cold-shock response of *Escherichia coli*. Mol. Microbiol. 19, 241–248.

Goldenberg, D., Azar, I., Oppenheim, A.B., Brandi, A., Gualerzi, C.O. and Pon, C.L. (1997). Role of *Escherichia coli cspA* promoter sequences and adaptation of translational apparatus in the cold shock response. Mol. Gen. Genet. 256, 282–290.

Gounot, A.M. and Russell, N.J. (1999). Physiology of cold-adapted microorganisms. In: Cold-adapted organisms—Ecology, physiology, enzymology and molecular biology. (Margesin, R. and Schinner, F., Eds.), pp. 33–56. Springer-Verlag, Heidelberg, Germany.

Goverde, R.L.J., Huis in 't Veld, J.H.J., Kusters, J.G. and Mooi, F.R. (1998). The psychrotrophic bacterium *Yersinia enterocolitica* requires expression of *pnp*, the gene for polynucleotide phosphorylase, for growth at low temperature (5°C). Mol. Microbiol. 28, 555–569.

Graumann, P.L. and Marahiel, M.A. (1994). The major cold shock protein of *Bacillus subtilis* CspB binds with high affinity to the ATTGG-and CCAAT sequences in single stranded nucleotides. FEBS Letters 338, 157–160.

Graumann, P.L. and Marahiel, M.A. (1996). Some like it cold: response of microorganisms to cold shock. Arch. Microbiol. 166, 293–300.

Graumann, P.L. and Marahiel, M.A. (1997). Effects of heterologous expression of CspB, the major cold-shock protein of *B. subtilis*, on protein synthesis in *Escherichia coli*. Mol. Gen. Genet. 253, 745–752.

Graumann, P.L. and Marahiel, M.A. (1998). A superfamily of proteins that contain the cold-shock domain. Trends Biochem. Sci. 23, 286–290.

Graumann, P.L. and Marahiel, M.A. (1999). Cold shock proteins CspB and CspC are major stationary-phase-induced proteins in *Bacillus subtilis*. Arch. Microbiol. 171, 135–138.

Graumann, P., Schröder K., Schmid, R. and Marahiel, M.A. (1996). Cold shock stress-induced proteins in *Bacillus subtilis*. J. Bacteriol. 178, 4611–4619.

Graumann, P., Wendrich, T.M., Weber, M.H.W., Schröder, K. and Marahiel, M.A. (1997). A family of cold shock proteins in *Bacillus subtilis* is essential for cellular growth and for efficient protein synthesis at optimal and low temperatures. Mol. Microbiol. 25, 741–756.

Hanna, M.H. and Liu, K. (1998). Nascent RNA in transcription complexes interacts with CspE, a small protein in *E. coli* implicated in chromatin condensation. J. Mol. Biol. 282, 227–239.

Hazeleger, W.C., Wouters, J.A., Rombouts, F.M. and Abee, T. (1998). Physiological activities of *Campylobacter jejuni* far below its minimal growth temperature. Appl. Environ. Microbiol. 64, 3917–3922.

Hesterkamp, T., Deuerling, E. and Bukau, B. (1997). The amino-terminal 118 amino acids of *Escherichia coli* trigger factor constitute a domain that is necessary and sufficient for binding to ribosomes. J. Biol. Chem. 272, 21865–21871.

Hillier, B.J., Rodriquez, H.M. and Gregoret, L.M. (1998). Coupling protein stability and protein function in *Escherichia coli* CspA. Folding Design 2, 87–93.

Hoch, J.A. and Silhavy, T.J. (1995). Two-component signal transduction. Washington DC: Am. Soc. Microbiol.

Jiang, W., Fang, L. and Inouye, M. (1996). The role of the 5′-end untranslated leader of the mRNA for CspA, the major cold-shock protein of *Escherichia coli*, in cold-shock adaptation. J. Bacteriol. 178, 4919–4925.

Jiang, W., Hou, Y. and Inouye, M. (1997). CspA, the major cold-shock protein of *Escherichia coli*, is an RNA chaperone. J. Biol. Chem. 272, 196–202.

Jiang, W., Jones, P.G. and Inouye, M. (1993). Chloramphenicol induces the transcription of the major cold shock gene of *Escherichia coli*, *cspA*. J. Bacteriol. 175, 5824–5828.

Jones, P.G. and Inouye, M. (1994). The cold-shock response—a hot topic. Mol. Microbiol. 11, 811–818.

Jones, P.G. and Inouye, M. (1996). RbfA, a 30S-ribosomal binding factor, is a cold-shock protein whose absence triggers the cold-shock response. Mol. Microbiol. 21, 1207–1218.

Jones, P.G., Cashel, M., Glaser, G. and Neidhardt, F.C. (1992a). Function of a relaxed-like state following temperature downshifts in *Escherichia coli*. J. Bacteriol. 174, 3903–3914.

Jones, P.G., Krah, R., Tafuri, S.R. and Wolffe, A.P. (1992b). DNA gyrase, CS7.4, and the cold shock response in *Escherichia coli*. J. Bacteriol. 174, 5798–5802.

Jones, P.G., Mitta, M., Kim, Y., Jiang, W. and Inouye, M. (1996). Cold shock induces a major ribosomal-associate protein that unwinds double stranded RNA in *Escherichia coli*. Proc. Natl. Acad. Sci. USA 93, 76–80.

Jones, P.G., VanBogelen, R. and Neidhardt, F.C. (1987). Induction of proteins in response to low temperature in *Escherichia coli.* J. Bacteriol. 169, 2092–2095.

LaTeana, A., Brandi A., Falconi, M., Spurio, R., Pon C.L. and Gualerzi, C.O. (1991). Identification of a cold shock transcriptional enhancer of the *Escherichia coli* major cold shock gene encoding nucleoid protein H-NS. Proc. Natl. Acad. Sci. USA 88, 10907–10911.

Loewen, P.C., Hu, B., Strutinsky, J. and Sparling, R. (1998). Regulation in the *rpoS* regulon of *Escherichia coli*. Can. J. Microbiol. 44, 707–717.

Mayr, B., Kaplan, T., Lechner, S. and Scherer, S. (1996). Identification and purification of a family of dimeric major cold shock protein homologs from the psychrotrophic *Bacillus cereus* WSBC10201. J. Bacteriol. 178, 2916–2925.

Mitta, M., Fang, L. and Inouye, M. (1997). Deletion analysis of *cspA* of *Escherichia coli*: requirement of the AT-rich UP element for *cspA* transcription and the downstream box in the coding for its cold shock induction. Mol. Microbiol. 26, 321–335.

Mizushima, T., Kataoka, K., Ogata, Y., Inoue, R. and Sekimizu, K. (1997). Increase in negative supercoiling of plasmid DNA in *Escherichia coli* exposed to cold shock. Mol. Microbiol. 23, 381–386.

Neuhaus, K., Francis, K.P., Rapposch, S., Görg, A. and Scherer, S. (1999). Pathogenic *Yersinia* species carry a novel, cold-inducible major cold shock protein tandem gene duplication producing both bicistronic and monocistronic mRNA. J. Bacteriol. 181, 6449–6455.

Neuhaus, K. Rapposch, S., Francis, K.P. and Scherer, S. (2000). Restart of exponential growth of cold-shocked *Yersinia enterocolitica* occurs after down-regulation of *cspA1/cspA2* mRNA. J. Bacteriol. 182, 3285–3288.

Newkirk, K., Feng, W., Jiang, W., Tejero, R., Emerson, S.D., Inouye, M. and Montelione, G.T. (1994). Solution NMR structure of the major cold shock protein (CspA) from *Escherichia coli*: identification of a binding epitope for DNA. Proc. Natl. Acad. Sci. USA 91, 5114–5118.

Phadtare, S. and Inouye, M. (1999). Sequence-selective interactions with RNA by CspB, CspC and CspE, members of the CspA family of *Escherichia coli*. Mol. Microbiol. 33, 1004–1014.

Prüß, B.M., Francis, K.P., Von Stetten, F. and Scherer, S. (1999). Correlation of 16S ribosomal DNA signature sequences with temperature-dependent growth rates of mesophilic and psychrotrophic strains of the *Bacillus cereus* group. J. Bacteriol. 181, 2624–2630.

Ray, M.K., Kumar, G.S. and Shivaji, S. (1994a). Phosphorylation of lipopolysaccharides in the Antarctic psychrotroph *Pseudomonas syringae*: a possible role in temperature adaptation. J. Bacteriol. 176, 4243–4249.

Ray, M.K., Kumar, G.S. and Shivaji, S. (1994b). Phosphorylation of membrane proteins in response to low temperature in an antarctic *Pseudomonas syringae*. Microbiology, 140, 3217–3223.

Russell, N.J. (1990). Cold adaptation of microorganisms. Phil. Trans. R. Soc. Lond. 326, 595–611.

Schindelin, H., Jiang, W., Inouye, M. and Heinemann, U. (1994). Crystal structure of CspA, the major cold shock protein of *Escherichia coli*. Proc. Natl. Acad. Sci. USA 91, 5119–5123.

Schindelin, H., Marahiel, M. A. and Heinemann, U. (1993). Universal nucleic acid-binding domain revealed by crystal structure of *B. subtilis* major cold-shock protein. Nature 364, 164–168.

Schindler, T., Graumann, P.L., Perl, D., Ma, S., Schmid, F.X. and Marahiel, M.A. (1999). The family of cold shock proteins of *Bacillus subtilis*. Stability and dynamics *in vitro* and *in vivo*. J. Biol. Chem. 274, 3407–3413.

Schindler, T., Perl, D., Graumann, P., Sieber, V., Marahiel, M.A. and Schmid, F.X. (1998). Surface-exposed phenylalanines in the RNP1/RNP2 motif stabilize the cold-shock protein CspB from *Bacillus subtilis*. Proteins: Struct. Funct. Gen. 30, 401–406.

Schröder, K., Graumann, P., Schnuchel, A., Holak, T.A. and Marahiel, M.A. (1995). Mutational analysis of the putative nucleic acid binding surface of the cold-shock domain, CspB, revealed an essential role of aromatic and basic residues in binding of single stranded DNA containing the Y-box motif. Mol. Microbiol. 16, 699–708.

Tanabe, H., Goldstein, J., Yang, M. and Inouye, M. (1992). Identification of the promoter region of the *Escherichia coli* major cold shock gene, *cspA*. J. Bacteriol. 174, 3867–3873.

Tomb, J.-F., White, O., Kerlavage, A.R., Clayton, R.A., Sutton, G.G., Fleischmann, R.D., et al. (1997). The complete genome sequence of the gastric pathogen *Helicobacter pylori*. Nature 388, 539–547.

VanBogelen, R.A. and Neidhardt, F.C. (1990). Ribosomes as sensors of heat and cold shock in *Escherichia coli*. Proc. Natl. Acad. Sci. USA 87, 5589–5593.

Welch, T.J., Farewell, A., Neidhardt, F.C. and Bartlett, D.H. (1993). Stress response of *Escherichia coli* to elevated hydrostatic pressure. J. Bacteriol. 175, 7170–7177.

Willimsky, G., Bang, H., Fischer, G. and Marahiel, M.A. (1992). Characterization of *cspB*, a *Bacillus subtilis* inducible cold shock gene affecting cell viability at low temperatures. J. Bacteriol. 174, 6326–6335.

Wouters, J.A., Frenkiel, H., De Vos, W.M., Kuipers, O.P. and Abee, T. Multiple disruptions of cold-shock genes in *Lactococcus lactis* MG1363 show the direct involvement of cold-shock proteins in gene regulation. Submitted.

Wouters, J.A., Jeynov, B., Rombouts, F.M., De Vos, W.M., Kuipers, O.P. and Abee, T. (1999). Analysis of the role of 7 kDa cold-shock proteins of *Lactococcus lactis* MG1363 in cryoprotection. Microbiology 145, 3185–3194.

Wouters, J.A., Kamphuis, H.H., Hugenholtz, J., Kuipers,

O.P., De Vos, W.M. and Abee, T. (2000a). Changes in glycolytic activity of *Lactococcus lactis* induced by low temperature. Appl. Environ. Microbiol. 66, 3686–3691.

Wouters, J.A., Mailhes, M., Rombouts, F.M., De Vos, W.M., Kuipers, O.P. and Abee, T. (2000b). Physiological and regulatory effects of controlled overproduction of the five cold shock proteins of *Lactococcus lactis* MG1363. Appl. Environ. Microbiol. 66, 3756–3763.

Wouters, J.A., Sanders, J.-W., Kok, J., De Vos, W.M., Kuipers, O.P. and Abee, T. (1998). Clustered organization and transcriptional analysis of a family of five *csp* genes of *Lactococcus lactis* MG1363. Microbiology 144, 2885–2893.

Yamanaka, K. and Inouye, M. (1997). Growth-phase-dependent expression of *cspD*, encoding a member of the CspA family in *Escherichia coli*. J. Bacteriol. 179, 5126–5130.

Yamanaka, K., Fang, L. and Inouye, M. (1998). The CspA family in *Escherichia coli*: multiple gene duplication for stress adaptation. Mol. Microbiol. 27, 247–255.

Yamanaka, K., Mitani, T., Ogura, T, Niki, H. and Hiraga, S. (1994). Cloning, sequencing, and characterization of multicopy suppressors of a *mukB* mutation in *Escherichia coli*. Mol. Microbiol. 13, 301–312.

Yamanaka, K., Mitta, M. and Inouye, M. (1999). Mutation analysis of the 5′ untranslated region of the cold shock *cspA* mRNA of *Escherichia coli*. J. Bacteriol. 181, 6284–6291.

Protein Adaptations and Signal Transduction. Edited by K.B. Storey and J.M. Storey

CHAPTER 5

Hibernation: Protein Adaptations

Alexander M. Rubtsov
Department of Biochemistry, School of Biology, Lomonosov Moscow State University, 119899 Moscow, Russia

1. Introduction

Winter survival for many small mammals can only be assured by hibernation, a torpor response to cold temperatures and food shortages that is characterized by a strong metabolic rate depression and a reduction in body temperature to near ambient (for review, see Lyman et al., 1982; Kalabukhov, 1985; Wang, 1985). It has become clear that the depression of metabolism during hibernation cannot be explained by the simple effect of low temperatures on the rates of biochemical reactions. Rather, the fall in body temperature during entry into hibernation is the result of a well-coordinated reduction in the rates of all intracellular processes (Geiser, 1988; Heldmaier and Ruf, 1992). For most homeotherms a decrease in body temperature of even a few degrees Celcius can be lethal because of the metabolic imbalance that results due to differing thermal sensitivities of various biochemical processes in organs and tissues. However, hibernating mammals can keep their metabolism fully balanced over an extremely wide range of core body temperatures—from about 40°C to almost 0°C. Such a unique ability of hibernators has attracted the attention of researchers for many years, but only during last two decades have the first steps been made to understand the molecular mechanisms that allow hibernators to survive under severe environmental conditions.

At present, many of the metabolic and regulatory adaptations that support the preparation for, maintenance of, and awakening from torpor in different organs of hibernating species remain unknown. However, some of the key strategic mechanisms involved in the adjustment of enzyme and other protein functions during hibernation seem to be conserved among hibernating mammals (Hochachka, 1986). This short review will focus mainly on an analysis of the molecular mechanisms that are involved in regulating the properties of enzymes responsible for the functioning of heart and skeletal muscles of hibernators during the winter. Hearts of hibernating mammals continue to contract at low body temperatures whereas those of nonhibernating mammals fail to work at only a few degrees Celcius below normal body temperatures (37–40°C) (Kondo and Shibata, 1984). Some skeletal muscles must also continue to work during hibernation including the intercostal muscles and diaphragm that support the breathing of hibernating animals, and the tonic muscles responsible for the maintenance of posture during torpor. Other skeletal muscles are inactive during hibernation. However, because winter hibernation consists of repeated bouts of torpor followed by brief arousals, skeletal muscles must readily regain their active state during these periodic awakenings. In addition, skeletal muscles play a very important role in heat production via shivering during rewarming from torpor, and muscle metabolism needs to be adjusted for this purpose (Lyman et al., 1982; Block, 1994).

2. Adjustment of energy metabolism for needs of hibernators

Metabolic rate is strongly reduced during hibernation (often to only 1–5% of normal resting state) and this allows hibernators to save 80–90% of the total energy (including the cost of periodic arousals) that would otherwise be needed to remain euthermic over a winter season at the same ambient temperatures (Panteleev, 1983; Wang, 1985). The reduction in energy metabolism during hibernation is connected with a suppression of both glycolysis (El Hachimi et al., 1990; Brooks and Storey, 1992) and oxidative phosphorylation (Fedotcheva et al., 1985; Bronnikov et al., 1990). This suppression is accomplished via a number of different mechanisms which include changes in enzyme isoform spectrum, differential temperature effects on kinetic and allosteric properties of enzymes, changes in enzyme association with intracellular organelles and/or structural proteins, and reversible phosphorylation of key regulatory enzymes by protein kinases. Nevertheless, energy metabolism is coordinated and regulated to cover all animal needs for ATP during torpor bouts, periodical arousals, and spring awakening from hibernation.

The general strategy for adjustment of energy metabolism during hibernation includes a near total suppression of glycolysis, a switch to the use of fat as the main fuel source during torpor, and a low consumption of protein that primarily fuels gluconeogenesis (for review, see Storey, 1997). Carbohydrate metabolism is suppressed by a number of factors, which probably work in concert during hibernation. For example, total activities of key glycolytic enzymes including hexokinase, phosphofructokinase (PFK), and pyruvate kinase (PK) in skeletal muscles of hibernating bat *Eptesicus fuscus* and jerboa *Jaculus orientalis* are strongly decreased (Yacoe, 1983; El Hachimi et al., 1990). The inhibition of PFK may be connected with disaggregation of the active tetrameric form of the enzyme into inactive dimers following the protonation of dissociable groups of the enzyme. This is so because although hibernators allow their body temperature to fall by as much as almost 40°C, blood pH is maintained practically constant. Since both the neutral pH of water and the pKs of the major dissociable groups on enzymes increase as temperature decreases, enzymes become more protonated as temperature decreases at constant pH. This mechanism of PFK inhibition was suggested for PFK from another hibernator, the ground squirrel *Citellus beecheyi* (Hand and Somero, 1983). Another way to control PFK activity is via reversible binding of the enzyme to skeletal muscle microfilaments. It was shown that PFK binding to myosin filaments from skeletal muscle of summer active ground squirrels *Citellus undulatus* is 1.5-fold higher than to myosin filaments from hibernating animals (Lukoyanova et al., 1996b). These changes in PFK binding are probably connected mainly with changes in the properties of contractile proteins. Nevertheless, this can also contribute to the modulation of PFK activity. Furthermore, because the ATP/AMP ratio in skeletal muscles increases significantly during hibernation, the activity of PFK is also inhibited via allosteric mechanisms (El Hachimi et al., 1990).

During hibernation, significant changes in enzyme isoform spectrum often take place. For example, in skeletal muscle and liver of the bat, *Myotis lucifugus*, activity of PK increases 1.5- to 2-fold during hibernation in contrast with the decreased activity of this enzyme in many other species (Borgmann and Moon, 1976). Because upon arousal a switch from lipid to carbohydrate utilization occurs, an active PK would be essential for this transition and glycolysis is blocked at a step prior to PK in this particular species of hibernator. PK from muscle of euthermic versus hibernating bats also had significantly different temperature sensitivities: Arrhenius plots showed a sharp break at 17°C versus 5°C, respectively. Moreover, the K_m value for ADP for PK from skeletal muscles of hibernating bats did not change with assay temperature whereas that of euthermic animals increased about 4-fold at low temperatures (below 17°C). All these changes in enzyme properties are probably connected with changes in the PK isozyme spectrum during hibernation. Polyacrylamide gel electrophoresis shows that bat skeletal muscle PK is present in nine isozymic forms in euthermic tissue but only in five isoforms in the tissue of hibernating animals and, based on their electrophoretic mobility, none of these five isoforms are identical to the isozymes present in euthermic muscles. Five electrophoretically distinct isoforms were also found in liver of hibernating and euthermic bats, but the positioning of these forms was also slightly altered (Borgmann and Moon, 1976).

In skeletal muscle of the hibernating jerboa, *Jaculus orientalis*, activity of another glycolytic enzyme, glyceraldehyde-3-phosphate dehydrogenase (GPDH), is 3–4-fold lower than in the euthermic animals (Soukri et al., 1995). Such a significant loss of enzyme activity is connected not only to a decrease in the amount of enzyme protein expressed in the muscle but also to the changes in GPDH isozyme composition. With the use of polyacrylamide gel electrophoresis and chromatofocusing, three identical GPDH isoforms were found in skeletal muscles of both hibernating and euthermic animals. However, whereas in euthermic jerboa GPDH I isozyme (pI 8.1–8.2) provides more than 90% of total enzyme activity, in hibernating animals the GPDH II isozyme (pI 7.8–7.9) reaches up to 65% of total activity. Enzyme isoforms prevailing under hibernating conditions also exhibited a decreased catalytic efficiency when compared with the corresponding major isoforms from euthermic animals. The main differences in their kinetic properties were relatively higher K_m and lower V_{max} values of GPDH isozymes from hibernating animals (Soukri et al., 1995). More detailed analysis has shown that the decrease of GPDH activity in skeletal muscle is mainly due to a suppression of protein expression, but in liver the lower enzyme activity is a result of protein covalent modification, probably via mono-ADP-ribosylation (Soukri et al., 1996a; 1996b). These data indicate that GPDH in the hibernating jerboa is regulated by different mechanisms in different tissues, that is, at a transcriptional level in muscle and a posttranslational level in liver.

One of the widespread mechanisms for changing enzyme properties during hibernation is reversible phosphorylation via the actions of protein kinases and phosphatases (for review, see Storey, 1997). A study of the kinetic properties of key regulatory enzymes of glycolysis in liver of the meadow jumping mouse, *Zapus hudsonius*, showed that the activity of glycogen phosphorylase is strongly depressed during hibernation as a result of a decrease in both the total amount of enzyme protein expressed and amount of enzyme in the active (phosphorylated) *a* form. Kinetic properties of PFK also indicated phosphorylation inactivation of this enzyme. In comparison with euthermic controls, PFK from hibernating animals showed a 2.5-fold increase in K_a for fructose-2,6-P_2 and an approximate 4-fold decrease in I_{50} values for ATP and citrate. Changes in the kinetics of PK (K_a for fructose-1,6-P_2 increased 4.4-fold and I_{50} for L-alanine decreased 6.3-fold) similarly indicated enzyme phosphorylation (Storey, 1987). In other tissues of *Z. hudsonius*, as well as in tissues of other species of hibernators, phosphorylation of glycolytic enzymes is probably less important in metabolic arrest. For example, no evidence of hibernation-induced PFK or PK covalent modification was found in liver, heart, kidney, and skeletal muscles of the ground squirrel, *Spermophilus richardsonii* (Brooks and Storey, 1992). However, in this case an inhibition of pyruvate dehydrogenase (PDH), the enzyme that gates carbohydrate entry into the citric acid cycle, occurs during hibernation as a result of enzyme phosphorylation. The amount of active PDH in kidney and heart of hibernating ground squirrels was reduced to only 3 and 4% of the corresponding euthermic values, respectively.

Energy production by mitochondria is also significantly reduced during hibernation. One of the reasons for this reduction is an accumulation in mitochondria of particular metabolites. For example, succinate oxidation by liver mitochondria from the ground squirrel *Citellus parryi* is strongly inhibited during hibernation by oxaloacetic acid and can be easily removed by an addition of glutamic and isocitric acids (Fedotcheva et al., 1985). In addition to an inhibition of respiration at the level of dehydrogenases, a significant decrease (more than 3-fold) in K^+ transport was observed. The changes in K^+

transport into mitochondria during hibernation are connected with changes in the respiratory chain. Presumably, they are the result of a decrease in both energy supply and membrane permeability, the former being not the main cause for the decrease of K^+ transport. Moreover, because internal K^+ concentration in mitochondria controls the rate of respiration and its movement across the mitochondrial membrane also regulates the influx of substrates and thus the rate of respiration, it can be suggested that the decrease in K^+ transport into mitochondria during hibernation makes some contribution to the inhibition of respiration. It is interesting to note that inhibition of both mitochondrial respiration and K^+ transport are spontaneously removed during arousal from torpor (Fedotcheva et al., 1985). With the use of liver mitochondria of another hibernator, the ground squirrel *Citellus undulatus*, it was shown that H-ATPase is a rate-limiting step in oxidative phosphorylation in hibernating animals (Bronnikov et al., 1985).

The strong decrease in the activity of this enzyme is connected with a lower affinity of the H-ATPase regulatory site for adenine nucleotides. As a result the total content of intramitochondrial ATP and ADP is decreased during hibernation by approximately 5-fold. A significant decrease in O_2 consumption by liver mitochondria from hibernating ground squirrels, *Spermophilus richardsonii,* was observed by Pehowich and Wang (1984). In addition, Arrhenius plots for succinate oxidase-linked H^+ ejection, Ca^{2+} uptake, and O_2 consumption exhibited discontinuities near 21°C in summer active and at 11–13°C in warm (20°C)- or cold (4°C)-acclimated winter ground squirrels, but were linear in hibernating and aroused animals.

This observation clearly shows that the differences in thermal behavior of mitochondrial membrane functions have a seasonal character and are independent of ambient and body temperature (Pehowich and Wang, 1984). Therefore, specific molecular mechanisms should be "switched on" during the preparation of the animals for winter hibernation. As a result of this, a number of key enzymes will change their properties independently of whether animals enter torpor or not.

However, it should be noted that in many cases mitochondrial functions in skeletal muscles are held at a relatively high level during hibernation. In homogenates of pectoralis muscles of the bat, *Eptesicus fuscus*, the maximal respiratory rate supported by palmitate oxidation does not vary seasonally (Yacoe, 1983). Suppression of pyruvate oxidation by physiological concentrations of palmitoyl-carnitine explains why maximal respiratory rates using pyruvate as a substrate are significantly greater in homogenates prepared from summer bats as compared with those from hibernating animals. This again demonstrates that skeletal muscle metabolism during hibernation is adjusted for the use of lipids but not carbohydrates as a fuel. A significant increase in the oxidative capacity in the gastrocnemius (by 65%) and semitendiosus (by 37%) muscles and 1.2-fold activation of citrate synthase in heart of hibernating ground squirrels, *Spermophilus lateralis,* were also found (Wickler et al., 1991). This activation of oxidative capacity in gastrocnemius and semitendinosus muscles occurs despite their moderate atrophy (by 14% and 42%, respectively) in contrast with many examples of disuse atrophy (inactivity, bed rest, space flight) when loss of skeletal muscle mass is accompanied by a decrease of oxidative capacity. The importance of oxidative metabolism during hibernation is supported by the observation that the content of myoglobin in skeletal muscles of the ground squirrel, *Citellus undulatus,* is 2.5–3-fold higher in hibernating than in summer active animals (Postnikova et al., 1997). This is probably due to the high oxygen demand of skeletal muscles during the first stage of arousal from torpor when body temperature rises from 0°C to 10–14°C using non-shivering thermogenesis and thermoregulatory tonus. At this stage the oxygen-dependent processes in muscles proceed under conditions of a blocked peripheral blood flow (Lyman et al., 1982).

Therefore, enzyme phosphorylation and the changes in isozyme spectrum, in combination with other mechanisms of enzyme control, provide a significant block on the use of carbohydrates as an energy source, contribute to the overall reduction in oxidative metabolism that suppresses thermogenesis, and facilitate the use of lipids as the major oxidative fuel in hibernation (Storey, 1997). Nevertheless, despite this coordinated suppression, the metabolism of hibernating animals provides a sufficient amount of energy to maintain all intracellular processes during torpor.

3. Molecular mechanisms of excitation–contraction coupling in heart and skeletal muscles of mammals

Contraction of the heart and skeletal muscles of hibernators as well as of all other mammals is controlled by the well-coordinated operation of a number of molecular machines: (1) voltage-gated plasma membrane Ca-channels and sarcoplasmic reticulum (SR) Ca-release channels that provide the transient increase in free Ca^{2+} concentration in cytoplasm necessary for contraction, (2) Ca-binding and contractile proteins of muscle filaments that convert Ca^{2+} signals into mechanical response, and (3) SR and plasma membrane Ca-ATPases which remove Ca^{2+} from the cytoplasm resulting in relaxation. In addition, the functional activity of these molecular machines is controlled and regulated by a plethora of accessory proteins including different protein kinases and Ca-binding proteins. In cardiac and skeletal muscles of mammals different mechanisms of excitation–contraction (E-C) coupling exist but in both tissues the same proteins participate in this process (for review, see Meissner and Lu, 1995; Rubtsov and Batrukova, 1997).

Contraction of cardiac and skeletal muscles begins from a depolarization of the plasma membrane that activates voltage-gated L-type Ca-channels, also called dihydropyridine receptors (DHPR). In both tissues DHPRs are related oligomeric protein complexes that are made up of five different subunits ($\alpha1$, $\alpha2$, β, γ, and δ). The $\alpha1$ subunit is capable of forming a Ca-channel in the absence of the other subunits, binds Ca-channel agonists and antagonists, and belongs to a voltage-sensitive ion channel superfamily which also includes Na- and K-channels (for review, see Catterall, 1995). All of these channels have four homologous domains each containing six transmembrane segments and a cytoplasmic pore-lining region and undergo conformational changes as a result of plasma membrane depolarization. The $\alpha2$, β, γ, and δ subunits of DHPRs do not participate directly in the formation of the voltage-gated conductance pathway, but they assist in the functional expression and contribute to the kinetics of Ca-channel function. The $\alpha1$ and β subunits are substrates for phosphorylation by numerous protein kinases, and phosphorylation-dephosphorylation seems to regulate Ca-channel function and its interaction with other proteins, including the SR Ca-release channel (Catterall, 1995; Mackrill, 1999).

The next participants in the E–C coupling process are cardiac and skeletal muscle isoforms of the SR Ca-release channel or ryanodine receptor (RyR). Both RyR isoforms have a number of common properties. They are homotetramers consisting of four identical monomers with a molecular mass of about 560 kDa, contain cytoplasmic and transmembrane domains, are activated by caffeine, micromolar Ca^{2+}, and adenine nucleotides and inhibited by Mg^{2+}, high Ca^{2+} concentrations, and ruthenium red (for review, see Rubtsov and Batrukova, 1997; Wagenknecht and Rademacher, 1997). In comparison with the skeletal muscle isoform, the cardiac RyR isoform is more sensitive to activation by low Ca^{2+} concentrations, less sensitive to inhibition by all of the above mentioned inhibitors, and binds the plant alkaloid, ryanodine, with higher affinity.

During the action potential in cardiac muscle, DHPRs located in the plasma membrane and its T-tubules provide an influx of extracellular Ca^{2+} into the cytoplasm by functioning as voltage-dependent Ca-channels. The rise in intracellular Ca^{2+} concentration triggers a massive release of Ca^{2+} from the SR lumen by opening closely apposed cardiac RyR (Meissner and Lu, 1995; Rubtsov and Batrukova, 1997). It is believed that up to 30% of the Ca^{2+} necessary for cardiac muscle contraction enters the cytoplasm through DHPRs, whereas the rest is released from SR via RyRs; this explains why mammalian hearts fail to work in Ca^{2+}-free solutions. Such a mechanism of E–C coupling was named Ca-induced Ca^{2+} release (CICR). By contrast, Ca^{2+} currents through DHPR Ca-channels do not seem to be required for E–C coupling in mammal skeletal muscles because removal of extracellular Ca^{2+} does not prevent muscle contraction. This led to the formulation of the "mechanical coupling" mechanism that suggests that voltage-sensing T-tubule DHPRs open SR RyR Ca-release channels through direct protein-protein interactions. It was shown that about 50% of SR RyRs in fact are directly linked to DHPRs. The Ca^{2+} released through these channels then opens RyR molecules that are not directly linked to DHPRs (Lamb, 2000).

It is now clear that the functioning of RyRs in cardiac and skeletal muscles and their interactions with DHPRs in skeletal muscles are regulated by a number of accessory proteins including the glycolytic enzymes GPDH and aldolase, the cytoplasmic Ca-binding protein calmodulin (CaM), the SR membrane Ca-binding proteins calsequestrin (CS), calreticulin (CR), sarcalumenin (SL), and histidine-rich Ca-binding protein (HCBP), as well as triadin, annexin VI, sorcin, different protein kinases, and many others (for review, see Mackrill, 1999). Depending on the intracellular concentration of Ca^{2+} and other second messengers these proteins can either activate or inhibit RyR Ca-channel activity providing fine regulation of its function.

An increase in Ca^{2+} concentration in the cytoplasm leads to the saturation by Ca^{2+} of microfilament Ca-binding regulatory proteins, first of all, cardiac and skeletal muscle isoforms of troponin C. Under these conditions the steric inhibition of actomyosin ATPase by troponin–tropomyosin complex is removed as a result of conformational changes in the complex. This allows the interaction of myosin heads with actin filaments that leads to muscle contraction (for review, see Gordon et al., 2000). At the same time, immediately after the rise of cytoplasmic Ca^{2+} concentration, the Ca-pumping proteins (SR and plasma membrane Ca-ATPases) start to work. These enzymes remove Ca^{2+} from the cytoplasm either transporting it out of cells (plasma membrane Ca-ATPase) or into the SR lumen (SR Ca-ATPase). The operation of Ca-ATPases decreases Ca^{2+} concentration in cytoplasm and this is followed by muscle relaxation. It should be noted that the activity of Ca-pumping ATPases is also controlled via interactions with a number of accessory proteins. Plasma membrane Ca-ATPase is activated by CaM (Carafoli, 1991). In SR membranes of cardiac cells Ca-ATPase is inhibited by the integral SR membrane protein phospholamban, but phosphorylation of phospholamban by cAMP- and/or Ca/CaM-dependent protein kinases removes this inhibition (Simmerman and Jones, 1998). In skeletal muscle SR, the modulation of Ca-ATPase activity by accessory proteins (including CR,CS, and SL) is also suggested (Krause and Michalak, 1997; Froemming and Ohlendieck, 1998). Therefore, the contractile activity of heart and skeletal muscles is controlled by the very well coordinated functioning of a number of proteins and protein complexes. Probably, some changes in these protein complexes and their accessory regulatory proteins should take place during hibernation to allow heart and skeletal muscle to operate under extremely unusual conditions.

4. Changes in the properties of enzyme systems responsible for the functional activity of heart and skeletal muscles during hibernation

4.1. Ca-channels of plasma membrane

To date, the peculiarities of the operation of skeletal muscle plasma membrane DHPRs during hibernation are largely unknown, but the properties of these channels in cardiac tissue have been studied in some detail. Electrophysiological experiments have shown that the electrical and mechanical characteristics of heart muscle are different in hibernating versus euthermic Asian chipmunks, *Tamias sibiricus* (Kondo and Shibata, 1984). The amplitude of the early plateau phase of the action potential of isolated papillary muscles from hibernating animals was much less than that from euthermic chipmunks. Such a depression may be attributed to the reduced activation of the slow inward current through DHPRs resulting in less Ca^{2+} influx across the cell membrane. The force-frequency relationship was also different in the two preparations. A stepwise increase in the driving frequency increased the contractile force of cardiac muscle from euthermic chipmunks but reduced that of the hibernating animals. These results suggest that in hibernating animals Ca^{2+} release from SR may play a more important role in contraction than does trans-sarcolemmal Ca^{2+} influx during the action potential. Probably, the lower Ca^{2+} influx is sufficient to trigger the release of Ca^{2+} from SR via CICRs. It is interesting to note that the changes in DHPR properties were already found in the heart of pre-hibernating chipmunks (Kondo, 1987). Therefore, changes in the mechanism of Ca^{2+} exchange in cardiac muscles are not a result of low temperature effects on plasma membrane proteins but are developed by still unknown intracellular processes during the preparation of animals for the hibernating season. A more detailed study of the action of the SR Ca-release channel modulators, caffeine and ryanodine, and a cardiotonic agent, isoprenaline, on the contractile activity of papillary muscle from hibernating chipmunks showed that SR not only plays a more important role in the release of the Ca^{2+} necessary for heart contraction but probably also has increased Ca-accumulating ability (Kondo, 1986; 1988). The same changes in the properties of DHPR Ca-channels during hibernation were found in the heart of the ground squirrel, *Citellus undulatus* (Alekseev et al., 1996). Using a perforated patch-clamp method it was shown that the potential-dependent Ca^{2+} current in isolated cardiocytes from hibernating animals is strongly inhibited in comparison with that in euthermic ground squirrels. A detailed analysis of experimental current traces allowed the authors to develop a kinetic model by which the conductance regulation of Ca-channels can be described by three processes: activation, and slow and fast inactivation of the channel (Alekseev et al., 1997; Kokoz et al., 1997a). Activation is related to the movement of the gating charge. Slow inactivation is associated with the movement of the gating charge and is current-dependent. Fast inactivation is a more complex process and cannot be represented as a single-stage conformational transition induced by the gating charge movement. Rather, it is regulated by cAMP-dependent phosphorylation. It was also shown that isoproterenol-induced cAMP-dependent phosphorylation partially restored the kinetic parameters of Ca^{2+} currents in cardiocytes of hibernating animals to values close to the parameters of active ground squirrels. However, the authors suggested the involvement of another, cAMP-independent, unidentified protein kinase in the regulation of DHPR Ca-channels in the heart of hibernators. During the active state this protein kinase phosphorylates an unknown site on one of subunits of DHPR, but during hibernation the phosphorylation of this site is blocked (Kokoz et al., 1999). Therefore, different protein kinases are probably involved in changing the characteristics of cardiac plasma membrane Ca-channels during hibernation.

4.2. Sarcoplasmic reticulum proteins

The increased ability of SR to pump Ca^{2+} that was suggested by Kondo (1986; 1988) for the heart of hibernating chipmunks was also seen in the heart of the hibernating ground squirrel, *Spermophilus richardsonii* (Belke et al., 1987). It was shown that SR preparations isolated from the heart of hibernating ground squirrels had the highest rate of Ca^{2+} uptake in comparison with SR preparations isolated from spring, fall, and winter active animals over a wide range of temperatures (from 4°C to 37°C). The increase in Ca-accumulating ability of the SR had a seasonal character: fall animals showed an intermediate rate of Ca^{2+} uptake in between that of spring and hibernating animals. This shows that fall ground squirrels are undergoing biochemical changes necessary for winter hibernation. The Ca-accumulating ability of SR preparations isolated from heart of winter active animals was lower than that of fall ground squirrels and close to the parameters of SR Ca-uptake of spring animals. Therefore, during the torpor state some additional changes occur in SR membranes resulting in a further increase of SR Ca-accumulating ability. In heart SR of winter active animals these changes not only did not occur but changes that had taken place during the fall (in preparation for hibernation) were actually reversed when the animals did not hibernate. However, the molecular mechanism of these changes is still unknown (Belke et al., 1987).

A detailed analysis of the protein composition of SR preparations isolated from heart of two hibernating ground squirrels, *Spermophilus richardsonii* and *Spermophilus columbianus*, has shown that cardiac SR of hibernators contains a unique isoform of the Ca-binding protein, CS (Milner et al., 1991). Its electrophoretic mobility significantly differed from that of both rabbit skeletal muscle and canine cardiac CS isoforms. Because CS plays an important role in SR RyR Ca-release channel operation (Yano and Zarain-Herzberg, 1994; Mackrill, 1999), the presence of this novel CS isoform in cardiac SR of hibernators is probably connected with the ability of hibernator heart to work at extremely low temperatures. It is interesting to note that in addition to this new CS isoform, cardiac SR preparations of hibernators also contain very atypical RyRs (Milner et al., 1991). Ground squirrel cardiac SR exhibits a nearly 6-fold higher number of ryanodine binding sites compared with bovine and sheep cardiac SR, but the affinity for ryanodine is about 3-fold lower in cardiac SR of hibernators. Therefore, the parameters of ryanodine binding by ground squirrel cardiac SR were much closer to those of rabbit skeletal muscle SR than to any known cardiac SR of non-hibernating mammals. The atypical properties of RyR and CS in cardiac SR of hibernators may explain the unusual type of E-C coupling in the heart of hibernating chipmunks where Ca^{2+} release from SR plays a more important role than Ca^{2+} influx through plasma membrane DHPR Ca-channels (Kondo and Shibata, 1984; Kondo, 1986). However, detailed analysis of chipmunk cardiac SR protein composition should be carried out.

An increased ability of SR to accumulate Ca^{2+} was also found in skeletal muscles of hibernating European hamsters, *Cricetus cricetus* (Agostini et al., 1991). In comparison with summer active animals, the maximum rate of Ca^{2+} uptake and the calcium storing capacity of SR were increased by 43% and 17%, respectively, in hibernating hamsters, but the affinity of SR Ca-ATPase for Ca^{2+} was decreased during hibernation. Moreover, the hydrolytic activity of SR Ca-ATPase was not increased during hibernation which indicates a tighter coupling between Ca-dependent ATP hydrolysis and Ca^{2+} transport. It should be noted also that the same changes in Ca^{2+} transport characteristics were found in winter active animals kept 22°C under a natural photoperiod. This indicates that during the preparation for hibernation a modulation of the properties of skeletal muscle SR Ca-ATPase occurs to better support the ability of SR to accumulate Ca^{2+} at low temperatures. However, the molecular mechanism of this modulation remains unknown.

A detailed study of the properties of SR preparations isolated from skeletal muscles of the ground squirrel, *Spermophilus undulatus,* in our laboratory has shown that, in contrast with hamster skeletal muscle, the activity of ground squirrel SR Ca-ATPase is significantly lower in hibernating animals than that in summer active ones (Table 5.1). The same decrease in Ca-ATPase hydrolytic activity was found in SR preparations of winter active animals. This probably indicates that the changes in Ca-ATPase activity are connected with a modulation of enzyme properties during the preparation of the animals for the hibernating period and are not a result of low temperature effects on SR membranes.

One of the reasons for a low Ca-ATPase activity in SR preparations from hibernating animals could be a low content of the enzyme protein in membranes. To check this possibility polyacrylamide gel electrophoresis was carried out and the protein composition of SR membranes from summer active and winter hibernating ground squirrels were compared. The data obtained show clearly that the content of Ca-ATPase protein in SR membranes of both winter hibernating and winter active animals is actually lower than that in SR preparations of summer active ground squirrels (Table 5.2). However, this difference is not large enough to fully explain the differences in enzyme activities between summer and winter groups. The specific activity of Ca-ATPase calculated using a

Table 5.1. Ca-ATPase activity (at 37°C) and phospholipid content in SR preparations isolated from skeletal muscles of summer active (SA), winter hibernating (WH), and winter active (WA) ground squirrels *Spermophilus undulatus* (mean ± SD). Skeletal muscles were collected during the middle of a torpor bout (WH1, body temperature 3-5°C) or at the conclusion of spontaneous arousal (WH2, body temperature 37°C); WA animals were kept at 20°C and natural photoperiod (body temperature 37°C). In all cases the values for SA preparations are significantly different from those for WH and WA preparations ($P < 0.05$). For details of experiments, see Shutova et al. (1999).

	SA ($n = 6$)	WH1 ($n = 8$)	WH2 ($n = 3$)	WA ($n = 3$)
Ca-ATPase activity (μmol/min/mg SR protein)	5.2 ± 0.4	2.4 ± 0.1	2.8 ± 0.2	2.0 ± 0.4
Ca-ATPase activity (μmol/min/mg enzyme protein)	14.2 ± 0.4	7.8 ± 0.9	9.2 ± 0.1	7.0 ± 1.4
Phospholipids (μmol P_i/mg protein)	1.04 ± 0.07	0.75 ± 0.05	0.74 ± 0.1	0.62 ± 0.07

Table 5.2. The relative content of major protein components of SR preparations isolated from skeletal muscles of summer active (SA), winter hibernating (WH), and winter active (WA) ground squirrels *Spermophilus undulatus* (mean ± SD). Skeletal muscles were collected during the middle of a torpor bout (WH1, body temperature 3–5°C) or at the conclusion of spontaneous arousal (WH2, body temperature 37°C); WA animals were kept at 20°C and natural photoperiod (body temperature 37°C). Data obtained from scanning gels stained by the cationic carbocyanine dye Stains-All are expressed relative to the mean intensity for the peak area of each protein in SA animals. Data obtained after staining gels with Coomassie Brilliant Blue R-250 are expressed as percentages of the total SR protein peak areas. In all cases the values for SA preparations are significantly different from those for WH and WA preparations ($p < 0.05$). For details of experiments, see Shutova et al. (1999).

	SA ($n = 6$)	WH1 ($n = 7$)	WH2 ($n = 3$)	WA ($n = 3$)
Stains-All scans:				
165 kDa HCBP	100 ± 9.1	39.8 ± 6.7	35.6 ± 6.1	36.7 ± 4.7
130 kDa SL	100 ± 6.4	39.9 ± 4.8	35.7 ± 8.1	33.4 ± 5.4
63 kDa CS	100 ± 5.0	28.7 ± 2.8	22.7 ± 5.2	21.2 ± 4.5
Coomassie Blue scans:				
450 kDa RyR	2.0 ± 0.4	1.0 ± 0.4	0.6 ± 0.1	1.5 ± 0.2
165 kDa HCBP	1.6 ± 0.1	0.6 ± 0.1	0.6 ± 0.2	0.6 ± 0.3
130 kDa SL	2.0 ± 0.1	0.6 ± 0.1	0.9 ± 0.3	0.8 ± 0.3
105 kDa Ca-ATPase	42.0 ± 3.1	29.4 ± 2.8	31.3 ± 3.8	25.0 ± 1.2
63 kDa CS	10.3 ± 1.1	3.3 ± 0.5	2.8 ± 0.7	3.0 ± 0.3
55 kDa protein	5.7 ± 0.9	8.7 ± 0.9	9.2 ± 0.8	8.9 ± 0.6
30 kDa protein	11.0 ± 1.9	17.0 ± 0.9	15.0 ± 1.1	17.0 ± 0.9
22 kDa protein	4.7 ± 1.6	10.6 ± 2.7	11.5 ± 1.5	12.4 ± 2.7

correction for the enzyme protein content in SR membranes is still ~2-fold lower in SR of winter, compared with summer, animals (Table 5.1).

On the other hand, the decrease in Ca-ATPase activity is probably connected with a significant increase in the protein to phospholipid ratio in SR membranes (Table 5.1) as well as with probable changes in the lipid composition of SR membranes during the winter period because lipid-protein interactions play significant role in SR Ca-ATPase operation (Lee et al., 1995), and the lipid composition of ground squirrel skeletal muscle SR membranes changes seasonally (Pehowich, 1994). However, in addition to the changes in enzyme hydrolytic activity, we have found significant changes in SR Ca-ATPase kinetic parameters during hibernation. The substrate dependence of Ca-ATPase from summer active ground squirrels is described by a curve with an undefined intermediary plateau, but the comparable relationship for

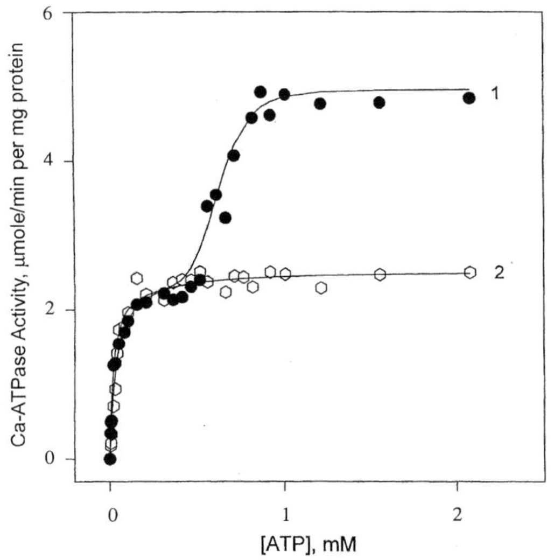

Fig. 5.1. ATP concentration dependence of skeletal muscle SR Ca-ATPase activity from summer active (1) and hibernating (2) ground squirrels, *S. undulatus*. Enzyme activity was measured at 37°C and expressed in µmol P_i released/min per mg protein. Data are means, $n = 3$.

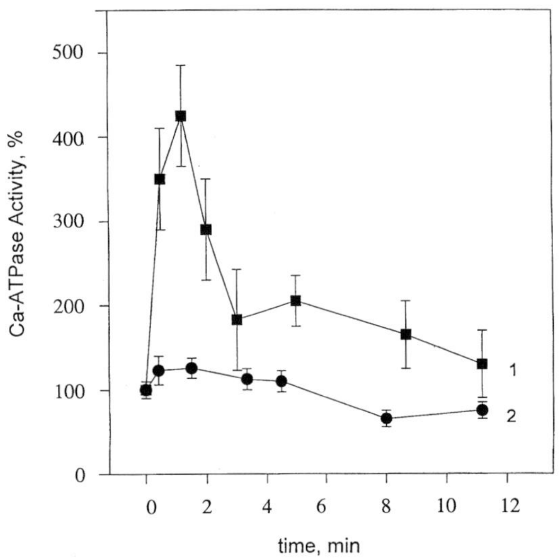

Fig. 5.2. Effect of endogenous protein kinase activation on Ca-ATPase hydrolytic activity in SR preparations isolated from skeletal muscles of hibernating (1) and summer active (2) ground squirrels. SR preparations were incubated in a medium containing 2 mM ATP, 2 mM $MgCl_2$, 1 mM EGTA, 1 mM DTT, 1 mM PMSF, and 10 mM imidazole, pH 7.0, at 25°C. Enzyme activity was measured at 37°C and expressed relative to the Ca-ATPase activity of each SR preparation without incubation. Data are means ± SD, $n = 3$. For details of the Ca-ATPase activity measurement, see Shutova et al. (1999).

the enzyme from hibernating animals is described by a simple hyperbolic curve (Fig. 5.1). The same changes in kinetic parameters were found for plasma membrane Na,K-ATPase from skeletal muscle of the ground squirrel, *Spermophilus lateralis* (MacDonald and Storey, 1999), and were linked with the phosphorylation of Na,K-ATPase by an unknown protein kinase(s).

To check for the possible involvement of endogenous protein kinase(s) in the regulation of skeletal muscle SR Ca-ATPase activity, we incubated SR membranes isolated from summer active and hibernating ground squirrels under conditions that stimulate protein kinase activation (Fig. 5.2). Incubation of SR membranes in the presence of ATP resulted in a significant transient increase in Ca-ATPase activity in preparations isolated from hibernating animals but had virtually no affect on the enzyme in SR preparations from summer active ground squirrels. The transient character of Ca-ATPase activation suggests that SR membranes probably also contain active protein phosphatase(s), which reverse the effect of protein kinase(s). Therefore, in contrast with Na,K-ATPase, phosphorylation of which inhibits enzyme activity (MacDonald and Storey, 1999), phosphorylation of Ca-ATPase or some other SR protein(s) results in enzyme activation and probably changes Ca-ATPase kinetic behavior.

Phosphorylation of skeletal muscle SR proteins was recently studied in our laboratory. As shown in Fig. 5.3, an incubation of SR membranes in the presence of $\gamma[^{32}P]$-ATP resulted in incorporation of phosphate onto a number of proteins in both summer active and hibernating animals, but the level of phosphorylation and the pattern of phosphorylated proteins were significantly different between the two groups. In general, SR preparations from skeletal muscles of summer active ground squirrels had about 2-fold higher phosphorylation ability in comparison with preparations from hibernating animals. The addition of exogenous cAMP had little effect on the protein phosphorylation pattern in both SR preparations, but addition of Ca^{2+} and CaM significantly inhibited phosphate incorporation into SR proteins especially in preparations isolated from summer active ground squirrels. It

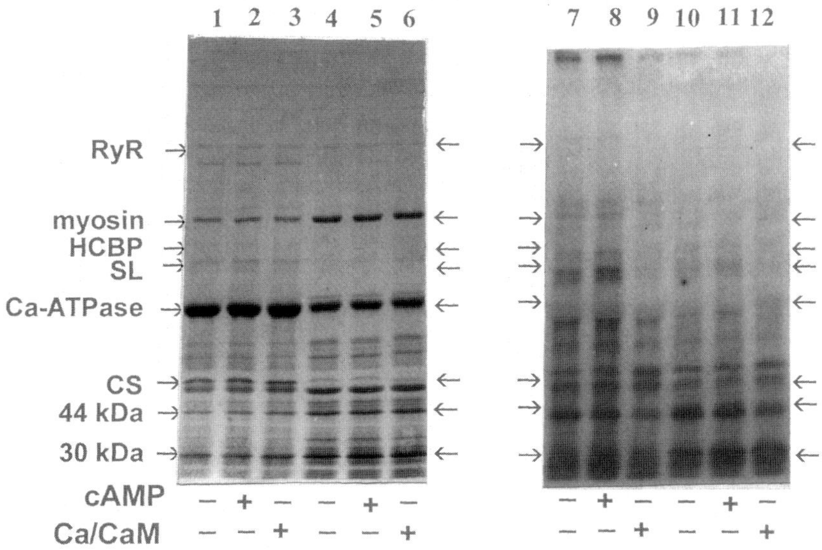

Fig. 5.3. Phosphorylation of SR proteins by endogenous protein kinase(s) in ground squirrel *S. undulatus* SR preparations. Electrophoregram (lanes 1–6) stained with Coomassie Brilliant Blue R-250 and autoradiogram of the same gel (lanes 7–12) are shown. For details of polyacrylamide gel electrophoresis, see Shutova et al. (1999). Each lane was loaded with 30 μg of SR protein from summer active (lanes 1–3 and 7–9) or hibernating (lanes 4–6 and 10–12) animals. Prior to electrophoresis, samples were incubated for 30 min at 30°C in a medium containing 250 μM $\gamma[^{32}P]$-ATP (1000 cpm/nmol), 5 mM $MgCl_2$, 1 mM EGTA, 1 mM DTT, 5 mM NaF, 0.1 mM vanadate, and 15 mM Tris-HCl, pH 7.4. Where indicated, 1 μM cAMP or 1 mM $CaCl_2$ and 10 nM CaM were added. Arrows indicate the position of the major SR proteins.

should be noted that the incorporation of phosphate into Ca-ATPase protein was not found. Therefore, skeletal muscle SR membranes of ground squirrels possess relatively high endogenous protein kinase activity and this activity is decreased during hibernation. Phosphorylation of some SR protein(s) but not Ca-ATPase itself seems to result in an increase in enzyme activity.

Detailed analysis of SR membrane protein composition has shown that skeletal muscle SR of ground squirrels (*S. undulatus*) contains atypical isoforms of the Ca-binding proteins HCBP and SL (Shutova et al., 1999). These differ from the corresponding isoforms found in rabbit and rat skeletal muscle SR in their electrophoretic mobility. The contents of these proteins as well as the content of another Ca-binding protein, CS, estimated with the use of the cationic carbocyanine dye Stains-All (which stains Ca-binding proteins in a dark blue color) was decreased dramatically (3–4-fold) in SR membranes of winter active and hibernating animals (Table 5.2). All these proteins interact with the skeletal muscle RyR Ca-release channel and are involved in its regulation (Mackrill, 1999), and phosphorylation of HCBP and SL by endogenous SR protein kinase inhibits RyR activity in rabbit skeletal muscle SR (Orr and Shoshan-Barmatz, 1996; Shoshan-Barmatz et al., 1996). We have demonstrated that ground squirrel isoforms of HCBP and SL can also be phosphorylated by endogenous SR protein kinase(s) (Fig. 5.3).

A decrease in the content of Ca-ATPase and Ca-binding proteins in ground squirrel skeletal muscle SR membranes during hibernation is accompanied by an increase in the content of a number of other SR proteins including unidentified proteins with molecular masses of 55, 30, and 22 kDa (Table 5.2; Fig. 5.3). There are a few known SR proteins with the same molecular masses. One possibility for the 55 kDa protein is the high-affinity Ca-binding protein, CR (Krauze and Michalak, 1997). Its content is known to change during skeletal muscle development. CR is the dominant SR Ca-binding protein in the early stages of development but is replaced by CS in mature myotubes (Koyabu et al., 1994; Froemmig and Ohlendieck, 1998). The opposite process may occur during hibernation. The 30 kDa protein is probably a CS-binding protein involved in the regulation of RyR function (Kagari et al., 1996). The protein(s) responsible for the large increase in the 22 kDa protein band in SR of hibernators is not clear at present. Nevertheless, it is evident that the content of many skeletal muscle SR proteins, at least some of them involved in the regulation of RyR and/or Ca-ATPase activity, as well as their phosphorylation state, are changed during hibernation. However, further studies will be necessary to identify which of these proteins are responsible for the adjustment of SR membranes for optimal functioning during hibernation.

4.3. Contractile proteins

Significant changes in cardiac and skeletal muscle contractile proteins also occur during hibernation. It was found that two types of myosin heavy chains (MHC) are expressed in heart ventricle of the European hamster, *Cricetus cricetus*: α-MHC with high and β-MHC with low enzymatic activity (Morano et al., 1992). These types of MHC have different temperature dependences: the activity of β-myosin is less dependent on temperature than that of α-myosin. Summer active animals expressed predominantly β-MHC (79% of total myosin) but during hibernation the expression of α-MHC increased up to 53% of total myosin. Winter active animals kept at room temperature showed a MHC isozyme pattern similar to that of summer active hamsters. Two forms of myosin light chains (MLC) were also found in all groups of animals. It was demonstrated that the endogenous level of protein phosphorylation of the MLC2 was decreased from 45% in summer active hamsters to 23% in hibernating animals. Therefore, the endogenous protein kinase(s) activity in the heart of the European hamster seems to change during hibernation.

Changes in the structural organization of myosin filaments were also observed in skeletal muscles of the ground squirrel, *Citellus undulatus,* at different stages of the hibernation cycle (Lukoyanova et al., 1996a). Reconstituted myosin filaments from hibernating, winter active, and aroused animals exhibit the ordered structure of myosin heads arranged regularly on the filament surface. However, the filaments reconstituted from myosin isolated from animals at the beginning of arousal from hibernation bouts (rectal temperature 12°C) have an irregular structure with a random arrangement of the myosin head clusters alternating with regions of different lengths without the heads. The authors ascribe these changes in myosin filaments to the appearance of new MHC isoforms during arousal. The changes in the structure of myosin filaments were accompanied by significant changes in the MLC spectrum. Although the amount of MLC2 was the same among all groups of animals, the content of MLC3 was about 3-fold lower in skeletal muscle of hibernating ground squirrels than in winter active animals and was increased by 1.5–2-fold during the first hours of arousal (Lukoyanova et al., 1996a; 1997). This suggested that rapid changes in MHC and MLC spectrum can result in a lowering of muscle contraction efficiency and an increase in heat production that can contribute to the nonshivering thermogenesis in skeletal muscles during arousal.

A significant decrease in actin-activated ATPase activity of skeletal muscle myosin from hibernating animals (by 60% in comparison with winter active ground squirrels) was also found (Lukoyanova et al., 1997). During arousals, myosin ATPase activity was increased about 2-fold. In the absence of the troponin-tropomyosin complex, the actomyosin from hibernating animals also demonstrated significantly lower Ca-sensitivity. All of these changes are connected with the differences in MHC and MLC properties because no change in the properties of actin were found during hibernation (Lukoyanova et al., 1998). Changes in myosin properties are probably connected with changes in the activity of endogenous protein kinase(s) in skeletal muscle cells. Thus, in myosin preparations from hibernating ground squirrels a high level of endogenous MLC2 phosphorylation (30–50% of the total MLC2) was found, whereas in all other preparations MLC2 was completely dephosphorylated (Lukoyanova et al., 1996a). This observation demonstrates once again the important role of phosphorylation–dephosphorylation processes in the protein adaptation to hibernation.

Recently, a significant increase in the production of mRNA transcripts encoding the ventricular MLC1 isoform was reported during hibernation in heart (about 2.5-fold) and skeletal muscle (about 3-fold) of the ground squirrel, *Spermophilus lateralis* (Fahlman et al., 2000). At the same time the level of expression of this gene in kidney and liver was unchanged during hibernation. This suggests that changes in the composition of the alkali MLC (MLC1 and MLC3) as well as a change in the level of phosphorylation of the regulatory MLC2 play important roles in the adjustment of the contractile apparatus of both heart and skeletal muscles for function at low body temperatures.

5. Concluding remarks

Even a brief review of recent investigations in the field of mammal hibernation shows that the entrance of animals into torpor is accompanied by numerous significant changes in the properties of enzymes and proteins. Very often these changes are connected with protein phosphorylation and/or dephosphorylation by endogenous protein kinase(s) and protein phosphatase(s). The expression of many proteins or their particular isoforms also changes during hibernation or in the pre-hibernating period. In addition, some proteins are represented in the tissues of hibernators by specific isoforms, which are absent in non-hibernating species. All of these mechanisms provide fine adjustment of metabolic processes for operation at the low body temperatures that characterize hibernation. However, very little is yet known about the initial triggering mechanisms that "switch on" the numerous intracellular processes that are necessary to prepare the animal to deal with the changes in environmental conditions. The cells of all organs and tissues of hibernators should receive specific signals from the central nervous and/or endocrine systems to start metabolic rearrangements. Such triggering signals are probably small molecules that can be easily delivered by the blood to any place in the animal body. In addition to hormones, small peptides produced by brain seem to play a very important role in the regulation of cell functions during hibernating-arousal cycles (see Kokoz et al., 1997b, and references therein). Therefore, despite significant progress in the analysis of the properties of enzymes and proteins from hibernating animals, a lot more remains to be done to obtain a full and detailed picture of the molecular mechanisms that are responsible for the adjustment of metabolism for hibernation on many different levels, ranging from the whole organism and its individual organs to the cellular and intracellular compartments.

References

Agostini, B., De Martino, L., Soltau, B. and Hasselbach, W. (1991). The modulation of calcium transport by skeletal muscle sarcoplasmic reticulum in the hibernating European hamster. Z. Naturforsch. 46c, 1109–1126.

Alekseev, A.E., Markevich, N.I., Korystova, A.F., Lankina, D.A. and Kokoz, Yu.M. (1997). The kinetic characteristics of the L-type calcium channels in cardiocytes of hibernators. 1. Development of kinetic model. Membr. Cell. Biol. 11, 31–44.

Alekseev, A.E., Markevich, N.I., Korystova, A.F., Terzic, A. and Kokoz, Yu.M. (1996). Comparative analysis of kinetic characteristics of L-type calcium channels in cardiac cells of hibernators. Biophys. J. 70, 786–797.

Belke, D.D., Pehowich, D.J. and Wang, L.C.H. (1987). Seasonal variation in calcium uptake by cardiac sarcoplasmic reticulum in a hibernator, the Richardson's ground squirrel. J. Therm. Biol. 12, 53–56.

Block, B.A. (1994). Thermogenesis in muscle. Ann. Rev. Physiol. 56, 535–577.

Borgmann, A.I. and Moon, T.W. (1976). Enzymes of the normothermic and hibernating bat, *Myotis lucifugus*: temperature as a modulator of pyruvate kinase. J. Comp. Physiol. B 107, 185–199.

Bronnikov, G.E., Vinogradova, S.O. and Mezentseva, V.S. (1990). Changes in kinetics of ATP-synthase and in concentration of adenine nucleotides in ground squirrel liver mitochondria during hibernation. Comp. Biochem. Physiol. B 97, 411–415.

Brooks, S.P.J. and Storey, K.B. (1992). Mechanism of glycolytic control during hibernation in the ground squirrel *Spermophilus lateralis*. J. Comp. Physiol. B 162, 23–28.

Carafoli, E. (1991). Calcium pump of plasma membrane. Physiol. Rev. 71, 129–153.

Catterall, W.A. (1995). Structure and function of voltage-gated ion channels. Annu. Rev. Biochem. 64, 493–531.

El Hachimi, Z., Tijuane, M., Boissonnet, G., Benjouad, A., Desmadril, M. and Yon, J.M. (1990). Regulation of the skeletal muscle metabolism during hibernation of *Jaculus orientalis.* Comp. Biochem. Physiol. B 96, 457–459.

Fahlman, A., Storey, J.M. and Storey, K.B. (2000) Gene up-regulation in heart during mammalian hibernation. Cryobiology 40, 332–342.

Fedotcheva, N.J., Sharyshev, A.A., Mironova, G.D. and Kondrashova, M.N. (1985). Inhibition of succinate oxidation and K^+ transport in mitochondria during hibernation. Comp. Biochem. Physiol. B 82, 191–195.

Froemming, G.R. and Ohlendieck, K. (1998). Oligomerization of Ca^{2+}-regulatory membrane components involved in excitation–contraction–relaxation cycle during postnatal development of rabbit skeletal muscle. Biochim. Biophys. Acta 1387, 226–238.

Geiser, F. (1988). Reduction of metabolism during hibernation and daily torpor in mammals and birds: temperature effect or physiological inhibition? J. Comp. Physiol. B 158, 25–37.

Gordon, A.M., Homsher, E. and Regnier, M. (2000). Regulation of contraction in striated muscle. Physiol. Rev. 80, 853–924.

Hand, S.C. and Somero, G.N. (1983). Phosphofructokinase of the hibernator, *Citellus beecheyi*: temperature and pH regulation of activity via influences on the tetramer–dimer equilibrium. Physiol Zool. 56, 380–388.

Heldmaier, G. and Ruf, T. (1992). Body temperature and metabolic rate during natural hypothermia in endotherms. J. Comp. Physiol. B 162, 696–706.

Hochachka, P.W. (1986). Defense strategies against hypoxia and hypothermia. Science 231, 234–241.

Kagari, T., Yamaguchi, N. and Kasai, M. (1996). Biochemical characterization of calsequestrin-binding 30-kDa protein in sarcoplasmic reticulum of skeletal muscle. *Biochem. Biophys. Res. Commun.* 227, 700–706.

Kalabukhov, N.I. (1985). Hibernation of Mammals. Nauka, Moscow.

Kokoz, Yu.M., Grichenko, A.S., Korystova, A.F., Lankina, D.A. and Markevich, N.I. (1999). Effect of isoproterenol on the L-type Ca^{2+} current in cardiac cells from rats and hibernating ground squirrels. Biosci. Rep. 19, 17–25.

Kokoz, Yu.M, Markevich, N.I., Korystova, A.F., Lankina, D.A. and Alekseev, A.E. (1997a). The kinetic characteristics of the L-type calcium channels in cardiac cells of hibernators. 2. Analysis of the kinetic model. Membr. Cell. Biol. 11, 213–234.

Kokoz, Yu.M., Zenchenko, K.I., Alekseev, A.E., Korystova, A.F., Lankina, D.A., Ziganshin, R.H., Mikhaleva, I.I. and Ivanov, V.T. (1997b). The effect of some peptides from the hibernating brain on Ca^{2+} current in cardiac cells and on the activity of septal neurons. FEBS Lett. 411, 71–76.

Kondo, N. (1986). Excitation-contraction coupling in the myocardium of hibernating chipmunks. Experientia 42, 1220–1222.

Kondo, N. (1987). Identification of a pre-hibernating state in myocardium from nonhibernating chipmunks. Experientia 43, 873–875.

Kondo, N. (1988). Comparison between effects of caffeine and ryanodine on electromechanical coupling in myocardium of hibernating chipmunks: role of internal Ca stores. Br. J. Pharmacol. 95, 1287–1291.

Kondo, N. and Shibata, S. (1984). Calcium source for excitation–contraction coupling in myocardium of nonhibernating and hibernating chipmunks. Science 225, 641–643.

Koyabu, S., Imanaka-Yoshida, K., Ioshi S.O., Nakano, T. and Yoshida, T. (1994). Switching of the dominant calcium sequestering protein during skeletal muscle differentiation. Cell Motil. Cytoskeleton 29, 259–270.

Krause, K.-H. and Michalak, M. (1997). Calreticulin. Cell 88, 439–443.

Lamb, G.D. (2000). Excitation–contraction coupling in skeletal muscle: comparison with cardiac muscle. Clin. Exp. Pharmacol. Physiol. 27, 216–224.

Lee, A.G., Dalton, K.A., Duggleby, R.C., East, J.M. and Starling, A.P. (1995). Lipid structure and Ca^{2+}-ATPase function. Biosci. Rep. 15, 289–298.

Lukoyanova, N.A., Ignat'ev, D.A., Kolaeva, S.G. and Podlubnaya, Z.A. (1997). Study of ATPase and regulatory properties of skeletal muscle myosin of ground squirrels (*Citellus undulatus*) at different stages of hibernation. Biofizika 42, 343–348.

Lukoyanova, N.A., Malyshev, C.L., Shpagina, M.D., Udal'tsov, S.N. and Podlubnaya, Z.A. (1998). The absence of seasonal changes in the isoform composition and functional properties of actin and thin filament associated proteins from skeletal muscle of hibernating ground squirrels. Biofizika, 43, 1134–1136.

Lukoyanova, N.A., Shpagina, M.D., Udal'tsov, S.N., Ignat'ev, D.A., Kolaeva, S.G. and Podlubnaya, Z.A. (1996a). Changes in structural organization of reorganized myosin filaments from skeletal muscles of the winter-hibernating suslik *Citellus undulatus* during awakening. Biofizika 41, 116–122.

Lukoyanova, N.A., Udal'tsov, S.N. and Podlubnaya, Z.A. (1996b). Binding of rabbit skeletal muscle phosphofructokinase to the filaments formed by skeletal muscle myosins from ground squirrel at different stages of hibernation. Biophys. J. 70, MP047.

Lyman, C.P., Willis, J.S., Malan, A. and Wang, L.C.H. (1982). Hibernation and Torpor in Mammals and Birds. Academic Press, New York.

MacDonald, J.A. and Storey, K.B. (1999). Regulation of ground squirrel Na^+K^+-ATPase activity by reversible phosphorylation during hibernation. Biochem. Biophys. Res. Commun. 254, 424–429.

Mackrill, J.J. (1999). Protein–protein interactions in intracellular Ca^{2+}-release channel function. Biochem. J. 337, 345–361.

Meissner, G. and Lu, X. (1995). Dihydropyridine receptor–ryanodine receptor interactions in skeletal muscles excitation–contraction coupling. Biosci. Rep. 15, 399–408.

Milner, R.E., Michalak, M. and Wang, L.C.H. (1991). Altered properties of calsequestrin and the ryanodine receptor in the cardiac sarcoplasmic reticulum of hibernating mammals. Biochim. Biophys. Acta 1063, 120–128.

Morano, I., Adler, K., Agostini, B., and Hasselbach, W. (1992). Expression of myosin heavy and light chains and phosphorylation of the phosphorylatable myosin light chain in the heart ventricle of the European hamster during hibernation and in summer. J. Muscle Res. Cell Motil. 13, 64–70.

Orr, I. and Shoshan-Barmatz, V. (1996). Modulation of the skeletal muscle ryanodine receptor by endogenous phosphorylation of 160/150-kDa proteins of the sarcoplasmic reticulum. Biochim. Biophys. Acta 1283, 80–88.

Panteleev, P.A. (1983). Bioenergetics of Small-Sized Mammals: Adaptation of Rodents and Insectivores to Temperature Conditions of the Environment. Nauka, Moscow.

Pehowich, D.J. (1994). Modification of skeletal muscle sarcoplasmic reticulum fatty acyl composition during arousal from hibernation. Comp. Biochem. Physiol. B 109, 571–578.

Pehowich, D.J. and Wang, L.C.H. (1984). Seasonal changes in mitochondrial succinate dehydrogenase activity in a hibernator, *Spermophilus richardsonii*. J. Comp. Physiol. B 154, 495–501.

Postnikova, G.B., Tselikova, S.V., Ignat'ev, D.A. and Kolaeva, S.G. (1997). Seasonal changes in myoglobin content in muscles of hibernating Yakutian ground squirrel. Biochemistry (Moscow) 62, 141–144.

Rubtsov, A.M. and Batrukova, M.A. (1997). Ca-release channels (ryanodine receptors) of sarcoplasmic reticulum: structure and properties. A review. Biochemistry (Moscow) 62, 933–945.

Shoshan-Barmatz, V., Orr, I, Well, S., Meyer, H., Varsanyi, M. and Heilmeyer, L.M. (1996). The identification of the phosphorylated 150/160-kDa proteins of sarcoplasmic reticulum, their kinase and their association with the ryanodine receptor. Biochim. Biophys. Acta 1283, 89–100.

Shutova, A.N., Storey, K.B., Lopina, O.D. and Rubtsov, A.M. (1999). Comparative characteristics of sarcoplasmic reticulum preparations from skeletal muscles of the ground squirrel *Spermophilus undulatus*, rats, and rabbits. Biochemistry (Moscow) 64, 1250–1257.

Simmerman, H.K.B. and Jones, L.R. (1998). Phospholamban: protein structure, mechanism of action, and role in cardiac contraction. Physiol. Rev. 78, 921–946.

Soukri, A., Valverde, F., Hafid, N., Elkebbaj, M.S. and Serrano, A. (1995). Characterization of muscle glyceraldehyde-3-phosphate dehydrogenase isoforms from euthermic and induced hibernating *Jaculus orientalis*. Biochim. Biophys. Acta 1243, 161–168.

Soukri, A., Valverde, F., Hafid, N., Elkebbaj, M.S. and Serrano, A. (1996a). Evidence for a posttranslational covalent modification of liver gliceraldehyde-3-phosphate dehydrogenase in hibernating jerboa (*Jaculus orientalis*). Biochim. Biophys. Acta 1292, 177–187.

Soukri, A., Valverde, F., Hafid, N., Elkebbaj, M.S. and Serrano, A. (1996b). Occurrence of a differential expression of the glyceraldehyde-3-phosphate dehydrogenase gene in muscle and liver from euthermic and induced hibernating jerboa (*Jaculus orientalis*). Gene 28, 139–145.

Storey, K.B. (1987). Regulation of liver metabolism by enzyme phosphorylation during mammalian hibernation. J. Biol. Chem. 262, 1670–1673.

Storey, K.B. (1997). Metabolic regulation in mammalian hibernation: enzyme and protein adaptations. Comp. Biochem. Physiol. A 118, 1115–1124.

Wagenknecht, T. and Rademacher, M. (1997). Ryanodine receptors: structure and macromolecular interactions. Curr. Opin. Struct. Biol. 7, 258–265.

Wang, L.C.H. (1985). Life at low temperature: mammalian hibernation. Cryo-Letters 6, 257–274.

Wickler, S.J., Hoyt, D.F. and van Breukelen, F. (1991). Disuse atrophy in the hibernating golden-mantled ground squirrel, *Spermophilus lateralis*. Am. J. Physiol. 261, R1214–R1217.

Yacoe, M.E. (1983). Adjustment of metabolic pathways in the pectoralis muscles of the bat, *Eptesicus fuscus*, related to carbohydrate sparing during hibernation. Physiol. Zool. 56, 648–658.

Yano, K. and Zarain-Herzberg, A. (1994). Sarcoplasmic reticulum calsequestrins: structural and functional properties. Mol. Cell. Biochem. 135, 61–70.

Protein Adaptations and Signal Transduction. Edited by K.B. Storey and J.M. Storey

CHAPTER 6

Aquaporins and water stress

Alfred N. Van Hoek, Yan Huang and Pinke Fang
Renal Unit and Program in Membrane Biology, Massachusetts General Hospital and Harvard Medical School, Boston, MA 02114, USA

1. Rationale

One of the evolutionary challenges that some animals have had to endure is adaptation to an environment that provides only limited access to water. Those that live in an arid environment have obviously coped successfully and do not seem to be highly dependent on the availability of drinking water. These animals have been a subject of considerable interest to renal physiologists. The increased urinary concentration capacities found in these animals are enormous and are correlated with relatively large extended papillae of the kidney. For example, the gerbil (or "sand rat") *Psammomys obesus* can concentrate urine to about 6500 mOsm or higher (Jamison et al., 1979). Recently, we discovered that *Dipodomys merriami merriami*, or Merriam's desert kangaroo rat, has unique distribution patterns of some aquaporins or water channels in the kidney and liver (Huang et al., 2001). This species is closely related to rats, mice and humans, and the interstitium of the papilla contains high amounts of urea. The different aquaporin distributions found in this species may provide important insights into the role and function of water channels, and into water transport physiology in the kidney and other organs in general.

2. Introduction

In all organisms, water (and solute) transport is critical in maintaining homeostasis under a variety of environmental conditions. Correlations between water transport physiology and environmental stresses such as water deprivation have contributed to understanding the mechanisms of urinary concentration. Early micropuncture experiments provided necessary data for the elucidation of many transport processes in the nephron. For example 60–70% of the filtered fluid was found to be reabsorbed in proximal tubules. Permeability experiments using proximal tubules indicated the existence of facilitated water transport, and that transport of water and solute occurred at near iso-osmotic conditions (Valtin, 1983). The characteristics of nephron function rest not only on micropuncture data, but also on electrical and chemical potentials in the various "compartments", and on biochemical data on Na^+-K^+-ATPase levels in different cell types and tubule segments. For example, the thick ascending limb of Henle has active sodium transport into the interstitium, whereas the thin descending limb is highly permeable to salts, urea and water, but there is no active transport of ions. In contrast, the ascending thin limbs are not permeable to water, but highly permeable to sodium and chloride and moderately to urea. All of these characteristics as well as many others contribute to the complex nature of fluid and electrolyte homeostasis in the kidney (Valtin, 1983; Valtin and Edwards, 1987; Bennet and Gardiner, 1987).

2.1. Identification of Aquaporins

There is substantial information available in the earlier literature about the existence of a common pore for water and solutes (reviewed, e.g., by

Finkelstein, 1987). The presence of common pores for solute and water was and still is a matter of considerable discussion, because the outcome of experiments was variable among different laboratories and among experimental approaches. With the advent of novel biochemical, biophysical, and molecular techniques, research has more recently focused on transport physiology at the cellular and subcellular levels. The discovery of water channels as integral membrane proteins occurred relatively recently (Van Hoek et al., 1991; Preston et al., 1991), long after the isolation and cloning of other transporters, because of technical and experimental limitations that delayed their identification. Once the first aquaporin was characterized by functional expression in *Xenopus* oocytes (Preston et al., 1992), and reconstitution into proteoliposomes (Van Hoek and Verkman, 1992; Zeidel et al., 1992), many other homologous water channels were subsequently identified. Numerous reviews have appeared on this subject to which the reader is referred (Agre et al., 1995; Agre and Nielsen, 1996; Beitz and Schulz, 1999; Brown et al., 1995; Knepper et al., 1996; Van Os et al., 1994; Verkman et al., 1995, 1996, 1999; Nielsen et al., 1999). Mouse knockout models (Verkman, 1999) and species dependent expression of aquaporins (Van Hoek et al., 2000; Huang et al., 2001) and other transporters have allowed a reevaluation of our understanding of renal physiology as well as water transport physiology in other organs, such as the liver and the stomach.

Aquaporins are small hydrophobic intrinsic membrane proteins of ~30 kDa. Analysis of aquaporin sequences indicated homology between the first and second halves of the molecule, suggesting the origin of aquaporin genes from an intragenic duplication event. Within the mammalian aquaporin family, two subgroups have been defined. One group selectively transports water (aquaporins) while the other group (aquaglyceroporins) also allows for passage of glycerol, and in some instances, urea and other uncharged solutes (Tsukaguchi et al., 1998). The unique feature of these promiscuous aquaporins is their gene structure, which is different from the water-selective aquaporins. Two additional peptide domains (Fig. 6.1) that are highly conserved characterize this subgroup. It is believed that these amino acids alter the pore-selective properties.

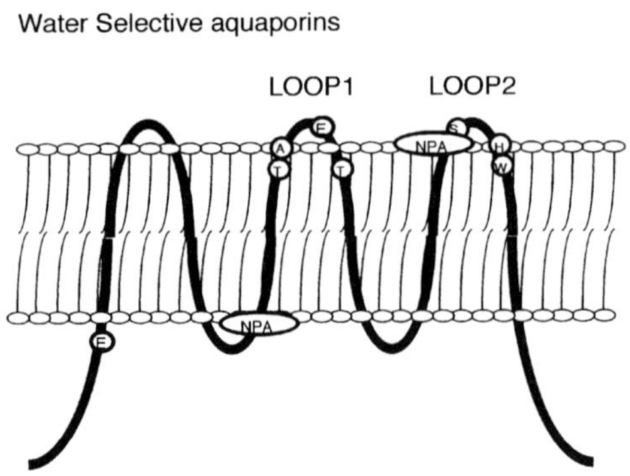

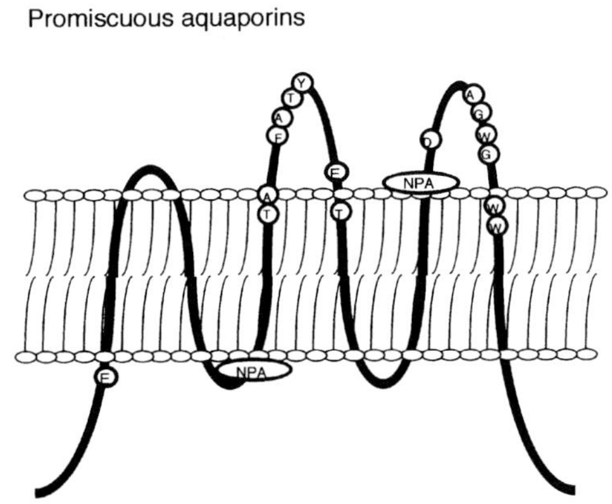

Fig. 6.1. Topology of two classes of aquaporins. By hydropathy analysis of the aquaporins, six putative membrane α helices are proposed, which is corroborated by structural studies, including the 3-D reconstruction in 2-D crystallography of AQP1 (Cheng et al., 1997). The aquaporins are also characterized by a tandemly repeating motif with two highly conserved NPA boxes, which are located on either side of the membrane. Thus, there is two-fold symmetry that was observed as a non-crystallographic pseudo-two-fold symmetry in 2-D crystals. The presence of extended loops (indicated as 1 and 2) in promiscuous aquaporins renders the protein less symmetrical. It is proposed that the extended loops should fold back into the lipid bilayer to functionally alter the pore characteristics.

In mammals, 10 homologous aquaporins have already been identified, and each has a particular tissue distribution (Table 6.1). The presence of different, but homologous, water channels or pores in

Table 6.1. Aquaporin distributions in mammals

Tissue/Aquaporin	0	1	2	3*	4	5	6	7*	8	9*
Brain		X			X					
Eye	X	X		X	X	X				
Ear					X					
Lacrimal gland						X				
Sweat gland		X								
Heart		X								
Liver		X							X	X
Stomach					X					
Intestine				X	X					
Kidney		X	X	X	X		X			
Reproductive organs		X	X					X	X	X
Bladder		X								X
Lung		X			X					
Skeletal muscle					X					
Smooth muscle									X	
Erythrocytes		X								
Leukocytes										X
Salivary gland		X								

The list is not exhaustive, but indicates the major tissue distribution of each of the water channels. The asterisk (*) indicates a promiscuous pore.

membranes in a variety of tissues suggests different functions or roles of each of the aquaporins. However, little is known yet about whether water channels are critical to sustain homeostasis in the various organs, and in addition, there are some observations that may require a reconsideration of the driving forces underlying osmotic water flow. In this respect, it becomes important to re-visit earlier experiments and assumptions that have led to our somewhat limited understanding of the physiology of water transport. This re-evaluation is partly imposed by the results obtained from functional measurements of promiscuous aquaporins, from aquaporin distribution studies of tissues from the desert rodent (Huang et al., 2001, and unpublished results), and from mouse knockout models (Verkman, 1999). This review will briefly discuss the functional properties of water transport, provide a rationalization for a major controversy that still exists, and perhaps allow for a physiological description of the highly adapted kidney found in kangaroo rats.

3. Osmosis, diffusion and functional properties of aquaporins

The advances in transport physiology measurements and continued identification of novel molecular water transporters have demonstrated the occurrence of common pores for solute and water. Osmotically induced water transport is associated with AQP3-mediated glycerol and urea transport (Echevarria et al., 1996; Tsukaguchi et al., 1998; Zeuthen and Klaerke, 1999), with AQP9-mediated polyol, carbamide and other uncharged solute transport (Tsukaguchi et al., 1998), with the CFTR cystic fibrosis transmembrane conductance regulator (Hasegawa et al., 1992), with the UT3 urea transporter (Yang and Verkman, 1998) and Na-solute transporters (Loo et al., 1999; Meinild et al., 2000). The low membrane density of the non-aquaporin transporters compared to the high membrane density of aquaporins suggest that only aquaporins make a quantitatively significant contribution to the overall plasma membrane water permeability. With respect to the solute permeabilities of the promiscuous aquaporins (aquaglyceroporins) it can be envisaged that these aquaporins also make a significant contribution to solute permeation, provided they constitute a common pore for water and solute. Interestingly, Wright and coworkers (Loo et al., 1999; Meinild et al., 2000) provide support for a hypothesis that uphill water transport (i.e., transport in the absence of, and even against, osmotic gradients) can occur by a cotransport mechanism in which one turnover of a cotransporter moves single Na and solute molecules accompanied by hundreds of water molecules. It remains to be elucidated whether this mechanism is relevant to the functioning of promiscuous aquaporins, because it would be expected that water molecules should move solutes (see below), not vice-versa.

The precise details that constitute osmotic transport as well as renal function as it involves urinary concentration are controversial. This is due mainly to current understanding of osmotic forces (see for instance Hammel (1999)), osmolality and thermodynamics. More than a century ago osmosis was described (Van't Hoff, 1887; Vegard, 1908) and it deals with the concept of different solutions that are separated by a membrane; the membrane is considered to be permeable only to water. A difference in solute concentration between the 2 compartments leads to volume (water) movement in the direction of the higher concentration of solute (Van't Hoff, 1887; Vegard, 1908; Mauro, 1957). Osmolality is based on the properties of any substance (solute) that is dissolved in water to lower the freezing point of water in proportion to the amount of substance dissolved. This also provides for measurement of osmolality by determination of the freezing point of the solution. Biological membranes are different from semipermeable membranes, because they allow permeation of water, as well as ions and other molecules. The presence of such a "leaky" membrane complicates the description of water and solute transport, which either proceeds through common or distinct pathways, and this has led to confusion and controversies. Diffusional "forces" driving solute movement down a gradient is one of the hallmarks of transport physiology, together with the osmotic response of volume flow. In renal physiology, any change in these processes is supposed to be a result of solute diffusion or of active transport of the solute against its chemical and/or electrical potential.

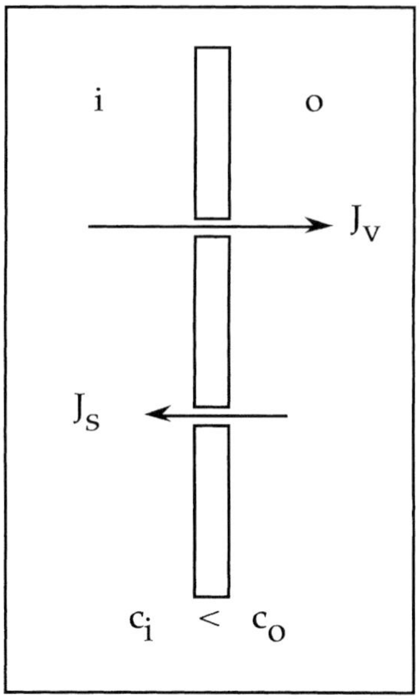

Fig. 6.2. Water and solute permeability across a semipermeable barrier separating two compartments. It is assumed that the concentration of a solute in compartment (o) is greater than in compartment (i). Volume flow will occur from (i) to (o) because of the properties of the barrier. Solute diffusion may occur from (o) to (i) if a pathway exists, making the barrier leaky.

3.1. Physiological relevance of solvent drag

The discovery of promiscuous aquaporins that do not discriminate between water molecules and certain solutes, and increased expression of these water channels (e.g. AQP3 and AQP9) in the kangaroo rat (Huang et al., 2001; unpublished results), provide support for the physiological relevance of solvent drag, i.e., that water flow provides the driving force for net solute flow. This mechanism allows for massive water plus solute movement by osmosis (induced by an impermeable solute, i.e., by a partial osmotic pressure difference) as shown by experiment, yet which by thermodynamic consideration leads to an uphill flow of water. The flow and osmotic pressure relationship is summarized by the Kedem and Katchalsky (1958) equations (Eqs. 6.1 and 6.2). The reader is referred to an excellent treatise for the application of these equations in current studies on water permeability measurements (Verkman, 2000). For a common pore for water and a solute, the drag term in Eq. 6.2, $J_v\ (1 - \sigma)$, contributes to the transport rate of the solute. The reflection coefficient σ occurs in both equations; in Eq. 6.1 it describes the efficacy of a certain solute to set up an osmotic gradient, and in Eq. 6.2 the same reflection coefficient determines whether solute movement occurs by volume flow. σ-values for a solute range from zero (no contribution to the osmotic gradient, i.e. common pore for water and the solute) to unity (100% contribution to the osmotic gradient, separate pathways for water and the solute). A mechanistic meaning for this factor, *a priori*, does not exist.

Equation 6.1 describes the volume flow (J_v) from compartment (i) to compartment (o) through a membrane (Fig. 6.2), while Eq. 6.2 describes a solute flux (J_s) from compartment (o) to compartment (i). For ease of explaining the flow-force relation, it is assumed that the concentration of a solute in compartment (o) is greater than the concentration in compartment (i) (Fig. 6.2). The mathematical description of the flow-force relation is:

$$J_v = P_f A\, V_w\{(P_i - P_o)/RT + \sigma(\Pi_o - \Pi_i)\} \qquad (6.1)$$

$$J_s = P_s(c_o - c_i) + J_v(1 - \sigma)\,(c_o + c_i)/2 \qquad (6.2)$$

where J_v is the volume flow (cm^3/s), and P_f is the osmotic water permeability coefficient (cm/s). A is the membrane area (cm^2), V_w, is the specific volume of water (18 cm^3/mol), and $P_i - P_o$, is the hydrostatic pressure difference from left (i) to right (o). R is the gas constant, T is the absolute temperature, and σ the reflection coefficient. $\Pi_o - \Pi_i$, is the osmotic pressure difference from right to left across the membrane. J_s is solute flux (cm^3/s), P_s is the solute diffusional permeability coefficient (cm/s), and $c_o - c_i$, is the solute concentration difference across the membrane from right to left. In most cases σ equals unity and the solvent drag term in Eq. 6. 2 is essentially zero and a solute flux is solely determined by its concentration gradient across the lipid bilayer. In experiments, the hydrostatic pressure difference is avoided and the two equations reduce to ($\sigma = 1 = \Pi = -RTc$): $J_v = P_f A\, V_w\, RT(c_i - c_o)$ and $J_s = P_s\,(c_o - c_i)$. It should be noted that Π_o and Π_i are determined by the sum of each of the solute concentrations in compartment (o), respectively compartment (i), and that each solute has its own σ.

3.2. Measurement of P_f, P_s and σ

For measurements of P_f and P_s, the oocyte expression system has been widely used to determine the biophysical properties of a water channel. The water permeability coefficient, P_f, is determined by placing aquaporin-expressing oocytes in a hypotonic solution (e.g. plain water) that leads to swelling. Since the main constituent of solutes inside the oocyte is NaCl (Barth's medium), the osmotic difference is determined by the NaCl gradient from inside to outside. Because NaCl is considered impermeable, it implies that J_s in Eq. 6.2 is zero. Thus from the initial slope (Fig. 6.3A and B, traces marked with asterisks) P_f is calculated. When the oocyte is placed in an isotonic solution of NaCl (Barth's medium) to which 1 mM labeled test solute is added, volume flow does not occur and J_v in Eq. 6.1 equals zero. J_s will equal the first part of Eq. 6.2 and this provides for the diffusional solute permeability coefficient (Tsukaguchi et al., 1998). Inhibitor studies with mercurial sulfhydryl compounds as well as measurements on mock-injected oocytes, in most cases, complete the study. To determine whether a solute and water share a common pathway, determination of the reflection coefficient for a test solute is performed with two solutions of equal osmolality but different compositions, i.e. one test (permeant) and one reference (impermeant) solution. Because P_f and P_s are determined in separate experiments, the reflection coefficient can be determined unambiguously, by using the two equations (Eqs. 6.1 and 6.2) and solving them numerically (Tsukaguchi et al., 1998). Experimentally, the oocyte is placed in the test solution and the inward movement of water plus test solute will lead to swelling. In Fig. 6.3 experiments are shown using AQP3 (and AQP9) expressing oocytes that were placed in a glycerol solution. Rapid swelling occurred to an extent that it appeared that the glycerol concentration (Fig. 6.3, traces marked with crossed squares) was not effective in opposing the outwardly directed electrolyte gradient. Calculation of the sigma from numerical simulation, being essentially zero, confirmed this conclusion. Thus, it did not matter whether oocytes were placed in water or a glycerol solution (up to 1000 mOsm); the results were similar (Tsukaguchi et al., 1998). The inability of glycerol to reduce swelling of the oocyte and the value of sigma suggest that water and glycerol share a common pathway in this experiment (Fig. 6.3).

There are, however, reports that have indicated that AQP3 is a relatively poor water channel,

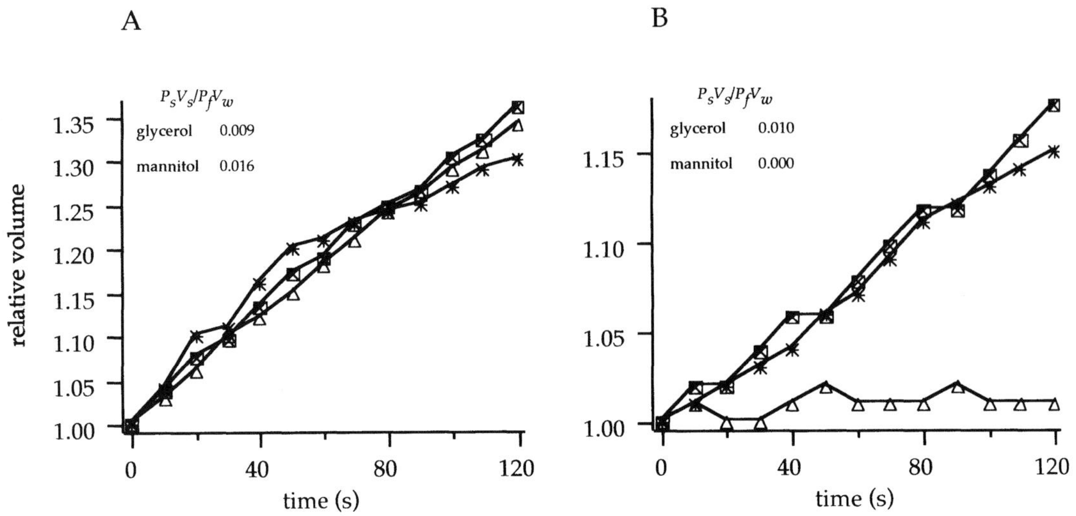

Fig. 6.3. Osmotic swelling of oocytes expressing AQP9 (A) and AQP3 (B). Asterisks depict osmotic water permeability when oocytes were placed in diluted Barth's solution, the hypo-osmotic response yielding P_f. The crossed-squares depict osmotic water permeability when the diluted Barth medium was adjusted to isotonicity with glycerol. Triangles depict osmotic water permeability when the outside medium (diluted Barth's solution) was adjusted with mannitol. Note that in the left panel (A), oocytes expressing AQP9, it did not matter whether the outside hypotonic medium was adjusted with glycerol or mannitol, rapid flow occurred, suggesting that mannitol and glycerol are non-osmolytes and cannot setup an effective partial osmotic gradient to oppose the partial osmotic gradient induced by NaCl. In AQP3 expressing oocytes, right panel (B), mannitol, but not glycerol, effectively induced an osmotic gradient to oppose the NaCl osmotic gradient. Note that Tsukaguchi et al (1998) showed that P_s-values were of the order of 20×10^{-6} cm/s, and water permeability coefficients were between 0.05 and 0.2 cm/s. Evaluation of $\sigma = 1 - P_sV_s/P_fV_w - f$ (see text) gave $f > 0.9$ for mannitol (AQP9), glycerol (AQP3, AQP9) and urea (AQP3, AQP9).

whereas it has a substantial permeability to urea and glycerol (Yang and Verkman, 1997). Also, it was claimed that AQP3 accommodates separate pathways for water and glycerol (Echevarria et al., 1996). In general, permeability measurements in aquaporin expressing *Xenopus* oocytes should be interpreted cautiously because it is possible that endogenous pathways for solutes are more pronounced when a facilitated pathway for water is created (Yool et al., 1996). This could explain some of the discrepancies reported in the literature. The common practice to determine whether an aquaporin possesses a common pore for solute and water is to evaluate the reflection coefficient. However, there are technical and experimental limitations that could lead to underestimation of the reflection coefficient sigma. Sigma is alternatively expressed as a frictional coefficient, f, plus the ratio of the solute diffusion coefficient and the water permeability coefficient (corrected by their partial specific volumes), $\sigma = 1 - P_sV_s/P_fV_w - f$. In the absence of a common pore, $f = 0$, the reflection coefficient can be substantially less than unity if the solute diffusion coefficient is relatively large (Kleinhans, 1998; Katkov, 2000). The zero sigma–values reported by Tsukaguchi et al (1998) for urea and glycerol in AQP3 expressing *Xenopus* oocytes included evaluation of P_sV_s/P_fV_w, which were less than 0.1 (Fig. 6.3), giving values for $f > 0.9$. This supports the notion that diffusion of a permeable solute, i.e. P_s, had a minor role in the rapid volume–solute responses (Fig. 6.3). Thus, the diffusional term in Eq. 6.2 did not contribute significantly to the overall flow J_s.

4. Uphill flow of water

In terms of thermodynamics, osmosis in conjunction with a leaky pore (i.e., a common pore for solute and water) leads to the paradox of an uphill flow of water (Finkelstein 1987), since volume flow occurs in the absence or against osmotic gradients. This conflict arises when osmolality of a solution is compared to its inability to setup the osmotic pressure. Perhaps the best way to approach this apparent conflict is the definition of the "entropy of mixing" and a reformulation of this definition within the context of biological membranes. The entropy of mixing is the conceptually additional increase of disorder when two substances are mixed. That is, the entropy of a mixture is equal to the sum of the entropies, which the individual components would have if each were at the same temperature, pressure and volume as the mixture, plus a correction term. This correction term is called the entropy of mixing and is given by $\Delta S_{mixing} = -N(1) \ln N(1)/N + -N(2) \ln N(2)/N$. The entropy of mixing is derived from considerations of particles of two kinds, $N(1)$ and $N(2)$ ($N = N(1) + N(2)$), the most probable distribution that each of the two kinds will have and be uniformly distributed in the whole volume. However, when an aquaporin renders a membrane equally permeable to water and glycerol, water and glycerol molecules appear to be indistinguishable (Fig. 6.3) and as such the entropy of mixing vanishes in the concept of thermodynamics, because $N = N(1) = N(2)$, and therefore ΔS_{mixing} looses its meaning. The "search" for a mechanistical description of sigma has been ongoing; Hill (1994, 1995) considered sigma a characteristic of molecular pore structure, giving reduced water permeabilities when the osmolyte could penetrate the pore. The explanation given here contributes to Finkelstein's (1987) evaluation of the uphill flow of water, observed when studies with a collodion membrane containing pores permeable to both water and urea were discussed (Meschia and Setnikar, 1958). Those earlier studies gave essentially the same results as studies with AQP3 and AQP9 expressing oocytes (Fig. 6.3) (Tsukaguchi et al., 1998). The implication of sigma expressing the degree of similarity of two kinds of molecules is essentially identical to the degree of efficacy of a solute to set up an osmotic pressure difference. In this context the following experiment, discussed by Finkelstein (1987), provides support for the "similarity of the two kinds". Consider the collodian membrane that has a common pore for water and urea. This membrane separates two compartments, one filled with 1 mM dextran, an impermeant, and one filled with either water or 100 mM urea (Meschia and Setnikar, 1958). In both cases, water moved at the same rate into the direction of the 1 mM dextran solution. There is no need to invoke the necessity that urea diffuses down its gradient at high rates towards the dextran side allowing water "to follow", which is an apparent prerequisite when "freezing point osmolalities" are considered. This high rate of diffusion, deduced from thermodynamic considerations, would also be in contradiction with the experimental values of Fig. 6.3, because $P_s V_s / P_f V_w < 0.1$.

The property that promiscuous aquaporins have of selectively defying a partial osmotic pressure may also be applicable to other transporters that have an aqueous pore. Moreover, water transport through aquaporins is nearly temperature insensitive, hence, solvent drag of solute through promiscuous aquaporins will also be relatively fast at low temperatures (Tsukaguchi et al., 1998). This then suggests a novel paradigm in transport physiology involving osmotically driven cotransport of water and solute by a small effective partial osmotic pressure difference. In addition, coexistence of a selective water channel and a promiscuous pore can provide for apparently complex responses to a small osmotic change.

5. Desert kangaroo rat and aquaporin distributions

To date, not much is known about the molecular distribution of transporters in desert animals. Only micropuncture data have provided data showing differences between species that live in an arid environment and species that do not. The desert kangaroo rat from the Sonoran Desert proved to be a good candidate for investigating aquaporin distributions. The species is phylogenetically closely related to mice and Wistar rats. Kangaroo rats are capable of concentrating their urine to 4500 mOsm and do not appear to be dependent on water to drink. Studies have been made of this species from two Arizona counties. In Yuma County annual precipitation is only 10.6 cm (occurring mostly during the winter months or the summer monsoon season), mean annual daily maximum and minimum temperatures are 31.9 and 14.7°C, respectively, and average daily temperatures during the hottest month (July) are 26–43°C. Comparable data for Gila County at 1200 m elevation are a yearly average precipitation of 43.6 cm, mean daily maximum and minimum temperatures of 23.5 and 6.2°C, respectively, average daily temperatures in July of 24–40°C. In the laboratory *D. merriami* could be held for six months at 25°C, if not longer, with only dry seeds and rolled oats to eat and no drinking water. Also, the urinary output is very low, such that the housing did not require much maintenance, if any. A habit that possibly helps in the preservation of water is that the kangaroo rat undergoes daily cycles of torpor (Yousef and Dill, 1971).

The major aquaporins that are thought to be important for water transport in the kidney are aquaporins 1, 2, 3, and 4 (Table 6.1). Several studies have shown that the cellular location of AQP1 and AQP2 in the kidneys of mice closely resembles that seen in other species (Breton et al., 1995; Knepper et al., 1996; Nielsen et al., 1999; Van Os et al., 1994; Verkman et al., 1995). AQP1 is localized to proximal tubules, thin descending limbs of Henle, and the descending and ascending limbs of the vasa recta. AQP2 is expressed in apical membranes and in subcellular vesicles lining the apical membranes of collecting duct principal cells. AQP2 is targeted to the cell plasma membrane by hormonal stimulation via vesicle trafficking and insertion into the membrane (Brown and Nielsen, 2000). AQP3 and AQP4 are known to localize to the basolateral membranes of collecting duct principal cells (Frigeri et al., 1995; Terris et al., 1995), thus principal cells are remarkable because they express three water channels.

In knockout studies it was shown that deletion of AQP1 results in a defect of the countercurrent system that reduced the urinary concentration capacity in the antidiuretic mouse (Ma et al., 1998; Chou et al., 1999). Thus, AQP1 is important in proximal tubule reabsorption and in the creation of a hypertonic medullary interstitium by countercurrent multiplication. Not surprisingly, AQP1 was also abundantly present in proximal tubule, thin limbs of Henle, and vasa recta of the desert kangaroo rat (Huang et al., 2001). Likewise, the apical membranes of collecting duct principal cells of the desert rodent contain large amounts of AQP2 (Huang et al., 2001). Point mutations in human AQP2 are associated with nephrogenic diabetes insipidus, a form of polyurea that affects the urinary concentration capacity (Deen et al., 1994). Thus, AQP2 is important for water reabsorption in collecting ducts. Because AQP2 is mainly localized to the apical side of principal cells, it is conceptually clear that the basolateral membrane requires other water channels to explain its high water permeability. The reason(s) for the co-localization of the selective water channel AQP4 and the aquaglyceroporin AQP3 remain(s) apparently unclear, although some insight has been provided by deletion of AQP3 and AQP4 (Chou et al., 1998; Ma et al., 2000). More information came from comparative immunocytochemistry, because differences in occurrence of these two water channels among species were noted (Van Hoek et al., 2000). In mice, but not rats, additional expression of AQP4 was found in basolateral membranes of proximal tubule of the S3 segment (Van Hoek et al., 2000), while surprisingly AQP4 could not be detected in the kidney of the desert rodent (Huang et al., 2001). In this species AQP3 is localized to cortical and medullary collecting ducts and it does not show a decrease from cortex to medulla as was found in Wistar rats (Ecelbarger et al., 1995) and mice (Huang et al., 2001). Collecting duct AQP4 immunocytochemistry in mice and Wistar rats showed in contrast an increase in staining from cortex to papilla, opposite to the gradient of AQP3 staining in these species.

AQP4-knockout studies in mice suggested that the urinary concentration capacity was reduced by ~10%. This small effect may also be related to AQP4 deletion from S3 proximal tubules in addition to deletion from the collecting ducts (Van Hoek et al., 2000). With respect to the role of AQP4 in collecting ducts, AQP4 appears to be dispensable for the urinary concentrating capacity, although the transepithelial flow in isolated perfused inner medullary collecting ducts was reduced four fold in AQP4 knockout mice (Chou et al., 1998). Thus, based on knockout studies alone the role of AQP4 is unclear. The absence of AQP4 from the kidneys of the desert rodent clearly indicates that AQP4 is not required for urinary concentration to occur. Interestingly, the absence of AQP4 was correlated with increased expression of AQP3 in these rodents. AQP3-knockout studies in mice highlighted the importance of this water channel, because deletion led to nephrogenic diabetes insipidus (Ma et al., 2000). The apparent explanation in mice was that the majority of distally-delivered fluid in the antidiuretic kidney is extracted from the lumen in cortical collecting products, and most of the AQP3 protein is constitutively expressed in cortical and outer medullary collecting ducts in the mouse. However, the role of AQP4 in collecting ducts remains to be elucidated.

6. Physiology of AQP3 and AQP4

It is of special interest that AQP3 expression is reduced in the inner medulla of mice and rats, where interstitial urea concentration is the highest, whereas it is abundantly expressed in this region in kangaroo rats. Because of the nature of this promiscuous aquaporin, the basolateral membrane of principal cells of the desert rodent, containing only AQP3, is a membrane highly permeable to water and urea. In mice, collecting ducts are in osmotic equilibrium with urea in the interstitium, but this does not necessarily mean that collecting ducts would contain urea in isosmolar amounts with respect to the interstitium. Isosmolar amounts of urea may be the found in kangaroo rats, because of abundant expression of AQP3. In mice, the presence of a water selective channel AQP4, in addition to AQP3, suggests that any change that could occur with the cortical-medullary urea gradient, and subsequently a change of the trans-epithelial urea gradient, will invoke osmotic flow of water, induced by urea, into the lumen. The absence of AQP4 would prevent this from happening in the kangaroo rat, again assuming that AQP3 is the major channel for water and urea in this part of the nephron. In other words, it is possible that the presence of AQP4 would render basolateral membranes more sensitive to lateral osmotic changes in urea gradients, and could affect the flows of water and urea across the apical membrane. The role of AQP4 in this part of the kidney is then perhaps to allow for facilitation of the collapse of the urea gradient. An indication for this possibility was given by a pilot study designed to "trigger" expression of AQP4 in the kidney of the desert rodent. These animals had free access to water. After four weeks, the animals were sacrificed and immunostaining did not reveal significant changes in aquaporin staining patterns, despite the fact that the rodents produced copious amounts of water. However, their bloated appearance may indicate an inability to reach a state of normal fluid homeostasis, providing support for an involvement of AQP4 in water clearance. Alternatively, the bloating could be the result of decreased glomerular filtration rate and low luminal flow in collecting ducts (Bennet and Gardiner, 1987; Valtin and Edwards, 1987).

7. Water transport in liver and stomach

Liver and stomach water transport physiology has not been given much attention, because only in recent years has the presence of aquaporins in these organs been appreciated. Aquaporin knockout models have not led to substantial new insight with respect to liver and stomach. However, in the kangaroo rat remarkable immunological differences have been observed that could provide novel insight into the role of certain aquaporins, including AQP1, AQP4 and AQP9.

AQP1 is a selective water channel and appears to be abundantly expressed in the sinusoidal membranes of the endothelium lining hepatic cells of the liver (Fig. 6.4A). Its occurrence in these membranes suggests that facilitated and selective water transport is required in the liver. Yet the absence of AQP1 in the endothelium in kangaroo rats (Fig. 6.4B) suggests that selective water transport is not necessary, or should be avoided when water limitation determines survival of the animal. Moreover,

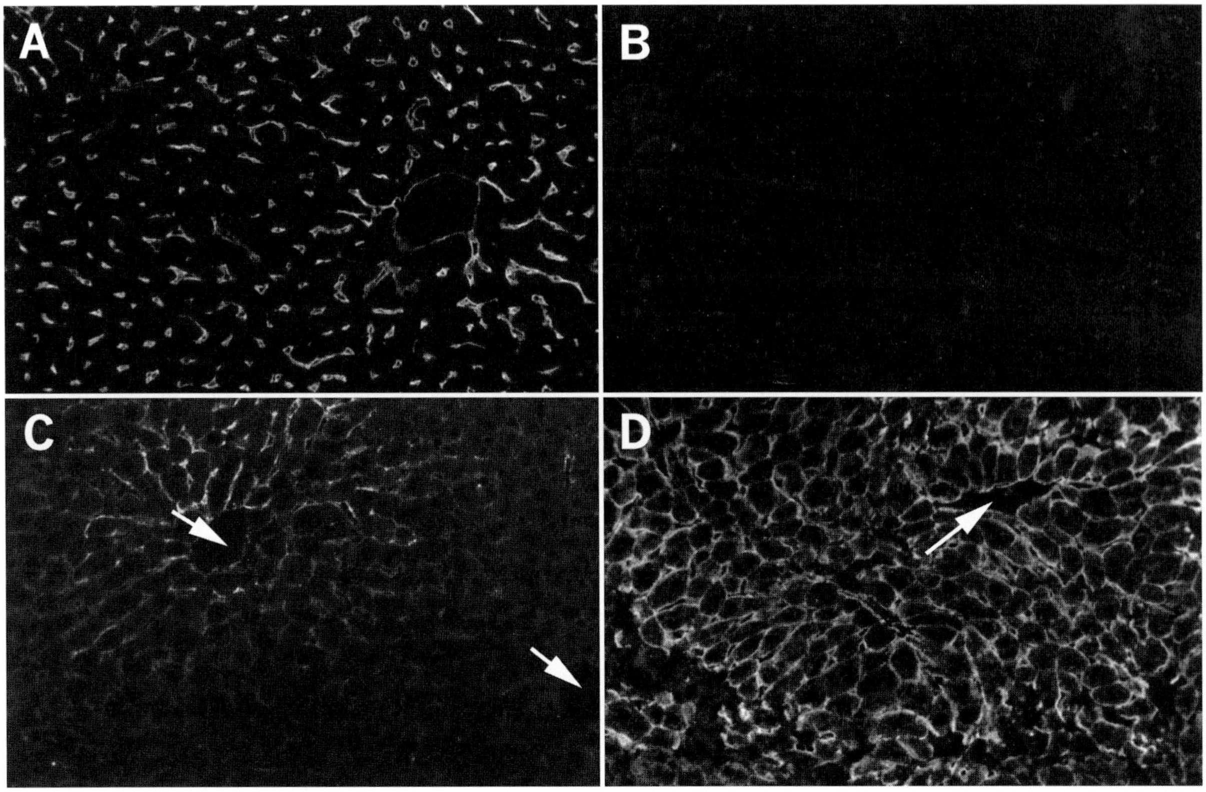

Fig. 6.4. Aquaporin immunocytochemistry of liver of mouse (A, C) and *Dipodomys merriami merriami* (B, D). (A, B): anti-AQP1 antibody immunostaining of sinusoidal membranes. (C, D): Immunostaining with anti-AQP9 antibody of hepatic cell membranes. (C) Left arrow: portal vein; right arrow: central vein. (D): Arrow: portal vein.

the lack of AQP1 is correlated with increased expression of AQP9 in the membranes of hepatocytes. In mice and rats, AQP9 appears to be abundantly expressed in hepatocytes that surround the portal vein and is considerably reduced or absent in hepatocytes surrounding the central vein (Fig. 6.4C). This lateral gradient of AQP9 staining resembles gradients of oxygen and metabolites from portal to central veins. The implication of increased immuno-staining with AQP9 antibodies in the liver of the kangaroo rat (Fig. 6.4D) is best illustrated by the nature of this promiscuous aquaporin, defying partial osmotic gradients. Expression of AQP9 in hepatocytes relates to rendering the membrane unable to distinguish between water and uncharged solutes. The consequences and advantages are that metabolites cannot induce cell swelling of hepatocytes (Tsukaguchi et al., 1998), while potentially such a membrane also provides for solvent drag of those metabolites as a result of a small effective osmotic difference by an impermeable solute.

The occurrence of AQP4 in the basolateral membranes of parietal cells of gastric mucosa, while the apical membranes (*cananiculi*) contain the important H^+-K^+ATPase that delivers acid in the lumen of the stomach, may also be important for delivering water for the digestion of nutrients. There was a remarkable higher number of parietal cells noted in the stomach of the kangaroo rat containing the AQP4 protein when compared to gastric mucosa of the mouse (Huang et al., 2001). The greater number of parietal cells may be related to environmental pressure, but it remains an enigma as to what could be responsible for this elevated expression of AQP4.

8. Adaptation

The finding that the kangaroo rat has different aquaporin distributions compared to other rodents illustrates the potentially diverse roles that aquaporins may possess. Additional *in situ* hybridization experiments with an anti-sense AQP4 probe suggested the presence of AQP4 message in kidneys of the kangaroo rat (Huang et al., 2001). AQP4 mRNA was confined to collecting duct principal cells and proximal tubule cells, similar to findings in mice. This suggests a down-regulation of AQP4 at the post transcriptional level. The apparent down-regulation of AQP1 and AQP4, and the up-regulation of AQP3 and AQP9 may be connected to the functional differences between selective water channels and the promiscuous pores AQP3 and AQP9. The ineffectiveness of particular solutes to contribute to the osmotic pressure when the membrane contains promiscuous pores may be related to the "dehydrated" state of the organism. This state is reflected also in morphological adaptations. In the kidney the papillae are relatively larger. Earlier micropuncture data have shown that oily substances from the interstitium were abundant in animals that have a high urinary concentration capacity. Later studies (Kaissling et al., 1975; Bohman and Jensen, 1976, 1978) showed that a correlation between oily substances and concentrated urine is more complex, but lipid droplets of the interstitial cells were smaller in gerbils and rabbits compared to rats, while their fine structure was similar. Interestingly, water-loaded rats showed a considerable increase in the number of lipid droplets when compared to dehydrated or untreated animals. In contrast, the interstitial cells of water loaded gerbils and rabbits were depleted of lipid droplets. In the light of the data presented here, those substances in the interstitium may exert the osmotic driving force, although the evidence is circumstantial. This idea (Hammel, 1999) may also have consequences for the apparent paradox in renal physiology with respect to oxygen consumption. Presence of substances in the interstitium, and the "thickening" of the blood just after the glomerulus may provide for near isosmotic reabsorption of water and solute in proximal tubule, without the requirement for active transport. For the desert rodent, and other species that have survived arid environments, it seems to be more economic to have a water-preserving organ that does not require much energy.

9. Concluding remarks

The increased occurrence of promiscuous aquaporins (AQP3, AQP9) in renal and hepatic tissues of the desert kangaroo rat at the apparent expense of water-selective aquaporins (AQP1, AQP4) provide support for a novel paradigm in water transport physiology. A promiscuous pore selectively defies a partial osmotic pressure that could give rise to volume-driven movement of the osmotic-inactive solute by an osmotic active osmolyte.

Acknowledgements

This work was supported by National Institutes of Health Research Grant RO1 DK55864. The authors wish to thank Dr. D. Brown for critical reading of the manuscript.

References

Agre, P., Brown, D. and Nielsen, S. (1995). Aquaporin water channels, unanswered questions and unresolved controversies. Curr. Opin. Cell Biol. 7, 472–83.

Agre, P. and Nielsen, S. (1996). The aquaporin family of water channels in kidney. Nephrology 17, 409–415.

Beitz, E. and Schultz, J.E. (1999). The mammalian aquaporin water channel family, A promising new drug target. Curr. Med. Chem. 6, 457–467.

Bennett, T. and Gardiner, S.M. (1987). Water deprivation, effects on fluid and electrolyte handling and plasma biochemistry in Long-Evans and Brattleboro rats. J. Physiol. 385, 35–48.

Bohman, S.O. and Jensen, P.K. (1976). Morphometric studies on the lipid droplets of the interstitial cells of the renal medulla in different states of diuresis. J. Ultrastruct. Res. 55, 182–192.

Bohman, S.O. and Jensen, P.K. (1978). The interstitial cells in the renal medulla of rat, rabbit, and gerbil in different states of diuresis. Cell Tissue Res.189, 1–18.

Breton, S., Alper, S.L., Gluck, S.L., Sly, W.S., Barker, J.E. and Brown, D. (1995). Depletion of intercalated cells from collecting ducts of carbonic anhydrase II-deficient (CAR2 null) mice. Am. J. Physiol. 269, F761–F774.

Brown, D., Katsura, T., Kawashima, M., Verkman, A.S. and Sabolic, I. (1995). Cellular distribution of the aquaporins, a family of water channel proteins. Histochem. Cell Biol. 104, 1–9.

Brown, D., Katsura, T. and Gustafson, C.E. (1998). Cellular mechanisms of aquaporin trafficking. Am. J. Physiol. 275, F328–F331.

Brown, D. and Nielsen S. (2000). The cell biology of vasopressin action. In: The Kidney, 6th edition. (Brenner, B., Ed.), pp. 575–594. W.B. Saunders Co, Orlando.

Cheng, A., Van Hoek, A.N., Yeager, M., Verkman, A.S. and Mitra, A.K. (1997). Three-dimensional organization of a human water channel. Nature 387, 627–630.

Chou, C.L., Ma, T., Yang, B., Knepper, M.A. and Verkman, A.S. (1998). Fourfold reduction of water permeability in inner medullary collecting duct of aquaporin-4 knockout mice. Am. J. Physiol. 274, C549–554.

Chou, C.L., Knepper, M.A., Hoek, A.N., Brown, D., Yang, B., Ma, T. and Verkman, A.S. (1999). Reduced water permeability and altered ultrastructure in thin descending limb of Henle in aquaporin-1 null mice. J. Clin. Invest. 103, 491–496.

Deen, P.M., Verdijk, M.A., Knoers, N.V., Wieringa, B., Monnens, L.A., van Os, C.H. and van Oost, B.A. (1994). Requirement of human renal water channel aquaporin-2 for vasopressin-dependent concentration of urine. Science 264, 92–95.

Ecelbarger, C.A,. Terris, J., Frindt, G., Echevarria, M., Marples, D., Nielsen, S. and Knepper, M.A. (1995). Aquaporin-3 water channel localization and regulation in rat kidney. Am. J. Physiol. 269, F663–F672.

Echevarria, M., Windhager, E.E. and Frindt, G. (1996). Selectivity of the renal collecting duct water channel aquaporin-3. J. Biol. Chem. 271, 25079–25082.

Finkelstein, A. (1987). Water movement through lipid bilayers, pores, plasma membranes. (Distinguished lecture series of the Society of General Physiologists; v. 4), Wiley-Interscience.

Frigeri, A., Gropper, M.A., Turck, C.W. and Verkman, A.S. (1995). Immunolocalization of the mercurial-insensitive water channel and glycerol intrinsic protein in epithelial cell plasma membranes. Proc. Natl. Acad. Sci. U.S.A. 92, 4328–4331.

Hammel, H.T. (1999). Evolving ideas about osmosis and capillary fluid exchange. FASEB J. 13, 213–231.

Hasegawa, H., Skach, W., Baker, O., Calayag, M.C., Lingappa, V. and Verkman, A.S. (1992). A multifunctional aqueous channel formed by CFTR. Science 258, 1477–1479

Hill, A.E. (1994). Osmotic flow in membrane pores of molecular size. J. Membr. Biol. 137, 197–203.

Hill, A.E. (1995). Osmotic flow in membrane pores. Int. Rev. Cytol. 163, 1–42.

Huang, Y., Tracy, R., Walsberg, G.E., Makkinje, A., Fang, P., Brown, D. and Van Hoek, A.N. (2001). Absence of aquaporin-4 water channels from kidneys of the desert rodent *Dipodomys merriami merriami*. Am. J. Physiol. 280, F794–F802.

Jamison, R.L., Roinel, N. and de Rouffignac C. (1979). Urinary concentrating mechanism in the desert rodent Psammomys obesus. Am. J. Physiol. 236, F448–F453.

Kaissling, B., de Rouffignac, C., Barrett, J.M. and Kriz, W. (1975). The structural organization of the kidney of the desert rodent, *Psammomys obesus*. Anat. Embryol. (Berl). 148, 121–143.

Katkov, I.I. (2000). A two-parameter model of cell membrane permeability for multisolute systems. Cryobiology 40, 64–83.

Kedem, O. and Katchalsky, A. (1958). Thermodynamic analysis of the permeability of biological membranes to non-electrolytes. Biochim. Biophys. Acta 27, 229–246.

Kleinhans, F.W. (1998). Membrane permeability modeling, Kedem-Katchalsky vs a two parameter formalism. Cryobiology 37, 271–289.

Knepper, M.A., Wade, J.B., Terris, J., Ecelbarger, C.A., Marples, D., Mandon, B., Chou, C.L., Kishore, B.K. and Nielsen, S. (1996). Renal aquaporins. Kidney Int. 49, 1712–1717.

Loo, D.D., Hirayama, B.A., Meinild, A.K., Chandy, G., Zeuthen, T. and Wright, E.M. (1999). Passive water and ion transport by cotransporters. J. Physiol. 518, 195–202.

Ma, T., Yang, B., Gillespie, A., Carlson, E.J., Epstein, C.J. and Verkman, A.S. (1998). Severely impaired urinary concentrating ability in transgenic mice lacking aquaporin-1 water channels. J. Biol. Chem. 273, 4296–4299.

Ma, T., Song, Y., Yang, B., Gillespie, A., Carlson, E.J., Epstein, C.J. and Verkman, A.S. (2000). Nephrogenic diabetes insipidus in mice lacking aquaporin-3 water channels. Proc. Natl. Acad. Sci .U. S. A. 97, 4386–4391.

Mauro, A. (1957). Nature of solvent transfer in osmosis. Science 126, 252–253.

Meinild, A.K., Loo, D.D., Pajor, A.M., Zeuthen, T. and Wright, E.M. (2000). Water transport by the renal Na(+)-dicarboxylate cotransporter. Am. J. Physiol. Renal. Physiol. 278, F777–F783.

Meschia G. and Setnikar I. (1958). Experimental study of osmosis through a colloidon membrane. J. Gen. Physiol. 42, 429–444.

Nielsen, S., Kwon, T.H., Christensen, B.M., Promeneur, D., Frokiaer, J. and Marples, D. (1999). Physiology and pathophysiology of renal aquaporins. J. Am. Soc. Nephrol. 10, 647–663.

Preston, G.M. and Agre, P. (1991). Isolation of the cDNA for erythrocyte integral membrane protein of 28-kilodaltons—member of an ancient channel family. Proc. Natl. Acad. Sci. USA 88, 11110–11114.

Preston, G.M., T.P. Carroll, W.B. Guggino and Agre, P. (1992). Appearance of water channels in *Xenopus* oocytes expressing red cell CHIP28 protein. Science 256, 385–387.

Terris, J., Ecelbarger, C.A., Marples, D., Knepper, M.A. and Nielsen, S. (1995). Distribution of aquaporin-4 water channel expression within rat kidney. Am. J. Physiol. 269, F775–F785.

Tsukaguchi, H., Shayakul, C., Berger, U.V., Mackenzie, B., Devidas, S., Guggino, W.B., Van Hoek, A.N. and Hediger, M.A. (1998). Molecular characterization of a broad selectivity neutral solute channel. J. Biol. Chem. 273, 24737–24743.

Valtin H., (1983). Renal Function. Little, Brown & Company, 2nd Edition, Boston, Toronto.

Valtin, H. and Edwards, B.R. (1987). GFR and the concentration of urine in the absence of vasopressin. Berliner-Davidson re-explored. Kidney Int. 31, 634–640.

Van Hoek, A.N., Hom, M.L., Luthjens, L.H., De Jong, M.D., Dempster, J.A. and Van Os, C.H. (1991). Functional unit of 30-kiloDalton for proximal tubule water channels as revealed by radiation inactivation. J. Biol. Chem. 266, 16633–16635.

Van Hoek, A.N. and Verkman, A.S. (1992). Functional reconstitution of the isolated erythrocyte water channel CHIP28. J. Biol. Chem. 267, 18267–18269.

Van Hoek, A.N., Ma, T., Yang, B., Verkman, A.S. and Brown, D. (2000). Aquaporin-4 is expressed in basolateral membranes of proximal tubule S3 segments in mouse kidney. Am. J. Physiol. 278, F310–F316.

Van 'T Hoff, J.H. (1887). Die Rolle des osmotischen Druckes in der Analogie zwischen Lösungen und Gase. Z. für Chemie 1, 481–508.

Van Os, C.H., Deen, P.M.T. and Dempster, J.A. (1994). Aquaporins, water selective channels in biological membranes. Molecular structure and tissue distribution. Biochim. Biophys. Acta 1197, 291–309.

Vegard, L. (1908). On the free pressure in osmosis. Proc. Camb. Phil. Soc. 15, 13–23.

Verkman, A.S., Shi, L.-B., Frigeri, A., Hasegawa, H., Farinas, J., Mitra, A., Skach, W., Brown, D., Van Hoek, A.N. and Ma, T. (1995). Structure and function of kidney water channels. Kidney Int. 48, 1069–1081.

Verkman, A.S., Van Hoek, A.N., Ma, T., Frigeri, A., Skach, W. R., Mitra, A., Tamarappoo, B.K. and Farinas, J. (1996). Mechanisms of water transport across mammalian cell membranes. Am. J. Physiol. 270, C12–C30.

Verkman, A.S. (1999). Lessons on renal physiology from transgenic mice lacking aquaporin water channels. J. Am. Soc. Nephrol. 10, 1126–1135.

Verkman, A.S. (2000). Water permeability measurement in living cells and complex tissues. J. Membr. Biol. 173, 73–87.

Yamamoto, T and Sasaki, S. (1998). Aquaporins in the kidney, emerging new aspects. Kidney Int. 54, 1041–1051.

Yang, B. and Verkman, A.S. (1997). Water and glycerol permeabilities of aquaporins 1–5 and MIP determined quantitatively by expression of epitope-tagged constructs in *Xenopus* oocytes. J. Biol. Chem. 272, 16140–16146.

Yang, B. and Verkman, A.S. (1998). Urea transporter UT3 functions as an efficient water channel. Direct evidence for a common water/urea pathway. J. Biol. Chem. 273, 9369–9372.

Yool, A.J., Stamer, W.D. and Regan, J.W. (1996). Forskolin stimulation of water and cation permeability in aquaporin 1 water channels. Science 273, 1216–1218.

Yousef M.K. and Dill D.B. (1971) Daily cycles of hibernation in the kangaroo rat, *Dipodomys merriami*. Cryobiology 8, 441–446.

Zeidel, M.L., Ambudkar, S.V., Smith, B.L. and Agre, P. (1992). Reconstitution of functional water channels in liposomes containing purified red cell CHIP28 protein. Biochemistry 31, 7436–7434.

Zeuthen, T. and Klaerke, D.A. (1999). Transport of water and glycerol in aquaporin 3 is gated by H(+). J. Biol. Chem. 274, 21631–21636.

Protein Adaptations and Signal Transduction. Edited by K.B. Storey and J.M. Storey

CHAPTER 7

Gene Expression Associated with Muscle Adaptation in Response to Physical Signals

Geoff Goldspink and Shi Yu Yang
Anatomy and Developmental Biology, Royal Free and UCL Medical School, University of London, Rowland Hill Street, London NW3 2PF, UK

1. Introduction

In this era of molecular biology many people think in terms of genetic programming. However, a prerequisite for the survival of animal species is adaptability. Muscle is one of the best examples of a tissue that has an inherent adaptability. It provides the power not only for locomotion but a number of life sustaining processes. Therefore its ability to function efficiently and economically over a range of conditions is crucial to survival. Genetic programming is important during embryological development, but after this, tissue mass and phenotype are to a large extent determined by environmental signals. It has been shown that changes in contractile function can be brought about quite rapidly by switching on one subset and repressing another subset of genes (Goldspink et al., 1992). In this way the tissue can be optimized for power output or fatigue resistance. These contractile characteristics are determined mainly by the predominant type of myosin cross bridge, i.e., the type of molecular motor of the individual muscle fibers. Different molecular motors are coded for by different myosin heavy chain (hc) isogenes which in the vertebrates comprise a family of separate genes. It is the regulation and switching of expression of the different myosin hc isoform genes that we have studied in response to physical signals during growth and adaptation in both mammals and fish.

Muscle mass as well as muscle phenotype determine muscle contractile performance as the maximum power output is directly related to force development (cross-sectional area of the fibers) as well as the speed of shortening (intrinsic rate of shortening times the number of sarcomeres in series). Muscle has other functions as well as producing movement; in particular, it acts as a store of metabolites. As well as being a major provider of glucose to fuel muscle contraction, it provides amino acids (e.g. glutamine) that are required for the synthesis of neurotransmitters and for acid base balance. In migrating animals and animals that experience a period of inanition, e.g. low food supply during the winter months, muscle is the major long term supplier of metabolites. It is now appreciated that in man, a marked loss of muscle mass during disease or ageing, is a major cause of death as the individual is unable to survive a traumatic experience which requires the increased provision of metabolites such as glutamate. We have studied the regulation of muscle mass and phenotype concomitantly to see if these two processes involve similar transduction mechanisms whereby physical signals result in the up-regulation of expression of specific genes.

Possibly with the exception of aquatic animals, there is a "trade-off" between the need to have sufficient muscle to provide for effective locomotion, particularly for burst speeds to capture prey or escape prey, and the energetic cost of transporting that muscle bulk. This is exemplified by human athletes in that long distance runners are very lean whilst sprinters are heavily muscled. This type of adaptation clearly has relevance to other terrestrial vertebrates in which the same sort of mechanisms exist to ensure extra muscle bulk is not being carried if it is not needed for peak power generation.

The increase in muscle mass in individuals who lift heavy weights is one of the most obvious examples of a tissue responding to mechanical stress. As described below we have cloned the cDNA of two isoforms of insulin-like growth factor I (IGF-I) which are derived from the IGF-I gene by alternate splicing. The expression of one of these was only detectable after mechanical stimulation. For this reason this has been called mechano growth factor (MGF). This splice variant has different exons, is not glycosylated, is smaller, has a shorter half life in the unbound state than the systemic liver type IGF-I. The other splice variant is expressed in muscle during rest but is also up-regulated by exercise, and is similar to the systemic liver type IGF-I. The evidence suggests that MGF has a high potency for inducing local protein synthesis and preventing apoptosis and therefore has an important role in local tissue repair and remodelling. Our physiological experiments show that stretch, and particularly stretch combined with electrical stimulation, rather than stimulation per se are important in inducing MGF expression (McKoy et al., 1999). The mechanotransduction mechanism involved is believed to involve the muscle cytoskeleton. During ageing, the production of growth hormone and IGF-I by the liver declines markedly. The discovery of MGF and muscle IGF-I provides a link between physical activity and gene expression and underlines the need for the elderly to remain active as the locally produced growth factors supplement the circulating IGF-I levels.

2. Mechanical factors that influence myosin heavy chain gene expression in mammalian muscle

Stretch has been shown to be a very powerful stimulant of muscle growth and muscle protein synthesis. During post-natal growth, skeletal muscles fibers elongate by adding new sarcomeres (Goldspink, 1964) serially to the ends of existing myofibrils (Griffin et al., 1971). Even mature muscles have been shown to be capable of adapting to a new functional length by adding or removing sarcomeres in series (Williams and Goldspink, 1973). In this way sarcomere length is adjusted back to the optimum for force generation, velocity and hence power output. The stretch effect and the adaptation to an increased functional length is known to be associated with increased protein synthesis (Goldspink and Goldspink, 1986). We studied the way gene expression in muscle is influenced by stretch by casting a limb with the muscle either in the shortened or lengthened position (Loughna et al., 1990). Several interesting findings emerged from this study including the fact that a slow soleus muscle which does not normally express fast type 2b myosin hc genes, begins to transcribe the fast myosin hc gene after only a day if its muscle fibers are not subjected to stretch or are not producing force. Stretch combined with electrical stimulation was also found to induce very rapid hypertrophy of the tibialis anterior muscle in the adult animal. Both force generation and stretch are major factors in activating protein synthesis and the combination of these stimuli apparently has a pronounced additive effect. Associated with this very significant increase in muscle size (over 35% in one week) was a marked increase (up to 250%) in RNA content of the muscles which peaked two days after the commencement of stretch. This rapid increase in total RNA, which was mainly ribosomal RNA, indicates that muscle fibre hypertrophy may be controlled mainly at the level of translation and that the rapid increase in the number of ribosomes means that more message can be translated into protein. There are situations when an abundant message is present but the fibers are still undergoing atrophy, e.g. lack of stretch (with and without electrical stimulation). This again indicates that muscle size, unlike muscle phenotype, may be regulated mainly at the level of translation.

A marked phenotypic change also occurred when both stretch and electrical stimulation are combined. The fast tibialis anterior of the rabbit

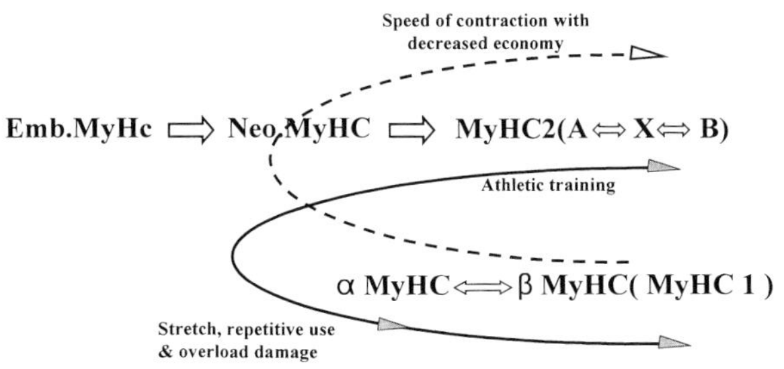

Fig. 7.1. This figure represents the family of mammalian myosin heavy chain genes that encode for different molecular motors. The sequential expression during development up to and just after birth is shown by the embryonic and neonatal MyHC genes after which the adult genes are expressed. These are the fast type 2 genes A, X and B in order of increasing contraction speeds. The latter is not expressed in the long muscles of large mammals including humans as the mechanical constraints mean that the overall rate of shortening with many sarcomeres in series would be too great. The main conversion during athletic training is the switch from the 2X to the 2A which is more economical and has a more oxidative metabolism. However, prolonged activity results in conversion to the very slow myosin type 1 MyHC which is the same as the cardiac beta myosin which is very economical and thus designed for postural and slow repetitive activities.

apparently becomes completely reprogrammed for the transcription of slow myosin hc and represses the expression of the fast myosin hc gene within only four days (Goldspink et al., 1992). Although it is generally accepted that the different fiber types in mammalian muscle are interconvertible there is some dispute concerning the nature of the physical signal. At one stage it was thought that the frequency of stimulation was the important factor in determining fibre type transition. However, it was shown that higher stimulation frequencies were just as effective in producing the fast to slow switch (Sreter et al., 1982).

Indeed, using plaster cast limb immobilization stretch alone was found to induce fast fibers to lay down slow type sarcomeres (Williams et al., 1986) and under these conditions virtually no EMG signal can be detected (Hnik et al., 1985). Therefore, it is not likely that stimulation frequency is the primary determinant of muscle phenotype. Instead, it appears to be the particular mechanical activity rather than the electrical stimulation per se that causes the switch in myosin hc gene expression. In our experiments more complete reprogramming of the muscle was obtained when stretch was combined with high frequency stimulation. This again suggests that the signal for the fast to slow change is mechanical strain. This makes physiological sense as it can be argued that the muscle cells, by responding to isometric overload, are adapting to an increased postural role (Goldspink et al., 1992).

3. Metabolic adaptation in relation to activity

In addition to the genetic reprogramming of muscle with respect to altering the predominant type of molecular motor, there are also metabolic changes. Muscle fibres expressing the slow type 1 myosin have a high mitochondrial content and a predominantly oxidative metabolism. The type 2a in human or the 2b in rodents are the next most oxidative fibers. Indeed, the most recent data on the effects of athletic training shows that the switch to very slow, oxidative type 1 fibers is difficult but that type 2x convert to 2a fibers relatively rapidly and vice versa. The evidence indicates that the metabolic changes are not coordinated with the switches in myosin type expression and can be dissociated under certain experimental conditions (Edgerton et al., 1980; Bigard et al., 2000). Also the time course for up-regulation of mitochondrial genes that are encoded by the nuclear genome is different from those encoded in the mitochondrial genome (Williams et al., 1986). For example, following electrical stimulation at 10 hz aldolase a which is a glycolytic enzyme encoded in the nuclear genome was reduced by five days but cytochrome b was unchanged. By 21 days aldolase a was decreased by five times and cytochrome b increased by four times. Recent research in our laboratory has shown that aldolase a binds to the 3′ untranslated region of myosin heavy chain genes and this may be the way the glycolytic/oxidative metabolic profile of muscle is linked to the expression of the myosin and other muscle genes (Kiri and Goldspink, unpublished data). However, it must be borne in mind that stimulation of a muscle in the unstretched position results in atrophy

(Tabary et al., 1981). Hence, the role of mechanical signals in changing the metabolism of individual muscle fibers is not understood. It is likely, for example, that the appropriate genes are activated by periods of hypoxia. These may be caused by periods of intensive physical activity but the primary effect is hypoxia.

Acclimation to lower temperatures in some fish species also results in an increase in mitochondrial density (Johnston and Maitland, 1980). It has been suggested that this is an adaptation to the longer diffusion times at the lower temperature (Johnston, 1982). A similar explanation was given for the increase in the ratio of surface area of the sarcotubular system to myofibrillar area that is associated with cold acclimation in carp (Penney and Goldspink, 1981a). Hypoxia is a topic that is described elsewhere but it should be borne in mind that it is often related to changes in the physical environment by the cells.

4. Switches in myosin gene expression in response to environmental temperature in fish muscle

One of the best examples of adaptation at the gene level is provided by fish. Some species of fish have the ability to rebuild their myofibrillar system (Johnston et al., 1973; Heap et al., 1985). This is achieved by expressing a different set of contractile myosin genes at low temperature to that expressed at warm environmental temperatures (Gerlach et al., 1990). Species that can acclimate in this way to low environmental temperatures develop more force (Heap et al., 1987) and a higher power output (Rome et al., 1985) at these temperatures than those acclimated to warmer environmental temperatures. The myofibrils of Antarctic fish have been shown to have a greater specific ATPase activity at low temperatures than tropical or temperate water fish but this is lost much more rapidly when the temperature is raised compared to fish from warmer water. Thus, there seems to have been a trade-off between catalytic activity and thermal stability. This change in thermal stability also occurs during seasonal acclimation of the carp (Penney and Goldspink, 1981b). Interestingly, this type of adaptation does not occur at low levels of food intake (Heap et al., 1986) indicating that starvation prevented gene expression and that adaptation must have involved synthesis of new proteins and the expression of different genes.

To elucidate this mechanism we made a carp genomic library and screened this for myosin hc genes using mammalian cDNA sequences under moderate stringency conditions. The clones were restriction mapped which resulted in 28 non-overlapping sequences. This indicated that the carp had a reasonably large family of myosin heavy chain genes that is about twice the size of that in mammals (Gerlach et al., 1990). Rather fortuitously the first sequence to be identified was from the gene that is predominantly expressed in white muscle at warm temperatures. This was done by extracting the RNA from red and white muscle of fish adapted to 25°C, 18°C or 8°C and carrying out a Northern analysis using the gene fragments as the probe. When carp maintained at a low temperature were acclimated to a warm temperature the time course for the expression of this gene was slightly in advance of the change in myofibrillar ATPase which suggested that this type of adaptation is regulated at the transcriptional level (Gerlach et al., 1990; Turay and Goldspink, 1991). These changes in myosin heavy chain genes in the carp have also been demonstrated at the protein level (Hirayama and Watabe, 1997). Hence, these species of fish adapt to seasonal changes in temperature by expressing different member genes of the myosin heavy chain isoform gene family at low as compared to warm temperatures. As well as different myosin hc genes for red, white and pink muscle we have identified quite a number of developmental myosin genes. Here again there is a scaling problem because as the fish increases in size the tail beat frequency decreases and, therefore, different genes have to be expressed at different times in ova and post hatching stages (Ennion et al., 1999). The species of fish studied are tetraploid and, therefore, presumably have at least twice the number of myosin isoform genes as compared to a mammal. Thus, evolutionary experiments by fish have resulted in some myosins being adapted for warm versus cold temperature swimming as well as for the different developmental stages and muscle types.

We have recently partly characterized the 5′ regulatory (promoter) sequence of carp FG2 myosin gene to see how a temperature switch may operate. Using engineered genes which consisted of the gene promoter sequence attached to a reporter cDNA we introduced these into cells in culture and into muscle *in vivo* by direct injection and looked at the effect of temperature. In this way we were able to determine which upstream regulatory sequences were necessary for the temperature effect (Gauvry et al., 1996). It is clear that some of the regulatory sequences are temperature sensitive but the detailed molecular mechanism has to be elucidated.

5. Molecular motor switching in response to muscle activity

The inherent ability of skeletal muscle to adapt to mechanical signals is related to its ability to "switch on" or "switch off" different isoform genes and to alter the general levels of expression of different subsets of genes. The fact that there are different myosin hc isoforms in fish as well as mammals means that a muscle fibre can alter its contractile properties by rebuilding its myofibrils using myosin hcs with the required slow or fast cross bridge cycling rate and thus changing its intrinsic velocity of contraction (V_{max}). In our work on muscle adaptation in mammals we also focused our attention on the myosin hc genes because of the strong correlation between the V_{max} of the muscle fibers and their fast and slow myosin hc content (Reiser et al., 1985). Our group have paid particular attention to the influence of physical signals (stretch and force generation) that determine the expression of the fast and slow myosin hc genes (Goldspink et al., 1992).

It is possible to chemically remove the cross-bridge heads (S1 fragments of the myosin heavy chains) and coat these onto microscope slides and show that these have the ability to move actin filaments if ATP is available in the medium. This shows that the S1 region of the myosin heavy chain is the molecular motor that generates the contractile force for muscular contraction (Spudich, 1994). Therefore, it was appropriate to study the regions of the myosin heavy chain genes that encode the molecular motors to see how they differ between different the fast and slow myosin isoforms. A major advance in muscle biology was made when the crystallographic structure of the myosin S1 (myosin cross-bridge head) for avian pectoralis muscle myosin was published. The possibility now exists for elucidating the mechanism of force development and of understanding the structure and function of the different types of myosin.

In order to study the molecular motors of different muscle fibre types we cloned the ATPase region of different myosin hc genes from the dog and also from species of fish that live at different temperatures (Rayment et al., 1993a; 1993b). These were then sequenced and compared using computer graphics. Although the ATP binding site is highly conserved there is a part of the ATPase pocket that differs between the different myosin hc isoforms. Part of this forms a loop, called the hypervariable loop, that projects over the enzymic pocket (Bobkov et al., 1996; Goodson et al., 1999). This differs in length, charge and distribution of charge and we believe that it acts as an electrostatic latch (Gauvry et al., 1997; 2000). The rate determining step in the myosin crossbridge cycle is the release of ADP. The longer this is delayed the longer the myosin attachment time. The length of time the latch covers the ATPase pocket will be determined by the charge difference between the loop and the body of the S1. The length and presumably the flexibility of the loop also differs between the myosin hc isoforms. As mentioned, the ability to rebuild the contractile system, including changing the type of molecular motor, enables muscle to respond to different physical environments.

6. Local control of muscle mass and phenotype

For some time it has been appreciated that there is local as well as the systemic control of tissue

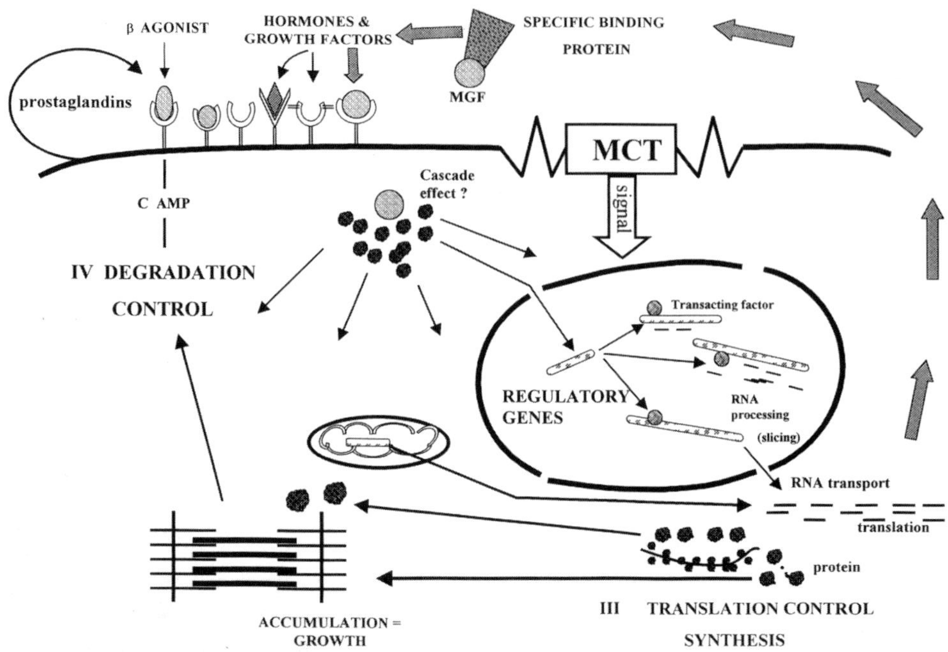

Fig. 7.2. This figure shows the autocrine/paracrine system that operates in muscle in response to mechanical strain and which plays a major role in controlling tissue mass. Accumulation of muscle mass is a balance between synthesis and breakdown of proteins. The latter is increased by the cytokines produced in several diseases in which muscle wasting is a problem. However, an increase in synthesis rate is brought about by growth factors of which the IGF-I s are the main ones. These include the systemic type of IGF-I that is produced by the liver and the local form of IGF-I called Mechano Growth Factor (MGF) which is expressed in response to stretch and overload. The nature of the mechanochemical transducer (MGT) and the initial second messenger is not known but this results in up-regulation of the IGF-I gene. The primary transcript is then spliced and translated into the MGF peptide or autocrine isoform. This then binds to the specific binding protein in the extracellular matrix which stabilizes the MGF and acts as a time release mechanism. MGF appears to use the same receptor as for the systemic IGF-I from the liver. After binding to the receptor a cascade effect occurs that results in transcriptional factors being produced which activate the structural genes such as actin and myosin that are required for muscle repair and remodelling.

growth. The post-natal growth spurt which occurs early in life is believed to be regulated to a large extent by growth hormone produced by the pituitary and which causes the release of IGF-Is from the liver and probably other tissues including muscle. However, it is well known that there are a good number of cell types which respond to mechanical signals and posses a mechanism for local control of growth remodelling and repair. As mentioned, tissue size and shape is not strictly pre-programmed but regulated to a large extent by mechanical factors. Cells that have an inherent ability to respond to mechanical factors have been termed mechanocytes and include fibroblasts, keratinocytes, osteoblasts, skeletal, cardiac and smooth muscle cells. For cardiac muscle this has been studied by Izumo's group (Sadoshima et al., 1997) and for skeletal muscle by our group (Yang et al., 1996; McKoy et al., 1999). We are now attempting to define the upstream signalling events that results in MGF production as well as the downstream effects of its using different molecular biology approaches (Kemp et al., 2000). Certainly, muscle offers one of the best models for studying this type of mechanotransduction as the mechanical activity generated by and imposed upon muscle tissue can be accurately controlled and measured in both *in vitro* and *in vivo* systems.

As mentioned, we have recently identified and cloned a growth factor which is expressed in muscle only when it is subjected to activity (Yang et al., 1996). Both the rabbit and human cDNA for this growth factor have now been sequenced and it is apparent that this is derived from the IGF-I gene by alternative splicing. The structure of the cDNA of this isoform indicates that it has different exons to the liver types and that it is not glycosylated. Therefore, it is expected to be smaller and have a shorter half-life than the liver IGF-Is. Thus, it is designed to act in an autocrine/paracrine rather than in a systemic fashion. It is possible that MGF is the end product of mechanotransduction signalling pathways in muscle and other cell types. Questions such as whether MGF is upregulated before membrane damage occurs or whether membrane damage initiates the production of the growth factor can be answered. Experiments are currently being performed to determine the mechanisms via which cells respond to mechanical stimuli and the link between the mechanical stimulus and gene expression as this represents a new and important area of physiology (Goldspink and Booth, 1992). As far as skeletal muscle is concerned, it has long been appreciated that there is local control over growth because if a muscle is exercised it is only that muscle which undergoes hypertrophy and not all the muscles of the limb.

Preliminary experiments have shown that in mdx and dydy dystrophic mice the mechanotransduction system is defective; MGF is not upregulated as it is in normal mice when the muscle is stretched (Goldspink et al., 1996). Dystrophin is associated with the membrane in normal muscle but it is absent in Duchenne dystrophy and in the mdx mouse. In autosomal dystrophies and the dydy mouse it is one of the proteins that connect the dystrophin to the membrane and to the extracellular matrix that is missing. This suggests that the dystrophin/dystrophin associated glycoproteins and extracellular proteins are part of a mechanotransduction system. Also nNOS in muscle has been shown to be associated with dystrophin and therefore dystrophin seems to have a more important role than just stabilizing the membrane. Recombinant (basic) IGF-I has been shown to have a beneficial effect on the muscles of the dystrophic dydy mouse (Zdanowicz et al., 1995) but the IGF-I1 used is probably an inappropriate form for maximum effectiveness. In IGF-I gene knockout experiments the transgenic offspring did not survive long after birth and examination of their muscles revealed that they were dystrophic (Powell-Braxton et al., 1993). Transgenic experiments in which the IGF-I gene is overexpressed (Coleman et al., 1995) have shown, however, that this results in muscle fibre hypertrophy. Hence, it appears that inadequate IGF-I and the inability of muscle to repair and adapt may be the causal mechanism of dystrophies.

7. Action of MGF in inducing muscle hypertrophy

In order to determine if the splice variant transcript that is detected in muscle following stretch or exercise has biological activity, we introduced by intramuscular injections a plasmid gene construct containing the MGF cDNA under the control of muscle regulatory elements. We obtained up to a 25% increase in muscle mass within two weeks following a single injection into the mouse tibialis anterior muscle. It has been reported from transgenic experiments that over-expression of the IGF-I gene results in increased muscle mass and injection of the liver type IGF-I cDNA into muscle produces a 20% increase in mass over a period of 4 months (Barton-Davis et al., 1998). We were therefore surprised at the potency of MGF which we have shown is associated with higher levels of the peptide in the muscle. In order to determine if the increase in wet tissue mass was of physiological significance we prepared cryostat sections of the muscles and measured muscle fibre size. Mean fibre size was increased but what was noticeable was that this was due to there being more large fibers in the injected muscle rather than an increase in all the fibers. With the intramuscular injection method first described by Wolff's group (Wolff et al., 1990) only some of the fibers took up and expressed the introduced construct. Therefore, it seems that it is these fibers that have undergone hypertrophy, the rapidity and extent of which is impressive.

In order to study the modus operandi of the muscle IGF-Is, particularly the autocrine splice variant (MGF), we generated the peptides using a peptide synthesizer and by introducing the cDNA into a expression system. A polyclonal antibody to MGF was shown to be specific by using its peptide to block its reaction. However, when a tissue section was treated with the MGF peptide before the antibody was added, the tissue lit up as it appears that the peptide binds to a binding protein and these sites are then detected by the MGF antibody. As well as cardiac and skeletal muscle this effect was particularly noticeable in brain tissue indicating the presence of specific MGF binding proteins in neuronal tissue.

8. Binding protein and local action of growth factors

The action of the IGF-I growth factors is an example of pre-receptor regulation. This involves binding proteins which act as a time release mechanism (Jones and Clemmons, 1995; Clemmons, 1997). As well as alternative splicing of the primary transcript of the IGF-I gene to produce more than one isoform of the peptide, there is also post-translational processing of the peptides. The pre-pro form is the first translation product and this is cleaved to yield the pro-peptide which is secreted by the activated cell. In the case of the IGF-I these bind to specific binding proteins which stabilise them and target their action. There are about at least 6 different IGF-I binding proteins (IGF-I BPs) described of which BP1 and BP3 are produced by the liver and are found in the circulation; the others are produced by specific tissues. Using Western blotting we have found that there is a particular IGF-I BP in muscle to which MGF binds and that it seems to be related to BP3. The systemic form of IGF-I does not bind to the MGFBP and in this way the local control differs from the systemic control of tissue growth.

9. Mechanotransduction mechanisms

The discovery of this growth factor provides a link between the mechanical stimulus and gene expression although the nature of the mechanochemical coupling process is not known. Recently, however there have been a number of studies on other cell types that implicate the cytoskeleton in mechanochemical transduction. The cytoskeleton is believed to be involved in the transduction of mechanical strain into chemical factors and then transcriptional factors that induce expression of specific genes. This may involve Ca^{2+} signalling and there is some evidence that Ca^{2+} fluxes induce fast to slow transition in primary skeletal muscle cultures (Meissner et al., 2000). Furthermore, the calcineurin-dependent transcriptional pathway is involved in the determination muscle fibre type (Chin et al., 1998) and it has been shown that IGF-I activates this pathway (Musaro et al., 1999). Although we have gone some way in understanding the regulation of muscle mass, the signalling pathways involved in determining muscle phenotype and the role of physical factors still needs to be resolved. The role of mechanical and physical stress in regulating genes is of considerable interest as it represents a new type of physiology.

10. Summary and conclusions

Muscle provides some striking examples of how a tissue can adapt to different types of physical signals. Adaptation to a different work regime is brought about by changes in fibre type and cross-sectional area which result in altered fatigue resistance and power output. This process involves quantitative and qualitative changes in gene expression including the myosin heavy chain isogenes which encode different types of molecular motors. Using a similar process some species of fish respond to seasonal temperate change and are able to rebuild their myofibrillar systems for warm and cold temperature swimming by selective myosin gene expression. In mammals it is apparent that muscle tissue is influenced by mechanical signals and that there must be local as well as systemic regulation of tissue mass and phenotype. The nature of the chemical link between the physical signal and the up-regulation of certain muscle genes involved in muscle mass determination has, to some extent, been elucidated by the cloning of a new growth factor that is only expressed in muscles subjected to stretch and/or exercise and which is designed for an autocrine/paracrine action. Experiments indicate that this is one of the ways muscle cells adapt to mechanical stress and how local induction of repair, remodelling and hypertrophy are induced. We now need to identify and understand the mechanochemical mechanisms involved in regulating expression of specific genes.

References

Barton-Davis, E.R., Shorturma, D.I., Musaro, A., Rosenthal, N. and Sweeney, H.L. (1998). Viral mediated expression of insulin-like growth factor I blocks the ageing-related loss of skeletal muscle function. Proc. Natl. Acad. Sci USA 95, 15603–15607.

Bigard, A.X., Mateo, P.H., Sanchez, H., Serrurier, B. and Ventura-Clapier, R. (2000). Lack of coordination in metabolic enzymes and myosin heavy chain isoforms in regenerated muscles of trained rats. J. Muscle Res. Cell Motility 21, 267–278.

Bobkov, A.A., Bobkova, E.A., Lin, S.H. and Reisler, E. (1996). The role of surface loops (residues 204–216 and 627–646) in the motor function of the myosin head. Proc. Natl. Acad. Sci USA 93, 2285–2289.

Chin, E.R., Olson, E.N., Richardson, J.A., Yang, Q., Humphries, C., Shelton, J.M., Wu, H., Zhu, W., Bassel-Duby, R. and Williams, R.S. (1998). A calineurin-dependent transcriptional pathway controls skeletal muscle fibre type. Genes Devel. 12, 2499–2500.

Clemmons, D.R. 1997. Insulin-like growth factor binding proteins and their role in controlling IGF actions. Cytokine & Growth Factor Rev. 8, 45–62.

Coleman, M.E., Demayo, F., Yin, K.C., Lee, H.M., Geske, R., Montgomery, C. and Schwartz, R.J. (1995). Myogenic vector expression of insulin-like growth factor 1 stimulates muscle cell differentiation and myofiber hypertrophy in transgenic mice. J. Biol. Chem. 270, 12109– 12116.

Edgerton, V.R., Goslow, G.E. Jr., Rasmussen, S.A. and Spector, S.A. (1980). Is resistance of a muscle to fatigue controlled by its motoneurones? Nature 285, 589–90.

Ennion, S., Wilkes, D., Gauvry, L., Alami-Durante, H. and Goldspink, G. (1999). Identification and expression analysis of two developmentally regulated myosin heavy chain gene transcripts in carp. J. Exp. Biol. 202, 1081–1092.

Gauvry, L., Ennion, S., Hansen, E., Butterworth, P. and Goldspink, G. (1996). The characterization of the 5′ regulatory region of a temperature-induced myosin-heavy-chain gene associated with myotomal muscle growth in the carp. Eur. J. Biochem. 236, 887–894.

Gauvry, L., Mohan-Ram, V., Ettelaie, C., Ennion, S. and Goldspink, G. (1997). Molecular motors designed for different tasks and to operate at different temperatures. J. Therm. Biol. 22, 367–373.

Gauvry , L., Ennion, S., Ettelaie, C. and Goldspink, G. (2000). Characterization of red and white muscle myosin heavy chain gene coding sequences from Antarctic and tropic fish. Comp. Biochem. Physiol. Part B 127, 575–588.

Gerlach, G.F., Turay, L., Malik, K.T., Lida, J., Scutt, A. and Goldspink, G. (1990). Mechanisms of temperature acclimation in the carp: a molecular biology approach. Am. J. Physiol. 259, R237–R244.

Goldspink, D.F. and Goldspink, G. (1986). The role of passive stretch in retarding muscle atrophy. In: Electrical stimulation and neuromuscular disorders. (Nix, W.A. and Vrbova, G. Eds.), pp 91–100. Springer Verlag, Berlin and Heidelberg.

Goldspink, G. (1964). Increase in length of skeletal muscle during normal growth. Nature 204, 1095–1096.

Goldspink, G. and Booth, F. (1992). General remarks—Mechanical signals and gene expression in muscle. Am. J. Physiol. 262, R327–R328.

Goldspink, G., Scutt, A., Loughna, P., Wells, D., Jaenicke, T. and Gerlach, G-F. (1992). Gene expression in skeletal muscle in response to mechanical signals. Am. J. Physiol. 262, R326–R363.

Goldspink, G., Yang, S.Y., Skarli, M. and Vrbova, G. (1996). Local growth regulation is associated with an isoform of IGF-I that is expressed in normal rabbit, mouse and human muscles but not in dystrophic mouse muscles when subjected to stretch. J. Physiol. 162, 496P.

Goodson, H.V., Warrick, H.M. and Spudich, J.A. (1999). Specialized conservation of surface loops of myosin: evidence that loops are involved in determining functional characteristics. J. Mol. Biol. 287, 173–185.

Griffin, G., Williams, P.E. and Goldspink, G. (1971). Region of longitudinal growth in striated muscle fiibres. Nature New Biology 232, 28–29.

Heap, S.P., Watt, P.W. and Goldspink, G. (1985). Consequences of thermal change on the myofibrillar ATPase of five freshwater teleosts. J. Fish Biol. 26, 733–738.

Heap, S.P., Watt, P.W. and Goldspink, G. (1986). Myofibrillar ATPase activity in the carp *Cyprinus carpio*: interactions between starvation and environmental temperature. J. Exp. Biol. 123, 373–382.

Heap, S.P., Watt, P.W. and Goldspink, G. (1987). Contractile properties of goldfish fin muscles following temperature acclimation. J. Comp. Physiol. B 157, 219–225.

Hirayama, Y. and Watabe, S. (1997). Structural differences in the crossbridge head of temperature-associated myosin subfragment-1 isoforms from carp fast skeletal muscle. Eur. J. Biochem. 246, 380–387.

Hnik, P., Vejsada, J., Goldspink, D.F., Kasiciki, S. and Krekule I. (1985). Quantitative evaluation of electromyogram activity in rat extensor and flexor muscles immobilized at different lengths. Exp. Neurol. 88, 515–528.

Ingebar, D.E. (1997). Tensegrity architectural basis of cellular mechano-transduction. Ann. Rev. Physiol. 59, 575–599.

Johnston, I.A., Frearson, N. and Goldspink, G. (1973). The effects of environmental temperature on the properties of myofibrillar adenosine triphosphatase from various species of fish. Biochem. J. 133, 735–738.

Johnston, I.A., Davison, W. and Goldspink, G. (1975). Adaptations in Mg^{++}-activated myofibrillar ATPase induced by temperature acclimation. FEBS Lett. 50, 293–295.

Johnston, I.A. and Maitland, B. (1980) Temperature acclimation in crucian carp, *Carassius carassius* L., morphometric analyses of muscle fibre ultrastructure. J. Fish Biol. 17, 113–125.

Johnston, I.A. (1982). Capillarisation, oxygen diffusion distances and mitochondrial content of carp muscles following acclimation to summer and winter temperatures. Cell Tissue Res. 222, 325–337.

Jones, I.J. and Clemmons, D.R. (1995). Insulin like growth factors and their binding proteins: biological actions. Endocr. Rev. 16, 3–34.

Kemp, T.J., Sadusky, T.J., Saltisi, F., Carey, N., Moss, J., Yang, S.Y., Sassoon, D.A., Goldspink, G. and Coulton, G.R. (2000). Identification of *Ankrd2*, a novel skeletal muscle gene coding for a stretch-responsive ankyrin-repeat protein. Genomics 66, 229–241.

Loughna, P.T., Izumo, S., Goldspink, G. and Nadal-Ginard, B. (1990). Disuse and passive cause rapid alterations in expression of development and adult contractile protein genes in adult skeletal muscle. Development 109, 217–223.

McKoy, G., Ashley, W., Mander, B.J.,Yang, S.Y., Williams, N., Russell, B. and Goldspink, G. (1999). Expression of IGF-l splice variants and structural genes in rabbit skeletal muscle are induced by stretch and stimulation. J. Physiol. 516, 583–592.

Meissner, J.D., Kubis, H.P., Scheibe, R.J. and Gros, G. (2000). Reversible Ca^{2+}-induced-fast- to-slow transition in primary skeletal muscle culture cells at the mRNA level. J. Physiol. 523, 19–28.

Musaro, A., McCullagh, K.J., Naya, F.J., Olson, E.N. and Rosenthal, N. (1999). IGF-I induces skeletal myocyte hypertrophy through calcineurin in association with GATA-2 and NF-ATc1. Nature 400, 581–585.

Penney, R.K. and Goldspink, G. (1981a). Temperature adaptation by the myotomal muscle of fish. J. Therm Biol. 6, 297–306.

Penney, R.K. and Goldspink, G. (1981b). Regulatory proteins and thermostability of myofibrillar ATPase in acclimated goldfish. Comp. Biochem. Physiol. B 69, 577–583.

Powell-Braxton, I., Hollingshead, P., Wartburton, C., Dowd, M., Pitts-Meek, S., Dalton, D., Gillett, H. and Stewart, T.A. (1993). IGF-I is required for normal embryonic growth in mice. Genes and Devel. 7, 2609–2617.

Rayment, I., Rypniewski, W.R., Schmidt-Base, K., Smith, R., Tomchick, D.R., Benning, M.M., Winkelmann, D.A., Wesenberg, G. and Holden, H.M. (1993a). Three-dimensional structure of myosin subfragment-1: a molecular motor. Science 261, 50–58.

Rayment, I., Holden, H.M., Whittaker, C.B., Yohn, C.B., Lorenz, M., Holmes, K.C. and Milligan, R.A. (1993b). Structure of the actin–myosin complex and its implications for muscle contraction. Science 261, 58–65.

Reiser, P.J., Moss, R.L., Giulian, G.G. and Greaser, M.L. (1985). Shortening velocity and myosin heavy chains of developing rabbit muscle fibres. J. Biol. Chem. 260, 14403–5.

Sadoshima, J. and Izumo, S. (1997). The cellular and molecular response of cardiac myocytes to mechanical stress. Ann. Rev. Physiol. 59, 551–71.

Spudich, J.A. (1994). How molecular motors work. Nature 372, 515–518.

Sreter, F.A., Pinter, K., Jolesz, F. and Mabauchi, K. (1982). Fast to slow transformation of fast muscles in response to long-term phasic stimulation. Exp. Neurol. 75, 95–102.

Tabary, J.C., Tabary, G. and Tabary, C. (1981). Experimental rapid sarcomere loss with concomitant hypoextensibility. Muscle and Nerve 43, 198–203.

Turay, L., Gerlach, G.-F. and Goldspink, G. (1991). Changes in myosin h.c. gene expression in the common carp during acclimation to warm environmental temperatures. J. Physiol. 435, 102P.

Williams, P.E and Goldspink, G. (1973) The effect of immobilization on the longitudinal growth of stretch and muscle fibres. J. Anat. 116, 45–55.

Williams, R.S., Salmons, S., Newsholme, A., Kaufman, R.E. and Mellor, J. (1986). Regulation of nuclear and mitochondrial gene expression by contractile activity in skeletal muscle. J. Biol. Sci. Chem. 261, 376–380.

Wolff, J.A., Malone, R.W., Williams, P., Chong, W., Ascadi, G., Jani, A. and Felgner, P.L. (1990). Direct-gene transfer into mouse muscle in vivo. Science 247, 1465–1468.

Yang, S.Y., Alnaqeeb, M., Simpson, H. and Goldspink, G. (1996). Cloning and characterisation of an IGF-1 isoform expressed in skeletal muscle subjected to stretch. J. Muscle Res. Cell Motility 17, 487–495.

Yang, S.Y., Alnaqeeb, M., Simpson, H. and Goldspink, G. (1997). Changes in muscle fibre type, muscle mass and IGF-1 gene expression in rabbit skeletal muscle subjected to stretch. J. Anat. 190, 613–622.

Zdanowicz, M.M., Moyse, J., Wingertzahn, M.A., O'Connor, M., Teichberg, S. and Slonim, A.E. (1995). Effect of insulin-like growth factor 1 in murine muscular dystrophy. Endocrinology 136, 4880–4886.

Protein Adaptations and Signal Transduction. Edited by K.B. Storey and J.M. Storey

CHAPTER 8

Early Responses to Mechanical Stress: From Signals at the Cell Surface to Altered Gene Expression

Matthias Chiquet and Martin Flück
M.E. Müller-Institute for Biomechanics, University of Bern, P.O. Box 30, CH-3010 Bern, Switzerland

1. Introduction

Our bodies are constantly subjected to mechanical stress, both of external origin (gravity and movements) and generated within the organism (contractile forces produced and hemodynamic stress). Since mechanosensation is essential for our ability to interact with the environment, this is a traditional field in neurophysiology (Sachs, 1993). In addition, however, mechanical stresses directly affect the function of cells and tissues. Although this has been known for a century (Wolff, 1892), the molecular mechanisms by which various cells translate mechanical stimuli directly into specific adaptive responses have only been studied for little more than a decade. Protocols for exposing cells or tissues to mechanical stresses vary, thus it is often difficult to compare results. Moreover, primary (cell-autonomous) responses need to be distinguished from secondary effects caused, for example, by stretch-induced growth factor release (Sadoshima and Izumo, 1997). Nevertheless, common themes seem to emerge from different systems (Chiquet, 1999).

In this chapter, we will concentrate on the direct, short term responses of cells to controlled mechanical stimulation. Examples are the effects of fluid shear stress on endothelial cells (Ishida et al., 1997), or of defined strain (deformation) on fibroblasts or myocytes grown on elastic substrates (Sadoshima and Izumo, 1997; MacKenna et al., 1998). For more comprehensive information, the reader is referred to recent reviews (Ingber, 1994; Sadoshima and Izumo, 1997; Ishida et al., 1997; Shyy and Chien, 1997; Galbraith and Sheetz, 1998; Chiquet, 1999).

Here, we will start by outlining the importance of mechanical stresses for cell differentiation and tissue homeostasis. Then, we will describe what is known so far about the mechanisms of mechanosensation by nonneuronal cells, as well as about the early signaling events elicited at the cell surface and within the cell. Finally, we will describe a few selected examples showing how transcription of "early response" genes is regulated at the promoter level by mechanical stimuli.

2. Mechanical stress and tissue homeostasis

Mechanical stresses have profound effects on cell and tissue homeostasis. For example, trabeculae in loaded bone exactly follow the calculated tension and compression lines, and healing trabecular bone is remodeled according to the altered stress patterns (Wolff, 1892). Hence, cells must sense the magnitude and direction of stresses within the bone matrix, and adapt its structure and composition to changing loads (Webb et al., 1997; Pavalko et al., 2000). Many other instances demonstrate the influence of mechanical stress on tissue differentiation and function. During embryogenesis, immobilization of limb muscles inhibits joint formation (Mikic et al., 2000). Hypertrophy is induced in vascular smooth muscle cells by high blood pressure (Mackie et al., 1992) and in cardiac myocytes by

Fig. 8.1. Induction of tenascin-C expression in chicken skeletal muscle by load *in vivo*. Chicken anterior latissimus dorsi (ALD) muscle was loaded by applying a weight to the left wing, and tenascin-C mRNA expression was analyzed. (A) Northern blot detecting *tenascin-C* transcripts (arrow) in total RNA from ALDs loaded for 0–24 hours. (B) and (C): In situ hybridization with digoxygenin-labeled *tenascin-C* antisense RNA probe of cryosections from control (B) and loaded (C) ALD muscle from the same animal. The polygonal cross-sections of individual muscle fibers are outlined by the endomysium, i.e., the thin sheet of extracellular matrix between adjacent fibers. Note the strong expression of *tenascin-C* mRNA after only four hours of loading in fibroblasts present within the endomysium. Bar, 100 μm.

mechanical overload (Sadoshima and Izumo, 1997). Muscles, tendons, and bones suffer atrophy under microgravity (Johnson, 1998).

Many effects of mechanical stress eventually result in systemic changes in cell and tissue metabolism. However, in various cell types specific genes are activated by defined mechanical stimuli, and hence cell differentiation is affected. For example, mechanically stressed fibroblasts in contracting skin wounds start to express α-smooth muscle actin (Grinnell, 1994). Endothelial cells under increased shear stress rapidly upregulate a defined set of genes (Khachigian, 1996). In connective tissues, altered expression of extracellular matrix (ECM) components is part of the adaptive response to mechanical stress (Chiquet, 1999). For example, tendon fibroblasts can differentiate into fibrocartilage at sites where the tendon is bent over a bone pulley (Benjamin and Ralphs, 1998). A prominent ECM component regulated by mechanical forces *in vivo* and *in vitro* is tenascin-C (Chiquet-Ehrismann et al., 1994; Flück et al., 2000). Tenascin-C is normally associated with collagen fibrils in tissues bearing high tensile stress, e.g. tendons and ligaments (Chiquet, 1999). In hypertensive rats, tenascin-C is upregulated in arterial smooth muscle cells (Mackie et al., 1992). When load is applied to the ulnae of living rats, tenascin-C mRNA and protein are increased in bone-forming periosteum (Webb et al., 1997). Conversely, tenascin-C expression is turned off in the leg tendons of immobilized chick embryos (Mikic et al., 2000). By fitting a small weight to the wings of young chickens, we found that tenascin-C mRNA is strongly induced in muscle fibroblasts that do not normally express this ECM component (Flück et al., 2000) (Fig. 8.1). A massive upregulation of tenascin-C mRNA and protein was observed only four hours after applying the load *in vivo*. Interestingly, the mRNA level for tenascin-Y, a related protein, decreased within the same time period, whereas the levels of other ECM proteins changed only later after applying load (Flück et al., 2000).

Thus, depending on the cell type and the kind of mechanical stimulus, expression of a limited number of genes is rapidly altered. It is therefore an important question how changes in mechanical stress are sensed by cells and translated into specific responses.

3. Mechanosensation at the cell surface

3.1. Force transduction between the extracellular matrix and the cytoskeleton

For normal growth and differentiation, most cells need to adhere to extracellular matrix (ECM).

Cell–matrix adhesion contacts (e.g. focal contacts) are prominent structures that form a physical link from the ECM across the cell membrane to the cytoskeleton (Miyamoto et al., 1995). Such contacts are essential for transducing mechanical forces from the outside to the inside of the cell and vice versa (Ingber, 1994). This is best exemplified by skeletal muscle where mechanical coupling between muscle fibers and tendons is essential for function (Mayer et al., 1997). In myotendinous junctions as in all cell–ECM contacts, the transmembrane link is primarily formed by integrins, a class of heterodimeric cell surface receptors (Shyy and Chien, 1997). With their extracellular domain, integrins bind to specific ECM molecules, whereas their cytoplasmic tail is tightly linked to the cytoskeleton. In skeletal muscle, integrin mutations result in muscle fiber rupture and dystrophy (Mayer et al., 1997).

Because of their involvement in mechanical coupling and their strategic location, transmembrane components within cell–matrix contacts are good candidates for translating mechanical stimuli into chemical or electrical signals. Primarily two classes of membrane proteins have been implicated in mechanosensation: ion channels and, more recently, integrins. They will be discussed in the next two sections.

3.2. Stretch-activated ion channels

In animal organs specialized for mechanosensation (e.g. the skin or the ear), stimuli are recorded by neurons or other excitable cells, triggering a change in their membrane potential. Specific cation channels are responsible for mechanically gated ion fluxes across the cell membrane, and they seem to be the actual strain gauges (Sachs, 1993). The primary structure of several stretch-activated channel protein subunits has been elucidated. Expectedly, they all contain additional domains suited to physically link them to either the ECM or the cytoskeleton. For example, unc-105, a cation channel of the degenerin family involved in stretch-induced muscle contractions in *C. elegans*, has an extracellular domain functionally coupled to collagen IV in the basement membrane (Liu et al., 1996). Another class of stretch-activated cation channel subunits of *Drosophila* possesses intracellular ankyrin repeats found in cytoskeleton-associated proteins (Walker et al., 2000). Most likely, strains arising within the ECM or the cytoskeleton are directly propagated to these stretch sensors, resulting in conformational changes and opening of the channel.

In mechanosensory cells, stretch-activated ion channels usually trigger the release of a neurotransmitter or another chemical mediator. This secondary signal then acts on target cells, thus indirectly evoking an adaptive response to mechanical stress.

3.3. Integrins as mechanosensory molecules

Several non-excitable cell types (e.g. endothelial cells, cardiac myocytes, fibroblasts, smooth muscle cells) are known to respond directly to mechanical stimulation by an altered pattern of gene expression, although they usually require quite high strains to be activated (Ishida et al., 1997; Sadoshima and Izumo, 1997; MacKenna et al., 1998; Zou et al., 1998). Presumably, this less sensitive but direct response to mechanical stress is not mediated primarily by ion channels, but by alternative mechanosensory molecules in the cell membrane. Since integrins are transmembrane components coupling the ECM to the cytoskeleton, and since their role in cell signaling is well established (Miyamoto et al., 1995), it has been speculated that they might also serve as mechanosensory molecules (Shyy and Chien, 1997). During the process of cell adhesion, initial binding of integrins to their ECM ligands leads to their activation and clustering, and to the assembly of focal adhesion contacts linked to the cytoskeleton. These protein complexes serve as “assembly lines” for signaling pathways (Miyamoto et al., 1995). However, integrins can only act as mechanotransducers if they are able to trigger signals in response to *changes* in the strength of their interaction with ECM. This is indeed the case. When fibronectin-coated microbeads are bound to fibroblast surfaces, they recruit focal adhesion and cytoskeletal components to these sites. As a consequence of cytoskeletal traction, bound beads are retrogradely transported on the cell surface (Choquet et al., 1997). When a laser trap is used to arrest a moving bead, the cell immediately reacts by increasing the tractional force, and within seconds pulls the bead out of the trap. To stop this bead again, a higher trapping force is now needed. Hence, *existing* integrin-mediated contacts are able to sense changes in the local force, and consequently trigger a very rapid local increase in the cellular traction at this site (Choquet et al., 1997). Similar experiments with ferromagnetic beads showed that direct mechanical stressing of integrin-mediated cell-ECM contacts triggers intracellular tyrosine phosphorylation and activation of MAP kinases (Schmidt et al., 1998). In accordance with a role in mechanosensation, integrins undergo conformational changes upon ligand binding, resulting in their activation (Miyamoto et al., 1995).

Despite this evidence, however, it is still possible that the actual strain gauges in ECM contacts are other signaling molecules associated with integrins. Growth factor receptors clustered in ECM contacts may be directly activated by strain in the absence of ligand (Hu et al., 1998). Alternatively, integrins might be mechanically coupled to ion channels (Chen and Grinnell, 1995). In *C. elegans*, in addition to a stretch-activated ion channel (Liu et al., 1996) an intracellular LIM domain protein colocalizing with integrin has been implicated in the response to stretch (Hobert et al., 1999). In all these cases, however, mechanosensation is physically linked to integrins in cell-ECM contacts.

4. Early generation of chemical signals at the cell surface

The earliest events in response to mechanical stress occur at the cell surface in the vicinity of sites of mechanosensation. They include opening of stretch-activated cation channels, induction of enzymatic activity of integrin-associated signaling molecules, induction of heterotrimeric G-proteins and production of reactive oxygen species (Sadoshima and Izumo, 1997). Here we will focus on the early mechanotransduction events at the cell surface, and discuss in the next sections how the rapid generation of chemical messengers translates, via signaling cascades, mechanical stress into transcriptional activation.

4.1. Calcium influx through stretch-activated cation channels

In endothelial and other cells, cyclic strain induces a transient increase in intracellular calcium within 12–30 seconds, through activation of gadolinium-sensitive, stretch-activated cation channels in the cell membrane (Rosales et al., 1997). Increases in intracellular Ca^{2+} can stimulate enzymatic activities (PKC, phospholipase A2, NO synthase), secretion of factors (Section 4.2) and transcription factor activity, and thus lead to changes in gene expression (Santella and Carafoli, 1997). The involvement of stretch-activated cation channels in direct gene activation was questioned by earlier studies showing that their inhibition with gadolinium did not prevent induction of immediate early genes *c-jun*, *c-fos*, and *c-myc* in cardiac myocytes by stretch (Sadoshima et al., 1992). Only recently, such channels were implicated in the fluid shear stress-induced increase of TGF-β1 mRNA in osteoblast-like cells (Sakai et al. 1998).

4.2. Secretion of autocrine/paracrine mediators

Various modes of mechanical stress can induce the rapid secretion of soluble factors, e.g. of platelet-derived growth factor (PDGF) by endothelial cells, angiotensin II by cardiac myocytes, or interleukin-4 by chondrocytes (Ishida et al., 1997; Sadoshima and Izumo, 1997; Millward-Sadler et al., 1999).

A variety of secretory processes are triggered by a rise in intracellular Ca^{2+} (Burgoyne and Morgan, 1998). Indeed, mechanically induced release of factors or neurotransmitters was shown to depend on a raise in intracellular Ca^{2+} (Chen and Grinnell 1995; Sadoshima and Izumo, 1997). Thus, the Ca^{2+}-mediated release of autocrine/paracrine factors by mechanically stressed cells might be a general mechanism by which secondary cellular responses are evoked.

4.3. Integrin-dependent events at the focal adhesion complex

Integrins provide a nucleation center for the assembly of cytoskeletal and signaling molecules (Miyamoto et al., 1995). The resulting focal adhesion complexes appear to be a major site where mechanical stress is translated directly into an intracellular response (Ishida et al., 1997; Sadoshima and Izumo, 1997; Galbraith and Sheetz, 1998). Many signalling molecules, including tyrosine kinases c-src and focal adhesion kinase (FAK), adapter proteins (e.g. Shc, Grb-2, Crk), small GTPases (Ras-, Rho- and Rac-family), phosphatidylinositol 3-kinase (PI3K), and phospholipase-C (PLC) are associated with integrins in focal adhesions, and are activated within minutes after application of mechanical stress (Miyamoto et al., 1995; Sadoshima and Izumo, 1997).

Protein phosphorylation in focal adhesions
Application of mechanical stress to many cell types causes within minutes the phosphorylation of signalling molecules which are an integral part of focal adhesion complexes (Shyy and Chien, 1997). An increased phosphotyrosine content and activity of protein tyrosine kinases c-src and FAK is observed rapidly after application of various kinds of mechanical stress (Jalali et al., 1998; Li et al., 1997; Ishida et al., 1997). Tyrosine phosphorylation creates src homology 2 (SH2) binding sites on FAK for recruitment of adaptor molecules (e.g. Shc, Grb-2). Presumably via activation of the GTPase Ras, these events can trigger mitogen-activated protein kinase (MAPK) signalling cascades in shear stressed endothelial cells (Shyy and Chien, 1997). As for c-src, there is good evidence for its involvement in regulating gene transcription following mechanical stress. In bovine aortic endothelial cells, fluid shear stress rapidly activated c-src, which was required for inducing the promoter activity of monocyte chemotactic protein-1 and c-fos genes via MAPK pathways (Jalali et al., 1998).

Small GTPases
By binding to adapter proteins (e.g. Shc, Grb-2) in focal adhesions, small GTPases (e.g. Rho, Ras, Rac) function as relays in the transduction of signals originating from membrane receptors, among them integrins. They regulate focal adhesion formation, cytoskeletal reorganisation and gene expression (Ishida et al., 1997; Berk et al., 1995). It has been recognized that small GTPases play an important role in mechanotransduction pathways upstream of phospholipase-C and MAPK signalling cascades (Ishida et al., 1997; Berk et al., 1995; Chen et al., 2000). Activation of Rho was required for stretch-induced activation of the ERK-1/2 signalling cascade and stretch-induced promoter activity of skeletal *α-actin* and *c-fos* genes in cardiac myocytes (Aikawa et al., 1999). Similarly, Rho was involved in control of fluid shear-induced cyclooxygenase (*COX-2*) and *c-fos* gene expression in an osteoblast cell line (Pavalko et al., 1998).

Phosphatidyl inositol metabolism
Several enzymes and metabolites of phosphatidyl inositol metabolism have been implicated in the cellular response to mechanical stress. Phosphatidyl inositol-3 kinase (PI3Kγ) has been shown to be activated by shear stress in bovine aortic endothelial cells (Go et al., 1998), leading to the activation of a MAPK pathway (JNK) which controls transcription of shear stress-induced genes (Go et al., 1998). A transient increase in inositol 1,4,5-trisphosphate (IP_3) and diacylglycerol (DAG) due to activation of membrane-bound phospholipase-C (PLC) appears to be a robust early response of many cells to mechanical stimuli. Activation of PLC was required for fluid shear stress-induced *c-fos* and *COX-2* mRNA expression in osteoblast-like cells (Chen et al., 2000). Cyclic strain in endothelial cells caused a transient elevation in IP_3 and DAG levels within seconds (Evans et al., 1997). IP_3 (via a transient increase in intracellular Ca^{2+}) and DAG serve as direct activators of protein kinase C (PKC), linking this enzyme to induced gene transcription in response to mechanical stress (see Section 5.3).

Heterotrimeric G proteins
Heterotrimeric guanine nucleotide-binding proteins (G proteins) transduce signals by coupling diverse cell membrane receptors to effectors (Jo et

al., 1997; Berk et al., 1995; Malek et al., 1999). Heterotrimeric G-proteins have been known for some time to be instantly activated in endothelial cells responding to increased fluid flow (Berk, 1995; Shyy and Chien, 1997). In endothelial cells, heterotrimeric G-proteins regulate shear stress induced MAPK signalling pathways (JNK and ERK) and are required for shear stress-induction of nitric oxide synthase (*NOS*) mRNA (Jo et al., 1997; Malek et al., 1999). Although focal adhesion complexes, receptor tyrosine kinases and/or cytokine receptors have been proposed to mediate heterotrimeric G-protein dependent signal transduction upon mechanical stimulation, other, yet unidentified shear stress receptors might be responsible (Berk et al., 1995; Pan et al., 1999).

4.5. Generation of intracellular reactive oxygen species

Reactive oxygen species (ROS; i.e. O_2^-, H_2O_2, and NO) influence the covalent modification of critical protein sulphydryl residues, an important parameter modulating cell growth and gene expression. Intracellular ROS also affect phosphatase and ubiquitin activity, and thus act as messengers (Finkel, 1998). Cellular production of these small diffusible molecules is stimulated by diverse ligands, including cytokines and growth factors. NO is synthesized by calcium-inducible and constitutively active forms of nitric oxide synthase (NOS). The enzymatic source of oxygen-derived free radicals O_2^- and H_2O_2 remain unclear but may involve a membrane bound NADH/NADPH-dependent oxidase which is present in a variety of cells and controlled by small GTPases (Yeh et al., 1999).

Recent findings indicate a role of intracellular ROS in modulating mechanical stress induced responses in various cell types. Exposure of endothelial cells to cyclic or shear stress produced increases in oxygen-derived ROS and oxidation of proteins within minutes (De Keulenaer et al., 1998; Hsieh et al., 1998; Yeh et al., 1999). Furthermore, stretching of osteoblasts lead to rapid release of NO (Pitsillides et al., 1995). Several reports imply a role of ROS in mechanically induced gene expression. Pretreatment of endothelial cells with

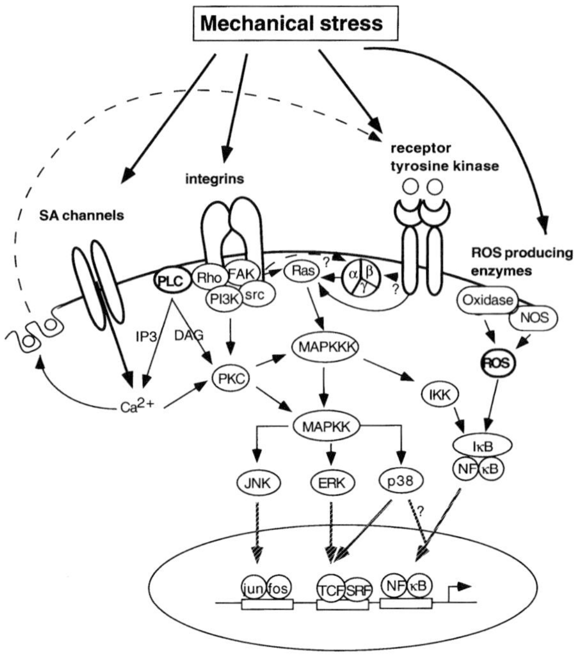

Fig. 8.2. Scheme indicating the signalling pathways involved in transcriptional responses to mechanical stress. Mechanical stress induces early events at the cell surface: Opening of stretch-activated (SA) cation channels; integrin-dependent activation of phospholipase-C (PLC), Phosphatidylinositol 3-kinase (PI3K), small GTPases (Rho,Ras), and protein tyrosine kinases in focal adhesions (FAK, src); activation of reactive oxygen species (ROS) producing enzymes (Oxidase, NOS); and activation of receptor tyrosine kinases. The chemical messengers trigger diverse signalling cascades. Increases in intracellular Ca^{2+} cause the secretion of autocrine/paracrine mediators. Activation of the small GTPase Ras and of protein kinase-C (via PLC and PI3K) lead to phosphorylation of upstream enzymes of the mitogen-activated protein kinase (MAPK) pathways (MAPKKK and MAPKK). These in turn activate the MAPKs ERK, JNK and p38, as well as I-kB kinase (IKK). Furthermore stretch-mediated activation of heterotrimeric G-proteins, through unknown events, can cause activation of MAPKs. Increased production of ROS and activated IKK lead to degradation of I-kB which enables nuclear translocation of transcription factor NF-kB. Similarly, activated ERK, JNK and p38 MAPK translocate to the nucleus and phosphorylate transcription factors TCF and AP1 (jun/fos), respectively. Activated transcription factors induce transcription of immediately early genes. For more details see text.

antioxidants markedly inhibited shear stress-induced transcription of several genes (De Keulenaer et al., 1998; Wung et al., 1999); the same was found for stretch-stimulated tenascin-C transcription in cardiac myocytes (Yamamoto et al. 1999). In summary, the results suggest that ROS are important intracellular messengers for the effects of mechanical stress on cells.

5. Triggering of intracellular signalling cascades

Various signalling cascades are used to translate the chemical messengers generated in response to mechanical stress into altered gene expression. Not surprisingly, they are the same as those triggered by various other extracellular signals. Only very recently, the janus kinase/signal transducers and activators of transcription (JAK/STAT) pathway has been implicated in mechanotransduction (Pan et al., 1999). More data exist for an essential function of mitogen-activated protein kinase (MAPK) and nuclear factor-kappa B (NF-κB) pathways, as well as for a regulatory function of protein kinase C (PKC), in the cellular responses to mechanical stress. The evidence is summarised below.

5.1. Mitogen-activated protein kinase (MAPK) pathways

Members of the MAPK family, ERK (extracellular signal regulated kinase), JNK (Jun N-terminal kinase)/SAPK (stress activated protein kinase) and p38 MAPK, play a major role in inducing the transcription of immediate early genes (Karin and Hunter, 1995). Three distinct cascades of upstream MAPK kinases selectively phosphorylate MAPK members and stimulate their serine/threonine phosphotransferase activity. Activated MAPKs translocate to the nucleus and regulate, by phosphorylation, the activity of numerous nuclear factors such as Elk-1 and c-jun. Thereby, they control the transcription of immediate early genes such as *c-fos*, *c-jun* and *egr-1* (Karin and Hunter, 1995; Wung et al., 1999).

Stimulation of MAPK cascades appears to involve translocation of upstream MAPK kinases to focal adhesions, and their selective association with activated small GTPase Ras. In addition to classical receptor tyrosine kinases, FAK and c-src in focal adhesions, as well as PKC and the small GTPase Rho have been shown to be upstream of MAPK pathways (Jalali et al., 1998, Shyy and Chien, 1997; Karin and Hunter, 1995). Similarly, activation of Ras has been shown to couple heterotrimeric G-proteins to MAPK triggering (Jo et al., 1997). Thus multiple extracellular stimuli that are known to induce homeostatic perturbations are integrated to activate MAPK pathways.

It is well documented that physical forces regulate JNK, ERK, and p38 MAPK activity within a matter of minutes. Static and cyclic stretch in cardiac myocytes, as well as fluid-shear stress in endothelial cells activate members of the ERK and JNK signaling cascade (Sadoshima and Izumo, 1997; Shyy and Chien, 1997). Furthermore, p38 MAPK was shown to be activated in a time-dependent fashion by cyclic strain in rat myocytes (Liang and Gardner, 1999). In vascular smooth muscle cells, the triggering of MAPK pathways lead to enhanced DNA binding activity of transcription factor AP-1 (Hu et al., 1998). Interestingly, recent data suggest that various MAPK are differentially induced depending on the mode of mechanical stress, as well as on the cells' ECM substrate and integrins. In biaxially stretched fibroblasts, for example, JNK1 was activated when the cells were plated on fibronectin, vitronectin or laminin. In contrast, ERK2 activity was only stimulated when the cells were plated on fibronectin, and neither MAPK was induced on a collagen substrate (MacKenna et al., 1998). In the case of rat fibroblasts, application of shear forces to their integrins by using collagen-coated magnetic beads activated p38 MAPK but not JNK or ERK1/2 (Lew et al., 1999).

Using specific inhibitors, several studies now link MAPK activation to immediate early gene induction in response to mechanical stress. For example, cyclic strain-induced increases in *egr-1* mRNA levels and enhanced *egr-1* promoter activity in endothelial cells are mediated mainly via the ERK pathway (Wung et al., 1999). In contrast, stretch-induced promoter activity of the brain natriuretic peptide gene in myocytes depends on activation of p38 MAPK (Liang and Gardner,

1999). Currently no data are available on the role of JNK in mechanically induced gene transcription. The prominent activation of ERK and p38 and their essential role in inducing transcription of immediate early genes suggests that MAPKs act as general switches required for the expression of "late" (e.g. Egr-1 dependent) mechano-responsive genes.

5.2. Nuclear factor-kappa B (NF-κB) pathway

Transcription factors of the NF-κB family form the distal elements of signalling cascades that are activated particularly in response to damaging stress, such as UV light, osmotic shock, ROS, or certain cytokines. NF-κB proteins bind as homo- or heterodimers to the promoter sequences of a variety of stress induced genes (Mercurio, 1999). Activity of NF-κB is controlled through the inhibitory subunit I-κB which retains NF-κB in the cytoplasm and thus prevents its interaction with gene promoters. Upon phosphorylation by I-κB kinase (IKK), I-κB is degraded, and active NF-κB translocates to the nucleus. IKK was shown to be regulated by an upstream MAPK kinase of the JNK pathway (Mercurio, 1999). Furthermore, p38 MAPK seems to cooperate with the machinery that activates NF-κB (Liang and Gardner, 1999).

Several reports point to the importance of NF-κB for gene activation in response to mechanical stress. For example, cyclic strain induced transcription of the platelet-activating factor receptor (*PAF-R*) gene was accompanied by activation of NF-κB and required four NF-κB binding sites in its promoter (Chaqour, 1999). Activation of NF-κB and DNA synthesis in stretched arterial smooth muscle cells was abolished by antioxidants (Hishikawa et al., 1997). In the promoters of certain genes (discussed in Section 6), NF-kB seems to interact functionally with a distinct shear-stress responsive element (SSRE) (Khachigian et al., 1995). Activation of IKK and NF-κB in endothelial cells by shear stress was shown to be mediated by integrins (Bhullar et al., 1998). This signalling pathway thus seems well suited to translate mechanical forces sensed at focal adhesions into a transcriptional response.

5.3. Protein kinase C

Protein kinase C (PKC) enzymes are a family of lipid-activated serine/threonine kinases that have been grouped into conventional (Ca^{2+}/IP_3 dependent), and novel and atypical (Ca^{2+}-independent; activated by products of PI3K) isoforms (Toker, 1998). Whereas the conventional PKC isoforms (α, β, γ) are mainly involved in cytoskeleton and focal adhesion assembly, other isoenzymes participate in various signalling processes. For example, PKC-ε may be involved in activation of ERK1/2 (Traub, 1997), whereas PKC-ζ was suggested to control NF-κB-mediated gene transcription through activation of an I-κB kinase (Ishida et al., 1997).

Several reports indicate that mechanical stress can induce rapid changes in PKC activity. In endothelial cells, cyclic stretch resulted in an early transient translocation of PKC activity from the cytosol to the membrane fraction at 10 seconds followed by a sustained elevation in PKC activity in the membrane at 100 seconds (Rosales and Sumpio, 1992); in keratinocytes, PKC translocation/activation was isoform-specific (Takei 1997). Several PKC isozymes are also induced by flow shear stress (Ishida et al., 1997).

A few observations indicate that distinct PKC enzymes may be critical for mechanically induced gene transcription. Accumulation of *c-fos* mRNA in stretched rat cardiac myocytes was markedly inhibited by down-regulation of protein kinase C (Yazaki et al., 1993). Also the induced transcription of platelet-activating factor receptor (*PAF-R*) observed after 1 h exposure of arterial smooth muscle cell to cyclic stretch was suppressed by PKC inhibitors. PKC seems to be required for enhanced promoter activity of the *PAF-R* gene in stretched arterial smooth muscle cells through a mechanism involving activation of NF-κB (Chaqour et al., 1999).

6. Transcriptional activation of mechano-responsive genes: examples

Despite the multiple signalling pathways triggered by mechanical stress, specific sets of genes are regulated depending on the cell type. On the level of transcription, only few of these genes respond directly. In many cases, the primary mechanical stimulus activates an "early response" gene leading to the rapid synthesis of a transcription factor, which then transactivates "secondary" mechano-responsive genes (Khachigian at al., 1996). Other genes are controlled via an auto- or paracrine feedback loop, involving a growth factor released in response to the mechanical stimulus (Sadoshima and Izumo, 1997). In this section, we will concentrate on a few genes for which the evidence points to a direct activation by mechanical signals. The corresponding gene products might in turn be important for the regulation of secondary response genes.

6.1. A transcription factor: Egr-1 (early growth response-1)

Like *c-jun* and *c-fos*, *egr-1* is a typical "immediate early" gene induced by various extracellular signals. Its gene product Egr-1 is a zinc-finger containing transcription factor which, upon phosphorylation via MAPK pathways in the cytoplasm, translocates to the nucleus and transactivates the promoters of target genes (Gashler et al., 1993). Because Egr-1 binding sequences are required for the activation of various gene promoters by mechanical signals, it has been speculated that *egr-1* might be a "master gene" for this type of response, and that it might be itself activated directly by mechanical stimulation (Khachigian at al., 1996). Indeed, *egr-1* mRNA levels were shown to be increased 10-fold within 30 min after applying fluid shear stress to endothelial cells (Schwachtgen et al., 1998) or cyclic strain to vascular smooth muscle cells (Morawietz et al., 1999); induction did not require protein synthesis. When endothelial cells were transfected with a reporter plasmid containing 1.2 kB of *egr-1* promoter sequence, the reporter gene was induced several-fold by shear stress. Five serum response elements (SREs) are found in the *egr-1* promoter sequence. Mutational studies showed that two of them, together with flanking Ets binding sites, are required and sufficient for inducibility of this promoter by shear stress (Schwachtgen et al., 1998). Composite SRE/Ets sites are recognized by serum response factor (SRF) and ternary complex factor (TCF) (Karin and Hunter, 1995). In fact, shear stress triggered the rapid phosphorylation/activation of the TCF component Elk-1 via the ras-ERK1/2 pathway in endothelial cells (Schwachtgen et al., 1998). Therefore, this is the most likely pathway by which the *egr-1* gene is activated by mechanical stress, and then in turn regulates secondary mechano-responsive genes. Interestingly, the *egr-1* gene promoter itself contains an Egr-1 binding site that acts as a silencer: after a brief burst of transcription, Egr-1 downregulates its own expression, stopping the response (Schwachtgen et al., 1998).

6.2. A growth factor: PDGF (platelet derived growth factor)

In endothelial cells, fluid shear stress leads to the production and release of platelet-derived growth factor (PDGF) (Khachigian et al., 1996). PDGF has two types of subunits (A and B) which form homo- or heterodimers. Interestingly, the two *PDGF* genes are regulated differently by shear stress. For *PDGF-A* the mechanism is indirect, requiring the prior synthesis of transcription factor Egr-1 (Khachigian et al., 1996). In contrast, the *PDGF-B* gene seems to be induced immediately. In its upstream promoter, a "shear stress responsive element" (SSRE) with the core nucleotide sequence GAGACC was identified by mutational analysis (Khachigian et al., 1995). The nuclear factor transactivating this SSRE was reported to be a member of the NF-κB family, although the element itself does not correspond to a canonical NF-κB binding sequence (Khachigian et al., 1995).

Recent data indicate that GAGACC-containing SSRE's in mechano-responsive genes are regulated by an integrin-triggered crosstalk between MAPK and NF-κB pathways. In heart myocytes subjected to cyclic stretch, the gene for brain natriuretic peptide (BNP) was reported to be upregulated by a pathway involving integrin signalling (Liang et al., 2000), followed by p38 MAPK-dependent activation of NF-κB (Liang and Gardner, 1999). Also in this case, GAGACC sequence elements in the *BNP* gene promoter seemed to be required for NF-κB-mediated transcriptional activation (Liang and Gardner, 1999).

However, in the case of the *PDGF-B* promoter not all types of mechanical signals seem to act via the GAGACC-containing SSRE. Recently, a different, more upstream sequence was found to be required for induction of this promoter by cyclic stretch in endothelial cells (Sumpio et al., 1998). Interestingly, this second mechano-responsive promoter region contains a canonical NF-κB binding sequence. These data indicate that various modes of mechanical stress (shear vs. cyclic stress) might act on distinct regions of this gene promoter.

6.3. An extracellular matrix protein: Tenascin-C

The adaptive response to mechanical stress involves remodeling of the extracellular matrix and coordinate changes in the expression of ECM components and proteinases. Many ECM components seem to be part of the secondary response to mechanical stress (Chiquet, 1999). Surprisingly, however, tenascin-C is a structural ECM component which fulfils some criteria of an immediate early gene. In various cell types both *in vivo* (Flück et al., 2000) and *in vitro* (Trächslin et al., 1999; Yamamoto et al., 1999), its mRNA is rapidly and transiently induced by mechanical stress. Induction is not blocked by cycloheximide, i.e., does not require prior synthesis of a transcription factor (Yamamoto et al., 1999). By culturing chick embryo fibroblasts in stretched or relaxed collagen gels, we have demonstrated that tenascin-C promoter-reporter plasmids have a high transcriptional activity in stretched cells, and low activity in relaxed cells. Deletion analysis revealed that the upstream promoter contains two separable regions, one induced by serum, the other by static stretch. The latter (100 bp) region conferred stretch inducibility to a heterologous promoter. Intriguingly, this "stretch-responsive" enhancer region contains a GAGACC sequence (GAGATC in the chick) like the one present in the SSRE of the *PDGF-B* gene (Chiquet-Ehrismann et al., 1994). Recently, we found a similar control region in the first intron of the gene for collagen XII (Chiquet et al., 1998), another stretch-regulated ECM component (Trächslin et al., 1999). The transcription factors binding to these regions in the *tenascin-C* and in the *collagen XII* gene remain to be identified.

As in the case of *PDGF-B*, however, other promoter regions seem to be involved as well in the regulation of the *tenascin-C* gene by mechanical stress. In heart myocytes cultured on fibronectin-coated silicon membranes, *tenascin-C* mRNA was rapidly and directly upregulated by cyclic strain (Yamamoto et al., 1999). A short region in the tenascin-C promoter was required for induction by cyclic stretch. This sequence includes a canonical NF-κB site, but differs from the previously identified GAGACC-containing enhancer region that responds to static stretch. Cotransfection of myocytes with dominant negative I-κB kinase blocked induction of *tenascin-C* promoter constructs by cyclic stretch, indicating that this site is indeed activated via the NF-κB pathway, presumably via ROS (Yamamoto et al., 1999). In conclusion, like for the *PDGF-B* gene, it is possible that different modes of mechanical stress regulate the *tenascin-C* gene via distinct sites in its promoter.

7. Conclusions and perspectives

The fundamental role of mechanical stress not only in various diseases, but also in the homeostasis of healthy cells and tissues has been recognized for a long time. However, there is still a large discrepancy between the empirical use of mechanical forces, e.g. in orthopedics, dentistry and plastic surgery, and the theoretical understanding of the cellular and molecular mechanisms involved. During the last decade, various tools have been developed to apply defined mechanical stress to cultured cells, and to observe their responses on the molecular level. Some mechanosensory molecules at the cell surface have been identified. Progress has been made in elucidating the conversion of mechanical into chemical signals, the triggering of intracellular signalling pathways, and the mechanisms of mechanically induced gene activation. A rough picture has emerged; however, many important detail questions remain unsolved.

For example, the biophysical mechanism by which any mechanosensory molecule converts a mechanical impulse into a chemical signal is still unknown. The hypothesis that mechanical forces induce conformational changes (and hence activity) in these molecules might eventually be validated using the laser trap technology, which allows studies of the effect of defined small forces on single protein domains.

Another large problem concerns the specificity of the cellular response to externally applied mechanical stress. Depending on the mode of the mechanical stimulus (tension, compression, shear; static vs. dynamic), as well as on the cell type, a bewildering array of intracellular signalling pathways seems to be triggered. Since forces propagate throughout a cell, any vectorial stress will always be sensed as tension in one part and as compression in another part of the cell. Thus, a single mechanical stimulus might trigger distinct signals in different locations at the cell surface. It is therefore not surprising that the array of signalling pathways elicited by various mechanical stresses is complex. The activation of MAP kinase and NF-kB signalling pathways seems to be a common theme in many cases. These pathways are of course triggered by other external stimuli as well. Nevertheless, mechanical stress clearly induces the transcription of distinct sets of genes, and specific mechano-responsive cis-acting elements exist in the promoters of these genes. It will be a challenge to find out how a cell is able to sort out mechanical from other signals using common signalling pathways, and to translate specific information on the mode, magnitude and direction of a mechanical stimulus into activation of distinct genes.

Acknowledgements

Work by the authors was supported by the Swiss National Science Foundation, the M.E. Müller-Foundation, and the Swiss Foundation for Research on Muscle Diseases.

References

Aikawa, R., Komuro, I., Yamazaki, T., Zou, Y., Kudoh, S., Zhu, W., Kadowaki, T. and Yazaki, Y. (1999). Rho family small G proteins play critical roles in mechanical stress-induced hypertrophic responses in cardiac myocytes. Circ. Res. 84, 458–466.

Benjamin, M. and Ralphs, J.R. (1998). Fibrocartilage in tendons and ligaments—an adaptation to compressive load. J. Anat. 193, 481–494.

Berk. B.C., Corson, M.A., Peterson, T.E. and Tseng, H. (1995). Protein kinases as mediators of fluid shear stress stimulated signal transduction in endothelial cells: a hypothesis for calcium-dependent and calcium-independent events activated by flow. J. Biomech. 28, 1439–1450.

Bhullar, I.S., Li, Y.S., Miao, H., Zandi, E., Kim, M., Shyy, J.Y. and Chien, S. (1998). Fluid shear stress activation of I-kappaB kinase is integrin-dependent. J. Biol. Chem. 273, 30544–30549.

Burgoyne, R.D. and Morgan, A. (1998). Calcium sensors in regulated exocytosis. Cell Calcium 24, 367–376.

Chaqour, B., Howard, P.S., Richards, C.F. and Macarak, E.J. (1999). Mechanical stretch induces platelet-activating factor receptor gene expression through the NF-kappaB transcription factor. J. Mol. Cell. Cardiol. 31, 1345–1355.

Chen, B.-M. and Grinnell, A.D. (1995). Integrins and modulation of transmitter release from motor nerve terminals by stretch. Science 269, 1578–1580.

Chen, N.X., Ryder, K.D., Pavalko, F.M., Turner, C.H., Burr, D.B., Qiu, J. and Duncan, R.L. (2000). Ca^{2+}regulates fluid shear-induced cytoskeletal reorganization and gene expression in osteoblasts. Am. J. Physiol. 278, C989–997.

Chiquet, M. (1999). Regulation of extracellular matrix gene expression by mechanical stress. Matrix Biol. 18, 417–426.

Chiquet, M., Mumenthaler, U., Wittwer, M., Jin, W. and Koch, M. (1998). The chick and human α1(XII) gene promoter: activity of highly conserved regions around the first exon and in the first intron. Eur. J. Biochem. 257, 362–371.

Chiquet-Ehrismann, R., Tannheimer, M., Koch, M., Brunner, A., Spring, J., Martin, D., Baumgartner, S. and Chiquet, M. (1994). Tenascin-C expression by fibroblasts is elevated in stressed collagen gels. J. Cell Biol. 127, 2093–2101.

Choquet, D., Felsenfeld, D.P. and Sheetz, M. P. (1997). Extracellular matrix rigidity causes strengthening of integrin-cytoskeleton linkages. Cell 88, 39–48.

De Keulenaer, G.W., Chappell, D.C., Ishizaka, N., Nerem, R.M., Alexander, R.W. and Griendling, K.K. (1998). Oscillatory and steady laminar shear stress differentially affect human endothelial redox state. Circ. Res. 82, 1094–1101.

Flück, M., Tunç-Civelec, V. and Chiquet, M. (2000). Rapid and reciprocal regulation of tenascin-C and tenascin-Y expression by loading of skeletal muscle. J. Cell Sci. 113, 3583–3591.

Evans, L., Frenkel, L., Brophy, C.M., Rosales, O., Sudhaker, C.B., Li, G., Du, W. and Sumpio, B.E. (1997). Activation of diacylglycerol in cultured endothelial cells exposed to cyclic strain. Am. J. Physiol. 272, C650–656.

Finkel, T. (1998). Oxygen radicals and signaling, Curr. Op. Cell Biol. 10, 248–253.

Galbraith, C.G. and Sheetz, M.P. (1998). Forces on adhesive contacts affect cell function. Curr. Opin. Cell Biol. 10, 566–571.

Gashler, A.L., Swaminathan, S. and Sukhatme, V.P. (1993). A novel repression module, an extensive activation domain, and a bipartite nuclear localization signal defined in the immediate-early transcription factor Egr-1. Mol. Cell Biol. 13, 4556–4571.

Go, Y.M., Park, H., Maland, M.C., Darley-Usmar, V.M., Stoyanov, B., Wetzker, R. and Jo, H. (1998). Phosphatidylinositol 3-kinase gamma mediates shear stress-dependent activation of JNK in endothelial cells. Am. J. Physiol. 275, H1898–H1904.

Grinnell, F. (1994). Fibroblasts, myofibroblasts, and wound contraction. J. Cell Biol. 124, 401–404.

Hishikawa, K., Oemar, B.S., Yang, Z. and Luscher, T.F. (1997). Pulsatile stretch stimulates superoxide production and activates nuclear factor-kappa B in human coronary smooth muscle. Circ. Res. 81, 797–803.

Hobert, O., Moerman, D.G., Clark, K.A., Beckerle, M.C. and Ruvkun, G. (1999). A conserved LIM protein that affects muscular adherens junction integrity and mechanosensory function in *Caenorhabditis elegans*. J. Cell Biol. 144, 45–57.

Hsieh, H.J., Cheng, C.C., Wu, S.T., Chiu, J.J., Wung, B.S. and Wang, D.L. (1998). Increase of reactive oxygen species (ROS) in endothelial cells by shear flow and involvement of ROS in shear-induced c-fos expression. J. Cell Physiol. 175, 156–162.

Hu, Y., Bock, G., Wick, G. and Xu, Q. (1998). Activation of PDGF receptor alpha in vascular smooth muscle cells by mechanical stress. FASEB J. 12, 1135–1142.

Ishida, T., Takahashi, M., Corson, M.A. and Berk, B.C. (1997). Fluid shear stress-mediated signal transduction: how do endothelial cells transduce mechanical force into biological responses? Ann. N.Y. Acad. Sci. 811, 12–23.

Ingber, D. (1994). Cellular tensegrity: exploring how mechanical changes in the cytoskeleton regulate cell growth, migration, and tissue pattern during morphogenesis. Int. Rev. Cytol. 150, 173–224.

Jalali, S., Li, Y.S., Sotoudeh, M., Yuan, S., Li, S., Chien, S. and Shyy, J.Y. (1998). Shear stress activates p60src-Ras-MAPK signaling pathways in vascular endothelial cells. Arterioscler. Thromb. Vasc. Biol. 18, 227–234.

Jo, H., Sipos, K., Go, Y.M., Law, R., Rong, J. and McDonald, J.M. (1997). Differential effect of shear stress on extracellular signal-regulated kinase and N-terminal Jun kinase in endothelial cells. J. Biol. Chem. 272, 1395–1401.

Johnson, R.B. (1998). The bearable lightness of being: bones, muscles, and spaceflight. Anat. Rec. 253, 24–27.

Karin, M. and Hunter, T. (1995). Transcriptional control by protein phosphorylation: signal transmission from the cell surface to the nucleus. Curr. Biol. 5, 747–757.

Khachigian, L.M., Resnick, N., Gimbrone Jr., M.A. and Tucker, C. (1995). Nuclear factor-kB interacts functionally with the platelet-derived growth factor-B chain shear-stress response element in vascular endothelial cells exposed to shear stress. J. Clin. Invest. 96, 1169–1175.

Khachigian, L.M., Lindner, V., Williams, A.J. and Collins T. (1996). Egr-1-induced endothelial gene expression: a common theme in vascular injury. Science 271, 1427–1431.

Lew, A.M., Glogauer, M. and Mculloch, C.A., (1999). Specific inhibition of skeletal alpha-actin gene transcription by applied mechanical forces through integrins and actin. Biochem. J. 341, 647–653.

Li, S., Kim, M., Hu, Y.L., Jalali, S., Schlaepfer, D.D., Hunter, T., Chien, S. and Shyy, J.Y. (1997). Fluid shear stress activation of focal adhesion kinase. Linking to mitogen-activated protein kinases. J. Biol. Chem. 272, 30455–30462.

Liang, F., Atakilit, A. and Gardner, D.G. (2000). Integrin dependence of brain natriuretic peptide gene promoter activation by mechanical strain. J. Biol. Chem. 275, 20355–20360.

Liang, F. and Gardner, D.G. (1999). Mechanical strain activates BNP gene transcription through a p38/NF-kappaB-dependent mechanism. J. Clin. Invest. 104, 1603–1612.

Liu, J., Schrank, B. and Waterston, R.H. (1996). Interaction between a putative mechanosensory membrane channel and a collagen. Science 273, 361–364.

MacKenna, D.A., Dolfi, F., Vuori, K. and Ruoslahti, E. (1998). Extracellular signal-regulated kinase and c-jun NH_2-terminal kinase activation by mechanical stretch is integrin-dependent and matrix-specific in rat cardiac fibroblasts. J. Clin. Invest. 101, 301–310.

Mackie, E.J., Scott-Burden, T., Hahn, A.W., Kern, F., Bernhardt, J., Regenass, S., Weller, A. and Bühler, F.R. (1992). Expression of tenascin by vascular smooth muscle cells. Alterations in hypertensive rats and stimulation by angiotensin II. Am. J. Pathol. 141, 377–388.

Malek, A.M., Jiang, L., Lee, I., Sessa, W.C., Izumo, S. and Alper, S.L. (1999). Induction of nitric oxide synthase mRNA by shear stress requires intracellular calcium and G-protein signals and is modulated by PI 3 kinase. Biochem. Biophys. Res. Commun. 254, 231–242.

Mayer, U., Saher, G., Bornemann, A., Echtermeyer, F., von

der Mark, H., Miosge, N., Pöschl, E. and von der Mark, K. (1997). Absence of integrin alpha 7 causes a novel form of muscular dystrophy. Nat. Genet. 17, 318–323.

Mercurio, M. and Manning, A.M. (1999). Multiple signals converging on NF-κB. Curr. Opin. Cell Biol. 11, 226–232.

Mikic, B., Wong, M., Chiquet, M. and Hunziker, E.B. (2000). Mechanical modulation of tenascin-C and collagen-XII expression during avian synovial joint formation. J. Orthop. Res. 18, 406–415.

Millward-Sadler. S.J., Wright, M.O., Lee, H., Nishida, K., Caldwell, H., Nuki, G. and Salter, D,M. (1999). Integrin-regulated secretion of interleukin 4: A novel pathway of mechanotransduction in human articular chondrocytes. J. Cell Biol. 45, 183–189.

Miyamoto, S., Teramoto. H., Coso, O.A., Gutkind, J.S., Burbelo, P.D., Akiyama, S.K. and Yamada, K.M. (1995). Integrin function: molecular hierarchies of cytoskeletal and signaling molecules. J. Cell Biol. 131, 791–805.

Morawietz, H., Ma, Y.H., Vives, F., Wilson, E., Sukhatme, V.P., Holtz, J. and Ives, H.E. (1999). Rapid induction and translocation of Egr-1 in response to mechanical strain in vascular smooth muscle cells. Circ. Res. 84, 678–687.

Pan, J., Fukuda, K., Saito, M., Matsuzaki, J., Kodama, H., Sano, M., Takahashi, T., Kato, T. and Ogawa, S. (1999). Mechanical stretch activates the JAK/STAT pathway in rat cardiomyocytes. Circ. Res. 84, 1127–1136.

Pavalko, F.M., Chen, N.X., Turner, C.H., Burr, D.B., Atkinson, S., Hsieh, Y.F., Qiu. J. and Duncan, R.L. (1998). Fluid shear-induced mechanical signaling in MC3T3-E1 osteoblasts requires cytoskeleton-integrin interactions. Am. J. Physiol. 275, C1591–1601.

Pitsillides, A.A., Rawlinson, S.C., Suswillo, R.F., Bourrin, S., Zaman, G. and Lanyon, L.E. (1995). Mechanical strain-induced NO production by bone cells: a possible role in adaptive bone (re)modeling? FASEB J. 9, 1614–1622.

Rosales, O.R., Isales, C.M., Barrett, P.Q., Brophy, C. and Sumpio, B.E. (1997). Exposure of endothelial cells to cyclic strain induces elevations of cytosolic Ca^{2+} concentration through mobilization of intracellular and extracellular pools. Biochem J. 326, 385–392.

Rosales, O.R. and Sumpio, B.E. (1992). Protein kinase C is a mediator of the adaptation of vascular endothelial cells to cyclic strain in vitro. Surgery 112, 459–466.

Sachs, F. (1993). Ion channels as mechanical transducers. In: Cell Shape Determinants: Regulation and Regulatory Role. (Stein, W.D. and Bronner, F., Eds.), pp. 63–92. Academic Press, San Diego.

Sadoshima, J. and Izumo, S. (1997). The cellular and molecular response of cardiac myocytes to mechanical stress. Annu. Rev. Physiol. 59, 551–557.

Sadoshima, J., Takahashi, T., Jahn, L. and Izumo, S. (1992). Roles of mechano-sensitive ion channels, cytoskeleton, and contractile activity in stretch-induced immediate-early gene expression and hypertrophy of cardiac myocytes. Proc. Natl. Acad. Sci. USA 89, 9905–9909.

Sakai, K., Mohtai, M. and Iwamoto, Y. (1998). Fluid shear stress increases transforming growth factor beta 1 expression in human osteoblast-like cells: modulation by cation channel blockades. Calcif. Tissue Int. 63, 515–520.

Santella, L. and Carafoli, E. (1997). Calcium signaling in the cell nucleus. FASEB J. 11, 1091–1109.

Schmidt, C., Pommerenke, H., Dürr, F., Nebe, B. and Rychly, J. (1998). Mechanical stressing of integrin receptors induces enhanced tyrosine phosphorylation of cytoskeletally anchored proteins. J. Biol. Chem. 273, 5081–5085.

Schwachtgen, J.L., Houston, P., Campbell, C., Sukhatme, V. and Braddock, M. (1998). Fluid shear stress activation of *egr-1* transcription in cultured human endothelial and epithelial cells is mediated via the extracellular signal-related kinase 1/2 mitogen-activated protein kinase pathway. J. Clin. Invest. 101, 2540–2549.

Shyy, J.Y.-J. and Chien, S. (1997). Role of integrins in cellular responses to mechanical stress and adhesion. Curr. Opin. Cell Biol. 9, 707–713.

Sumpio, B.E., Du, W., Galagher, G., Wang, X., Khachigian, L.M., Collins, T., Gimbrone, M.A. Jr. and Resnick, N. (1998). Regulation of PDGF-B in endothelial cells exposed to cyclic strain. Arterioscler. Thromb. Vasc. Biol. 18, 349–355.

Takei, T., Han, O., Ikeda, M., Male, P., Mills, I. and Sumpio, B.E. (1997). Cyclic strain stimulates isoform-specific PKC activation and translocation in cultured human keratinocytes. J. Cell. Biochem. 67, 327–337.

Trächslin, J., Koch, M. and Chiquet, M. (1999). Rapid and reversible regulation of collagen XII expression by mechanical stress. Exp. Cell Res. 247, 320–328.

Traub, O., Monia, B.P., Dean, N.M. and Berk, B.C. (1997). PKC-epsilon is required for mechano-sensitive activation of ERK1/2 in endothelial cells. J. Biol. Chem. 272, 31251–31257.

Toker, A. (1998). Signaling through protein kinase C. Front. Biosci. 3, D1134–1147.

Walker, R.G., Willingham, A.T. and Zuker, C.S. (2000). A *Drosophila* mechanosensory transduction channel. Science 287, 2229–2234.

Webb, C.M., Zaman, G., Mosley, J.R., Tucker, R.P., Lanyon, L.E. and Mackie, E.J. (1997). Expression of tenascin-C in bones responding to mechanical load. J. Bone Miner. Res. 12, 52–58.

Wolff, J. (1892). Das Gesetz der Transformation der Knochen. Verlag August Hirschwald, Berlin. English translation (1986): Springer-Verlag, Berlin.

Wung, B.S., Cheng, J.J., Chao, Y.J., Hsieh, H.J. and Wang, D.L., (1999). Modulation of Ras/Raf/extracellular signal-regulated kinase pathway by reactive oxygen species is involved in cyclic strain-induced early growth response-1 gene expression in endothelial cells. Circ. Res. 84, 804–812.

Yamamoto, K., Dang, Q.N., Kennedy, S.P., Osathanondh, R., Kelly, R.A., Lee, R.T. (1999). Induction of tenascin-C in cardiac myocytes by mechanical deformation. Role of reactive oxygen species. J. Biol. Chem. 274, 21840–21846.

Yazaki, Y., Komuro, I., Yamazaki, T., Tobe, K., Maemura, K., Kadowaki, T. and Nagai, R. (1993). Role of protein kinase system in the signal transduction of stretch-mediated protooncogene expression and hypertrophy of cardiac myocytes. Mol. Cell. Biochem. 119, 11–16.

Yeh, L.H., Park, Y.J., Hansalia, R.J., Ahmed, I.S., Deshpande, S.S., Goldschmidt-Clermont, P.J., Irani, K. and Alevriadou, B.R. (1999). Shear-induced tyrosine phosphorylation in endothelial cells requires Rac1-dependent production of ROS. Am. J. Physiol. 276, C838–847.

Zou, Y., Hu, Y., Metzler, B. and Xu, Q. (1998). Signal transduction in arteriosclerosis: mechanical stress-activated MAP kinases in vascular smooth muscle cells. Int. J. Mol. Med. 1, 827–834.

CHAPTER 9

Fasting and Refeeding: Models of Changes in Metabolic Efficiency

Stephen P.J. Brooks
Nutrition Research Division, Health Products and Food Branch, Health Canada, 2203C Banting Research Centre, 1 Ross Ave., Ottawa, ONT, Canada K1A 0L2

1. Introduction

Most rodents do not possess extensive metabolic energy reserves and, therefore, must rely on small fuel stores when food availability is limited or absent. These animals are in serious danger of death if a source of food is not found quickly. There are, however, certain metabolic adaptations that help to prolong survival during fasting. One such adaptation—metabolic depression—is a universal mechanism whereby body energy expenditure is reduced and body weight is actively defended. This strategy holds many advantages for fasting animals including prolongation of available fuel reserves and a reduced rate of protein utilization for it is protein depletion in critical organs that ultimately results in death. Profound metabolic depression is a well known phenomenon among hibernating mammals where it permits survival over extended periods of winter dormancy. Hibernating mammals can reduce metabolic rates to at least 30%, and often less than 5%, of corresponding euthermic rates (Geiser, 1988; Ruf and Heldmaier, 1992; Heldmaier et al., 1999). Because of metabolic depression, energy savings over the winter can be as much as 88% compared with the energy that would be required to maintain normal euthermic metabolism over the winter. These savings mean that the amount of stored body fat required to survive this prolonged period of fasting can be maintained at a reasonable level. This is especially important in smaller animals that have larger surface area/lean body mass ratios and could not physically store enough fat to survive the winter under euthermic conditions (French, 1988).

Although metabolic depression is common to hibernating and torpid species, it is not confined to these animals and there are many similarities between hibernation and fasting in non-hibernating mammals. Many non-hibernating mammals also enter a state of reduced metabolic activity when food intake is restricted and weight loss occurs. For example, grey seal pups reduce their mass-specific metabolic rate by 45% in response to a period of fasting after weaning (Nørdoy et al., 1990) and the spiny desert mouse can reduce its resting metabolic rate to 50% of normal during a fast (Merkt and Taylor, 1994). Common laboratory rodents, like mice and rats have also shown evidence for metabolic depression during a fast (Corbett et al., 1986; Dulloo and Girardier, 1990; Keesey and Corbett, 1990; Dulloo and Calokatisa, 1991; Dulloo and Girardier, 1992) although the metabolic rate depression is not as profound as that seen during hibernation (Panemangalore et al., 1989; Geiser and Ruf, 1995). The underlying physiological signal for metabolic depression may be similar in both hibernators and non-hibernating (fasting) animals. For example, although changes in photoperiod are a primary signal inducing hibernation, food and water restriction can also act as induction signals in facultative hibernators (Folk, 1974; Davis, 1976; Mrosovsky, 1978). In addition, hibernation is characterized by a prolonged fasting period since most species exhibit little or no food caching during the winter. If food is present, hibernating animals eat negligible amounts during cyclical bouts of arousal (Mrosovsky and Barns, 1974).

Physiological similarities between fasting, energy restricted and hibernating animals suggest that the underlying biochemical processes are similar. Other reviews have dealt with changes associated with hibernation. The present chapter details the biochemical changes associated with food deprivation or energy restriction in non-hibernating animals. This chapter also describes the biochemical and physiological events that accompany refeeding in these animals. These studies were originally carried out to better understand dieting and weight relapse in humans and so the final chapter deals with metabolic depression associated with dieting.

2. Biochemical and physiological changes associated with fasting and energy restriction

Both lean and obese rodent models have been used to study fasting and energy restriction. Each model offers advantages. Because of their large fat deposits, obese rodents can survive prolonged periods of fasting because proteolysis is kept to a minimum (Cherel et al., 1992) and hepatic glycogen stores are less depleted (Koubi et al., 1991). This makes them ideal as models for obesity-associated dieting and permits long term fasting and energy restriction studies. Lean rodents, on the other hand, have low fat stores and utilize stored glycogen reserves quickly. These animals are useful for studying enzyme-level changes associated with starvation and energy restriction since the enzyme changes are rapid and profound.

2.1. The biochemical controls on fasting gluconeogenesis: demands on muscle protein

Fasting animals have no external source of energy and so must quickly re-organize metabolism to ensure survival. However, the pattern and extent of the metabolic changes are dependent on the original metabolic state of the animal. Experiments have shown that the ability of fasting animals to use fat stores over protein "stores" (muscle protein) directly depends on the size of the body lipid stores (Goodman et al., 1980). Lipid use is advantageous because protein is needed to maintain organ, and especially cardiac, function (van Itallie and Yang, 1984). This means that obese animals can survive for prolonged periods without food because protein utilization is reduced. However, even with protein sparing, the low, constant nitrogen loss in fasting animals over an extended period of time will eventually limit the length of time that an obese animal can survive without food by causing irreversible organ damage (Cherel et al., 1992). Obesity, therefore, is advantageous in animals that are dependent on seasonal sources of food. This is descriptive of most hibernators, some of which accumulate up to 50% of their weight in body fat prior to hibernation (Bauman, 1992). This logic also apparently applies to small mammals, such as rats and mice that do not accumulate large body fat reserves but may experience periodic bouts of fasting in the wild; in these, prolonged survival times correlate with increased fat stores (Goodman et al., 1980).

The exact mechanism for the fat-associated protein sparing effect has not been elucidated but it is thought to be the result of two independent processes. Firstly, amino acids are used as fuel when fatty acid oxidation is slowed by depletion of fat reserves. This can be brought about through many different routes: a diversion of amino acids from gluconeogenesis to oxidative pathways to provide cellular energy, an increased rate of glucose oxidation brought about through a decreased availability of fatty acids as fuel, and an increased rate of glucose utilization by neural tissues resulting from a decrease in ketone body production. The latter two processes will result in an increased rate of amino acid utilization for gluconeogenic purposes. In all cases, the result is the same: increased proteolysis. The relationship between fatty acid oxidation and proteolysis has been shown by several different experiments. Food deprivation in newborn rats was associated with a profound hypoglycaemia resulting from a low gluconeogenic rate. This occurred because these rats had no white adipose tissue and, consequently, no fatty acids to fuel fasting metabolism (Ferré et al., 1978). In starved adult rats, administration of an inhibitor of free fatty acid oxidation or an antilipolytic agent also led to a hypoglycemic state. This was associated with a rise in skeletal muscle proteolysis and an increase in urine N excretion (Lowell and Goodman, 1987). The cellular biochemical controls governing these processes are thought to be exerted mainly at the substrate level and are numerous. For example: (1) reduced ATP levels (and increased ADP levels) stimulate glycolysis and amino acid catabolism, (2) increased Pi levels (brought about through a relative rise in ADP levels) stimulate glycolysis at the phosphofructokinase locus, and (3) lower acetyl-CoA concentrations activate pyruvate dehydrogenase leading to a stimulation of glycolysis (Denton, 1996).

Differences exist in the cellular and systemic signals for increased amino acid utilization. Cellular signals involve changes in energy levels to allosterically regulate gluconeogenesis and amino acid utilization. However, changes in available systemic energy (such as plasma glucose concentrations) do not, apparently, regulate the increased proteolysis on the systemic level. This is shown by near constant plasma glucose values throughout a fast (Cherel et al., 1992). In addition, plasma glucagon and insulin concentrations are unlikely to change significantly during a fast, even if glucose levels fall slightly and muscle tissue (the source of amino acids available for gluconeogenesis) might be less permeable to amino acids during a fast (Warner et al., 1989). Recent studies suggest that increased proteolysis is concurrent with an increase in plasma corticosterone levels (Fig. 9.1). This may be related to decreased β-hydroxybutyrate levels that accompany the exhaustion of lipid reserves. Corticosterone levels are also known to be involved in modulating the efficiency of energy utilization during and after a fast (Dulloo et al., 1990).

Another important mechanism of protein sparing in obese animals involves using the glycerol released during lipolysis to fuel gluconeogenesis (Lin, 1977). This provides extra carbon units for gluconeogenesis and, therefore, reduces the amino acid requirement. The biochemical basis for this effect must lie in processes external to the liver since a consideration of the regulatory aspects of the hepatic gluconeogenic pathway offers no insights. During a fast, glutamine and alanine account for 60 to 80% of the amino acids released from skeletal muscle and taken up by the liver (Young, 1991). Glutamine is derived from glutamate as well as directly from muscle protein. Alanine probably comes from amino acid catabolism to pyruvate followed by transamination to alanine using glutamate or aspartate as the nitrogen donor. Most of the amino acid-derived flux into liver gluconeogenesis, therefore, occurs via glutamate dehydrogenase (GDH) and alanine aminotransferase (Ala-AT). GDH is a highly regulated allosteric enzyme located in the mitochondrion; glutamate oxidation is inhibited by GTP and NADH and is activated by ADP and NAD^+ (Smith et al., 1975). The mitochondrial concentrations of GTP, ADP, NADH, and NAD^+ are unlikely to be dramatically influenced by an increased glycerol concentration since fatty acid oxidation generates NADH at a fairly high rate in fasted animals. In addition, only part of the glycerol is oxidized by the mitochondrial glycerol 3-phosphate dehydrogenase (G3PDH, see below). Both alanine and glutamine metabolites must eventually pass through the phosphoenolpyruvate kinase (PEPCK) locus to form phosphoenolpyruvate (PEP). Once PEP has been formed, the carbon skeletons are pushed up the glycolytic pathway to the phosphofructokinase/fructose 1,6-bisphosphatase and glucokinase/glucose 6-phosphatase loci. All three enzyme loci are important control points for gluconeogenesis. PEPCK is the primary enzyme that controls amino acid entry into gluconeogenesis (Rognstad, 1979) and is primarily controlled by changes in total activity (Hers and Hue, 1983); PEPCK activity increases 2–3 fold in fasted animals (Remmen and Ward, 1994). Fructose 1,6-bisphosphatase (FBPase) and phosphofructokinase (PFK) are regulated by hormones and dietary status largely through the concentration of fructose 2,6-bisphosphate (fru 2,6-P_2; Pilkis and Granner, 1992). Glucokinase (GK) and glucose 6-phosphatase

(G6Pase) are regulated primarily through changes in gene transcription to reduce the amount of enzyme. GK is also regulated by translocation from the nucleus (where it is bound to an inhibitory protein and inactive) to the cytosol (where it is active; Toyoda et al., 1995). These latter enzymes are primarily involved in long-term regulation of hepatic gluconeogenesis.

Unlike the amino acids, glycerol enters the gluconeogenic pathway via glycerol kinase and is then oxidized by glycerol 3-phosphate oxidoreductase (soluble) or by glycerol 3-phosphate dehydrogenase (mitochondrial) to dihydroxyacetone phosphate. This bypasses the PEPCK locus but still leaves glycerol utilization under the control of the PFK/F6Pase and GK/G6Pase loci. Glycerol utilization is also dependent on NAD^+ concentrations (oxidoreductase) or FAD concentrations (dehydrogenase), which may be limiting because of the significant rate of fatty acid oxidation in starving animals. Regulation of glycerol utilization is apparently confined to the PFK/FBPase and GK/G6Pase loci. This is assumed since neither glycerol kinase nor glycerol 3-phosphate oxidoreductase appear to be highly regulated. Glycerol 3-phosphate dehydrogenase is regulated by changes in activity but its activity is increased when the tissues are exposed to thyroid hormones such as thyroxine and triiodothyronine (Lin, 1977). Since this is exactly opposite to that which occurs during starvation when metabolism decreases (Goodman et al., 1980; Rothwell et al., 1982), it suggests that glycerol 3-phosphate dehydrogenase does not regulate glycerol utilization during fasting.

Despite the fact that glycerol is regarded as highly gluconeogenic (Lin, 1977), the rate of its utilization as a substrate for *de novo* glucose synthesis suggests otherwise. In isolated hepatocytes, the rates of gluconeogenesis with various substrates follow the order: fructose ≈ dihydroxyacetone phosphate > lactate ≈ pyruvate > glycerol ≈ alanine. The decrease for each step is approximately 50% (Hers and Hue, 1983). In experiments where hepatocytes from starved rats were incubated with a mixture of amino acids and lactate, glutamine contributed 9.8% and alanine contributed 10.8% to gluconeogenesis. However, lactate was the greatest contributor at 68% (Kaloyianni and Freedland, 1990). It is interesting that, despite their different points of entry into the gluconeogenic pathway, the relative rates of glycerol and amino acid utilization are similar. This is due to a combination of kinetic factors that are coincidental rather than any commonality in control points. For amino acid utilization, PEPCK is a rate-controlling gluconeogenic step for the use of 4-carbon skeletons. For glycerol, the kinetic properties of glycerol kinase, glycerol 3-phosphate dehydrogenase and glycerol 3-phosphate oxidoreductase regulate glycerol utilization since dihydroxyacetone phosphate utilization is very rapid (see above). These considerations suggest that the control for the protein-sparing effect of glycerol lies outside the liver.

It is possible to calculate the contribution that glycerol makes to protein sparing in an obese starving animal using the values from Cherel et al. (1990). From their data, one can approximate 0.345 g/d of body protein and 1.93 g/d of body fat were oxidized by obese animals over the final 78 d portion of the fasting period (see Fig. 9.1). Assuming that the triglycerides are all triolein and using the reported value of 91% triglyceride as a percentage of total fat, one can calculate that approximately 0.18 g/d glycerol was released by the white adipose tissue. This translates into 1.5 kcal/d of carbohydrate. Over the same period, the animals used approximately 6.2 kcal/d of protein. Thus, glycerol contributed approximately 20% of the total carbohydrate derived from gluconeogenesis (assuming that all the glycerol went into gluconeogenesis). This shows the substantial protein sparing effect associated with the glycerol released by triglyceride hydrolysis.

2.2. *Lipid metabolism in starving animals*

From a whole body perspective, lipid metabolism is a simple process. Without food intake, lipids are mobilized from stores to supply energy needs. Non-protein respiratory quotients (RQ) fall to near 0.7, reflecting a most important reliance on fatty acid oxidation as fuel. Whole body compositional

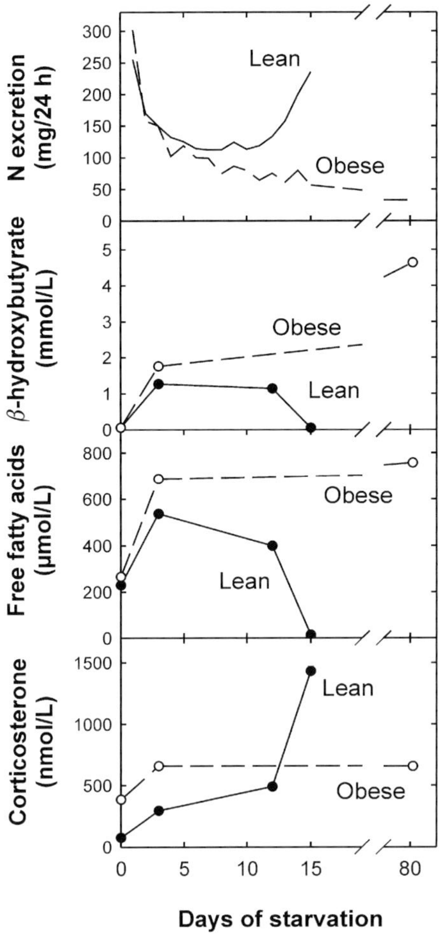

Fig. 9.1. Biochemical and physiological changes in fasting lean and obese animals. Nitrogen excretion was followed daily and plasma levels of β-hydroxybutyrate (β-HB), free fatty acids and corticosterone were measured on day 0, day 3, and day 15 (lean) or day 81 (obese). Note the rapid increase in N excretion (showing rapid protein utilization) that coincides with a decrease in plasma β-HB and free fatty acid as well as an increase in plasma corticosterone. The data are modified from Cherel et al. (1992).

analysis also shows this fact: approximately 80% and 90% of fasting energy was derived from fat oxidation in lean and obese rats, respectively (Hill et al., 1984; Cherel et al., 1992). The increased lipid mobilization is due to the action of glucagon on hormone-sensitive lipase (Bertrand et al., 1987). Interestingly, the sensitivity of fat cells to hormone stimulation appears to be a function of cell size so that fasting-associated decreases in fat cell size increase the sensitivity to glucagon (Bertrand et al., 1987). Glucose utilization by isolated white adipose tissue was similar in fasted and refed animals but large differences occurred in the proportions of glucose carbon incorporated into metabolic end products (Kather et al., 1972). It was apparent that fasting reduced carbon entry into lipogenesis and the tricarboxylic acid pathway. Calculations revealed that starving reduced the relative proportion used for fatty acid esterification and glycolysis (Grail and Davies, 1990).

Fat utilization during fasting is facilitated by increases in lipoprotein lipase activity in heart (Ruge et al., 2000) as well as increases in uncoupling proteins 2 and 3 in skeletal muscle (Samec et al., 1998). Fat is also used for thermogenesis by brown adipose tissue (BAT) in rats and mice but BAT activity is reduced during fasting. Non-hibernating animals must maintain body temperature throughout a fast, despite losing the thermogenic activity associated with digestive processes (and mediated by insulin and carbohydrate; Rothwell et al., 1983). The exact mechanism for this process is not clear but it appears that BAT plays only a limited role in fasting animals. BAT itself atrophies during a fast probably because of a reduction in fat deposits (Muralidhara and Desautels, 1994). Thyroxin and triiodothyronine levels decrease during a fast and this is expected to reduce metabolic rate (Rothwell et al., 1982) and lower BAT activity. Lower BAT activity was shown by a decreased basal oxygen consumption and decreased basal lipolysis rate during fasting (Nagashima et al., 1995). This decrease is accompanied by a marked decline in the mRNA for the uncoupling protein (Knott et al., 1992), a lower lipoprotein lipase activity (Fried et al., 1983b) and a reduction in guanosine diphosphate binding (Rothwell et al., 1984). Interestingly, these changes are opposite to those in energy-restricted animals where BAT lipoprotein lipase activity is greater than in control rats (Fried et al., 1993b). Thus, although BAT activity is reduced in fasted animals, it may be important site for thermogenesis in energy-restricted animals. The regulatory processes governing these processes have yet to be deciphered but may be related to changes in insulin receptors during fasting (Knott et al., 1992).

Fat utilization is not an arbitrary process and preferential retention of specific fatty acids can be observed after fasting or repeated cycles of fasting and refeeding. White adipose tissue fat deposits preferentially retain linoleic acid-enriched triacylglycerols, which may be the result of reduced mobilization (Raclot and Groscolas, 1995) or lower utilization of mobilized linoleic acid as fuel followed by increased re-esterification (Chen and Cunnane, 1992). This selectivity may be a function of the differential activity of outer mitochondrial carnitine palmitoyltransferase activity towards long-chain polyunsaturated fatty acids (Gavino and Gavino, 1991).

3. Biochemical changes associated with refeeding

Refed animals undergo a biochemical re-organization that promotes the repletion of carbohydrate reserves (liver glycogen), fat reserves (white adipose tissue) and muscle protein (skeletal muscle). A decrease in the levels of corticosterone is expected to accompany refeeding and would promote skeletal muscle repletion by reducing the rate of muscle proteolysis. Protein synthesis rates are also stimulated during refeeding by approximately 60% (Svanberg et al., 1998). This may be related to increased dietary protein during the refeeding period (Yoshizawa et al., 1998) and is independent of circulating growth hormone and insulin-like growth factor I (Svanberg et al., 1998). The reduced rate of proteolysis and increased rate of synthesis combine to reduce the rate of alanine release from muscle (Goodman et al., 1990) so that dietary glucose becomes an important source of glucose units for *de novo* glycogen synthesis in post-energy restricted animals. Thus, starved rats refed a high fat, low carbohydrate diet accumulated approximately 50% of the liver glycogen of animals fed a normal diet (Björntorp et al., 1983). White adipose tissue lactate is also an important of carbon for gluconeogenesis in refed animals (Newby et al., 1990).

Dietary carbohydrate provides much of the glucose units for *de novo* glycogen synthesis but gluconeogenesis is also important for glycogen repletion in refed rats. This was shown by a decreased hepatic glycogen content and rate of glycogen re-synthesis (Sugden et al., 1983) after treating refed rats with 3-mercaptopicolinate, a specific inhibitor of PEPCK. During feeding, gluconeogenesis is normally inhibited by insulin but in fasting-refed rats, glycogenesis continues at a rapid rate apparently unaffected by the increased insulin levels (Penicaud et al., 1985).

Glycogen levels continually increase in refed animals but the control of glycogen synthesis is complex. Liver cAMP values do not change appreciably even though changes in glycogen phosphorylase (GP) and glycogen synthase (GS) are apparent. GP transiently decreases during early refeeding and returns to normal by 10 h and GS continually decreases during refeeding. Increased hexose phosphate concentrations and fru 2,6-P_2 concentrations demonstrate increased entry of carbon into the glycolytic pathway and suggest an increased rate of glycolysis (stimulated via the activating effects of increased fru 2,6-P_2 on PFK). Thus, although glycolysis may be regulated by increased insulin and decreased glucagon levels in refed animals, it appears that glycogen synthesis is regulated by allosteric modifiers of the enzyme and not by insulin or glucagon-mediated changes in enzyme activity (van de Werve and Jeanrenaud, 1987). This conclusion is supported by studies showing a lack of effect of glucagon in stimulating glycogenolysis in fasted–refed rats (Blain and Kerbacher, 1986). Increased glycolysis may be partly mediated by a translocation of GK from the nucleus to the cytoplasm that also occurs during refeeding. This serves to activate GK as the regulatory protein that inhibits activity remains in the nucleus (Toyoda et al., 1995; Fernández-Novell et al., 1999).

Inhibition of G6Pase activity is also important for glycogen repletion since G6Pase would divert carbon away from phosphglucomutase and glycogen synthetase. A high G6Pase activity would also fuel futile cycling involving the PFK and G6Pase loci which would lead to excess energy loss. G6Pase activity is high at the end of a fast and decreases progressively with time (Minassian et al., 1995). This may reflect either the disappearance of an inhibitory metabolite or reversible phosphorylation of the enzyme. This seems to be part of the mechanism to suppress hepatic glucose output during post-prandial periods and like the gluconeogenic pathway, is also insensitive to insulin (Terrettaz et al., 1986).

Other changes in refed animals are governed by increased sympathetic nervous activity (Rothwell et al., 1982), thyroid hormone levels (Rothwell et al., 1982; Matsumura et al., 1982), insulin (Rothwell et al., 1983) and decreasing glucagon levels as well as by changes in AMP-activated protein kinase (Winder and Hardie, 1999). Corticoids also participate but their role is complex. Refeeding is expected to reduce cortisol levels but adrenalectomy prior to refeeding attenuates the rate of body fat increase (Dulloo et al., 1990). This observation suggests that corticoids many decrease slowly upon refeeding and their increased levels during a time of *ad lib* refeeding may drive the rapid rate of body energy gain that is apparent during this period (see below).

Refeeding is characterized by a rapid gain in body weight and body fat. The relative contribution of lipogenesis and dietary fat to fat accretion during refeeding is difficult to assess but it appears that dietary fat may be the most important factor in replenishing fatty tissue depots since animals fed a high fat, low carbohydrate diet had higher epididymal and perirenal fat pad weights (Björntorp et al., 1983). Liver lipogenesis increases upon refeeding fasted rats and this effect is especially apparent in animals refed a low fat, high carbohydrate diet (Nutrition Reviews, 1969). However, the exact contribution of lipogenesis to fat accretion is unknown. For example, refeeding with a high carbohydrate diet may increase fatty acid synthesis by 5–20 fold above the fed state (Horton et al., 1998) but this effect is not universally observed (Brooks and Lampi, 2001). Fat sparing is also apparent and is most likely due to alterations in lipid mobilization that accompany rises in insulin during refeeding. In energy restricted–refed rats, oxidation of dietary fatty acids was approximately 2.5 times lower than control animals demonstrating the profound degree of alteration in fat metabolism in refed animals.

It is clear that refeeding with a high carbohydrate based diet facilitates lipid storage possibly due to high insulin levels in refed animals (Björntorp et al., 1983). Increased rates of fatty acid synthesis are suggested by a dramatic induction of the enzymes of the fatty acid and triacylglycerol synthesis pathways such as fatty acid synthase (FAS) and mitochondrial glycero-3-phosphate acyltransferase (Sul and Wang, 1998) as well as ATP citrate lyase (Fukuda et al., 1992) but this may be misleading. The rates of fat synthesis in the rat correlate more closely with acetyl CoA carboxylase activity (Moir and Zammit, 1993; Brooks and Lampi, 1999) despite large changes in FAS activities; there are only modest changes in acetyl CoA carboxylase activity during refeeding (Moir and Zammit, 1990). FAS and glycero-3-phosphate acyltransferase are regulated by changes in genc transcription with insulin stimulating transcription and glucagon inhibiting transcription through *cis*-acting elements within the promotors (Horton et al., 1998; Sul and Wang, 1998). Changes in lipoprotein lipase activity may also be important in replenishment of fat stores. Lipoprotein lipase activity increases dramatically in refed animals (Doolittle et al., 1990) and returns to normal only after fat cell size returns to control values (Fried et al., 1983a). This is apparently due to changes in mRNA levels as well as changes in the rate of enzyme synthesis (Doolittle et al., 1990).

4. Metabolic depression and metabolic efficiency

Although it is commonly believed that laboratory rodents can only minimally depress their metabolic rates during food stress, this is not true. Studies with mice made obese by subcutaneous injection of glutamate have shown metabolic rate reductions of 25–30% during fasting (Dulloo and Calokatisa, 1991). Similar results had previously been observed with lean, fasted rats (Cumming and Morrison, 1960). Reductions in metabolic rate are also apparent in energy-restricted (dieting) animals. For example, a drop in metabolic rate of approximately 15% was observed in lean and

Table 9.1. Net energy for maintenance and net energetic efficiency in rats refed after calorie restriction[1].

Period	Group	ME intake (kcal/d)	ΔBody energy (kcal/d)	XE (kcal/d/kg LBM)	% of control value
Restriction (14 d)	ER	45.2	–2.04	822	67% of AM
	AM	111.7	12.81	1235	—
0–2 d refeeding	ER	103.5	24.6	985	55% of WM
	WM	104.5	1.05	1780	—
2–7 d refeeding	ER	109.2	18.68	1024	109% of WM 90% of AM
	WM	104.8	23.08	943	—
	AM	162.8	11.1	1137	—

[1]Male Sprague Dawley rats calorie restricted by feeding 50% of energy for 14 d.
LBM: lean body mass; ME: metabolizable energy; ER: energy restricted; AM: age matched; WM: weight matched. Data adapted from Brooks and Lampi (2001).

obese animals fed 55% of the calories required for maintenance of body weight (Keesey and Corbett, 1990). Similar results have been published by Boyle et al. (1981) and Forsum et al. (1980). Panemangalore et al. (1989) showed that rats made obese by feeding high fat diets maintained a weight in excess of that predicted from control rat growth curves when they were fed 80% or 60% of controls. Data from Walks et al. (1983) showed that previously obese rats required less energy to maintain body weight than controls.

Data from our lab showed a 33% reduction in the net energy for maintenance (XE) when lean rats were fed 50% of their maintenance calories (Table 9.1). XE is a measure of the energy required to perform necessary biochemical and physiological functions (the resting metabolic rate; RMR) but also includes the energy expended in locomotory function, shivering and non-shivering thermogenesis as well as the thermal effect of food. Resting metabolic rate forms a major portion of daily energy expenditure and so is a major fraction of the XE. Calculations in rats show that up to 85% of energy expenditure can be attributed to RMR (Iossa et al., 1999).

The fasting-associated and energy restriction-associated changes in XE might be due to changes in locomotory function or other non-essential energy expenditure but measurements by Boyle et al. (1981) suggest that rats do not alter their locomotory activity during energy restriction. Boyle et al. (1981) did, however, observe a lower heat increment in response to an intubated meal. A lower RMR is also expected since this is a function of metabolic mass (essentially the protein content of the animal); protein content decreases during fasting.

Hormonal control of energy expenditure during and after a fast is complex. Dulloo et al. (1990) provided suggestive evidence for corticoid involvement in the adaptive changes in energy expenditure during refeeding. In their adrenalecomized rats, they observed a reduced difference in energy expenditure between the refed and control groups from 18 to 8%. A sympathetic involvement in also apparent in energy sparing. Evidence for catecholamine-induced changes in thyroid metabolism and for a sympathetic involvement in thyroid-dependent responses to fasting and refeeding have been observed (Rothwell et al., 1982). The exact contribution of the individual hormonal systems is difficult to assess. For example, Young and Landsberg (1997) demonstrated that the suppression of sympathetic activity that occurs during fasting is completely reversed by 1 day of refeeding. Thus, it appears that sympathetic activity is rapidly restored during refeeding whereas the increase in metabolic efficiency can last as long as two weeks (Dulloo and Girardier, 1992). It is unlikely that thyroid hormones mediate the increase in energy efficiency since they are known to increase during refeeding (Rondeel et al., 1992) and this would be expected to increase, rather than decrease, energy expenditure.

4.1. Refeeding fasted and energy-restricted animals

The most striking consequence of fasting or energy-restricting animals is the rapid weight gain that occurs once food consumption returns to normal (Fig. 9.2). This is apparently due to the persistence of the fasting-associated metabolic rate depression after the re-introduction of food (Björntorp and Yang, 1982). This can be shown by a reduction in oxygen consumption during refeeding of energy restricted obese mice over a 3-week period (Dulloo and Calokatisa) or a lower

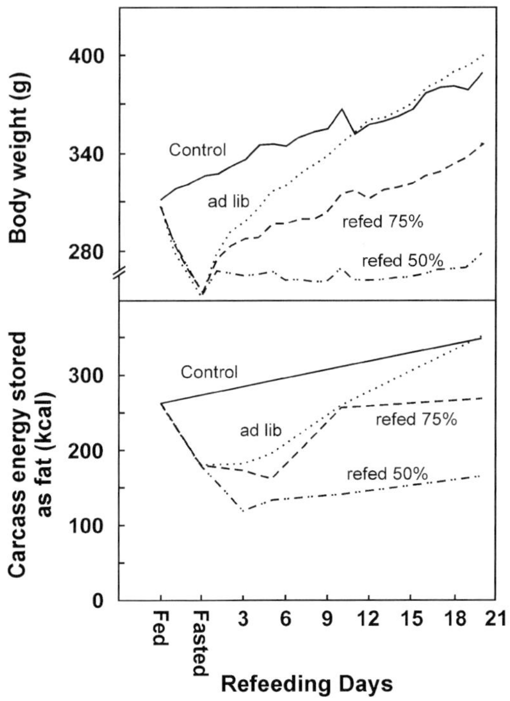

Fig. 9.2. Body weight and body fat gain in refed animals. Gain in live body weight (top, g) and carcass energy stored as fat (kcal) as a function of refeeding after fasting (3 d). The animals were male Wistar rats between 10–11 weeks of age prior to fasting. The animals were refed either *ad lib* or with 75% or 50% of pre-fast food intake. Control animals were not fasted. The data are modified from Hill et al. (1984).

XE value in refed normal rats that persisted for 2 d after refeeding had commenced (Table 9.1).

The effect of the persistent metabolic rate reduction is clearly evident because the refed animals eat the same amount of food as age-matched or weight-matched, non-fasted controls but gain weight and body energy much more rapidly. The metabolic rate depression can, therefore, manifest as an apparent increase in metabolic efficiency (i.e., a higher gain in body energy per kcal food consumed). The increased rate of weight gain can continue for a lengthy period and can lead to a final body weight greater than that of non-fasted control animals (Fig. 9.2). This period is characterized by a high rate of fat accretion, approximately three times that of control animals. This is expected since fat represents the body's only form of excess fuel storage. A significant protein accretion is also observed during refeeding, showing that actual growth rates may have been affected by fasting or energy restriction. Interestingly, increased protein synthesis is not associated with an increase in tissue DNA so that the protein/DNA ratio is increased in post-energy restricted animals (Young et al., 1989). The rate of protein energy gain is much slower than that of fat gain.

The increased metabolic efficiency can be quantified by measuring energy balance over defined experimental periods. This is done by sacrificing animals at the start and end of the refeeding time course and measuring body composition. Changes in metabolic efficiency can be expressed in several different ways. For example, energy efficiency is calculated as the increase in body energy per kcal of metabolizable energy (ME) intake.

$$\text{Energy efficiency} = \frac{\text{increase in body energy (kcal)}}{\text{ME intake (kcal)}} \tag{1}$$

where ME is defined as:

$$\text{ME} = \text{total ingested energy} - \text{fecal energy} - \text{urine energy} - \text{gas energy} \tag{2}$$

Thus, energy efficiency gives some idea of the ability of the animal to capture ingested energy and deposit it as fat and protein. However, it (or feed efficiency, an equivalent value found in the literature) is only a crude measure of metabolic efficiency because it does not take into account potential changes in the XE requirements of the animal.

Ideally, the metabolic efficiency should be a measure of the animal's ability to capture excess energy as body energy. Excess energy can be defined as any energy not required for basic physiological and biochemical processes. If a whole body calorimeter is available, one can obtain the RMR in post-absorptive, awake but quiescent animals. However, the majority of experiments have been carried out without the use of a whole body calorimeter and so only the XE value is available. Some authors have attempted to define a more theoretically sound measure of energy efficiency by subtracting the XE value from total ME intake:

$$\text{Net Energetic Efficiency (NEF)} = \frac{\text{increase in body energy (kcal)}}{\text{ME intake} - \text{XE(kcal)}} \qquad (3)$$

In absolute terms, the NEF is greater than the energy efficiency (Table 9.2) but this is of little practical consequence because only the relative difference in energy efficiency between metabolic states is important. However, as a measure of energetic efficiency, this calculation is suspect since XE can be defined as (see Livesey, 1993):

$$\text{XE} = \text{ingested ME} + E_{\text{fat loss}} - 1.36 \times E_{\text{fat gain}} - 2.25 \times E_{\text{LBM}} \qquad (4)$$

where $E_{\text{fat loss}}$ is the energy derived from the mobilization of body fat stores, $E_{\text{fat gain}}$ is the energy deposited as body fat and E_{LBM} is the energy deposited as lean body mass (protein). The factors 1.36 and 2.25 are the energy costs of depositing body fat and protein, respectively. These have been measured experimentally by Pullar and Webster (1977) under a variety of conditions. Thus, the denominator of Eq. 3 can be rewritten as:

$$\text{ME intake} - \text{XE} = 1.36 \times E_{\text{fat gain}} + 2.25 \times E_{\text{LBM}} \qquad (5)$$

Table 9.2. Estimates of metabolic efficiency during refeeding

	Refed[1]	Weight matched	Age matched
ME[2] intake (kcal)	1946	1954	2088
Increase in body energy (kcal)	460	280	202
EE (MEI – body energy)	1486	16745	1886
Energy efficiency[3] (%)	23.5	14.3	9.6
BMR[4] (kcal)	921	1039	1406
Net energetic efficiency[5]	45	31	30

Data adapted from Dulloo and Girardier (1990).
[1]Male Sprague Dawley rats refed *ad lib* after 10 d of energy restriction (approximately 50% of *ad lib* energy intake).
[2]Metabolizable energy (ME).
[3]Energy efficiency = Increase in body energy/ME intake × 100%
[4]Basal Metabolic Rate (BMR) estimated as 101 kcal × body wgt$^{-0.75}$ × d^{-1} (for the refed and age matched animals) or 103 kcal × body wgt$^{-0.75}$ × d^{-1} (for weight matched animals; see Pullar and Webster, 1977).
[5]Net energetic efficiency = Increase in body energy/(ME intake – Maintenance energy) × 100%.

with the term $E_{\text{fat loss}}$ equal to 0 because these animals are refed (Table 9.2). Thus, NEF is defined as:

$$\text{Net Energetic Efficiency (NEF)} = \frac{E_{\text{fat gain}} + E_{\text{LBM}}}{1.36 \times E_{\text{fat gain}} + 2.25 \times E_{\text{LBM}}} \qquad (6)$$

The differences reported in Table 9.2 are, therefore, most likely due to changes in XE (locomotive activity or other processes) as a function of the experimental condition since XE values were calculated from the data of Pullar and Webster (1977). Alternatively, they may reflect differences in the energy costs of depositing body fat and protein, but this is unlikely. As indicated in the previous section, several different factors contribute to the XE value but, as a rule, XE is a function of the metabolic mass of the animal, usually estimated as the protein content or (body weight)$^{0.75}$, because the RMR is related directly to metabolic mass. Its value, however, is not constant between experiments because it is extremely sensitive to environmental and experimental conditions. For example, any changes in room temperature or housing conditions will change the XE value.

4.2. Factors affecting metabolic efficiency

Several different factors can influence post-fasting weight gain. The extent of weight regain is a function of post-fasting caloric intake with greater gain at higher caloric intakes (Fig. 9.2). This is expected since weight gain depends on an excess intake of ME above RMR requirements. In agreement with this, increased exercise during the refeeding period reduces the extent of weight and fat regain (Presta et al., 1984). It is also evident that factors controlling fat metabolism are intimately involved in mediating the increase in metabolic efficiency. Yang et al. (1990) demonstrated that food efficiency correlated with the degree of adipocyte filling. In addition, the extent of weight gain was influenced by the fat content of the diet with much greater metabolic efficiency observed at very high fat intakes representing at least 40% of the total energy intake (Dulloo and Girardier, 1992).

These intakes are much higher than normal for laboratory rodents; the American Institute of Nutrition recommended fat intake for laboratory rodents is 16.7% of total energy (for growth) and 10.0% (for maintenance; see Reeves et al., 1993) and so represent non-physiological situations. Nevertheless, these results argue for a fat-sparing mechanism that must be operative in the post-fasting rodent. Evidence for this effect also comes from refeeding experiments using diets differing in their fat profiles. Data from Dulloo et al. (1995) showed that the highest energy efficiencies were observed in rats fed diets containing longer-chain fatty acids with a lower degree of polyunsaturation (lard or olive oil). Lower energy efficiencies occurred when rats were red diets high in polyunsaturated fats (safflower or fish oil) or diets high in shorter chain fatty acids (coconut oil). Measurements of the rates of fatty acid synthesis during refeeding of energy-restricted animals also provides evidence for fat sparing during refeeding. Using extrapolated rates of fatty acid synthesis, dietary fat intake and body fat loss it is possible to show a significant reduction in overall fat oxidation during refeeding (Brooks and Lampi, 2001).

4.3. Relevance to humans

Interest in metabolic depression during periods of food restriction is related to the human obsession with dieting to lose excess body weight even though this strategy is rarely successful (Bronwell and Rodin, 1994). The futility of current diet regimes has been demonstrated many times in the literature and a recent report showed that attempts at weight loss through dieting were significantly related to major weight gain in adult Finns after controlling for several confounders (Korkeila et al., 1999). It has been postulated that, by analogy with fasting rodents, part of the problem may be an apparent dieting-associated decrease in metabolic rate that could persist after resumption of normal caloric intake. As indicated above, there is good evidence for this in obese and non-obese laboratory rodents. However, controversy over the relationship between dieting, metabolic rate and obesity in humans has persisted for several years.

Evidence in favour of an increased metabolic efficiency in energy-restricted humans has come from many sources. Shetty (1984) measured the resting metabolic rate (in awake, resting, 12–14 h post-absorptive individuals) in control and chronically undernourished unskilled Indian labourers and found a 26% lower RMR in the undernourished labourers. Ravussin et al. (1988) showed that weight gain was associated with a lower metabolic rate in Pima Indians. Gingras et al. (2000) demonstrated that female chronic dieters with low resting energy expenditure had lower lean body mass at equivalent body mass indices and higher ratios of abdominal to gluteal fat. These results suggested that individuals with a lower metabolic rate had an increased risk of becoming obese. Studies with dieting obese subjects appeared to support this finding. In dieting obese women, de Boer et al. (1986) observed a lower energy requirement for maintenance of body weight after dieting; the energy requirement decreased approximately 10%

Table 9.3. Metabolic efficiency in humans during energy restriction.

Subjects	Condition	Eff_{rel}	Reference
Twelve overweight women (BMI > 25)	Before weight loss	0.88	De Boer et al. (1986)
	After loss of 7.3 kg (average) over 10 wk	0.79	
Eight moderately overweight male subjects	Before weight loss	0.94	Rumpler et al. (1991)
	After loss of 5.1 kg (average) over 27 d (fed 50% of maintenance calories)	0.81	

The relative efficiency of energy use (Eff_{rel}) is calculated as (de Boer et al., 1986):

$$Eff_{rel} = \frac{\left(\frac{RE_1}{W_1} - \frac{RE_2}{W_2}\right)}{\left(\frac{MEI_1}{W_1} - \frac{MEI_2}{W_2}\right)} = \frac{\text{efficiency of dietary energy use}}{\text{efficiency of body energy use}}$$

where metabolizable energy intake (MEI) and retained body energy (RE) are expressed in kcal/d (or kJ/d) and body weight (W) is measured in kg. Decreases in Eff_{rel} reflect a lower energy requirement for maintenance of fat free mass.

from 204 kJ/kg fat free mass to 185 kJ/kg fat free mass. Foster et al. (1999) showed a lower resting energy expenditure in post-dieting black and white women. The evidence for increased efficiency is best demonstrated by a reduction in Eff_{rel} values meaning that relatively less body energy is required to maintain the same fat free body mass (Table 9.3). As is the case in rodents, the increase in metabolic efficiency can also manifest as a increased capacity for storing fat. In weight-reduced men, carbohydrate balance tended to be lower and fat balance tended to be higher after weight loss suggesting a fat-sparing effect of weight loss that persisted after resumption of normal feeding (Rumpler et al., 1991). Changes in the respiratory quotient were also observed by Larson et al. (1995) in post obese individuals (males and females) that tended to favour fat storage. These studies complement other studies on changes in metabolic rates in obese and post-obese subjects (Geissler et al., 1987; Ravussin et al., 1988; Froidevaux et al., 1993). A recent report has provided a potential mechanism linking metabolic rate depression and starvation in humans. Yanovski et al. (2000) found a relationship between uncoupling protein 2, body composition and RMR demonstrating the potential for a relationship between RMR and the body's ability to store body fat.

The evidence for a relationship between metabolic rate, dieting and obesity is not absolute. Several studies have failed to find a relationship between dieting and metabolic efficiency. For example, de Peuter et al. (1992) failed to show any significant differences in the RMR of obese and post-obese women and Wyatt et al. (1999) found no change in RMR in previously obese men and women. In addition, the method used to calculate lean body mass (the basis for comparisons of metabolic rate) has been called into question by Ravussin and Bogardus (1989) who pointed out that an inherent bias may exist in the calculation of metabolically active body mass. Treuth et al. (2000) failed to find a significant difference in RMR or activity-related energy expenditure in girls with or without a familial predisposition to obesity. Weinsier et al. (2000) observed a reduced metabolic rate in 24 overweight, postmeonpausal women during feeding with a 800 kcal/d diet. However, the body composition-adjusted resting metabolic rate returned to normal when measured 3 weeks after a return to consuming an energy balanced diet. Zwiauer et al. (1992) reported a significant decrease in RMR during a 3-week weight reduction but this failed to be sustained 12 months after dieting. These latter two results may have missed a period of increased energy efficiency if there is a temporal component to any change in metabolic rate. This is suggested by the findings of Zauner et al. (2000) who showed that subjects starved for 84 hours had a 1.14 fold higher RMR than controls. This was apparently due to increased serum norepinephrine levels that accompanied declining serum glucose values during early starvation showing the importance of temporal changes in energy metabolism during calorie restriction and fasting.

If a change in metabolic efficiency exists, it is likely to be modest. In a recent review, Ravussin and Bogardus (2000) point out that variability in metabolic rates may only account for 12% of the variability in body mass index (BMI; Ravussin and Bogardus, 2000). Thus, as a predictor of obesity it is a relatively small component. This estimate is close to the increase in metabolic efficiency observed in some post-dieting individuals. As the human dieting studies were carried out with obese individuals, it is difficult to separate any putative changes in metabolic efficiency from actual differences in metabolic rates that may be present in these subjects.

Publication #549 of the Bureau of Nutritional Sciences.

References

Bauman W.A. (1992). Hibernation: a model of adaptive hyperlipogenesis and associated metabolic features. In: Obesity (Björntorp, P. and Brodoff, B.N., Eds.), pp. 206–219. J.B. Lippincott Company, Philadelphia, PA.

Bertrand, H.A., Anderson, W.R., Masoro, E.J. and Yu, B.P. (1987). Action of food restriction on age-related changes in adipocyte lipolysis. J. Gerentol. 42, 666–673.

Björntorp, P. and Yang, M.U. (1992). Refeeding after fasting in the rat: effects on body composition and food efficiency. Am. J. Clin. Nutr. 36, 444–449.

Björntorp, P., Yang, M.U. and Greenwood, M.R. (1983). Refeeding after fasting in the rat: effects of carbohydrate. Am. J. Clin. Nutr. 37, 396–402.

Blair, J.B. and Kerbacher, J.J. (1986). Inability of glucagon to regulate glycogen metabolism in rat hepatocytes isolated after fasting and refeeding high-carbohydrate diets. Arch. Biochem. Biophys. 251, 250–259.

Boyle, P.C., Storlein, L.H., Harper, A.E. and Keesey, R.E. (1981). Oxygen consumption and locomotor activity during restricted feeding and realimentation. Am. J. Physiol. 241, R392–R397.

Bronwell, K.D. and Rodin, J. (1994). The Dieting Maelstrom. Is it possible and advisable to lose weight? Am. Physiologist 49, 781–791.

Brooks, S.P.J. and Storey, K.B. (1992). Mechanisms of glycolytic control during hibernation of the ground squirrel *Spermophilus lateralis*. J. Comp. Physiol. B 162, 23–28.

Brooks, S.P.J. and Lampi, B.J. (1999). Effect of dietary fat on whole body fatty acid synthesis in weanling rats. J. Nutr. Biochem. 10, 291–298.

Brooks, S.P.J. and Lampi, B.J. (2001). Fatty acid oxidation and fatty acid synthesis in calorie restricted rats. J. Nutr. Biochem., submitted.

Chen, Z.-Y. and Cunnane, S.C. (1992). Preferential retention of linoleic-enriched triacylglycerols in liver and serum during fasting. Am. J. Physiol. 263, R233–R239.

Cherel, Y., Robin, J.-P., Heitz, A., Calgari, C. and Le Maho, Y. (1992). Relationships between lipid availability and protein utilization during prolonged fasting. J. Comp. Physiol. B 162, 305–313.

Corbett, S.W., Stern, J.S. and Keesey, R.E. (1986). Energy expenditure in rats with diet-induced obesity. Am. J. Clin. Nutr. 44, 173–180.

Cumming, M.C. and Morrison, S.D. (1960). The total metabolism of rats during fasting and refeeding. J. Physiol. London 154, 219–243.

Davis, D.E. (1976). Hibernation and circadian rhythms of food consumption in marmots and ground squirrels. Q. Rev. Biol. 51, 477–514.

Deavers, D.R. and Musacchia X.J. (1980). Water metabolism and renal function during hibernation and hypothermia. Fed. Proc. 39, 2969–2973.

De Boer, J.O., van Es, A.J.H., Roovers, L.A., van Raaij, J.M.A. and Hautvast, J.G.A.J. (1986). Adaptation of energy metabolism of overweight women to low-energy intake, studied with whole-body calorimeters. Am. J. Clin. Nutr. 44, 585–595.

De Peuter, R., Withers, R.T., Brinkman, M., Tomas, F.M. and Clark, D.G. (1992). No differences in rates of energy expenditure between post-obese women and their matched, lean controls. Int. J. Obesity 16, 801–808.

Denton, R.M. (1996). The hormonal regulation of pyruvate dehydrogenase complex. Adv. Enzyme Regul. 36, 183–198.

Doolittle, M.H., Ben-Zeev, O., Elovson, J., Martin, D. and Kirchgessner, T.G. (1990). The response of lipoprotein lipase to feeding and fasting. Evidence for post-translational regulation. J. Biol. Chem. 265, 4570–4577.

Dulloo, A.G. and Girardier, L. (1990). Adaptive changes in energy expenditure during refeeding following low-calorie intake: evidence for a specific metabolic component favoring fat storage. Am. J. Clin. Nutr. 52, 415–420.

Dulloo, A.G., Seydoux, J. and Girardier, L. (1990). Role of corticosterone in adaptive changes in energy expenditure during refeeding after low calorie intake. Am. J. Physiol. 259, E658–E664.

Dulloo, A.G. and Girardier, L. (1992). Influence of dietary

composition on energy expenditure during recovery of body weight in the rat: implications for catch-up growth and obesity relapse. Metabolism 41, 1336–1342.

Dulloo, A.G. and Calokatisa, R. (1991). Adaptation to low calorie intake in obese mice: contribution of a metabolic component to diminished energy expenditures during and after weight loss. Int. J. Obesity 15, 7–16.

Dulloo, A.G., Mensi, N., Seydoux, J. and Girardier, L. (1995). Differential effects of high-fat diets varying in fatty acid composition on the efficiency of lean and fat tissue deposition during weight recovery after low food intake. Metabolism 44, 273–279.

Fernández-Novell, J.M., Castel, S., Bellido, D., Ferrer, J.C., Vilaró, S. and Guinovart, J.J. (1999). Intracellular distribution of hepatic glucokinase and glucokinase regulatory protein during the fasted to refed transition in rats. FEBS Lett. 459, 211–214.

Ferré, P., Pégorier, J.P., Marliss, E.B. and Girard, J.R. (1978). Influence of exogenous fat and gluconeogenic substrates on glucose homeostasis in the newborn rat. Am. J. Physiol. 234, E129–E136.

Folk, E.G. (1974). Environmental physiology. Lea and Febiger, Philadelphia, PA.

Foster, G.D., Wadden, T.A., Swain, R.M., Anderson, D.A. and Vogt, R.A. (1999). Changes in resting energy expenditure after weight loss in obese African American and White women. Am. J. Clin. Nutr. 69, 13–17.

Forsum, E., Hillman, P.E. and Nesheim, M.C. (1980). Effect of energy restriction on total heat production, basal metabolic rate, and specific dynamic action of food in rats. J. Nutr. 111, 1691–1697.

French, A.R. (1988). The patterns of mammalian hibernation. Am. Sci. 76, 569–581.

Fried, S.K., Hill, J.O., Nickel, M. and DiGirolamo, M. (1983a). Prolonged effects of fasting-refeeding on rat adipose tissue lipoprotein lipase activity: influence of caloric restriction during refeeding. J. Nutr. 113, 1861–1869.

Fried, S.K., Hill, J.O., Nickel, M. and DiGirolamo, M. (1983b). Novel regulation of lipoprotein lipase activity in rat brown adipose tissue: effects of fasting and caloric restriction during refeeding. J. Nutr. 113, 1870–1874.

Froidevaux, F., Schutz, Y., Christin, L. and Jéquier, E. (1993). Energy expenditure in obese women before and during weight loss, after refeeding, and in the weight-relapse period. Am. J. Clin. Nutr. 57, 35–42.

Fukuda, H., Katsurada, A. and Iritani, N. (1992). Effects of nutrients and hormones on gene expression of ATP citrate-lyase in rat liver. Eur. J. Biochem. 209, 217–222.

Gavino, G.R. and Gavino, V.C. (1991). Rat liver outer mitochondrial carnitine palmitoyltransferase activity towards long-chain polyunsaturated fatty acids and their CoA esters. Lipids 26, 266–270.

Gingras, J.R., Harber, V., Field, C.J. and McCargar, L.J. (2000). Metabolic assessment of female chronic dieters with either normal or low resting energy expenditures. Am. J. Clin. Nutr. 71, 1413–1420.

Geiser, F. (1988). Reduction of metabolism during hibernation and daily torpor in mammals and birds: temperature effect or physiological inhibition. J. Comp. Physiol. B 158, 25–37.

Geiser, F. and Ruf, T. (1995). Hibernation versus daily torpor in mammals and birds: physiological variables and classification of torpor patterns. Physiol. Zool. 68, 935–966.

Geissler, C.A., Miller, D.S. and Shah, M. (1987). The daily metabolic rate of the post-obese and the lean. Am. J. Clin. Nutr. 45, 914–920.

Goodman, M.N., Larsen, P.R., Kaplan, M.M., Aoki, T.T., Vernon, R.Y. and Ruderman, N.B. (1980). Starvation in the rat. II. Effect of age and obesity on protein sparing and fuel metabolism. Am. J. Physiol. 239, E277–E286.

Goodman, M.N., Dietrich, R. and Luu, P. (1990). Formation of gluconeogenic precursors in rat skeletal muscle during fasted–refed transition. Am. J. Physiol. 259, E513–E516.

Grail, B.M. and Davies, J.I. (1990). Changes in adipose tissue glucose metabolism associated with the fasting–refeeding cycle. Biochem Soc. Trans. 18, 992.

Hand, S.C. and Somero, G.N. (1983). Phosphofructokinase of the hibernator *Citellus beecheyi*: temperature and pH regulation of activity via influences on the tetramer–dimer equilibrium. Physiol. Zool. 56, 380–388.

Hers, H.G. and Hue, L. (1983). Gluconeogenesis and related aspects of glycolysis. Ann. Rev. Biochem. 52, 617–653.

Heldmaier, G., Klingenspor, M., Werneyer, M., Lampi, B.J., Brooks, S.P.J. and Storey, K.B. (1999). Metabolic adjustments during daily torpor in the Djungarian hamster. Am. J. Physiol. 276, E896–E906.

Hill, J.O., Fried, S.K. and DiGrolamo, M. (1984). Effects of fasting and restricted refeeding on utilization of ingested energy in rats. Am. J. Physiol. 247, R318–R327.

Horton, J.D., Bashmakov, Y., Shimomura, I. and Shimano, H. (1998). Regulation of sterol regulatory element binding proteins in livers of fasted and refed mice. Proc. Natl. Acad. Sci. USA 95, 5987–5992.

Iossa, S., Lionetti. L., Mollica, M.P., Barletta, A. and Liverini, G. (1999). Energy intake and utilization vary during development in rats. J. Nutr. 129, 1593–1596.

Kaloyianni, M. and Freedland, R.A. (1990). Contribution of several amino acids and lactate to gluconeogenesis in hepatocytes isolated from rats fed various diets. J. Nutr. 120, 116–122.

Kather, H., Rivera, M. and Brand, K. (1972). Interrelationship and control of glucose metabolism and lipogenesis in isolated fat-cells. Effect of the amount of glucose uptake on the rates of the pentose phosphate cycle and of fatty acid synthesis. Biochem. J. 128, 1089–1096.

Keesey, R.E. and Corbett, S.W. (1990). Adjustments in daily energy expenditure to caloric restriction and weight loss by adult obese and lean Zucker rats. Int. J.

Obesity 14, 1079–1084.

Knott, R.M., Trayhurn, P. and Hesketh, J.E. (1992). Changes in insulin-receptor mRNA levels in skeletal muscle and brown adipose tissue of weanling rats during fasting and refeeding. Br. J. Nutr. 68, 583–592.

Korkeila, M., Rissanen, A., Kaprio, J., Sorensen, T.I. and Koskenvuo, M. (1999). Weight-loss attempts and risk of major weight gain: a prospective study in Finnish adults. Am. J. Clin. Nutr. 70, 965–975.

Koubi, H., Duchamp, C. and Géloën, A. (1991). Resistance of hepatic glycogen to depletion in obese Zucker rats. Can. J. Pharmacol. 69, 841–845.

Larson, D.E., Ferraro, R.T., Robertson, D.S. and Ravussin, E. (1995). Energy metabolism in weight-stable post-obese individuals. Am. J. Clin. Nutr. 62, 735–739.

Lin, E.C.C. (1977). Glycerol utilization and its regulation in mammals. Ann. Rev. Biochem. 46, 765–795.

Livesey, G. (1993). Comments on the methods used to determine the energy values of carbohydrates: dietary fibre, sugar alcohols and other bulking agents. Int. J. Food Sci. Nutr. 44, 221–241.

Lowell, B.B. and Goodman, M.N. (1987). Protein sparing in skeletal muscle during prolonged starvation. Dependence on lipid fuel availability. Diabetes 36, 14–19.

Matsumura, M., Kuzuya, N., Kawakami, Y. and Yamashita, K. (1982). Effects of fasting, refeeding, and fasting with T_3 administration on Na-K, ATPase in rat skeletal muscle. Metabolism 41, 995–999.

Merkt, J.R. and Taylor, C.R.. (1994). "Metabolic switch" for desert survival. Proc. Natl. Acad. Sci. 91, 12313–12316.

Minassian, C., Daniele, N., Bordet, J.-C., Zitoun, C. and Mithieux, G. (1995). Liver glucose 6-phosphatase activity is inhibited by refeeding in rats. J. Nutr. 125, 2727–2732.

Minassian, C., Montano, S. and Mithieux, G. (1999). Regulatory role of glucose 6-phosphatase in the repletion of liver glycogen during refeeding in fasted rats. Biochim. Biophys. Acta 1452, 172–178.

Moir, A.M. and Zammit, V.A. (1990). Changes in the properties of cytosolic acetyl-CoA carboxylase studied in cold-clamped liver samples from fed, starved and starved-refed rats. Biochem. J. 272, 511–517.

Moir, A.M. and Zammit, V.A. (1993). Rapid switch of hepatic fatty acid metabolism from oxidation to esterification during diurnal feeding of meal-fed rats correlates with changes in the properties of acetyl-CoA carboxylase, but not of carnitine palmitoyltransferase I. Biochem. J. 291, 241–246.

Mrosovsky, N. and Barnes, D.S. (1974). Anorexia, food deprivation and hibernation. Physiol. Behav. 12, 265.

Mrosovsky, N. (1978). Circannual cycles in hibernators. In: Strategies in cold: natural torpidity and thermogenesis (Wang, L.C.H. and Hudson, J.W., Eds.), pp. 21–65. Academic Press, New York, NY.

Muralidhara, D.V. and Desautels, M. (1994). Changes in brown adipose tissue composition during fasting and refeeding of diet-induced obese mice. Am. J. Physiol. 266, R1907–R1915.

Nagashima, Y., Ohno, T., Ogawa, K. and Kuroshima, A. (1995). Effects of fasting and refeeding on some metabolic characteristics of rat brown adipose tissue. Japan. J. Physiol. 45, 645–658.

Newby, F.D., Wilson, L.K., Thacker, S.V. and DiGirolamo, M. (1990). Adipocyte lactate production remains elevated during refeeding after fasting. Am. J. Physiol. 259, E865–E871.

Nørdoy, E.S., Ingebretsen, O.C. and Blix, A.S. (1990). Depressed metabolism in low protein catabolism in fasting gray seal pups. Acta Physiol. Scand. 139, 361–369.

Nutrition Reviews (1969). Changes in lipogenesis during refeeding of fasted mice. Nutr Revs 27, 56–59.

Panemangalore, M., Lee, C.J. and Wilson, K. (1989). Adaptive response of rats with diet-induced adiposity to energy restriction. Ann. Nutr. Metab. 33, 39–48.

Penicaud, L., Kande, J., Le Magnen, J. and Girard, J.R. (1985). Insulin action during fasting and refeeding in rat determined by euglycemic clamp. Am. J. Physiol. 249, E514–E518.

Pilkis, S.J. and Granner, D.K. (1992). Molecular physiology of the regulation of hepatic gluconeogenesis and glycolysis. Ann. Rev. Physiol. 54, 885–909.

Presta, E., Yang, M.U., Segal, K.R. and Björntorp, P. (1984). Energy depot replenishment in rats during refeeding after fasting: effect of exercise. Am. J. Clin. Nutr. 40, 1011–1016.

Pullar, J.D. and Webster, A.J.F. (1977). The energy cost of fat and protein deposition in the rat. Br. J. Nutr. 37, 355–363.

Raclot, T. and Groscolas, R. (1995). Selective mobilization of adipose tissue fatty acids during energy depletion in the rat. J. Lipid. Res. 36, 2164–2173.

Ravussin, E., Lillioja, S., Knowler, W.C., Christin, L., Freymond, D., Abbott, W.G., Boyce, V., Howard, B.V. and Bogardus, C. (1988). Reduced rate of energy expenditure as a risk factor for body-weight gain. N. Engl. J. Med. 318, 467–472.

Ravussin, E. and Bogardus, C. (1989). Relationship of genetics, age, and physical fitness to daily energy expenditure and fuel utilization. Am. J. Clin. Nutr. 49, 968–975.

Ravussin, E. and Bogardus, C. (2000). Energy balance and weight regulation : genetics versus environment. Br. J. Nutr. 83, S17–S20.

Reeves, P.G., Nielsen, F.H. and Fahey, G.C. (1993). AIN-93 purified diets for laboratory rodents: final report of the American Institute of Nutrition ad hoc writing committee on the reformulation of the AIN-76A rodent diet. J. Nutr. 123, 1939–1951.

Rothwell, N.J., Saville, M.E. and Stock, M.J. (1982). Sympathetic and thyroid influences on metabolic rate in fed, fasted, and refed rats. Am. J. Physiol. 243, R339–R346.

Rothwell, N.J., Saville, M.E. and Stock, M.J. (1983). Role of insulin in thermogenic responses to refeeding in 3-day-fasted rats. Am. J. Physiol. 245, E160–E165.

van Remmen, H. and Ward, W.F. (1994). Effect of age on induction of hepatic phosphoenolpyruvate carboxykinase by fasting. Am. J. Physiol. 267, G195–G200.

Rondeel, J.M., Heide, R., de Greef, W.J., van Toor, H, van Haasteren, G.A., Klootwijk, W. and Visser, T.J. (1992). Effect of starvation and subsequent refeeding on thyroid function and release of hypothalamic thyrotropin-releasing hormone. Neuroendocrinology 56, 348–353.

Rognstad, R. (1979). Rate-limiting steps in metabolic pathways. J. Biol. Chem. 254, 1875–1878.

Rothwell, N.J., Saville, M.E. and Stock, M.J. (1983a). Sympathetic and thyroid influences on metabolic rate in fed, fasted, and refed rats. Am. J. Physiol. 243, R339–R346.

Rothwell, N.J., Saville, M.E. and Stock, M.J. (1983b). Role of insulin in thermogenic responses to refeeding in 3-day-refed rats. Am. J. Physiol. 245, E160–E165.

Rothwell, N.J., Saville, M.E. and Stock, M.J. (1984). Brown fat activity in fasted and refed rats. Biosci. Rep. 4, 351–357.

Ruf, T. and Heldmaier, G. (1992). The impact of daily torpor on energy requirements in the Djungarian hamster, *Phodopus sungorus*. Physiol. Zool. 65, 994–1010.

Ruge, T., Bergo, M., Hultin, M., Olivecrona, G. and Olivecrona, T. (2000). Nutritional regulation of binding sites for lipoprotein lipase in rat heart. Am. J. Physiol. 278, E211–E218.

Rumpler, W.V., Seale, J.L., Miles, C.W. and Bodwell, C.E. (1991). Energy-intake restriction and diet-composition effects on energy expenditure in men. Am. J. Clin. Nutr. 53, 430–436.

Samec, S., Seydoux, J. and Dulloo, A.G. (1998). Interorgan signaling between adipose tissue metabolism and skeletal muscle uncoupling protein homologs: is there a role for circulating free fatty acids? Diabetes 47, 1693–1698.

Shetty, P.S. (1984). Adaptive changes in basal metabolic rate and lean body mass in chronic undernutrition. Human Nutr. Clin. Nutr. 38C, 443–451.

Smith, E.L., Austen, B.M., Glumenthal, K.M. and Nyc, J.F. (1975). Glutamate dehydrogenases. In: The Enzymes (Boyer, P.D., Ed.), pp. 294–367. Academic Press, New York, NY.

Snapp, B.D. and Heller, H.C. (1981). Suppression of metabolism during hibernation in ground squirrels (*Citellus lateralis*). Physiol. Zool. 54, 297–307.

Storey, K.B. (1987). Regulation of liver metabolism by enzyme phosphorylation during mammalian hibernation. J. Biol. Chem. 262, 1670–1673.

Sugden, M.C., Watts, D.I., Palmer, T.N. and Myles, D.D. (1983). Direction of carbon flux in starvation and after refeeding: *in vitro* and *in vivo* effects of 3-mercaptopicolinate. Biochem. Int. 7, 329–337.

Sul, H.S. and Wang, D., (1998). Nutritional and hormonal regulation of enzymes in fat synthesis: studies of fatty acid synthase and mitochondrial glycerol-3-phosphate acyltransferase gene transcription. Annu. Rev. Nutr. 18, 331–351.

Svanberg, E., Powell-Braxton, L., Ohlsson, C., Zachrisson, H. and Lundholm, K. (1998). The role of the growth hormone/insulin-like growth factor I axis in stimulation of protein synthesis in skeletal muscles following oral refeeding. Endocrinology 139, 4906–4910.

Terrettaz, J., Assimacopoulos-Jeannet, F. and Jeanrenaud, B. (1986). Inhibition of hepatic glucose production by insulin in vivo in rats: contribution of glycolysis. Am. J. Physiol. 250, E346–E351.

Toyoda, Y., Miwa, I., Kamiya, M., Ogiso, S., Nonogaki, T., Aoki, S. and Okuda, J. (1995). Tissue and subcellular distribution of glucokinase in rat liver and their changes during fasting–refeeding. Hisochem. Cell Biol. 103, 31–38.

Treuth, M., Butte, N. and Wong, W. (2000). Effects of familial predisposition to obesity on energy expenditure in multiethnic prepubertal girls. Am. J. Clin. Nutr. 71, 893–900.

van de Werve, G. and Jeanrenaud, B. (1987). The onset of liver glycogen synthesis in fasted–refed lean and genetically obese (*fa/fa*) rats. Diabetologia 30, 169–174.

van Itallie, T.B. and Yang, M.-U. (1984). Cardiac dysfunction in obese dieters : a potentially lethal complication of rapid, massive weight loss. Amer. J. Clin. Nutr. 39, 695–702.

Warner, B.W., Hummel, R.P., Hasselgren, P.O., James, J.H. and Fischer, J.E. (1989). Inhibited amino acid uptake in skeletal muscle during starvation. J. Parenter. Enteral. Nutr. 13, 344–348.

Walks, D., Lavau, M., Presta, E., Yang, M.U. and Bjorntorp, P. (1983). Refeeding after fasting in the rat : effects of dietary-induced obesity on energy balance regulation. Am. J. Clin. Nutr. 37, 387–395.

Winder, W.W. and Hardie, D.G. (1999). AMP-activated protein kinase, a metabolic master switch: possible roles in type 2 diabetes. AM. J. Physiol. 277, E1–E10.

Wyatt, H.R., Grunwald, G.K., Seagle, H.M., Klem, M.L., McGuire, M.T., Wing, R.R. and Hill, J.O. (1999). Resting energy expenditure in reduced-obese subjects in the National Weight Control Registry. Am. J. Clin. Nutr. 69, 1189–1193.

Yang, M.U., Presta, E. and Bjorntorp, P. (1990). Refeeding after fasting in rats : effects of duration of starvation and refeeding on food efficiency in diet-induced obesity. Am. J. Clin. Nutr. 51, 970–978.

Yanovski, J.A., Diament, A.L., Sovik, K.N., Nguyen, T.T., Li, H., Sebring, N.G. and Warden, C.H. (2000). Associations between uncoupling protein 2, body composition, and resting energy expenditure in lean and obese African American, white and Asian children. Am. J. Clin. Nutr. 71, 1405–1412.

Yoshizawa, F., Kimball, S.R., Vary, T.C. and Jefferson, L.S. (1998). Effect of dietary protein on translation initiation in rat skeletal muscle and liver. Am. J. Physiol. 275, E814–E820.

Young, E.A., Cantu, T.L. and Harris, M.M. (1989). Gastrointestinal and cardiac response to refeeding after low-calorie semistarvation. Am. J. Clin. Nutr. 50, 922–929.

Young, J.B. and Landsberg, L. (1997). Suppression of sympathetic nervous system during fasting. Obes. Res. 5, 646–649.

Young, V.R. (1991). Nutrient interactions with reference to amino acid oxidation and protein metabolism in non-ruminants: particular emphasis on protein–energy relations in man. J. Nutr. Sci. 30, 239–267.

Zauner, C., Schneeweiss, B., Kranz, A., Madl, C., Ratheiser, K., Kramer, L, Roth, E., Schneider, B. and Lenz, K. (2000). Resting energy expenditure in short-term starvation is increased as a result of an increase in serum norepinephrine. Am. J. Clin. Nutr. 71, 1511–1515.

Zwiauer, K.F.M., Mueller, T. and Widhalm, K. (1992). Resting metabolic rate in obese children before, during and after weight loss. Int. J. Obesity 16, 11–16.

Protein Adaptations and Signal Transduction. Edited by K.B. Storey and J.M. Storey

CHAPTER 10

Nutritional Regulation of Hepatic Gene Expression

Howard C. Towle
Department of Biochemistry, Molecular Biology and Biophysics, University of Minnesota, 6-155 Jackson Hall, 321 Church St. SE, Minneapolis, MN 55455, U.S.A.

1. Introduction—energy homeostasis

Maintenance of energy homeostasis is critical to the survival of all animals. For most species, a considerable portion of each day is spent in the processes of procuring, digesting, absorbing and distributing key energy nutrients to various cells and tissues of the organism. Frequent intake of metabolizable fuel substrates will allow an organism to meet ongoing requirements for energy utilization. However, environmental conditions will invariably occur in which energy sources are inadequate to meet energy needs. In addition, during times of prolonged inactivity due to sleep or hibernation or times of prolonged physical activity, such as bird migration, intake of energy ceases. Since energy is needed continually to maintain life processes, all organisms have evolved mechanisms for storing energy internally that can be drawn upon when demands exceed available sources. For mammals, these energy stores consist of triglycerides, glycogen and proteins with the former representing nearly 80% of the total available energy stores in humans.

Given the critical need for maintaining energy homeostasis, all organisms have evolved complex mechanisms to regulate the processes of energy utilization and storage. These regulatory circuits allow organisms to draw on energy stores during times of energy need due to high levels of energy expenditure or conditions of insufficient caloric uptake. Conversely, when presented with conditions in which energy consumption exceeds ongoing energy utilization, pathways favoring storage of excess calories are invoked and existing energy stores are preserved. These regulatory processes are highly complex and diverse. They involve, for example, neuronal signaling circuits that control appetite and satiety. They also involve mechanisms, currently only poorly understood, that effect efficiency of energy utilization. Finally, processes affecting the metabolic pathways of fuel oxidation and energy storage are highly regulated. Clearly, attempting to cover all of these regulatory processes adequately in this chapter is not practical. Instead, this chapter will focus on one important aspect of this regulatory circuitry: that involving changes in gene expression in response to altered energy uptake that occurs in the liver of mammals. These mechanisms generally provide for longer-term control of energy metabolism that functions in periods of hours rather than minutes. As such, these mechanisms can be viewed as adaptive responses that allow the organism to cope with changing environmental conditions.

2. Role of the liver in energy homeostasis

In mammals the liver plays a central role in the processes of interconversion, distribution, and storage of energy metabolites. Because of its privileged position in portal circulation, the liver is the first major organ to gain access to most incoming carbohydrate and amino acid nutrients from intestinal absorption. It is a major site of carbohydrate storage in the form of glycogen that can be broken down in times of need to derive glucose for tissues highly dependent on this energy metabolite, such as the brain. It is an important site for distributing fatty acids obtained from the diet or adipose stores to other tissues in the form of lipoproteins. In times of fasting, the liver is able to synthesize and export glucose from amino acid precursors derived from protein degradation through the process of gluconeogenesis. In times of severe energy shortage, the liver provides alternative sources of energy through the formation of ketone bodies. In contrast, when excess energy in the form of carbohydrates is consumed, the liver can convert these carbohydrate calories to the preferred energy storage form of triglycerides for transport and storage in adipose. This process of converting carbohydrate to fatty acids and triglycerides is termed lipogenesis.

The control of metabolic pathways in the liver involves many hormonal signals that function through specific receptors to alter hepatic function. For example, the counterregulatory hormones, insulin and glucagon, are critical for controlling glucose homeostasis. When blood glucose levels are low, such as during a period of fasting, glucagon secreted from the pancreatic alpha cell serves as a signal to turn on processes enhancing glucose production by the liver. These include increases in glycogenolysis and gluconeogenesis. Concomitantly, processes that utilize glucose in the liver, such as glycogen synthesis and glycolysis, are inhibited. In contrast, when blood glucose levels are elevated, insulin is secreted from the pancreatic beta cell. In the liver, insulin acts to decrease plasma glucose levels by stimulating glucose utilization for glycogen synthesis and glycolysis. Insulin also stimulates glucose uptake in muscle and adipose. Both glucagon and insulin function in part by altering the transcription of specific genes in the liver and other target tissues. For example, glucagon, via its intracellular second messenger cAMP, stimulates expression of the gene encoding phosphoenolpyruvate carboxykinase, the rate-limiting step in gluconeogenesis (Hanson and Reshef, 1997). Insulin activates glucokinase gene expression to promote glucose utilization by stimulating its phosphorylation to glucose-6-phosphate (O'Brien and Granner, 1996). In this manner, these critical hormones of glucose homeostasis can help to maintain a steady supply of glucose to peripheral tissues of the body that are highly dependent on this energy source.

While the essential role of hormones in controlling processes involved in energy homeostasis in mammals have been recognized for many years, the role of another important regulatory input has only been more recently recognized. These signaling molecules are the energy nutrients themselves or metabolites derived from these nutrients. In addition to providing substrates for oxidation in deriving energy, fuel substrates can also act as important intracellular signals to control cell function. Again, this control is exerted in part by changing the expression patterns of specific genes involved in metabolism. It is this topic on which the remainder of this chapter will focus.

3. Fatty acid oxidation and the peroxisome proliferator-activated receptor

3.1. The hepatic response to fasting

Mammals have evolved a metabolic response that allows them to survive long periods of energy deprivation. A prominent feature of this metabolic response involves a switch from carbohydrates and fatty acids for energy production in the fed state to a reliance on fatty acids and ketones in the fasted state. Most of the interconversions in energy substrates occur in the liver. In fasting conditions, high glucagon, glucocorticoids and epinephrine promote the hydrolysis of triacylglycerol stores from adipose tissue, resulting in increased free fatty acids in plasma. Free fatty acids are taken up by the liver where they can undergo several fates. First, free fatty acids can be oxidized in the mitochondria via β-oxidation to provide energy for supporting liver function. Second, they can be reesterified to triacylglycerols and secreted as VLDL particles to provide energy for other tissues capable of utilizing fatty acids. Finally, fatty acids can serve as precursors for the synthesis of ketone bodies that provide an alternate energy source for tissue such as the brain. These processes are all enhanced in the liver under fasting conditions by the activation and increased synthesis of rate-limiting enzymes in these pathways. Over recent years, one of the major players in the process of enzyme induction has been identified as a nuclear receptor that functions by directly binding to fatty acids and their derivatives.

The peroxisome proliferator-activated receptors (PPARs) are ligand-inducible transcription factors that are members of the nuclear hormone receptor superfamily (Desvergne and Wahli, 1999). Like all members of the family, PPARs contain a DNA binding domain consisting of two zinc-finger motifs and a carboxyl terminal region capable of binding to a hydrophobic ligand. PPARs function by binding to a specific DNA sequence as a heterodimer with the partner, retinoid X receptor. This PPAR response element is located within the transcriptional promoter or enhancer of genes regulated by the receptor. Binding to specific ligand generally converts the DNA-bound receptor complex from an inactive or repressing state to one capable of stimulating transcription of the proximal gene. PPARs were first identified as receptors for a diverse set of amphipathic carboxylates that include the widely used fibrate class of hypolipidemic drugs (Isseman and Green, 1990). When administered to rodents, these compounds trigger a characteristic response that includes hepatomegaly, increased number and size of peroxisomes and the induction of many enzymes induced in fatty acid oxidation. Three isoforms of the PPARs have been cloned: PPARα, which is expressed in metabolically active tissues such as liver, heart and kidney; PPARγ, which is expressed predominantly in white and brown adipose tissue; and PPARδ, which is expressed in virtually every tissue. All three receptor isotypes appear to play major and distinct roles in lipid homeostasis, although the precise physiological function of PPARδ remains to be defined.

3.2. Role of PPARα in the hepatic response to fasting

Several lines of evidence have implicated PPARα as playing a major role in regulating the process of fatty acid oxidation. An examination of 'target' genes that are transcriptionally induced by the PPARα is illuminating (Fig. 10.1). The first identified target genes were several enzymes implicated in the peroxisomal β-oxidation pathway (Dreyer et al., 1992). Subsequently, PPARα has been found to be involved in induction of many enzymes of

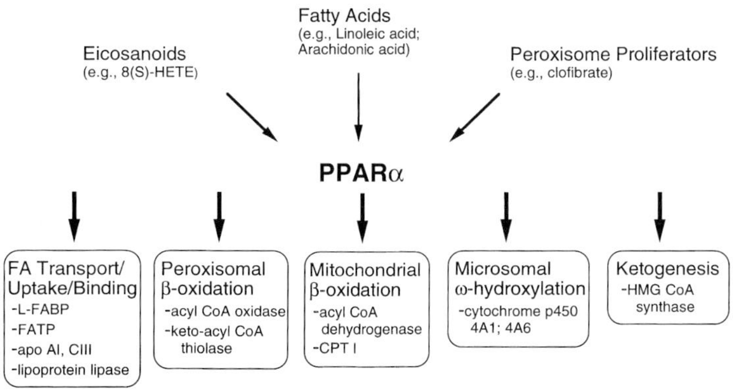

Fig. 10.1. PPARα functions to regulate various aspects of hepatic lipid metabolism by stimulating transcription of specific genes. The top line shows natural and synthetic ligands that can activate the transcriptional activity of PPARα. Various aspects of lipid metabolism that are stimulated by PPARα are shown in the boxes. For each process stimulated, examples of genes that are transcriptionally induced are indicated. (See Desvergne, 1999, 253, for references). HETE, hydroxyeicosatetraenoic acid; L-FABP, liver fatty acid binding protein; FATP, fatty acid transporter protein; apo AI, CIII, apolipoprotein AI, CIII; CPT, carnitine palmitoyl transferase; HMG-CoA, 3-hydroxy-3-methylglutaryl-CoA.

other key lipid metabolic pathways, including mitochondrial fatty acid oxidation, fatty acid transport and uptake, lipoprotein metabolism, ketogenesis and microsomal fatty acid ω-hydroxylation. All of these processes are consistent with a role in responding to increased needs for fatty acid utilization during fasting. Consistent with this idea, PPARα expression in liver is also known to be upregulated in conditions such as fasting or stress, which require energy mobilization (Lemberger et al., 1996).

A search for the natural ligand for the PPARs has revealed a remarkably diverse set of potential hydrophobic ligands that can interact with the various PPAR isoforms. Using transfection assays, PPARs were shown to be activated by micromolar concentrations of a large group of fatty acids. However, the potential metabolic activity of these fatty acids and their relatively low affinity for PPAR raised question as to whether they were indeed direct ligands in vivo. The availability of high affinity, synthetic radioligands for the PPARs provided a means of addressing this question. Using competition assays, a variety of fatty acids with chain lengths between 14 and 20 carbons were able to compete for radioligand binding to PPARα and PPARγ (Kliewer et al., 1997). In general, unsaturated fatty acids were more potent ligands for the PPARs than saturated fatty acids. Oleic, linoleic, arachidonic, and linolenic acids bound at concentrations in the 5 to 20 μM range that are found in nonesterified fatty acids in human serum. Thus, unsaturated fatty acids likely serve as natural ligands for the PPARα, coupling fatty acid oxidation and metabolism with expression of the genes whose products regulate these processes. In addition to fatty acids, certain eicosanoids function as PPAR activators. For example, leukotriene B4 and 8(S)-hydroxy-eicosatetraenoic acid were identified as relatively high affinity ligands for PPARα and may provide an alternative route of activation of PPARs in certain tissues (Michalik and Wahli, 1999).

The promiscuity of the PPARs in binding to a variety of natural and synthetic ligands is surprising in view of the high specificity of other members of the nuclear hormone receptor superfamily. The basis of this promiscuity has been revealed from structural studies of the ligand binding domain (Xu et al., 1999a). Although the general structure of the ligand binding domain is similar to other known nuclear receptors, the volume of the ligand-binding cavity is nearly three times that of other receptors and is only partially occupied by specific ligands. This larger ligand-binding pocket may have evolved to allow PPARs to interact with multiple ligands and function as general fatty acid sensors.

The key role of PPARα in the metabolic response to fasting has been recently confirmed by the analysis of mice deleted for the PPARα gene (Lee et al., 1995). PPARα-null mice that are subjected to 24 h fasting display a massive accumulation of lipid in the liver and severe hypoglycemia, hypoketonemia, hypothermia, and elevated free fatty acid levels (Kersten et al., 1999; Leone et al., 1999). This phenotype is consistent with a dramatic impairment of fatty acid oxidation. In fact, the phenotype of fasted PPARα-null mice resembles that of humans with genetic defects in mitochondrial fatty acid oxidation enzymes. These data, together with previous observations on PPARα, firmly establish a critical role for this transcription factor in responding to fasting and the maintenance of energy homeostasis.

3.2. Role of PPARγ in adipogenesis

The PPARγ isotype also clearly plays a major, but quite distinct, role in regulating lipid homeostasis (Auwerx, 1999; Spiegelman, 1998). PPARγ is expressed in highest levels in white and brown adipose tissues. One of the functions of PPARγ is in the process of adipogenesis. In models of adipocyte differentiation, the activation of PPARγ expression is an early step in the process of differentiation. In fact, retroviral expression of PPARγ in many fibroblast cell lines, followed by the addition of a PPAR ligand, gave abundant differentiation that included lipid accumulation and characteristic changes in cell morphology (Tontonoz et al., 1994). PPARγ in turn acts in conjunction with other transcription factors to activate expression of genes responsible for the adipocyte phenotype.

Many of these genes are involved in fatty acid conversion to triglycerides and hence PPARγ effects appear to be opposite those of PPARα in promoting deposition rather than oxidation of fatty acids. Consistent with this hypothesis, PPARγ has been found to reduce expression of the adipose signaling factor, leptin (Hollenberg et al., 1997). Leptin normally functions to signal from adipose tissue to the brain that adipose stores are adequate, leading to reduced appetite and increased energy expenditure. By inhibiting leptin production, PPARγ would promote adipose storage of triglycerides and energy deposition.

Like PPARα, PPARγ can bind to a diverse array of fatty acids and fatty acid derivatives, including unsaturated fatty acids and specific eicosanoids. One of the more interesting ligands, however, is a synthetic one—the thiazolidinediones, which are used pharmacologically for treatment of type II diabetes. Thiazolidinediones, such as troglitazone, enhance the actions of insulin to reduce plasma glucose levels and markedly lower circulating levels of fatty acids. How the actions of a receptor that promotes adipogenesis leads to reduced plasma glucose is not yet entirely clear. However, these data do support an important overall role for PPARγ in regulating glucose and lipid homeostasis.

4. Lipogenesis and the induction of lipogenic enzyme genes

During fasting conditions, mammals depend largely on triglyceride stores to provide the energy necessary for maintaining life functions. Consequently, it is critical that triglyceride stores are deposited during times when energy intake exceeds immediate energy needs. These stores are largely maintained by adipose tissues, particularly white adipose. Fatty acids derived from dietary triglycerides can be transported to adipose tissue via chylomicron particles. In the adipocyte, fatty acids released from the chylomicron are reesterified with glycerol to form triglycerides that are stored in lipid droplets in the cell. Dietary carbohydrates in excess of those needed to meet immediate energy needs and replenish glycogen stores can also be converted to triglycerides. This process is termed lipogenesis and occurs in many mammalian species predominantly in the liver. Triglycerides produced in the liver are packaged into VLDL particles for secretion and utilization by peripheral tissues. In particular, adipose tissue can again utilize fatty acids derived from VLDL particles to synthesis triglycerides.

Similar to the systemic response to fasting that allows the enhanced mobilization and utilization of triglyeride stores, a response occurs in mammals to allow them to more effectively store excess energy taken in during a meal. This response is mediated via both hormonal signals, principally insulin, and metabolic signals. Feeding of a high carbohydrate diet to rodents has been used as a model of this response. Following a high carbohydrate diet, the induction of a number of key lipogenic enzymes occurs in the liver (Hillgartner et al., 1995; Towle et al., 1997). These include key steps in converting glucose and other carbohydrate substrates into triglycerides (Fig. 10.2). Enzymes induced include those directly responsible for fatty acid synthesis: acetyl CoA carboxylase and fatty acid synthase. Several enzymes of glycolysis, such as phosphofructo-1-kinase and pyruvate kinase, that provide acetyl CoA substrate for the process, are also induced. In addition, enzymes that generate NADPH needed to drive lipogenesis, such as malic enzyme, are elevated. Finally, enzymes involved with triglyeride synthesis and packaging, such as acyl CoA synthetase, are induced.

While the set of enzymes induced by high carbohydrate diet have been worked out, far less is known about the pathways involved in induction of these gene products. One major question involves the respective roles of insulin and metabolic products, such as glucose, in the process of induction. Both of these factors increase following a high carbohydrate meal and provide potential effectors for altering liver function. A second major question is the nature of the transcriptional factors involved in mediating the response. Finally, the mechanism of activation of these transcription factors must be worked out to provide an understanding of this pathway. To date, two transcription factors have been implicated in the process. These two factors

Liver Gene Products Induced by High Carbohydrate Diet

Glucose Uptake/Glycolysis:
- —GLUT2 glucose transporter
- —Glucokinase
- —Phosphofructo-1-kinase
- —Aldolase
- —Pyruvate kinase

Fatty acid synthesis:
- —ATP citrate lyase
- —Acetyl CoA carboxylase
- —Fatty acid synthase

NADPH Generation:
- —Malic enzyme
- —Glucose-6-phosphate dehydrogenase
- —6-Phosphogluconate dehydrogenase

Triglyceride synthesis/maturation:
- —Glycerol-3-P dehydrogenase
- —Stearoyl CoA desaturase
- —Acyl CoA synthetase

Other:
- —S_{14}
- —Glucose-6-phosphatase
- —Microsomal triglycride tranfer protein
- —6-Phosphofructo-2-kinase/ fructose-2,6-bisphosphatase

Fig. 10.2. Liver genes that are induced by feeding of high carbohydrate diet to rats or mice. Each of the gene products indicated has been shown to be induced at the level of increased mRNA following the feeding of a high carbohydrate, fat-free diet to rats or mice. The gene products are grouped according to their function in the overall process of lipogenesis. (See Towle, 1997, 16 for references).

may function to mediate insulin and glucose signals separately, and work coordinately to control the overall process of lipogenic enzyme induction.

5. Lipogenesis and the sterol regulatory element binding protein

5.1. SREBP in the regulation of cholesterol homeostasis

The first transcription factors that were implicated in the regulation of lipogenesis were the sterol regulatory element binding proteins (SREBPs). SREBPs were identified by workers studying the expression of genes encoding cholesterol metabolic enzymes (Brown and Goldstein, 1997). Although cholesterol is not an energy nutrient, it does function as an essential membrane component. Levels of cholesterol must be tightly regulated to provide sufficient cholesterol for membrane biosynthesis, but to limit its excess accumulation, which can be toxic to cells. This regulation is accomplished in part by controlling the production of several key enzymes and proteins. When cholesterol levels in a cell are low, production of enzymes of cholesterol biosynthesis, such as HMG-CoA reductase and HMG-CoA synthase, are elevated. Similarly, increased production of the LDL receptor allows increased uptake of cholesterol from the plasma pools. These changes in protein production occur through the transcriptional activation of the corresponding genes. Conversely, when cholesterol levels in cells are adequate or elevated, the transcription of this same set of genes is reduced. In this manner, cholesterol acts to regulate its own synthesis and uptake in the cell.

The identification of SREBP as the major transcription factor involved in regulation of genes of cholesterol metabolism began with the use of the transfection assay to identify the critical regulatory sequences of these genes. In this approach, the 5′-flanking region of the gene of interest, such as the LDL receptor gene, is cloned upstream to a reporter gene in a plasmid vector. The reporter gene encodes an easily assayed marker enzyme not normally expressed in mammalian cells, such as chloramphenicol acetyl transferase or luciferase. The plasmid DNA is then transfected into cultured cells where the promoter sequences will be recognized by the cellular transcriptional machinery and used to synthesize reporter mRNA and protein. In this manner, reporter enzyme levels will provide a readout of promoter activity under various conditions of cell incubation. By growing cells in the presence or absence of cholesterol, gene regulatory sequences involved in the transcriptional response to cholesterol could be identified and mapped.

Using transfection analyses, a critical regulatory sequence in the genes of cholesterol metabolism was identified and termed the sterol regulatory element (SRE). This short 10 base pair sequence was conserved in several of the genes involved in cholesterol metabolism and postulated to serve as a binding site for a transcription factor that coordinately regulates these genes (Goldstein and Brown, 1990). Purification of a nuclear protein that specifically bound to the SRE led to the cloning and identification of the first SREBP isotype (Yokoyama et al., 1993). SREBPs are members of a family of DNA binding proteins known as the basic/helix–loop–helix/leucine zipper family. This large family of transcription factors function as dimers and regulate a diverse set of functions in development and tissue-specific expression of genes (Kadesch, 1993). Three forms of SREBP have been identified. SREBP-1a and 1c are derived from a common gene through the use of alternative transcription start sites, whereas SREBP-2 is the product of a distinct gene. SREBP-1a was independently isolated as a factor that binds to the fatty acid synthase gene and promotes adipocyte differentiation (Tontonoz et al., 1993).

Examination of the mechanism of activation of the SREBPs by cholesterol led to discovery of a novel pathway in the field of transcription activation. When the SREBP genes were sequenced, they were found to encode for products that were much larger than the purified nuclear SREBP transcription factor. These precursor forms were in the range of 125 kilodaltons, whereas the nuclear forms were only 66 to 68 kilodaltons and originated from the amino terminal segment of the larger form. Using antibodies, the 125 kilodalton form of SREBP was found to be located in the endoplasmic reticulum, anchored by two closely spaced membrane spanning segments such that the amino terminal and carboxyl terminal portions of the molecule were on the cytosolic face of the membrane (Wang et al., 1994). When cholesterol levels in the cell are sufficient, SREBP occurs predominantly as the precursor form. However, when cholesterol levels fall, a proteolytic cleavage of the precursor occurs to release the amino terminal portion. This segment then translocates to the nucleus where in binds to the SRE and turns on the expression of genes of cholesterol metabolism. The details of the proteolytic activation of the SREBP precursor are not yet known, but the large carboxyl terminal domain is clearly involved in the cholesterol-sensing mechanism (Sakai et al., 1997). Perhaps a change in membrane properties induced by altered levels of cholesterol or a cholesterol derivative are responsible for activating the process.

5.2. SREBP in regulation of lipogenesis

More recently, it has become apparent that SREBPs are more widely involved in controlling lipid metabolism and, in particular, are involved in activation of lipogenesis under conditions of energy excess (Fig. 10.3). This hypothesis initially arose from observations made on transgenic mice that produced the nuclear (constitutively active) form of SREBP-1a specifically in the liver (Shimano et al., 1996). The phenotype of these animals was an extremely enlarged liver that was engorged with both cholesterol and triglycerides. Examination of mRNA levels showed that overexpression of the nuclear SREBP-1a led to the increased accumulation of mRNAs for a number of enzymes involved in the process of lipogenesis, as well as cholesterol metabolism. These included key enzymes such as acetyl CoA carboxylase and fatty acid synthase. The rates of lipogenesis in these animals were greatly increased and the massive accumulation of triglycerides and cholesterol apparently blocked the normal pathways for lipoprotein secretion and led to lipid accumulation in the liver. Consistent with this finding was the observation that several of the lipogenic enzyme genes contain DNA regulatory elements similar to the SRE. In particular, fatty acid synthase and acetyl CoA carboxylase were found to bind to SREBP and to be inducible by sterol activation of SREBP (Lopez et al., 1996; Magana and Osborne, 1996).

In cultured cells, SREBP-1a and SREBP-2 are the principal isotypes expressed and cholesterol

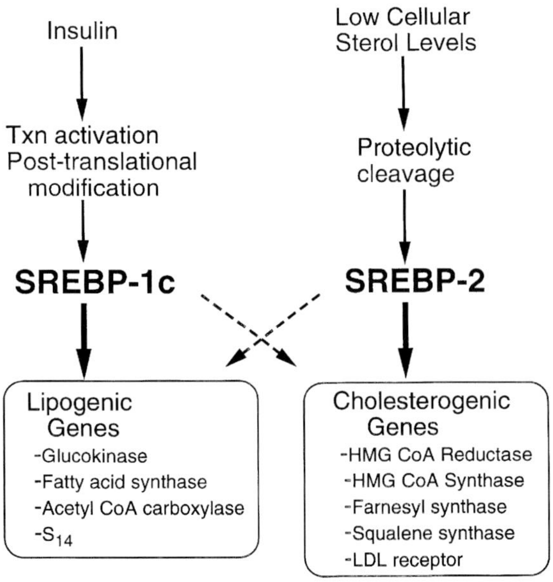

Fig. 10.3. Distinct roles of SREBP isotypes in regulating hepatic lipid metabolism by stimulating transcription of specific genes. SREBP-2 is activated in response to low sterol levels which activates its proteolytic cleavage from an endoplasmic reticulum-bound precursor to release the nuclear transcription factor. Genes regulated by SREBP-2 encode enzymes that are involved in aspects of cholesterol biosynthesis and uptake. SREBP-1c is transcriptionally activated in response to insulin and may also be post-translationally modified by phosphorylation. Genes regulated by SREBP-1c encode enzymes primarily involved in lipogenesis. Since the two SREBP isotypes share a similar DNA recognition site, some cross-regulation likely occurs, as indicated by dotted arrows.

limitation activates the cleavage of both these forms. However, in animals it was found that SREBP-1c is expressed at much higher levels than SREBP-1a in liver and adipose (Shimomura et al., 1997). SREBP-2 cleavage from the endoplasmic reticulum was stimulated by cholesterol depletion, but not SREBP-1c cleavage (Sheng et al., 1995). This led to the proposal that SREBP-2 was primarily associated with regulating genes involved in cholesterol metabolism, whereas SREBP-1c might be involved in lipogenic responses. SREBP-1c appears to be a less potent transcriptional activator than SREBP-1a due to a truncated amino–terminal activation domain (Shimano et al., 1997). When transgenic mice that overexpress the nuclear form of SREBP-1c were studied, increased lipogenesis and expression of lipogenic enzyme genes were observed, but to a much lower extent than with the SREBP-1a transgenics.

Consistent with the notion that SREBP-1c is involved in regulating lipogenesis, the expression of SREBP-1c is increased in response to insulin (Foretz et al., 1999a; Kim et al., 1998). While the pathway between the insulin receptor and SREBP-1c gene induction is unknown, this observation does provide a pathway for mediating insulin effects on the process of lipogenesis. It is also possible that insulin may act posttranslationally to increase the potency of SREBP-1c as a transcriptional activator. SREBP-1a has been shown to be a phosphoprotein and its phosphorylation is increased in response to insulin treatment in cultured cells (Kotzka et al., 1998). Whether a similar phenomenon occurs for SREBP-1c in vivo is not yet known, but certainly would not be unexpected. On the other hand, there is no evidence to suggest a change in the rate of proteolytic cleavage in the activation of SREBP-1c in response to carbohydrate feeding.

Taken together, these observations support a model in which SREBPs serve to coordinately regulate the processes of cholesterol and fatty acid biosynthesis. For fatty acid biosynthesis, the anabolic hormone insulin may be the key regulator that activates this pathway physiologically. Elucidation of the pathways leading to SREBP activation by insulin should be of interest in further understanding this signaling pathway.

6. Lipogenesis and the carbohydrate responsive transcription factor

6.1. Glucose metabolism generates an intracellular signal for inducing lipogenic enzyme genes

Evidence for a pathway of carbohydrate activation independent of insulin and SREBP arose from studies performed in primary cultured hepatocytes. Hepatocytes that are cultured in the presence of insulin and high glucose levels induce the same battery of lipogenic enzymes as observed in whole animals fed high carbohydrate diet (e.g. Mariash et al., 1981). On the other hand, if hepatocytes are cultured in low glucose conditions (5.5 mM), which approximate fasting blood glucose levels, the addition of insulin to the media does not by itself activate expression of most lipogenic enzyme genes. The exception is the glucokinase gene, which encodes the high Km hexokinase that is critical for initiating the glycolytic pathway in hepatocytes. Glucokinase gene expression responds directly to changes in insulin levels (Iynedjian et al., 1989) and recent data suggest that this regulation is mediated by SREBP (Foretz et al., 1999b). At present, the regulatory site responsible for insulin action on the glucokinase gene has not been mapped, so that this hypothesis awaits confirmation.

In contrast to glucokinase, all other lipogenic enzyme genes studied to date require elevated levels of glucose in addition to insulin for induction. In the face of a fixed insulin level, increasing glucose levels induce lipogenic gene expression with a $K_{0.5}$ of approximately 8 mM (Prip-Buus et al., 1995), comparable to that for the glucokinase reaction. In addition to glucose, other carbohydrate substrates that can be metabolized through the glycolytic pathway are capable of inducing gene expression (Mariash and Oppenheimer, 1983). However, non-metabolizable analogs of glucose do not stimulate this pathway. Consequently, it has been proposed that glucose metabolism is necessary to generate a cellular signal for transcriptional activation of lipogenic enzyme genes. Insulin plays a permissive role in this process due to the critical role for glucokinase expression to achieve high rates of glycolysis in the hepatocyte. However, the need for insulin can be circumvented to some degree by constitutively expressing glucokinase in cultured cells (Doiron et al., 1994).

The nature of the intracellular signal that is responsible for mediating the glucose response in primary hepatocytes is unknown. Several intermediates in glucose metabolism have been suggested as potential mediators. These include glucose-6-phosphate and ribulose-5-phosphate (Doiron et al., 1996; Girard et al., 1997). The data supporting these metabolites are at present largely correlative. For example, glucose-6-phosphate levels found in cells following treatment with various inducing and non-inducing metabolic fuels correlates with rates of pyruvate kinase gene expression. Furthermore, glucose-6-phosphate levels increase at an early time following glucose administration, preceding the earliest changes in transcriptional induction. However, conclusive evidence for the role of any specific intermediate in the process will require identification of the cellular machinery involved in transducing this signal into changes in gene transcription.

6.2. The carbohydrate response element

Evidence for an independent pathway of glucose regulation has also arisen from studies on the regulatory sites involved in mediating this response. Using transfection assays in hepatocytes, the carbohydrate (or glucose) response element (ChoRE) has been mapped in two genes known to be induced in response to glucose. One of these genes is L-type pyruvate kinase (L-PK). This enzyme catalyzes the final reaction of glycolysis, one of three essentially irreversible steps in the process. The other gene product is termed 'S_{14}'. This gene product responds to feeding a high carbohydrate meal with extremely fast kinetics—changes in mRNA levels can be observed by 30 min (Mariash et al., 1986). Although its physiological function is not known, S_{14} has been postulated to play a role in lipid metabolism. The S_{14} gene is expressed at high levels in liver and adipose in adults and encodes a 17 kilodalton nuclear polypeptide (Jump and Oppenheimer, 1985). Blocking the induction of S_{14} mRNA via antisense oligonucleotides inhibits the increased rate of lipogenesis normally observed in cultured hepatocytes upon addition of high glucose to the media (Kinlaw et al., 1995).

Mapping of the carbohydrate response elements of either the L-PK or S_{14} gene revealed a common regulatory site that was responsible for mediating the effects of glucose (Bergot et al., 1992; Liu et al., 1993; Shih and Towle, 1992). Mutation of this site blocked the normal glucose response of either promoter (Bergot et al., 1992; Kaytor et al., 1997). Furthermore, linking oligonucleotides containing this regulatory site to heterologous promoters that are not normally regulated by glucose can confer a glucose-responsive activity to these promoters (Shih et al., 1995). Thus, the ChoRE is both necessary and sufficient to support the glucose response. When the regulatory sites of rat L-PK and rat and mouse S_{14} genes were compared to each other, a distinct sequence similarity was noted (Koo and Towle, 2000). This similarity involved the presence of two motifs related to the consensus sequence (5′)CACG. The arrangement of these motifs was either as an inverted repeat with nine base pairs separating the motifs or as a direct repeat with seven base pair separation (Fig. 10.4). The spacing was found to be critical for the ability to respond to glucose (Shih et al., 1995). Thus, the DNA-binding factor that recognizes this site appears to make two contacts with the DNA helix. Based on the consensus sequence, it is reasonable to speculate that the factor involved is a member of a large family of "E box" binding proteins. These proteins all share a common DNA-binding motif composed of a basic region that contacts the DNA helix and a protein–protein dimerization domain that positions the basic region in the proper conformation to contact the DNA (Kadesch, 1993). This family commonly functions as dimers, either homodimers or heterodimers between family members. In the case of the ChoRE, each CACG motif would presumably contact one monomer subunit of the dimer.

rat S14 (-1442/-1420)	tctCACGTGgtggcccTGTGctt
rat L-PK (-171/-142)	gcgCACGgggcactccCGTGgtt
mouse S14 (-1450/-1425)	gtccatggCACGctggagtCAGCcct
rat ACC I (-95/-119)	tgcccCACGacgttttCACAtggaca
rat FAS (-7194/-7218)	gtgaCACGcctgtggCACAtgcagg

Fig. 10.4. ChoRE sequences from various lipogenic enzyme genes. Each of the oligonucleotide sequences has been shown to form the ChoRF complex on electrophoretic mobility shift assay. Numbers indicate the location of the sequences relative to the transcriptional start site. The arrows indicate the consensus motif (5′)CACG that is present in two copies in each oligonucleotide.

6.3. The carbohydrate responsive transcription factor

Since many basic/helix–loop–helix factors exist in any given cell, it is important to determine which of the family members is actually involved in regulation in response to glucose. To approach this problem, we carried out an in vitro mutagenic approach. Starting with the rat or mouse S_{14} ChoRE oligonucleotides, point mutations were introduced at various locations throughout the site (Kaytor et al., 1997; Koo and Towle, 2000). The effect of these point mutations were assessed by linking the oligonucleotides to a promoter that was not glucose-responsive. The constructs were then introduced into hepatocytes and tested for their ability to support a glucose response. Many of the point mutations, especially those within the conserved CACG motifs, led to a loss of function. Other mutations were without effect or in a few cases, actually gave an increased responsiveness to glucose. This same battery of mutated oligonucleotides was then tested for their ability to bind to nuclear proteins extracted from livers of animals treated with high carbohydrate diet. Using an electrophoretic mobility shift assay, several DNA–protein complexes were detected with the wild type rat or mouse S_{14} oligonucleotide. Of these, only one specific complex was found to invariably disappear when mutant oligonucleotides defective in supporting a glucose response were used. Furthermore, this complex was retained on all mutant oligonucleotides that remained functional for the glucose response. This correlation was observed with over 15 wild type and mutant oligonucleotides and strongly suggests that this complex contains the nuclear factor responsible for mediating the glucose response. This same complex has also been more recently detected with ChoRE sites from the fatty acid synthase and acetyl CoA carboxylase gene (see Fig. 10.4) (Koo, S.-H. and Towle, H.C., unpublished data). We have termed this nuclear protein ChoRF for carbohydrate responsive factor (Koo and Towle, 2000).

Given that SREBP itself is an E box binding protein, it was critical to determine whether ChoRF was related to SREBP. A number of lines of evidence suggest that ChoRF is indeed a novel and specific factor (Koo, S.-H. and Towle, H.C., unpublished data). First of all, the ChoRF complex does not comigrate with the SREBP complex on EMSA. Second, antibodies to SREBP did not block formation of the ChoRF complex, but did block SREBP binding to its recognition site. Third, an authentic SRE binding site was incapable of competing for binding of ChoRF to the S_{14} or L-PK ChoRE. Conversely, ChoRE-containing oligonucleotides did not compete for SREBP binding to the SRE site. Fourth, transfection of hepatocytes with an expression vector for nuclear forms of SREBP led to activation of SRE-containing reporter constructs, but not from constructs containing ChoREs. These data suggest that ChoRF is a novel factor distinct from SREBP. Efforts to purify and clone the ChoRF are currently underway. Clearly, elucidating the pathway by which glucose can modulate the activity of this factor will require identification of the factor.

7. Model for lipogenic enzyme gene regulation

These data lead us to suggest the following model for the regulation of lipogenic enzyme genes. Two transcription factors are involved in the regulation of lipogenic enzyme genes in response to dietary carbohydrate (Fig. 10.5). One of these factors is SREBP and this factor is regulated in response to insulin at the level of its expression and perhaps its activity. The second factor is ChoRF and it is regulated by glucose metabolism. No changes were found in the level of ChoRF binding in extracts prepared from animals treated with diets promoting lipogenesis. Hence, it would appear that the primary regulation of ChoRF occurs posttranslationally by modification of its ability to activate transcription. This idea is consistent with the very rapid response of S_{14} to glucose, which appears to initiate well within 30 min. We would suggest that the genes regulated by carbohydrate feeding contain either SREBP or ChoRF binding sites, or in many cases may contain both. Glucokinase appears to be an example of a gene regulated primarily by SREBP in response to insulin. The L-PK gene has not been found to contain SREBP-binding sites and is unaffected by SREBP expression, suggesting that it may be regulated predominantly by ChoRF in response to glucose. We suspect that for many of the other lipogenic enzyme genes that both factors may function coordinately to provide induction, accounting for the dual requirement for insulin and glucose observed in hepatocyte cultures. One case where this seems to be the case is the rat S_{14} gene. The ChoRE site responsible for ChoRF binding and glucose activation is located at approximately –1430 relative to the transcription start site. Recently, Jump and coworkers demonstrated that SREBP can bind to the S_{14} promoter at a site located at –170 (Mater et al., 1999). We postulate that these two sites are both required for optimal induction of S_{14} in response to high carbohydrate diet.

The involvement of two distinct factors in supporting the lipogenic enzyme gene response may have several advantages over a single factor. First, it may serve to assure that all physiological conditions are appropriate for turning on lipogenesis—

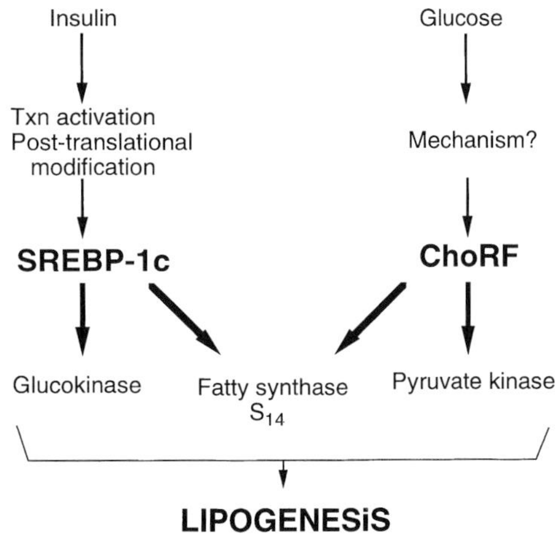

Fig. 10.5. Model for the dual role of SREBP-1c and ChoRF in regulating genes of hepatic lipogenesis. See text for description.

both insulin signal and increased substrate must be reaching the liver. Second, it may serve to provide a finer degree of control to the various gene products that must be coordinately produced at appropriate levels than could be afforded by a single factor. Third, these factors are each potentially subject to varying regulatory inputs that could influence their behavior and thus help to coordinate the output of lipogenesis to varying input signals. For example, polyunsaturated fatty acids in the diet are known to repress the induction of lipogenic enzyme genes by high carbohydrate diet. Recent data suggests that polyunsaturated fatty acids function by inhibiting the nuclear accumulation of SREBP and hence limiting transcriptional induction (Mater et al., 1999; Worgall et al., 1998; Xu et al., 1999b; Yahagi et al., 1999). Similarly, glucagon acts to inhibit expression of lipogenic enzyme genes, consistent with its role as a signal of the fasted state. This action appears to be mediated at least in part through the ChoRE site, suggesting that glucagon acts by controlling ChoRF activity. In this manner, various metabolic and stress inputs can be coordinated by the cell to provide the appropriate output for these enzymes involved in energy storage.

8. Conclusions

Several important themes have emerged from the studies concerning regulation of energy homeostasis in response to changing nutritional states in mammals. First, changes in gene expression, particularly at the level of gene transcription, are critical points of regulation for adaptive responses of the organism to its environmental conditions. These changes allow the organism to modify the enzyme repertoire of any cell or tissue to meet the present energy needs. The enzymes that are expressed are then subject to fine level control of their activity to provide the acute regulation of metabolic pathways.

Second, energy nutrients themselves, as well as hormonal cues, are important in promoting the intracellular signaling pathways responsible for controlling gene expression. While hormonal signals are critical for communication between tissues that must coordinate their activity, intracellular signaling in response to nutrients and their metabolites provides another level of regulatory inputs that must be met to ensure appropriate physiological responses. In this regard, it should be recalled that the regulation of the secretion and synthesis of many of critical hormones of energy homeostasis, such as insulin and glucagon, occur in response to nutrient and metabolite levels as well. Hence, the overall metabolic state of an organism is critical for dictating the responses that occur in tissues such as the liver.

Third, a variety of mechanisms are involved in controlling the activity of transcription factors. The PPARs are regulated by directly binding to fatty acid and eicosanoid ligands. The PPARs are representatives of a large set of so-called 'orphan' receptors related to the steroid receptor superfamily. While the endogenous ligands for PPARs have been identified, there are many other orphan receptors that are likely to be involved in regulating genes of metabolic importance awaiting identification of their ligands. Just recently, intracellular receptors for oxysterols and bile acids have been identified and one may logically expect many others to follow (Russell, 1999). In addition to direct ligand binding, transcription factors can be regulated by a variety of other mechanisms including post-translational modification and cellular localization, as observed for the SREBPs.

Fourth, it should be emphasized that all of the various transcription factors that function to regulate transcription of metabolically-important enzymes and proteins work coordinately with many other factors to regulate the final levels of transcription of any specific gene. Although this discussion has focused on individual factors and their regulation, the promoter and enhancer regions of genes invariably contain binding sites for many factors that can both activate and repress transcription in response to differing signals. In this manner, integration of a variety of inputs from varying environmental and internal sensors can provide the optimal output of enzyme production to meet the needs of the organism and ensure energy homeostasis is maintained.

References

Auwerx, J. (1999). PPARγ, the ultimate thrifty gene. Diabetologia 42, 1033–1049.

Bergot, M.-O., Diaz-Guerra, M.-J. M., Puzenat, N., Raymondjean, M. and Kahn, A. (1992). Cis-regulation of the L-type pyruvate kinase gene promoter by glucose, insulin and cyclic AMP. Nucl. Acids Res. 20, 1871–1878.

Brown, M.S. and Goldstein, J.L. (1997). The SREBP pathway: Regulation of cholesterol metabolism by proteolysis of a membrane-bound transcription factor. Cell 89, 331–340.

Desvergne, B. and Wahli, W. (1999). Peroxisome proliferator-activated receptors: Nuclear control of metabolism. Endocrine Rev. 20, 649–688.

Doiron, B., Cuif, M.-H., Chen, R. and Kahn, A. (1996). Transcriptional glucose signaling through the glucose response element is mediated by the pentose phosphate pathway. J. Biol. Chem. 271, 5321–5324.

Doiron, B., Cuif, M.-H., Kahn, A. and Diaz-Guerra, M.-J. (1994). Respective roles of glucose, fructose, and insulin in the regulation of the liver-specific pyruvate kinase gene promoter. J. Biol. Chem. 269, 10213–10216.

Dreyer, C., Krey, G., Keller, H., Givel, F., Helftenbein, G. and Wahli, W. (1992). Control of the peroxisomal beta-oxidation pathway by a family of nuclear hormone receptors. Cell 68, 879–887.

Foretz, M., Guichard, C., Ferre, P. and Foufelle, F. (1999a). Sterol regulatory element binding protein-1c is a major mediator of insulin action on the hepatic expression of glucokinase and lipogenesis-related genes. Proc. Natl. Acad. Sci. USA 96, 12737–12742.

Foretz, M., Pacot, C., Dugail, I., Lemarchand, P., Guichard, C., Liepvre, X.L., Berthelier-Lubrano, C., Spiegelman, B., Kim, J.B., Ferre, P. and Foufelle, F. (1999b). ADD1/SREBP-1c is required in the activation of hepatic lipogenic gene expression by glucose. Mol. Cell. Biol. 19, 3760–3768.

Girard, J., Ferre, P. and Foufelle, F. (1997). Mechanisms by which carbohydrates regulate expression of genes for glycolytic and lipogenic enzymes. Annu. Rev. Nutr. 17, 325–52.

Goldstein, J.L. and Brown, M. S. (1990). Regulation of the mevalonate pathway. Nature 343, 425–430.

Hanson, R.W. and Reshef, L. (1997). Regulation of phosphoenolpyruvate carboxykinase (GTP) gene expression. Annu. Rev. Biochem. 66, 581–611.

Hillgartner, F.B., Salati, L.M. and Goodridge, A.G. (1995). Physiological and molecular mechanisms involved in nutritional regulation of fatty acid synthesis. Phys. Rev. 75, 47–76.

Hollenberg, A.N., Susulic, V.S., Madura, J.P., Zhang, B., Moller, D.E., Tontonoz, P., Sarraf, P., Spiegelman, B.M. and Lowell, B.B. (1997). Functional antagonism between CCAAT/enhancer binding protein-α and peroxisome proliferator-activated receptor-γ on the leptin promoter. J. Biol. Chem. 272, 5283–5290.

Isseman, I. and Green, S. (1990). Activation of a member of the steroid hormone receptor superfamily by peroxisome proliferators. Nature 347, 645–650.

Iynedjian, P.B., Jotterand, D., Nouspikel, T., Asfari, M. and Pilot, P.R. (1989). Transcriptional induction of glucokinase gene by insulin in cultured liver cells and its repression by the glucagon-cAMP system. J. Biol. Chem. 264, 21824–21829.

Jump, D.B. and Oppenheimer, J. H. (1985). High basal expression and 3,5,3′-triiodothyronine regulation of messenger ribonucleic acid S_{14} in lipogenic tissues. Endocrinology 117, 2259–2266.

Kadesch, T. (1993). Consequences of heteromeric interactions among helix–loop–helix proteins. Cell Growth Differ. 4, 49–55.

Kaytor, E.N., Shih, H.-M. and Towle, H.C. (1997). Carbohydrate regulation of hepatic gene expression. Evidence against a role for the upstream stimulatory factor. J. Biol. Chem. 272, 7525–7531.

Kersten, S., Seydoux, J., Peters, J.M., Gonzalez, F.J., Desvergne, B. and Wahli, W. (1999). Peroxisome proliferator-activated receptor α mediates the adaptive response to fasting. J. Clin. Invest. 103, 1489–1498.

Kim, J.B., Sarraf, P., Wright, M., Yao, K.M., Mueller, E., Solanes, G., Lowell, B.B. and Spiegelman, B.M. (1998). Nutritional and insulin regulation of fatty acid synthetase and leptin gene expression through ADD1/SREBP1. J. Clin. Invest. 101, 1–9.

Kinlaw, W.B., Church, J.L., Harmon, J. and Mariash, C.N. (1995). Direct evidence for a role of the "spot 14" protein in the regulation of lipid synthesis. J. Biol. Chem. 270, 16615–16618.

Kliewer, S.A., Sundseth, S.S., Jones, S.A., Brown, P.J., Wisely, G.B., Koble, C.S., Devchand, P., Wahli, W., Willson, T.M., Lenhard, J.M. and Lehmenn, J. M. (1997). Fatty acids and eicosanoids regulate gene expression through direct interactions with peroxisome proliferator-activated receptors α and γ. Proc. Natl. Acad. Sci. USA 94, 4318–4323.

Koo, S.-H. and Towle, H.C. (2000). Glucose regulation of mouse S_{14} gene expression in hepatocytes: Involvement of a novel transcription factor complex. J. Biol. Chem. 275, 5200–5207.

Kotzka, J., Muller-Wieland, D., Koponen, A., Njamen, D., Kremer, L., Roth, G., Munck, M., Knebel, B. and Krone, W. (1998). ADD1/SREBP-1c mediates insulin-induced gene expression linked to the MAP kinase pathway. Biochem. Biophys. Res. Commun. 249, 375–379.

Lee, S.S., Pineau, T., Drago, J., Lee, E.J., Owens, J.W., Kroetz, D.L., Fernandez-Salguero, P.M., Westphal, H. and Gonzalez, F.J. (1995). Targeted disruption of the alpha isoform of the peroxisome proliferator-activated receptor gene in mice results in abolishment of the pleiotropic effects of peroxisome proliferators. Mol. Cell. Biol. 15, 3012–3022.

Lemberger, T., Saladin, R., Vazquez, M., Assimacopoulos, F., Staels, B., Desvergne, B., Wahli, W. and Auwerx, J. (1996). Expression of the peroxisome proliferator-activated receptor alpha gene is stimulated by stress and follows a diurnal rhythm. J. Biol. Chem. 271, 1764–1769.

Leone, T.C., Weinheimer, C.J. and Kelly, D.P. (1999). A critical role for the peroxisome proliferator-activated receptor α (PPARα) in the cellular fasting response: The PPARα mouse as a model of fatty acid oxidation disorders. Proc. Natl. Acad. Sci. USA 96, 7473–7478.

Liu, Z., Thompson, K.S. and Towle, H.C. (1993). Carbohydrate regulation of the rat L-type pyruvate kinase gene requires two nuclear factors: LF-A1 and a member of the c-myc family. J. Biol. Chem. 268, 12787–12795.

Lopez, J.M., Bennett, M.K., Sanchez, H.B., Rosenfeld, J.M. and Osborne, T.F. (1996). Sterol regulation of acetyl coenzyme A carboxylase: A mechanism for coordinate control of cellular lipid. Proc. Natl. Acad. Sci. USA 93, 1049–1053.

Magana, M.M. and Osborne, T.F. (1996). Two tandem binding sites for sterol regulatory element binding proteins are required for sterol regulation of fatty-acid synthase promoter. J. Biol. Chem. 271, 32689–32694.

Mariash, C.N., McSwigan, C.R., Towle, H.C., Schwartz, H.L. and Oppenheimer, J.H. (1981). Glucose and triiodothyronine both induce malic enzyme in the rat hepatocyte culture. J. Clin. Invest. 68, 1485–1490.

Mariash, C.N. and Oppenheimer, J.H. (1983). Stimulation of malic enzyme formation in hepatocyte culture by metabolites: Evidence favoring a nonglycolytic metabolite as the proximate induction signal. Metabolism 33, 545–552.

Mariash, C.N., Seelig, S., Schwartz, H.L. and Oppenheimer, J.H. (1986). Rapid synergistic interaction between thyroid hormone and carbohydrate on $mRNA_{S14}$ induction. J. Biol. Chem. 261, 9583–9586.

Mater, M.K., Thelen, A.P., Pan, D.A. and Jump, D.B. (1999). Sterol response element-binding protein 1c (SREBP-1c) is involved in the polyunsaturated fatty acid suppression of hepatic S_{14} gene transcription. J. Biol. Chem. 274, 32725–32732.

Michalik, L. and Wahli, W. (1999). Peroxisome proliferator-activated receptors: three isotypes for a multitude of functions. Curr. Op. Biotech. 10, 564–570.

O'Brien, R.M. and Granner, D.K. (1996). Regulation of gene expression by insulin. Phys. Rev. 76, 1109–1161.

Prip-Buus, C., Perdereau, D., Foufelle, F., Maury, J., Ferre, P. and Girard, J. (1995). Induction of fatty-acid-synthase gene expression by glucose in primary culture of rat hepatocytes. Dependency upon glucokinase activity. Eur. J. Biochem. 230, 309–315.

Russell, D.W. (1999). Nuclear orphan receptors control cholesterol catabolism. Cell 97, 539–542.

Sakai, J., Nohturfft, A., Cheng, D., Ho, Y.K., Brown, M.S. and Goldstein, J.L. (1997). Identification of complexes between the COOH-terminal domains of sterol regulatory element-binding proteins (SREBPs) and SREBP cleavage-activating protein. J. Biol. Chem. 272, 20213–20221.

Sheng, Z., Otani, H., Brown, M.S. and Goldstein, J.L. (1995). Independent regulation of sterol regulatory element binding proteins 1 and 2 in hamster liver. Proc. Natl. Acad. Sci. USA 92, 935–938.

Shih, H.-M., Liu, Z. and Towle, H.C. (1995). Two CACGTG motifs with proper spacing dictate the carbohydrate regulation of hepatic gene transcription. J. Biol. Chem. 270, 21991–21997.

Shih, H.-M. and Towle, H.C. (1992). Definition of the carbohydrate response element of the rat S_{14} gene: Evidence for a common factor required for carbohydrate regulation of hepatic genes. J. Biol. Chem. 267, 13222–13228.

Shimano, H., Horton, J.D., Hammer, R.E., Shimomura, I., Brown, M.S. and Goldstein, J.L. (1996). Overproduction of cholesterol and fatty acids causes massive liver enlargement in transgenic mice expressing truncated SREBP-1a. J. Clin. Invest. 98, 1575–1584.

Shimano, H., Horton, J.D., Shimomura, I., Hammer, R.E., Brown, M.S. and Goldstein, J.L. (1997). Isoform 1c of sterol regulatory element binding protein is less active than isoform 1a in livers of transgenic mice and in cultured cells. J. Clin. Invest. 99, 846–854.

Shimomura, I., Shimano, H., Horton, J.D., Goldstein, J.L. and Brown, M.S. (1997). Differential expression of exons 1a and 1c in mRNAs for sterol regulatory element binding protein-1 in human and mouse organs and cultured cells. J. Clin. Invest. 99, 838–845.

Spiegelman, B.M. (1998). PPARγ: Adipogenic regulator and thiazolidinedione receptor. Diabetes 47, 507–514.

Tontonoz, P., Hu, E. and Spiegelman, B.M. (1994). Stimulation of adipogenesis in fibroblasts by PPARγ2, a lipid-activated transcription factor. Cell 79, 1147–1156.

Tontonoz, P., Kim, J.M., Graves, R.A. and Spiegelman, B.M. (1993). ADD1: a novel helix–loop–helix transcription factor associated with adipocyte determination and differentiation. Mol. Cell. Biol. 13, 4753–4759.

Towle, H.C., Kaytor, E.N. and Shih, H.-M. (1997). Regulation of the expression of lipogenic enzyme genes by carbohydrate. Annu. Rev. Nutr. 17, 405–33.

Wang, X., Sato, R., Brown, M.S., Hua, X. and Goldstein, J.L. (1994). SREBP-1, a membrane-bound transcription factor released by sterol-regulated proteolysis. Cell 77, 53–62.

Worgall, T.S., Sturley, S.L., Seo, T., Osborne, T.J. and Deckelbaum, R.J. (1998). Polyunsaturated fatty acids decrease expression of promoters with sterol regulatory elements by decreasing levels of mature sterol regulatory element-binding protein. J. Biol. Chem. 273, 25537–25540.

Xu, H.E., Lambert, M.H., Montana, V.G., Parks, D.J., Blanchard, S.G., Brown, P.J., Sternbach, D.D., Lehmann, J.M., Wisely, G.B., Wilson, T.M., Kliewer, S.A. and Milburn, M.V. (1999a). Molecular recognition of fatty acids by peroxisome proliferator-activated receptors. Mol. Cell 3, 397–403.

Xu, J., Nakamura, M.T., Cho, H.P. and Clarke, S.D. (1999b). Sterol regulatory element binding protein-1 expression is suppressed by dietary polyunsaturated fatty acids. J. Biol. Chem. 274, 23577–23583.

Yahagi, N., Shimano, H., Hasty, A.H., Amemiya-Kuto, M., Okazaki, H., Tamura, Y., Iizuka, Y., Shionoiri, F., Ohashi, K., Osuga, J.-I., Harada, K., Gotada, T., Nagai, R., Ishibashi, S. and Yamada, N. (1999). A crucial role of sterol regulatory element-binding protein-1 in the regulation of lipogenic gene expression by polyunsaturated fatty acids. J. Biol. Chem. 274, 35840–35844.

Yokoyama, C., Wang, X., Briggs, M.R., Admon, A., Wu, J., Hua, X., Goldstein, J.L. and Brown, M.S. (1993). SREBP-1, a basic–helix–loop–helix–leucine zipper protein that controls transcription of the low density lipoprotein receptor gene. Cell 75, 187–197.

Protein Adaptations and Signal Transduction. Edited by K.B. Storey and J.M. Storey

CHAPTER 11

The AMP-activated/SNF1 Protein Kinases: Key Players in the Response of Eukaryotic Cells to Metabolic Stress

D. Grahame Hardie
Wellcome Trust Biocentre, School of Life Sciences, Dundee University, Dundee, DD1 5EH, Scotland, U.K.

1. Introduction

Single-celled heterotrophic eukaryotes such as the yeast *Saccharomyces cerevisiae* are constantly subject to the vagaries of their extracellular environment, especially variation in availability of a carbon source. They must have systems that allow them to monitor such situations and respond appropriately. Green plants are also subject to frequent stressful conditions, including heat, cold or water stress, variable cloud cover, shading, or grazing of leaves by animals. A plant in any situation where photosynthesis is inhibited is in a state of starvation akin to that of a yeast cell without a carbon source. By contrast, cells in a multicellular organism, particularly in homeothermic organisms such as mammals, are relatively cossetted. They are constantly bathed in a rich nutrient medium, which is normally held at a constant temperature, pH, ionic strength and glucose concentration. When a mammal dies of starvation, this probably happens not because its cells have been starved of glucose, but because it has succumbed to an infection or some other secondary problem.

Despite these apparent differences, it now appears that a signalling pathway (the AMPK/SNF1 protein kinase system), that may have evolved in single-celled eukaryotes as a mechanism to respond to starvation for a carbon source, has been highly conserved right across the animal, plant and fungal kingdoms. This pathway involves a kinase cascade in which one or more upstream kinases phosphorylate and activate the downstream kinase. In mammals and other metazoans, where the downstream kinase is called the AMP-activated protein kinase (AMPK), the cascade does not respond directly to the availability of a carbon source, but instead is activated by cellular stresses that deplete ATP. It achieves this by sensing the cellular levels of AMP and ATP, being activated by a rise in this ratio. Surprisingly, the related protein kinases in fungi and plants do not appear to respond to AMP and ATP in the same manner, and the key signals that regulate them remain unclear. Despite these apparent differences in regulatory properties, the mammalian, fungal and plant AMPK/SNF1 protein kinases can all be regarded as "metabolic master switches" and their downstream effects are similar although not identical.

2. Early studies of the AMPK/SNF1 protein kinases

2.1. Mammalian AMP-activated protein kinase

Although it did not receive its current name until 1988, the first observations that can with hindsight be attributed to the AMP-activated protein kinase (AMPK) were reported in 1973. Gibson and coworkers (Beg et al., 1973) reported that microsomal preparations of HMG-CoA reductase (a regulatory enzyme of cholesterol synthesis) were inactivated by incubation with MgATP and a soluble protein fraction. They proposed that phosphorylation and inactivation of HMG-CoA reductase (an integral protein of the smooth endoplasmic reticulum) was being catalyzed by a protein kinase in the soluble fraction. In independent work, Carlson and Kim (1973) had been studying acetyl-CoA carboxylase, a regulatory enzyme of fatty acid synthesis. They reported that crude preparations of the enzyme from rat liver were inactivated on incubation with MgATP and that this appeared to be due to phosphorylation of the enzyme. For many years it was not realized that Gibson's and Kim's groups were working with distinct functions of a single protein kinase. This connection was made by the author's laboratory who showed that a highly purified preparation of "acetyl-CoA carboxylase kinase-3" also inactivated HMG-CoA reductase (Carling et al., 1987), and that these two activities co-purified from rat liver extracts (Carling et al., 1989). Since it no longer seemed appropriate to name the kinase after one of its substrates, we renamed it AMP-activated protein kinase after its allosteric regulator, 5′-AMP (Munday et al., 1988a; Munday et al., 1988b; Sim and Hardie, 1988; Hardie et al., 1989).

2.2. The yeast SNF1 protein kinase

The yeast *Saccharomyces cerevisiae* grows on glucose as its preferred carbon source, and as long as it is available in the medium expression of a large number of genes is repressed (Gancedo, 1998). This phenomenon is known as *glucose repression*, and a manifestation of it was reported 100 years ago (Dienert, 1900) when it was found that yeast transferred from glucose to galactose medium required a period of adaptation before they would start to grow. Genes repressed by glucose include those required for metabolism of alternative fermentable carbon sources such as galactose and sucrose. Thus, glucose represses the *GAL* genes required to metabolize galactose, and the *SUC2* gene, encoding a secreted form of invertase that can break down sucrose. Glucose also represses genes required for oxidative metabolism and hence for growth on non-fermentable carbon sources such as glycerol and ethanol. Screening for mutants that would not grow on sucrose or non-fermentable carbon sources resulted in the isolation of *cat1* (Zimmermann et al., 1977), *ccr1* (Ciriacy, 1977) and *snf1* (Carlson et al., 1981) mutants, which turned out to be alleles of the same gene, now usually called *SNF1* (*sucrose non-fermenting-1*). Celenza and Carlson (1986) sequenced the gene and showed that it encoded a protein kinase. However it was not until complete or partial sequences for the catalytic subunit of AMPK were obtained (Carling et al., 1994; Mitchelhill et al., 1994), that it was realized that the *SNF1* gene product was the yeast homologue of the catalytic subunit of mammalian AMPK.

2.3. The higher plant SNF1-related protein kinases

In 1991 Halford and coworkers cloned a cDNA from rye that encoded a putative protein kinase closely related to the yeast *SNF1* gene product (Alderson et al., 1991). Subsequently, homologues were cloned from many plant species, and they have been termed the SNF1-related kinase-1 (SnRK1) group to distinguish them from other plant protein kinases more distantly related to Snf1p (Halford and Hardie, 1998). The rye gene was shown to be functionally related to *SNF1* in that expression of the DNA in *snf1* mutant yeast restored growth on non-fermentable carbon sources (Alderson et al., 1991). The following year the author's group (MacKintosh et al., 1992) reported that extracts of a variety of green plants contained protein kinase activities with biochemical properties very similar to those of mammalian AMPK. With the discovery that Snf1p was the yeast homologue of the AMPK catalytic subunit (Carling et al., 1994; Mitchelhill et al., 1994) it became very likely that the plant protein kinases described by the author's group was the products of the *SNF1*-related genes described by Halford (Alderson et al., 1991). This was subsequently confirmed (Ball et al., 1995; Barker et al., 1996).

3. Structure of the AMPK/SNF1 kinases

3.1. Structure of mammalian AMP-activated and yeast SNF1 protein kinases

The catalytic subunit of AMPK was identified by labelling with a reactive ATP analogue as a polypeptide of 63 kDa (Carling et al., 1989), and the kinase was subsequently purified to homogeneity and shown to be a heterotrimer of three subunits (Davies et al., 1994) now termed α, β and γ. DNAs encoding all of these subunits have been cloned and each has been found to exist as multiple isoforms (α1, α2, β1, β2, γ1, γ2, γ3) (Carling et al., 1994; Gao et al., 1995; Gao et al., 1996; Stapleton et al., 1996; Woods et al., 1996; Thornton et al., 1998; Cheung et al., 2000). Current indications are that particular isoforms of the α subunit do not selectively associate with particular isoforms of β and γ (Thornton et al., 1998; Cheung et al., 2000) so that all twelve possible combinations of the three subunits may exist. These isoform combinations exhibit subtle differences in tissue distribution, subcellular localization and regulatory properties, as is discussed further below.

It is becoming clear that the yeast SNF1 protein kinase is also a heterotrimer of α, β and γ subunits. In this review I will use the term “SNF1 complex” or “SNF1 protein kinase” to refer to the active heterotrimeric complex, and Snf1p (i.e. the product of the *SNF1* gene) to refer to the catalytic (α) subunit. The yeast homologue of the mammalian γ subunit is the product of the *SNF4* gene, which (as its name suggests) emerged from the same mutant screen as *SNF1*. Its gene product, Snf4p, forms a complex with Snf1p (Celenza et al., 1989) and they co-purify from yeast extracts (Wilson et al., 1996). Now that the sequence of the *S. cerevisiae* genome is complete it is clear that there are only single genes encoding the yeast α and γ subunits (*SNF1* and *SNF4*). However there are three homologues of the mammalian β subunits, i.e. Sip1p, Sip2p and Gal83p (Yang et al., 1994). Disruption of all three β subunit genes (*SIP1, SIP2, GAL83*) was reported not to cause a growth defect similar to those of *snf1* or *snf4* mutants, even though there appear to be no other β subunit genes in the genome (Yang et al., 1994). This is surprising, especially considering that in the related yeast *Kluyveromyces lactis* mutation of single genes encoding α and β subunit homologues (*FOG2* and *FOG1* respectively) causes the same *snf*-like phenotype (Goffrini et al., 1996).

The α (Snf1p) subunits of AMPK and SNF1 are the catalytic subunits, containing kinase domains at the N-terminus followed by regulatory domains of size approximately equal to that of the kinase domain (Fig. 1). The regulatory domains appear to contain “autoinhibitory” sequences that maintain the kinase in an inactive state by binding to the kinase domain in the absence of an activating signal (Jiang and Carlson, 1996; Crute et al., 1998). The autoinhibitory sequence may be a “pseudosubstrate” sequence that binds to the substrate-binding groove of the kinase domain as in other messenger-activated protein kinases (Kemp et al., 1994), although this has not yet been directly demonstrated. If this model is correct, AMP would activate the protein kinase by displacing the autoinhibitory sequence, as discussed further below.

The β subunit appears to be the “scaffold” on which the α and γ subunits assemble (Yang et al., 1994; Woods et al., 1996). When the sequence of mammalian β1 and β2 are compared with those of the yeast β subunit genes, there are two conserved regions called the *KIS* (*kinase interaction sequence*) and *ASC* (*association with SNF1 kinase*) domains (Fig. 11.1). Two-hybrid analysis in the yeast system shows that the *KIS* domain is responsible for interaction with α and the *ASC* domain for interaction with γ (Jiang and Carlson, 1997), and it seems likely that this will also be true of the mammalian system. There is also genetic evidence in yeast that the β subunits act as “adapter” subunits that direct the SNF1 complex to particular downstream targets (Hardie et al., 1998; Vincent and Carlson, 1999).

When they were first sequenced, it was reported that the mammalian γ subunits and yeast Snf4p were not related to any other proteins except to each other. However using more sensitive methods Bateman (1997) found that the mammalian and yeast γ subunits each contain four tandem repeats of a module of about 60 residues known as a CBS

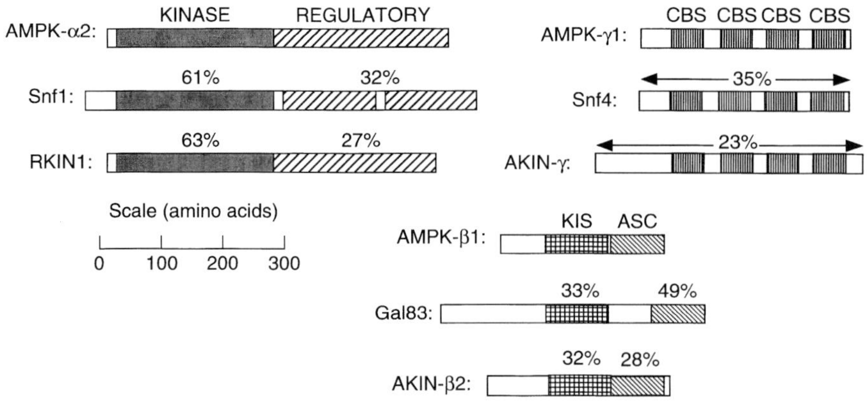

Fig. 11.1. Domain structures of subunits of the AMPK/SNF1 kinase subfamily. Subunits are drawn as linear bars approximately to scale, and with N-termini on the left. The plant sequences represented are from rye for the catalytic subunit (Alderson et al., 1991), and *Arabidopsis* for the putative β and γ subunits (Bouly et al., 1999). Related domains are represented as hatched or shaded boxes, and figures above these boxes for the yeast and plant subunits are % sequence identities with the related domain in the rat subunit shown. Assignment of CBS domains in the AMPK-γ/Snf4p sequences was as in Bateman (Bateman, 1997), and in this case only the overall sequence identity of the whole subunits are presented. Redrawn from (Hardie et al., 1998).

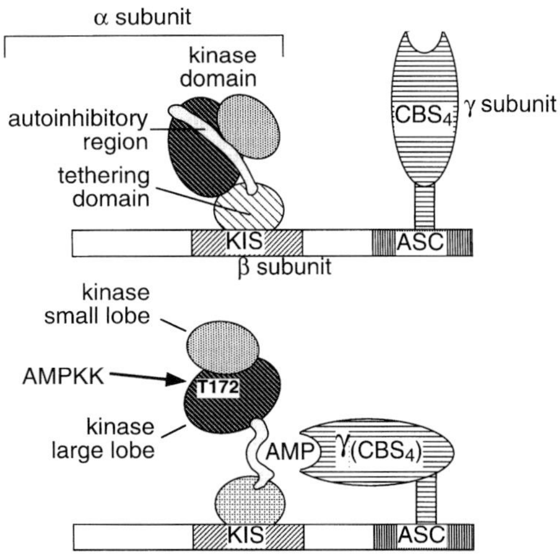

Fig. 11.2. Model for the regulation of the AMPK complex. In the absence of AMP, the complex exists predominantly in the inactive conformation shown in the top panel, where the α and γ subunits do not interact (except indirectly via the β subunit). In this state phosphorylation of Thr-172, and access to exogenous substrates, is blocked by interactions between the catalytic cleft of the kinase domain and the autoinhibitory region on the α subunit. In the active conformation (lower panel), interaction between the catalytic cleft and the autoinhibitory region is prevented by interaction of the latter with one or more of the CBS domains of the γ subunit. The interaction between the autoinhibitory region and the γ subunit is stabilized by binding of AMP, which involves residues on both subunits. Reproduced from (Cheung et al., 2000).

domain, which also occurs in a number of other proteins but whose function is unknown. Genetic evidence in yeast suggests that the γ subunit (Sn4p) is involved in the mechanism of activation of the SNF1 complex. In the absence of glucose in the medium, when the complex is active, a two-hybrid interaction between Snf4p and the regulatory domain of the catalytic subunit (Snf1p) can be demonstrated (Jiang and Carlson, 1996). The region of Snf1p responsible for this interaction overlaps with that responsible for the autoinhibition of the kinase domain under conditions (high glucose) where the kinase is inactive. This leads to a model (Fig. 11.2) where the γ subunit binds to the regulatory domain of the α subunit, displacing the kinase domain and activating the latter (Jiang and Carlson, 1996).

3.2. Structure of higher plant SNF1-related protein kinases

The structures of the high plant homologues of AMPK/SNF1 have not been studied in such detail, although by gel filtration the plant kinases appear to have native molecular masses of 150–200 kDa (Ball et al., 1994; Sugden et al., 1999b). This suggests that they are oligomers rather than monomeric forms of the 60 kDa catalytic subunit. DNAs encoding homologues of the β and γ subunits have also been cloned from higher plants (Bouly et al., 1999; Lakatos et al., 1999), and work is in progress in the author's laboratory to examine whether these subunits also occur in the 150–200 kDa complexes.

4. Regulation of the AMPK/SNF1 kinases

4.1. Regulation of mammalian AMP-activated protein kinase

As its name suggests, AMP-activated protein kinase is allosterically activated by 5′-AMP. Because this effect is antagonized by ATP, the $A_{0.5}$ (concentration of AMP giving half-maximal activation) depends on the ATP concentration used in the assay, and increases from 4 to 30 μM as the ATP concentration is increased from 0.2 to 4 mM (Corton et al., 1995). The kinase is therefore activated more dramatically by a rise in AMP that is coupled with a fall in ATP. Because of the very active adenylate kinase enzyme present in eukaryotic cells, AMP and ATP always vary in reciprocal directions. The adenylate kinase reaction (2ADP ↔ AMP + ATP) appears to be close to equilibrium in all eukaryotic cells, and as pointed out many years ago by Krebs (1963) this means that the AMP:ATP ratio is a much more sensitive indicator of a low cellular energy status than ADP:ATP. In fact if the adenylate kinase reaction is *at equilibrium*, it is easy to show AMP:ATP will vary as the *square* of the ADP:ATP ratio.

Interestingly, the degree of activation of AMPK complexes by AMP depends on the isoform composition. All combinations appear to be activated by AMP over the same concentration range, but the degree of activation varies. Complexes containing the α2 catalytic subunit are activated to a greater extent than those containing α1 (Salt et al., 1998a). Complexes containing the γ2 catalytic subunit are activated to a greater extent than those containing γ1 (Cheung et al., 2000), while those containing γ3 appear to be almost independent of AMP. A simple explanation for these results is that AMP binds at the interface between the α and γ subunits. Independent evidence from this comes from the use of two reactive ATP/AMP analogues (*p*-fluorosulphonylbenzoyl adenosine and 8-azido AMP) that both bind at the allosteric site but label the α and γ subunits respectively (Cheung et al., 2000). This also fits nicely with the model devised by Carlson's group for the yeast complex (Jiang and Carlson, 1996), where an interaction between the γ subunit (Snf4p) and the regulatory domain of the α subunit (Snf1p) displaces the autoinhibitory effect of the latter on the kinase domain and thus activates the complex. If the same mechanism operated for AMPK, then AMP would promote the active conformation by binding between α and γ and stabilizing their interaction (Fig. 11.2).

This allosteric activation mechanism is only part of the story: as mentioned in the introduction, the AMPK/SNF1 systems are protein kinase cascades. AMPK is activated by phosphorylation by at least one upstream kinase, AMPKK. This phosphorylates the α subunit at a site (Thr-172) within the "activation loop" where many other protein kinases are regulated by phosphorylation (Johnson et al., 1996). AMP stimulates this phosphorylation and activation as well as causing allosteric activation (Weekes et al., 1994). Since the allosteric effect is at most five-fold, whereas the effect of phosphorylation is >50-fold, the latter is quantitatively more important. At first it was not clear whether AMP affected phosphorylation by binding to the substrate (i.e. AMPK) or the enzyme (i.e. AMPKK). Eventually the author's laboratory was able to show that AMP has no less than four effects on the system:

1. allosteric activation of the downstream kinase, AMPK;
2. allosteric activation of the upstream kinase, AMPKK (Hawley et al., 1995);
3. binding to the downstream kinase, making it a *better* substrate for AMPKK (Hawley et al., 1995);
4. binding to the downstream kinase, making it a *worse* substrate for protein phosphatases (Davies et al., 1995).

With the possible exception of effect (2), all of these effects of AMP are antagonized by high concentrations of ATP, and may be due to binding at a single allosteric site on AMPK. Computer simulations in the author's laboratory suggested that this multi-step mechanism makes the system respond to AMP in an *ultrasensitive* manner, i.e. following a sigmoidal rather than a hyperbolic response curve. In the same study, we were able to demonstrate that the response to a change in the concentration of the activating nucleotide was sigmoidal in intact cells (Hardie et al., 1999). The AMPK cascade would therefore act as a sensor that monitors the cellular AMP:ATP ratio and is switched on by a small change in the ratio over a critical concentration range.

By the 1960s a small number of metabolic enzymes (e.g. muscle phosphorylase and phosphofructokinase, liver fructose-1,6-bisphosphatase) had been found to be regulated allosterically in reciprocal directions by AMP and ATP. Atkinson (Ramaiah et al., 1964) generalized from these findings and proposed that all branch-points between catabolism and anabolism might be regulated by AMP/ATP or ADP/ATP. He called this the *adenylate control* or *energy charge* hypothesis, the latter term coming from the analogy between ATP and ADP and the chemicals in an electrical cell or battery, with a high ATP:ADP ratio being equivalent to a fully charged battery. The idea caused much interest at the time and is still mentioned in many biochemistry textbooks. However, after the first examples very few other metabolic enzymes were subsequently found to respond directly to these nucleotides. In the author's view the discovery of the AMPK system represents a fulfilment of the energy charge hypothesis, although what Atkinson had not anticipated is that most of the effects of reduced energy charge would be mediated *indirectly* via a protein kinase cascade.

The idea that the AMPK system is sensor of cellular energy charge was reinforced by findings that AMPK is inhibited by physiological concentrations of phosphocreatine (Ponticos et al., 1998). This appears to be a purely allosteric effect, and the author's laboratory was unable to find any effects on phosphorylation by AMPKK or on dephosphorylation by protein phosphatases.

4.2. Regulation of yeast SNF1 protein kinase

Sudden removal of glucose from the medium of yeast in logarithmic growth results in a dramatic activation of the SNF1 complex due to phosphorylation (Woods et al., 1994; Wilson et al., 1996). Although not directly demonstrated, this is probably due to phosphorylation of Snf1p at Thr-210, equivalent to Thr-172 on mammalian AMPK. Thus, mammalian AMPKK (which phosphorylates AMPK at Thr-172) can also activate the SNF1 complex *in vitro* (Wilson et al., 1996), while mutation of Thr-210 abrogates the ability of Snf1p to confer growth on alternative carbon sources *in vivo* (Estruch et al., 1992). Along with activation of the SNF1 complex, removal of glucose from the medium of yeast in logarithmic growth causes dramatic increases in the cellular AMP:ATP ratio. This demonstrates that sudden glucose deprivation is a severe stress to yeast in logarithmic growth: under these conditions the organism has no carbohydrate reserves and is dependent on the continual supply of external glucose. It also suggested that a rise in AMP:ATP ratio might be the elusive signal that switched on gene derepression. Unfortunately, all attempts to demonstrate effects of AMP on activation of the SNF1 complex in cell-free systems have failed. Certainly AMP does not allosterically activate the SNF1 complex, although because the upstream kinases in yeast remain undefined, it has not been possible to examine all of the effects of AMP observed in the mammalian system. The signals that switch the SNF1 complex on and off in response to the availability of external glucose therefore remain elusive. Genetic evidence shows that a functional complex between Reg1p and Glc7p (the regulatory and catalytic subunits of a form of protein phosphatase-1) is required to maintain the SNF1 complex in an inactive state in high glucose. The Reg1p-Glc7p and SNF1 complexes interact *in vivo* by two hybrid analysis (Sanz et al., 2000), and an obvious hypothesis is that the Reg1p-Glc7p phosphatase is responsible for dephosphorylation of the SNF1 complex at Thr-210. Genetic evidence has also suggested that the PII isoform of hexokinase (encoded by the *HXK2* gene) has a role as a "glucose sensor" in addition to its known role in the initial metabolism of glucose to glucose-6-phosphate. It appears to be involved in maintaining the SNF1 complex in an inactive state in high glucose (Sanz et al., 2000). It may be that while the mammalian AMPK system has evolved to detect any cellular stress that causes ATP depletion, the SNF1 system is more specialized for monitoring the availability of glucose.

4.3. Regulation of higher plant SNF1-related protein kinases

The plant SnRK1 kinases are activated by phosphorylation at the threonine residue within the "activation loop" equivalent to Thr-172 in mammalian AMPK-α1/2 and Thr-210 in yeast Snf1p. A spinach SnRK1 is inactivated by mammalian protein phosphatases, and can then be reactivated by incubation with MgATP and mammalian AMPKK. During these experiments the activity correlates with the phosphorylation of the threonine residue (Thr-175) as assessed by probing blots with an antibody specific for the phosphorylated form of the "activation loop". Although AMP does not allosterically activate the plant SnRK1 kinases, the nucleotide has been found to inhibit their dephosphorylation, thus mirroring one of the four effects of AMP on the mammalian system (Sugden et al., 1999a). Unfortunately, estimation of AMP concentrations in plant cells is problematical because of compartmentation, and it remains unclear whether AMP is a physiological regulator in the plant system.

5. Cellular stresses that switch on the AMPK/SNF1 systems

5.1. Activation of AMPK in intact cells and in vivo

In normal, unstressed cells maintained under ideal conditions the AMP:ATP ratio is very low (of the order of 1:100) and the AMPK system has a very low basal activity. However the ratio is increased, and AMPK activated, by any cellular stress that either interferes with ATP production or increases ATP consumption. Stresses that interfere with ATP *production* and activate AMPK include:

1. *heat shock* (Corton et al., 1994);
2. *metabolic poisoning* with inhibitors of the TCA cycle such as arsenite (Corton et al., 1994), or inhibitors of respiration such as oligomycin (Marsin et al., 2000);
3. *hypoxia*, which has been studied in perfused heart muscle (Marsin et al., 2000);
4. *hypoglycaemia* (i.e. glucose deprivation, the stress that activates yeast SNF1) which activates AMPK in pancreatic β cells (Salt et al., 1998b; DaSilva-Xavier et al., 2000), but appears to be less effective in cells like liver that can rapidly mobilize glycogen stores;
5. *ischaemia*, which has been studied in perfused heart muscle (Kudo et al., 1995; Kudo et al., 1996; Marsin et al., 2000), and can be regarded as a combination of hypoxia and hypoglycaemia.

A stress that increases ATP *consumption* and activates AMPK is exercise in skeletal muscle *in vivo* (Winder and Hardie, 1996), a response that can be mimicked by electrical stimulation of muscle *in vitro* (Hutber et al., 1997; Vavvas et al., 1997). Exercise is of particular interest with respect to AMPK because it is a "physiological" stress that occurs on a regular basis, whereas most of the other stresses in the list above can be regarded as rarer "pathological" events.

In all of the cases listed above, activation of AMPK has been shown to correlate with increases in AMP:ATP ratio, and at present there is no reason to implicate any other regulators of AMPK. However it should be noted that a change in AMPK is only detectable in a cell extract if it involves a stable, covalent change, e.g. phosphorylation. As mentioned in the previous section, phosphocreatine does not appear to trigger changes in phosphorylation. This may be why, in the early stages of electrical stimulation of muscle, the AMPK target acetyl-CoA carboxylase became inactivated at time points when AMPK activation could not be detected (Hutber et al., 1997). In the early stages of exercise, phosphocreatine would become depleted before there was any change in

ATP or AMP, the effect on AMPK activity would be purely allosteric and would not be preserved on homogenization of the muscle.

5.2. Regulation of yeast SNF1 and plant SnRK1 kinases in vivo

As already discussed, the SNF1 protein kinase complex is rapidly activated by glucose deprivation due to phosphorylation (Woods et al., 1994; Wilson et al., 1996), and this remains the only stress that has been shown to activate the yeast system. The regulation of the plant kinases has as yet been little studied. However, Halford's group (Purcell et al., 1998) showed that potato plants expressing a SnRK1 DNA in leaves in antisense orientation were defective in the induction of sucrose synthase mRNA in high sucrose. Sucrose is of course the main form in which reduced carbon is transported around the plant, and despite its name sucrose synthase is thought to be responsible for the *degradation* of sucrose. This result seems somewhat paradoxical in that it indicates that the function of the SnRK1 kinases is required under conditions of carbon excess, rather than carbon deprivation as in yeast and mammals. Nevertheless it is consistent with findings that an *Arabidopsis thaliana* SnRK1 is activated by treatment with high sucrose (Bhalerao et al., 1999). In unpublished work the author's laboratory has also shown that removal of sucrose from *Arabidopsis* cells in suspension culture results in *inactivation* rather than activation of SnRK1.

6. Target pathways and proteins for AMPK/SNF1 systems

6.1. Recognition of targets by the AMPK/SNF1 protein kinase family

Studies with variant synthetic peptide substrates (Weekes et al., 1993; Dale et al., 1995b) and site-directed mutagenesis of recombinant protein substrates (Ching et al., 1996) have revealed that AMPK, SNF1 and higher plant SnRK1s recognize very similar motifs on their target proteins (Fig.

```
                -5 -4-3         ↓       +4
Animal AMPK:    M-(X-R)-X-X-S-X-X-X-L
                L   K       T           M
                I   H                   I
                F                       F
                V                       V

                -5  -3       ↓       +4
Yeast SNF1:     L-X-R-X-X-S-X-X-X-L
                M           T       I
                I                   F
                F                   M
                V                   V

                -5 -4-3         ↓       +4
Plant SnRK1:    M-(X-R)-X-X-S-X-X-X-L
                V   K       T           I
                L   H                   F
                F                       M
                I                       V
```

Fig. 11.3. Minimal recognition motifs around phosphorylation sites for the AMPK/SNF1 family. These motifs are based on studies of variant synthetic peptides (Dale et al., 1995b). Amino acids are shown using the single letter code with the phosphorylated amino acid indicated by a vertical arrow. Key features for recognition are hydrophobic residues at the –5 and +4 positions (numbering with respect to the phosphoamino acid), and a basic residue at the –3 position. The parentheses around the (–4, –3) positions for the animal and plant kinases indicate that in these cases the basic residue can be at either of these positions. For each kinase, the preferred amino acid is shown on the top line, with alternatives listed below in order of decreasing preference.

11.3). Further studies of the recognition of substrates by AMPK are continuing in the author's laboratory, and it is clear that recognition also involves determinants outside of the 9 residue span implied by Fig. 11.3. For a target protein to be phosphorylated by the kinase, it may only be necessary for a subset of these determinants to be present. However the hydrophobic residue at –5 (i.e. 5 residues N-terminal to the phosphoamino acid), and the basic residue at –3 or –4, do appear to be crucial. Of course, these determinants must not only be present, but they must also be in an accessible position at the surface of the protein.

An additional feature of recognition of target proteins is whether the target protein and the kinase are present at the same subcellular location. Ongoing studies reveal that the different isoform combinations of AMPK can be localized differently. For example, AMPK complexes containing the α2 isoform are partly located in the nucleus, whereas those containing the α1 isoform are not (Salt et al., 1998a; Turnley et al., 1999; DaSilva-Xavier et al., 2000). In yeast, the Gal83p isoform of the β subunit also appears to be directly responsible for targetting the SNF1 complex to Sip4p, a transcription factor that binds to the carbon-source responsive element in the promoter of gluconeogenic genes (Vincent and Carlson, 1999).

6.2. Targets for mammalian AMPK

An important development in the identification of targets pathways and proteins for AMPK was the development of 5-aminoimidazole-4-carboxamide (AICA) riboside as a method to activate the kinase in intact cells (Fig. 11.4). This nucleoside is taken up into cells and phosphorylated by adenosine kinase to the monophosphorylated AICA ribotide, usually referred to as ZMP. ZMP is an intermediate in the pathway of synthesis of the purine nucleotides IMP and AMP. However in most cells uptake of AICA riboside and phosphorylation to ZMP is rapid but its further metabolism is slow, so that ZMP accumulates to high levels without affecting cellular ATP, ADP or AMP (Corton et al., 1995). ZMP is an AMP analogue that mimics the effect of AMP on allosteric activation (Henin et al., 1996) as well as phosphorylation (Corton et al., 1995) of AMPK. AICA riboside has now been very widely used to identify processes regulated by AMPK in intact cells, although its specificity remains uncertain. Indeed ZMP has been reported to also mimic the effects of AMP on glycogen phosphorylase (Young et al., 1996) and fructose-1,6-bisphosphatase (Vincent et al., 1991). In addition, in some cell types such as cardiomyocytes (Javaux et al., 1995) incubation with AICA riboside does not lead to accumulation of ZMP and activation of AMPK, possibly because ZMP is rapidly metabolized. Other, molecular biological methods to manipulate AMPK in intact cells are now becoming available, such as expression of constitutively active or

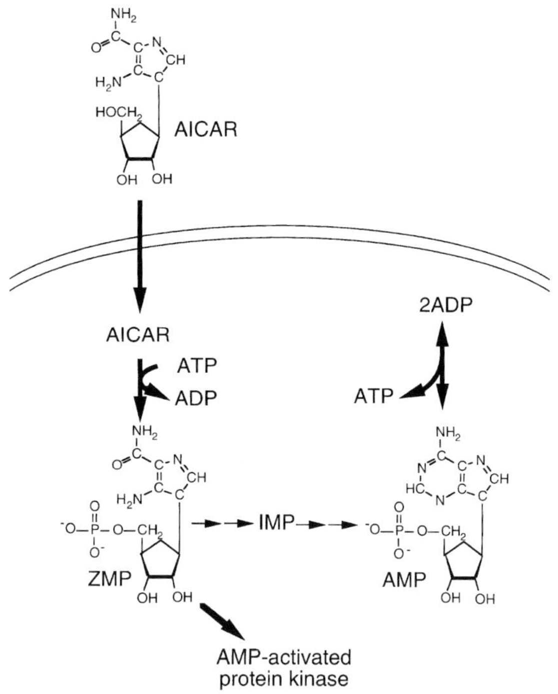

Fig. 11.4. Mechanism of action of AICA riboside in intact cells. The nucleoside is transported across the plasma membrane and phosphorylated to ZMP in the cytoplasm to ZMP, which mimics the effects of AMP on AMPK activation. Although ZMP can be metabolized to IMP and AMP by the pathway of purine nucleotide synthesis, in most cells this appears to be rather slow compared with its uptake and phosphorylation. ZMP therefore accumulates to the high concentrations necessary to activate AMPK, without changing the cellular levels of AMP. If the latter did occur the levels of ATP and ADP would also be affected by the adenylate kinase reaction, shown on the right. Redrawn from (Corton et al., 1995).

dominant negative AMPK mutants from adenovirus vectors (Woods et al., 2000). While these methods have their own associated drawbacks, if they confirm results obtained using AICA riboside then the evidence for a role of AMPK in control of the pathway becomes more convincing.

The pathways and processes for which there is evidence for regulation by AMPK are summarized in the following sections. In general, AMPK appears to switch on ATP-producing catabolic pathways, and switch off ATP-consuming processes including biosynthetic (anabolic) pathways. This makes physiological sense for a system that is switched on by stresses causing ATP depletion.

Where the actual target for AMPK is known, this is given in parentheses after the name of the pathway.

Isoprenoid/sterol synthesis (HMG-CoA reductase)

AMPK was originally discovered for its ability to inactivate HMG-CoA reductase (Beg et al., 1973), and this remains one of the best established substrates. AMPK phosphorylates the enzyme at Ser-871, close to the C-terminus, and this totally inactivates the enzyme probably because the phosphate forms an ionic interaction with His-865, a residue involved in the catalytic mechanism (Omkumar and Rodwell, 1994). Activation of AMPK causes phosphorylation of HMG-CoA reductase at Ser-871 and dramatic inhibition of sterol synthesis (Gillespie and Hardie, 1992; Corton et al., 1995), while mutation of Ser-871 to alanine prevents the inhibition of sterol synthesis caused by ATP depletion (Sato et al., 1993).

Acute effects on fatty acid synthesis/oxidation (acetyl-CoA carboxylase)

Another classical substrate for AMPK (Carlson and Kim, 1973), acetyl-CoA carboxylase (ACC) catalyzes a key regulatory step in fatty acid synthesis (conversion of acetyl-CoA to malonyl-CoA). The liver isoform (ACC1) is phosphorylated by AMPK at three sites, although phosphorylation at Ser-79 appears to be responsible for inactivation (Davies et al., 1990; Ha et al., 1994). This site is phosphorylated in isolated hepatocytes (Sim and Hardie, 1988) and in rat liver *in vivo* (Davies et al., 1992), and activation of AMPK in isolated hepatocytes by arsenite (Corton et al., 1994) or AICA riboside (Corton et al., 1995) leads to dramatic inhibition of fatty acid synthesis. Skeletal muscle expresses a different isoform (ACC2) that is also inactivated by AMPK (Winder et al., 1997). Since muscle does not express fatty acid synthase, ACC2 appears not to be involved in fatty acid synthesis but instead in regulation of fatty acid oxidation. Inactivation of ACC2 lowers the cellular concentration of its product malonyl-CoA, which is an inhibitor of fatty acid uptake into mitochondria (McGarry and Brown, 1997). Thus, exercise in rat skeletal muscle causes activation of AMPK, inactivation of ACC2, and a drop in malonyl-CoA (Winder and Hardie, 1996). Similarly, stimulation of AMPK by AICA riboside in perfused rat muscle causes inactivation of ACC2, a drop in malonyl-CoA, and consequent stimulation of fatty acid oxidation (Merrill et al., 1997). This makes sense because during endurance exercise the muscle must increase its metabolism of fatty acids to maintain ATP levels. AICA riboside also stimulates fatty acid oxidation in rat hepatocytes (Velasco et al., 1997).

Effect on gene transcription

As well as the acute effects on fatty acid synthesis just described, AMPK also has chronic effects on fatty acid synthesis by inhibiting gene expression. In primary hepatocytes AMPK activation inhibits the expression of several lipogenic genes, including acetyl-CoA carboxylase, fatty acid synthase, L-pyruvate kinase and S14 (Foretz et al., 1998; Leclerc et al., 1998; Woods et al., 2000). It also inhibits the expression of the L-pyruvate kinase and proinsulin genes in a pancreatic β cell line (DaSilva-Xavier et al., 2000). In H4IIE hepatoma cells, AMPK activation totally inhibits expression of phosphoenolpyruvate carboxykinase (a key enzyme of gluconeogenesis), as well as causing a partial inhibition of expression of glucose-6-phosphatase, which catalyzes the final common step in gluconeogenesis and glycogen breakdown (Lochhead et al., 2000).

The effects just described all involve *inhibition* of transcription. However in skeletal muscle, where AMPK is activated by exercise (Winder and Hardie, 1996), AICA riboside induces *increased* expression of several proteins, including GLUT4, hexokinase, several mitochondrial enzymes, and the mitochondrial uncoupling protein UCP3 (Winder et al., 2000; Zhou et al., 2000). This is intriguing because these same adaptations are seen in response to endurance training, implying that AMPK activation might be responsible for both the acute and the chronic effects of exercise on muscle metabolism.

The actual targets for phosphorylation by AMPK that explain the modulation of expression of these genes are not known, although their promoters are well characterized and work is in progress to test candidate transcription factors as AMPK substrates. As discussed elsewhere, complexes containing the $\alpha 2$ subunit of AMPK are partially localized in the nucleus, suggesting that this might be the form that regulates transcription.

Lipolysis (hormone-sensitive lipase)
Hormone sensitive lipase is phosphorylated by AMPK at Ser-565. Although this does not appear to have any direct effects on lipase activity, it completely prevents phosphorylation and activation by cyclic AMP-dependent protein kinase at the neighbouring site, Ser-563 (Garton et al., 1989). Activation of AMPK would therefore be expected to antagonize the lipolytic effects of cyclic AMP-elevating hormones, and this has been shown to be the case in isolated adipocytes using AICA riboside (Sullivan et al., 1994; Corton et al., 1995). At first sight it may appear that lipolysis is a catabolic pathway and that it might not make sense for it to be inhibited by AMPK. However in most cases the fatty acids derived from lipolysis are not oxidized in the same cell (this is certainly true of adipocytes). If free fatty acids derived from lipolysis are not immediately removed from the cell they will recycle into triacylglycerol, consuming 2 ATP equivalents in the process. Inhibition of the hormone-sensitive lipase by AMPK may therefore be a mechanism for preventing this "futile" cycling under conditions where removal of the fatty acids is restricted.

Glycolysis in heart (6-phosphofructo-2-kinase)
Hue's group (Marsin et al., 2000) has recently reported evidence for an intriguing new mechanism by which AMPK stimulates glycolysis in cardiac muscle, thus stimulating anaerobic ATP production by glycolysis during periods of ischaemia or hypoxia. AMPK phosphorylates 6-phosphofructo-2-kinase at Ser-466, leading to an increased V_{max} for the production of fructose-2,6-bisphosphate, an activator of the glycolytic enzyme 6-phosphofructo-1-kinase. This would combine with effects of AMPK on glucose uptake in the heart (see below) to stimulate glycolysis. Hue's group provided convincing evidence that this mechanism is operative in heart during hypoxia, as well as in cultured cells expressing the recombinant heart isoform of 6-phosphofructo-2-kinase when respiration was inhibited by oligomycin. The heart isoform is only expressed in cardiac muscle and in kidney, and this mechanism would not appear to operate in tissues expressing other isoforms such as skeletal muscle and liver.

Phosphocreatine-ATP interconversion (creatine kinase)
Ponticos et al (1998) reported that the muscle (MM) isoform of creatine kinase was phosphorylated and activated by AMPK, and that these two proteins are physically associated in muscle. Although creatine kinase catalyzes a reversible reaction (creatine + ATP $\leftrightarrow$ phosphocreatine + ADP) in skeletal muscle it is believed that the mitochondrial (Mi) isoform operates in the direction of phosphocreatine synthesis, and that the phosphocreatine then diffuses to the myofibril where the MM isoform operates in the direction of ATP synthesis. In exercising muscle, AMPK would only become activated to a large extent once the phosphocreatine had been depleted and the AMP:ATP ratio started to rise. Under these conditions the creatine:phosphocreatine ratio would be high, and phosphorylation of the MM creatine kinase may be a mechanism for switching it off and preventing its operation in reverse, converting ATP to ADP.

Nitric oxide production (NO synthase)
Kemp's group (Chen et al., 1999) reported that the endothelial isoform of NO synthase was phosphorylated at Ser-1177 by AMPK, leading to an increase in activity particularly at low concentrations of Ca^{2+}-calmodulin. They also provided evidence that this phosphorylation occurs in ischaemic heart. The phosphorylation site appears to be conserved in the neuronal isoform (nNOS) which is also phosphorylated by AMPK (Fryer et al., 2000) although in this case the effects on activity have not been studied. Nitric oxide has many effects, but its classical effect is to cause relaxation of vascular smooth muscle, thus increasing blood flow. This would provide an intriguing mechanism whereby ischaemia or hypoxia would activate

AMPK, triggering the localized release of NO, and thus automatically increasing the blood flow to the tissue affected.

Glucose transport (targets unknown)
Since many cells utilize glucose as their primary carbon source for ATP production, it makes sense for AMPK to activate glucose uptake. Activation of AMPK using AICA riboside in skeletal or cardiac muscle leads to a stimulation of glucose transport that is mediated by translocation of GLUT4 from intracellular sites to the plasma membrane (Merrill et al., 1997; Hayashi et al., 1998; Kurth-Kraczek et al., 1999; Russell et al., 1999; Hayashi et al., 2000). Since AMPK is activated by exercise (Winder and Hardie, 1996), there is considerable interest in the idea that AMPK might mediate the well-known effects of exercise on glucose uptake. There is also a small stimulation of glucose transport via GLUT4 in adipocytes. An intriguing difference between muscle and adipose tissue is that in the former the effects of insulin and AICA riboside are additive (Hayashi et al., 1998), whereas in adipocytes AICA riboside inhibits the insulin effect (Salt et al., 2000). Activation by AICA riboside in cells that only express GLUT1 also stimulates glucose transport via a mechanism that does not involve translocation (Abbud et al., 2000). In none of these cases is the mechanism completely understood, although it has recently been found that the effects of AICA riboside in skeletal muscle or a muscle cell line (Fryer et al., 2000), or in adipocytes (Salt et al., 2000), are blocked by inhibitors of NO synthase. In the muscle cell line the effects are also blocked by a guanyl cyclase inhibitor, suggesting that the sequence of events is AMPK → NO synthase → guanyl cyclase → cyclic GMP → GLUT4 translocation (Fryer et al., 2000).

Other effects of AICA riboside
AICA riboside has other interesting effects on cells, including protection against apoptosis (Stefanelli et al., 1998; Durante et al., 1999), and inhibition of autophagy (Samari and Seglen, 1998). The target for AMPK in these cases, or indeed whether the effects are truly mediated by AMPK activation, remain unclear at present.

6.3. Targets for the yeast SNF1 complex

Although the overall functions of the SNF1 protein kinase are well understood through genetic approaches, few actual target proteins have been identified. Acetyl-CoA carboxylase is phosphorylated and inactivated by the SNF1 complex both *in vitro* (Mitchelhill et al., 1994) and *in vivo* (Woods et al., 1994). This suggests that the SNF1 kinase (like AMPK in mammals) might switch off fatty acid synthesis under conditions of metabolic stress, i.e. glucose deprivation. Another very interesting target is the transcription factor Mig1p, which is phosphorylated by the SNF1 complex *in vitro* at four sites (Smith et al., 1999). Mig1p binds to the promoters of many glucose-repressed genes and represses their expression, and mutation of three of the four SNF1 sites causes Mig1 to become a constitutive repressor *in vivo* (Ostling and Ronne, 1998), suggesting that SNF1 derepresses these genes in part by direct phosphorylation of Mig1p. Phosphorylation appears to abolish the repressive effect of Mig1p by causing its translocation out of the nucleus (DeVit and Johnston, 1999). Another transcription factor that is a good candidate to be a direct target of SNF1 is Sip4p, which binds to the carbon-source responsive element of gluconeogenic genes (Vincent and Carlson, 1998) and is phosphorylated in response to glucose starvation *in vivo* in a SNF1-dependent manner (Lesage et al., 1996).

Although acetyl-CoA carboxylase and Mig1p are perhaps the only well established substrates for the SNF1 complex, many cellular processes are abnormal in *snf1* mutants. As well as the failure to derepress genes required for growth on alternative carbon sources (described above), *snf1* mutants have defects in gluconeogenesis, glycogen storage, sporulation, tolerance to heat stress, and peroxisome biogenesis (Hardie et al., 1998). The molecular targets for the SNF1 protein kinase responsible for most of these effects are not known.

6.4. Targets for the plant SnRK1 complexes

Higher plant SnRK1 complexes have been shown to phosphorylate and inactivate a number of important metabolic enzymes in cell-free assays, including HMG-CoA reductase (HMGR), sucrose phosphate synthase (SPS) and nitrate reductase (NR) (Dale et al., 1995a; Douglas et al., 1997; Sugden et al., 1999b). HMGR was in fact the first metabolic enzyme in which regulation by phosphorylation was shown to be conserved between plants and animals (Dale et al., 1995a). Although it has not yet been proved that these metabolic enzymes are physiological targets in plants *in vivo*, these findings imply that the plant kinases may regulate isoprenoid biosynthesis (HMGR), sucrose synthesis (SPS) and nitrate assimilation for amino acid synthesis (NR). Like the animal homologue, the plant kinases may therefore be global regulators of biosynthesis. It is also very likely that they regulate gene expression. The best evidence for this has been cited already, i.e. that potato plants expressing a SnRK1 DNA in antisense orientation in leaves were defective in the induction of sucrose synthase mRNA by high sucrose (Purcell et al., 1998). Despite its name, sucrose synthase is thought to catalyze intracellular sucrose *breakdown in vivo* (sucrose + UDP ↔ UDP-glucose + fructose). Invertase (encoded in yeast by *SUC2*, one of the first genes shown to be regulated by the SNF1 complex) catalyzes *extracellular* breakdown of sucrose by a different reaction (sucrose + H_2O ↔ glucose + fructose). Nevertheless, it is intriguing that the yeast and plant homologues of the SNF1 protein kinase both appear to regulate expression of enzymes involved in sucrose breakdown.

7. Future perspectives

Our understanding of the physiological functions of the AMPK/SNF1 family of protein kinases has progressed dramatically in the last decade, but a number of important questions remain. The mammalian AMPK system responds directly to the energy charge of the cell by sensing AMP, ATP and phosphocreatine, although the possibility that there are other inputs into the system cannot be ruled out. The regulation of the yeast and plant systems remains much less clear. They both seem to respond to the availability of a carbon source, although rather paradoxically the yeast SNF1 complex is activated by glucose deprivation, whereas the plant SnRK1s appear to be activated by addition of high sucrose. The yeast and plant kinases are not directly activated by AMP, although AMP does inhibit dephosphorylation of the plant kinases, and it has not yet been possible to test the effects of AMP on the upstream kinases from these kingdoms. An important difference between fungal and plant cells and mammalian cells is that the latter are bathed in a medium in which the glucose is maintained at a constant 5–10 mM by sophisticated endocrine mechanisms. Carbohydrate deprivation is therefore likely to be much less of an issue for a mammalian cell that for a yeast or plant cell. Possibly in the latter cases the systems respond in a more direct manner to the availability of carbohydrate, rather than only responding indirectly by sensing adenine nucleotides. The mechanisms by which the yeast and plant systems sense carbohydrate availability remain unclear, although in the former case it may involve hexokinase PII and the Reg1p–Glc7p protein phosphatase (Sanz et al., 2000). There is also some evidence that hexokinase is involved in sugar sensing in plants (Jang et al., 1997), although whether these effects are transmitted through the SnRK1 system remains unclear.

It is already clear that these protein kinase cascades have important effects on gene expression in animals, plants and yeast. Another important challenge for the future is the elucidation of the molecular mechanisms for these effects. The yeast system provides excellent models here, where the repressor protein Mig1p (Smith et al., 1999) (and perhaps also the transcription activator Sip4p (Vincent and Carlson, 1999)) appear to be direct targets for phosphorylation by the SNF1 complex.

Acknowledgements

Studies in the author's laboratory have been supported by the Wellcome Trust, The Biotechnology and Biological Sciences Research Council, the Medical Research Council and Diabetes UK.

References

Abbud, W., Habinowski, S., Zhang, J.Z., Kendrew, J., Elkairi, F.S., Kemp, B.E., Witters, L.A. and Ismail-Beigi, F. (2000). Stimulation of AMP-Activated Protein Kinase (AMPK) Is Associated with Enhancement of Glut1-Mediated Glucose Transport. Arch. Biochem. Biophys. 380, 347–352.

Alderson, A., Sabelli, P.A., Dickinson, J.R., Cole, D., Richardson, M., Kreis, M., Shewry, P.R. and Halford, N.G. (1991). Complementation of *snf1*, a mutation affecting global regulation of carbon metabolism in yeast, by a plant protein kinase cDNA. Proc. Natl. Acad. Sci. USA 88, 8602–8605.

Ball, K.L., Barker, J., Halford, N.G. and Hardie, D.G. (1995). Immunological evidence that HMG-CoA reductase kinase-A is the cauliflower homologue of the RKIN1 subfamily of plant protein kinases. FEBS Lett. 377, 189–192.

Ball, K.L., Dale, S., Weekes, J. and Hardie, D.G. (1994). Biochemical characterization of two forms of 3-hydroxy 3-methylglutaryl CoA reductase kinase from cauliflower (*Brassica oleracia*). Eur. J. Biochem. 219, 743–750.

Barker, J.H.A., Slocombe, S.P., Ball, K.L., Hardie, D.G., Shewry, P.R. and Halford, N.G. (1996). Evidence that barley HMG-CoA reductase kinase is a member of the SNF1-related protein kinase family. Plant Physiol. 112, 1141–1149.

Bateman, A. (1997). The structure of a domain common to archaebacteria and the homocystinuria disease protein. Trends Biochem. Sci. 22, 12–13.

Beg, Z.H., Allmann, D.W. and Gibson, D.M. (1973). Modulation of 3-hydroxy-3-methylglutaryl coenzyme: A reductase activity with cAMP and with protein fractions of rat liver cytosol. Biochem. Biophys. Res. Comm. 54, 1362–1369.

Bhalerao, R.P., Salchert, K., Bako, L., Okresz, L., Szabados, L., Muranaka, T., Machida, Y., Schell, J. and Koncz, C. (1999). Regulatory interaction of PRL1 WD protein with *Arabidopsis* SNF1-like protein kinases. Proc. Natl. Acad. Sci. USA 96, 5322–5327.

Bouly, J.P., Gissot, L., Lessard, P., Kreis, M. and Thomas, M. (1999). *Arabidopsis thaliana* proteins related to the yeast SIP and SNF4 interact with AKINalpha1, an SNF1-like protein kinase. Plant J. 18, 541–550.

Carling, D., Aguan, K., Woods, A., Verhoeven, A.J.M., Beri, R.K., Brennan, C.H., Sidebottom, C., Davison, M.D. and Scott, J. (1994). Mammalian AMP-activated protein kinase is homologous to yeast and plant protein kinases involved in the regulation of carbon metabolism. J. Biol. Chem. 269, 11442–11448.

Carling, D., Clarke, P.R., Zammit, V.A. and Hardie, D.G. (1989). Purification and characterization of the AMP-activated protein kinase. Copurification of acetyl-CoA carboxylase kinase and 3-hydroxy-3-methylglutaryl-CoA reductase kinase activities. Eur. J. Biochem. 186, 129–136.

Carling, D., Zammit, V.A. and Hardie, D.G. (1987). A common bicyclic protein kinase cascade inactivates the regulatory enzymes of fatty acid and cholesterol biosynthesis. FEBS Lett. 223, 217–222.

Carlson, C.A. and Kim, K.H. (1973). Regulation of hepatic acetyl coenzyme A carboxylase by phosphorylation and dephosphorylation. J. Biol. Chem. 248, 378–380.

Carlson, M., Osmond, B.C. and Botstein, D. (1981). Mutants of yeast defective in sucrose utilization. Genetics 98, 25–40.

Celenza, J.L. and Carlson, M. (1986). A yeast gene that is essential for release from glucose repression encodes a protein kinase. Science 233, 1175–1180.

Celenza, J.L., Eng, F.J. and Carlson, M. (1989). Molecular analysis of the *SNF4* gene of *Saccharomyces cerevisiae*: evidence for physical association of the SNF4 protein with the SNF1 protein kinase. Mol. Cell. Biol. 9, 5045–5054.

Chen, Z.P., Mitchelhill, K.I., Michell, B.J., Stapleton, D., Rodriguez-Crespo, I., Witters, L.A., Power, D.A., Ortiz de Montellano, P.R. and Kemp, B.E. (1999). AMP-activated protein kinase phosphorylation of endothelial NO synthase. FEBS Lett. 443, 285–289.

Cheung, P.C.F., Salt, I.P., Davies, S.P., Hardie, D.G. and Carling, D. (2000). Characterization of AMP-activated protein kinase γ subunit isoforms and their role in AMP binding. Biochem. J. 346, 659–669.

Ching, Y.P., Davies, S.P. and Hardie, D.G. (1996). Analysis of the specificity of the AMP-activated protein kinase by site-directed mutagenesis of bacterially expressed 3-hydroxy 3-methylglutaryl-CoA reductase, using a single primer variant of the unique site elimination (USE) method. Eur. J. Biochem. 237, 800–808.

Ciriacy, M. (1977). Isolation and characterization of yeast mutants defective in intermediary carbon metabolism and in carbon catabolite repression. Molec. Gen. Genet. 154, 213–220.

Corton, J.M., Gillespie, J.G. and Hardie, D.G. (1994). Role of the AMP-activated protein kinase in the cellular stress response. Current Biol. 4, 315–324.

Corton, J.M., Gillespie, J.G., Hawley, S.A. and Hardie, D.G. (1995). 5-Aminoimidazole-4-carboxamide ribonucleoside: a specific method for activating AMP-activated protein kinase in intact cells? Eur. J. Biochem. 229, 558–565.

Crute, B.E., Seefeld, K., Gamble, J., Kemp, B.E. and Witters, L.A. (1998). Functional domains of the alpha1 catalytic subunit of the AMP-activated protein kinase. J. Biol. Chem. 273, 35347–35354.

Dale, S., Arró, M., Becerra, B., Morrice, N.G., Boronat, A., Hardie, D.G. and Ferrer, A. (1995a). Bacterial expression of the catalytic domain of 3-hydroxy-3-methylglutaryl CoA reductase (isoform HMGR1) from *Arabidopsis thaliana*, and its inactivation by phosphorylation at serine-577 by *Brassica oleracea* 3-hydroxy-3-methylglutaryl CoA reductase kinase. Eur. J. Biochem. 233, 506–513.

Dale, S., Wilson, W.A., Edelman, A.M. and Hardie, D.G. (1995b). Similar substrate recognition motifs for mammalian AMP-activated protein kinase, higher plant HMG-CoA reductase kinase-A, yeast SNF1, and mammalian calmodulin-dependent protein kinase I. FEBS Lett. 361, 191–195.

DaSilva-Xavier, G., Leclerc, I., Salt, I.P., Doiron, B., Hardie, D.G., Kahn, A. and Rutter, G. (2000). Role of the AMP-activated protein kinase in the regulation by glucose of islet β-cell gene expression. Proc. Natl. Acad. Sci. USA 97, 4023–4028.

Davies, S.P., Carling, D., Munday, M.R. and Hardie, D.G. (1992). Diurnal rhythm of phosphorylation of rat liver acetyl-CoA carboxylase by the AMP-activated protein kinase, demonstrated using freeze-clamping. Effects of high fat diets. Eur. J. Biochem. 203, 615–623.

Davies, S.P., Hawley, S.A., Woods, A., Carling, D., Haystead, T.A.J. and Hardie, D.G. (1994). Purification of the AMP-activated protein kinase on ATP-γ-Sepharose and analysis of its subunit structure. Eur. J. Biochem. 223, 351–357.

Davies, S.P., Helps, N.R., Cohen, P.T.W. and Hardie, D.G. (1995). 5′-AMP inhibits dephosphorylation, as well as promoting phosphorylation, of the AMP-activated protein kinase. Studies using bacterially expressed human protein phosphatase-2Cα and native bovine protein phosphatase-$2A_C$. FEBS Lett. 377, 421–425.

Davies, S.P., Sim, A.T.R. and Hardie, D.G. (1990). Location and function of three sites phosphorylated on rat acetyl-CoA carboxylase by the AMP-activated protein kinase. Eur. J. Biochem. 187, 183–190.

DeVit, M.J. and Johnston, M. (1999). The nuclear exportin Msn5 is required for nuclear export of the Mig1 glucose repressor of *Saccharomyces cerevisiae*. Curr. Biol. 9, 1231–1241.

Dienert, F. (1900). Sur la fermentation du galactose et sur l'accoutumance des levures a ce sucre. Ann. Inst. Pasteur 19, 139–189.

Douglas, P., Pigaglio, E., Ferrer, A., Halford, N.G. and MacKintosh, C. (1997). Three spinach leaf nitrate reductase/3-hydroxy-3-methylglutaryl-CoA reductase kinases that are regulated by reversible phosphorylation and/or Ca^{2+} ions. Biochem. J. 325, 101–109.

Durante, P., Gueuning, M.A., Darville, M.I., Hue, L. and Rousseau, G.G. (1999). Apoptosis induced by growth factor withdrawal in fibroblasts overproducing fructose 2,6-bisphosphate. FEBS Lett 448(2–3), 239–43.

Estruch, F., Treitel, M.A., Yang, X. and Carlson, M. (1992). N-terminal mutations modulate yeast SNF1 protein kinase function. Genetics 132, 639–50.

Foretz, M., Carling, D., Guichard, C., Ferré, P. and Foufelle, F. (1998). AMP-activated protein kinase Inhibits the glucose-activated expression of fatty acid synthase gene in rat hepatocytes. J. Biol. Chem. 273, 14767–14771.

Fryer, L.G.D., Hardie, D.G. and Carling, D. (2000). Stimulation of glucose transport by AMP-activated protein kinase via activation of nitric oxide synthase. Diabetes, in press.

Gancedo, J.M. (1998). Yeast carbon catabolite repression. Microbiology and Molecular Biology Reviews 62, 334–361.

Gao, G., Fernandez, S., Stapleton, D., Auster, A.S., Widmer, J., Dyck, J.R.B., Kemp, B.E. and Witters, L.A. (1996). Non-catalytic β- and γ-subunit isoforms of the 5′-AMP-activated protein kinase. J. Biol. Chem. 271, 8675–8681.

Gao, G., Widmer, J., Stapleton, D., Teh, T., Cox, T., Kemp, B.E. and Witters, L.A. (1995). Catalytic subunits of the porcine and rat 5′-AMP-activated protein kinase are members of the SNF1 protein kinase family. Biochim. Biophys. Acta 1266, 73–82.

Garton, A.J., Campbell, D.G., Carling, D., Hardie, D.G., Colbran, R.J. and Yeaman, S.J. (1989). Phosphorylation of bovine hormone-sensitive lipase by the AMP-activated protein kinase. A possible antilipolytic mechanism. Eur. J. Biochem. 179, 249–254.

Gillespie, J.G. and Hardie, D.G. (1992). Phosphorylation and inactivation of HMG-CoA reductase at the AMP-activated protein kinase site in response to fructose treatment of isolated rat hepatocytes. FEBS Lett. 306, 59–62.

Goffrini, P., Ficarelli, A., Donnini, C., Lodi, T., Puglisi, P.P. and Ferrero, I. (1996). FOG1 and FOG2 genes, required for the transcriptional activation of glucose-repressible genes of Kluyveromyces lactis, are homologous to GAL83 and SNF1 of *Saccharomyces cerevisiae*. Curr. Genet. 29, 316–26.

Ha, J., Daniel, S., Broyles, S.S. and Kim, K.H. (1994). Critical phosphorylation sites for acetyl-CoA carboxylase activity. J. Biol. Chem. 269, 22162–22168.

Halford, N.G. and Hardie, D.G. (1998). SNF1-related protein kinases: global regulators of carbon metabolism in plants? Plant Mol. Biol. 37, 735–748.

Hardie, D.G., Carling, D. and Carlson, M. (1998). The AMP-activated/SNF1 protein kinase subfamily: metabolic sensors of the eukaryotic cell? Ann. Rev. Biochem. 67, 821–855.

Hardie, D.G., Carling, D. and Sim, A.T.R. (1989). The AMP-activated protein kinase—a multisubstrate regulator of lipid metabolism. Trends Biochem. Sci. 14, 20–23.

Hardie, D.G., Salt, I.P., Hawley, S.A. and Davies, S.P.

(1999). AMP-activated protein kinase: an ultrasensitive system for monitoring cellular energy charge. Biochem. J. 338, 717–722.

Hawley, S.A., Selbert, M.A., Goldstein, E.G., Edelman, A.M., Carling, D. and Hardie, D.G. (1995). 5′-AMP activates the AMP-activated protein kinase cascade, and Ca^{2+}/calmodulin the calmodulin-dependent protein kinase I cascade, via three independent mechanisms. J. Biol. Chem. 270, 27186–27191.

Hayashi, T., Hirshman, M.F., Fujii, N., Habinowski, S.A., Witters, L.A. and Goodyear, L.J. (2000). Metabolic stress and altered glucose transport: activation of AMP-activated protein kinase as a unifying coupling mechanism. Diabetes 49, 527–531.

Hayashi, T., Hirshman, M.F., Kurth, E.J., Winder, W.W. and Goodyear, L.J. (1998). Evidence for 5′ AMP-activated protein kinase mediation of the effect of muscle contraction on glucose transport. Diabetes 47, 1369–1373.

Henin, N., Vincent, M.F. and Van den Berghe, G. (1996). Stimulation of rat liver AMP-activated protein kinase by AMP analogues. Biochim. Biophys. Acta 1290, 197–203.

Hutber, C.A., Hardie, D.G. and Winder, W.W. (1997). Electrical stimulation inactivates muscle acetyl-CoA carboxylase and increases AMP-activated protein kinase activity. Am. J. Physiol. 272, E262–E266.

Jang, J.C., Leon, P., Zhou, L. and Sheen, J. (1997). Hexokinase as a sugar sensor in higher plants. Plant Cell 9, 5–19.

Javaux, F., Vincent, M.F., Wagner, D.R. and van den Berghe, G. (1995). Cell-type specificity of inhibition of glycolysis by 5-amino-4-imidazolecarboxamide riboside. Lack of effect in rabbit cardiomyocytes and human erythrocytes, and inhibition in FTO-2B rat hepatoma cells. Biochem. J. 305, 913–919.

Jiang, R. and Carlson, M. (1996). Glucose regulates protein interactions within the yeast SNF1 protein kinase complex. Genes Dev. 10, 3105–3115.

Jiang, R. and Carlson, M. (1997). The Snf1 protein kinase and its activating subunit, Snf4, interact with distinct domains of the Sip1/Sip2/Gal83 component in the kinase complex. Mol. Cell. Biol. 1, 2099–2106.

Johnson, L.N., Noble, M.E.M. and Owen, D.J. (1996). Active and inactive protein kinases: structural basis for regulation. Cell 85, 149–158.

Kemp, B.E., Parker, M.W., Hu, S., Tiganis, T. and House, C. (1994). Substrate and pseudosubstrate interactions with protein kinases: determinants of specificity. Trends Biochem Sci 19, 440–444.

Krebs, H. (1963). The Croonian lecture—gluconeogenesis. Proc. Roy. Soc. Lond. B 159, 545–564.

Kudo, N., Barr, A.J., Barr, R.L., Desai, S. and Lopaschuk, G.D. (1995). High rates of fatty acid oxidation during reperfusion of ischemic hearts are associated with a decrease in malonyl-CoA levels due to an increase in 5′-AMP-activated protein kinase inhibition of acetyl-CoA carboxylase. J. Biol. Chem. 270, 17513–17520.

Kudo, N., Gillespie, J.G., Kung, L., Witters, L.A., Schulz, R., Clanachan, A.S. and Lopaschuk, G.D. (1996). Characterization of 5′AMP-activated protein kinase activity in the heart and its role in inhibiting acetyl-CoA carboxylase during reperfusion following ischemia. Biochim. Biophys. Acta 1301, 67–75.

Kurth-Kraczek, E.J., Hirshman, M.F., Goodyear, L.J. and Winder, W.W. (1999). 5′ AMP-activated protein kinase activation causes GLUT4 translocation in skeletal muscle. Diabetes 48, 1667–1671.

Lakatos, L., Klein, M., Hofgen, R. and Banfalvi, Z. (1999). Potato StubSNF1 interacts with StubGAL83: a plant protein kinase complex with yeast and mammalian counterparts. Plant J. 17, 569–574.

Leclerc, I., Kahn, A. and Doiron, B. (1998). The 5′-AMP-activated protein kinase inhibits the transcriptional stimulation by glucose in liver cells, acting through the glucose response complex. FEBS Lett. 431, 180–184.

Lesage, P., Yang, X. and Carlson, M. (1996). Yeast SNF1 protein kinase interacts with SIP4, a C6 zinc cluster transcriptional activator: a new role for SNF1 in the glucose response. Mol. Cell. Biol. 16, 1921–1928.

Lochhead, P.A., Salt, I.P., Walker, K.S., Hardie, D.G. and Sutherland, C. (2000). 5-aminoimidazole-4-carboxamide riboside mimics the effects of insulin on the expression of the 2 key gluconeogenic genes PEPCK and glucose-6-phosphatase. Diabetes 49, 896–903.

MacKintosh, R.W., Davies, S.P., Clarke, P.R., Weekes, J., Gillespie, J.G., Gibb, B.J. and Hardie, D.G. (1992). Evidence for a protein kinase cascade in higher plants: 3-hydroxy-3-methylglutaryl-CoA reductase kinase. Eur. J. Biochem. 209, 923–931.

Marsin, A.S., Bertrand, L., Rider, M.H., Deprez, J., Beauloye, C., Vincent, M.F., Van den Berghe, G., Carling, D. and Hue, L. (2000). Heart 6-phosphofructo-2-kinase is a new substrate for AMP-activated protein kinase: role in the stimulation of glycolysis in ischaemic myocardium. Current Biol., in press.

McGarry, J.D. and Brown, N.F. (1997). The mitochondrial carnitine palmitoyltransferase system. From concept to molecular analysis. Eur. J. Biochem. 244, 1–14.

Merrill, G.M., Kurth, E., Hardie, D.G. and Winder, W.W. (1997). AICAR decreases malonyl-CoA and increases fatty acid oxidation in skeletal muscle of the rat. Am. J. Physiol. 273, E1107–E1112.

Mitchelhill, K.I., Stapleton, D., Gao, G., House, C., Michell, B., Katsis, F., Witters, L.A. and Kemp, B.E. (1994). Mammalian AMP-activated protein kinase shares structural and functional homology with the catalytic domain of yeast Snf1 protein kinase. J. Biol. Chem. 269, 2361–2364.

Munday, M.R., Campbell, D.G., Carling, D. and Hardie,

D.G. (1988a). Identification by amino acid sequencing of three major regulatory phosphorylation sites on rat acetyl-CoA carboxylase. Eur. J. Biochem. 175, 331–338.

Munday, M.R., Carling, D. and Hardie, D.G. (1988b). Negative interactions between phosphorylation of acetyl-CoA carboxylase by the cyclic AMP-dependent and AMP-activated protein kinases. FEBS Lett. 235, 144–148.

Omkumar, R.V. and Rodwell, V.W. (1994). Phosphorylation of Ser(871) impairs the function of His(865) of Syrian hamster 3-hydroxy-3-methylglutaryl-CoA reductase. J. Biol. Chem. 269, 16862–16866.

Ostling, J. and Ronne, H. (1998). Negative control of the Mig1p repressor by Snf1p-dependent phosphorylation in the absence of glucose. Eur. J. Biochem. 252, 162–168.

Ponticos, M., Lu, Q.L., Morgan, J.E., Hardie, D.G., Partridge, T.A. and Carling, D. (1998). Dual regulation of the AMP-activated protein kinase provides a novel mechanism for the control of creatine kinase in skeletal muscle. EMBO J. 17, 1688–1699.

Purcell, P.C., Smith, A.M. and Halford, N.G. (1998). Antisense expression of a sucrose non-fermenting-1-related protein kinase sequence in potato results in decreased expression of sucrose synthase in tubers and loss of sucrose-inducibility of sucrose synthase transcripts in leaves. Plant J. 14, 195–202.

Ramaiah, A., Hathaway, J.A. and Atkinson, D.E. (1964). Adenylate as a metabolic regulator. Effect on yeast phosphofructokinase kinetics. J. Biol. Chem. 239, 3619.

Russell, R.R., Bergeron, R., Shulman, G.I. and Young, L.H. (1999). Translocation of myocardial GLUT-4 and increased glucose uptake through activation of AMPK by AICAR. Am. J. Physiol. 277, H643–649.

Salt, I.P., Celler, J.W., Hawley, S.A., Prescott, A., Woods, A., Carling, D. and Hardie, D.G. (1998a). AMP-activated protein kinase—greater AMP dependence, and preferential nuclear localization, of complexes containing the α2 isoform. Biochem. J. 334, 177–187.

Salt, I.P., Connell, J.M.C. and Gould, G.W. (2000). 5-aminoimidazole-4-carboxamide ribonucleoside (AICAR) inhibits insulin-stimulated glucose transport in 3T3-L1 adipocytes. Diabetes 49, 1649–1656.

Salt, I.P., Johnson, G., Ashcroft, S.J.H. and Hardie, D.G. (1998b). AMP-activated protein kinase is activated by low glucose in cell lines derived from pancreatic β cells, and may regulate insulin release. Biochem. J. 335, 533–539.

Samari, H.R. and Seglen, P.O. (1998). Inhibition of hepatocytic autophagy by adenosine, aminoimidazole-4-carboxamide riboside, and N_6-mercaptopurine riboside. Evidence for involvement of AMP-activated protein kinase. J. Biol. Chem. 273, 23758–23763.

Sanz, P., Alms, G.R., Haystead, T.A. and Carlson, M. (2000). Regulatory interactions between the Reg1-Glc7 protein phosphatase and the Snf1 protein kinase. Mol. Cell. Biol. 20, 1321–1328.

Sato, R., Goldstein, J.L. and Brown, M.S. (1993). Replacement of Serine-871 of hamster 3-hydroxy-3-methylglutaryl CoA reductase prevents phosphorylation by AMP-activated protein kinase and blocks inhibition of sterol synthesis induced by ATP depletion. Proc. Natl. Acad. Sci. USA 90, 9261–9265.

Sim, A.T.R. and Hardie, D.G. (1988). The low activity of acetyl-CoA carboxylase in basal and glucagon-stimulated hepatocytes is due to phosphorylation by the AMP-activated protein kinase and not cyclic AMP-dependent protein kinase. FEBS Lett. 233, 294–298.

Smith, F.C., Davies, S.P., Wilson, W.A., Carling, D. and Hardie, D.G. (1999). The SNF1 kinase complex from *Saccharomyces cerevisiae* phosphorylates the repressor protein Mig1p *in vitro* at four sites within or near Regulatory Domain 1. FEBS Lett. 453, 219–223.

Stapleton, D., Mitchelhill, K.I., Gao, G., Widmer, J., Michell, B.J., Teh, T., House, C.M., Fernandez, C.S., Cox, T., Witters, L.A. and Kemp, B.E. (1996). Mammalian AMP-activated protein kinase subfamily. J. Biol. Chem. 271, 611–614.

Stefanelli, C., Stanic, I., Bonavita, F., Flamigni, F., Pignatti, C., Guarnieri, C. and Caldarera, C.M. (1998). Inhibition of glucocorticoid-induced apoptosis with 5-aminoimidazole-4-carboxamide ribonucleoside, a cell-permeable activator of AMP-activated protein kinase. Biochem. Biophys. Res. Commun. 243, 821–6.

Sugden, C., Crawford, R.M., Halford, N.G. and Hardie, D.G. (1999a). Regulation of spinach SNF1-related (SnRK1) kinases by protein kinases and phosphatases is associated with phosphorylation of the T loop and is regulated by 5′-AMP. Plant J. 19, 1–7.

Sugden, C., Donaghy, P.G., Halford, N.G. and Hardie, D.G. (1999b). Two SNF1-related protein kinases from spinach leaf phosphorylate and inactivate HMG-CoA reductase, nitrate reductase and sucrose phosphate synthase *in vitro*. Plant Physiol. 120, 257–274.

Sullivan, J.E., Brocklehurst, K.J., Marley, A.E., Carey, F., Carling, D. and Beri, R.K. (1994). Inhibition of lipolysis and lipogenesis in isolated rat adipocytes with AICAR, a cell-permeable activator of AMP-activated protein kinase. FEBS Lett. 353, 33–36.

Thornton, C., Snowden, M.A. and Carling, D. (1998). Identification of a novel AMP-activated protein kinase β subunit isoform which is highly expressed in skeletal muscle. J. Biol. Chem. 273, 12443–12450.

Turnley, A.M., Stapleton, D., Mann, R.J., Witters, L.A., Kemp, B.E. and Bartlett, P.F. (1999). Cellular distribution and developmental expression of AMP-activated protein kinase isoforms in mouse central nervous system. J. Neurochem. 72, 1707–16.

Vavvas, D., Apazidis, A., Saha, A.K., Gamble, J., Patel, A.,

Kemp, B.E., Witters, L.A. and Ruderman, N.B. (1997). Contraction-induced changes in acetyl-CoA carboxylase and 5′-AMP-activated kinase in skeletal muscle. J. Biol. Chem. 272, 13255–13261.

Velasco, G., Geelen, M.J.H. and Guzman, M. (1997). Control of hepatic fatty acid oxidation by 5′-AMP-activated protein kinase involves a malonyl-CoA-dependent and a malonyl-CoA-independent mechanism. Arch. Biochem. Biophys. 337, 169–175.

Vincent, M.F., Marangos, P.J., Gruber, H.E. and Van den Berghe, G. (1991). Inhibition by AICA riboside of gluconeogenesis in isolated rat hepatocytes. Diabetes 40, 1259–1266.

Vincent, O. and Carlson, M. (1998). Sip4, a Snf1 kinase-dependent transcriptional activator, binds to the carbon source-responsive element of gluconeogenic genes. EMBO J. 17, 7002–7008.

Vincent, O. and Carlson, M. (1999). Gal83 mediates the interaction of the snf1 kinase complex with the transcription activator sip4. EMBO J. 18, 6672–6681.

Weekes, J., Ball, K.L., Caudwell, F.B. and Hardie, D.G. (1993). Specificity determinants for the AMP-activated protein kinase and its plant homologue analysed using synthetic peptides. FEBS Lett. 334, 335–339.

Weekes, J., Hawley, S.A., Corton, J., Shugar, D. and Hardie, D.G. (1994). Activation of rat liver AMP-activated protein kinase by the upstream kinase kinase in a purified, reconstituted system. Effects of AMP and AMP analogues. Eur. J. Biochem. 219, 751–757.

Wilson, W.A., Hawley, S.A. and Hardie, D.G. (1996). The mechanism of glucose repression/derepression in yeast: SNF1 protein kinase is activated by phosphorylation under derepressing conditions, and this correlates with a high AMP:ATP ratio. Current Biology 6, 1426–1434.

Winder, W.W. and Hardie, D.G. (1996). Inactivation of acetyl-CoA carboxylase and activation of AMP-activated protein kinase in muscle during exercise. Am. J. Physiol. 270, E299–E304.

Winder, W.W., Holmes, B.F., Rubink, D.S., Jensen, E.B., Chen, M. and Holloszy, J.O. (2000). Activation of AMP-activated protein kinase increases mitochondrial enzymes in skeletal muscle. J. Appl. Physiol. 88, 2219–2226.

Winder, W.W., Wilson, H.A., Hardie, D.G., Rasmussen, B.B., Hutber, C.A., Call, G.B., Clayton, R.D., Conley, L.M., Yoon, S. and Zhou, B. (1997). Phosphorylation of rat muscle acetyl-CoA carboxylase by AMP-activated protein kinase and cAMP-dependent protein kinase. J. App. Physiol. 82, 219–225.

Woods, A., Azzout-Marniche, D., Foretz, M., Stein, S.C., Lemarchand, P., Ferre, P., Foufelle, F. and Carling, D. (2000). Characterization of the role of AMP-activated protein kinase in the regulation of glucose-activated gene expression using constitutively active and dominant negative forms of the kinase. Mol. Cell. Biol. 20, 6704–6711.

Woods, A., Cheung, P.C.F., Smith, F.C., Davison, M.D., Scott, J., Beri, R.K. and Carling, D. (1996). Characterization of AMP-activated protein kinase β and γ subunits: assembly of the heterotrimeric complex *in vitro*. J. Biol. Chem. 271, 10282–10290.

Woods, A., Munday, M.R., Scott, J., Yang, X., Carlson, M. and Carling, D. (1994). Yeast SNF1 is functionally related to mammalian AMP-activated protein kinase and regulates acetyl-CoA carboxylase *in vivo*. J. Biol. Chem. 269, 19509–19515.

Yang, X., Jiang, R. and Carlson, M. (1994). A family of proteins containing a conserved domain that mediates interaction with the yeast SNF1 protein kinase complex. EMBO J. 13, 5878–5886.

Young, M.E., Radda, G.K. and Leighton, B. (1996). Activation of glycogen phosphorylase and glycogenolysis in rat skeletal muscle by AICAR—an activator of AMP-activated protein kinase. FEBS Lett. 382, 43–47.

Zhou, M., Lin, B.Z., Coughlin, S., Vallega, G. and Pilch, P.F. (2000). UCP-3 expression in skeletal muscle: effects of exercise, hypoxia, and AMP-activated protein kinase. Am. J. Physiol. 279, E622–E629.

Zimmermann, F.K., Kaufmann, I., Rasenberger, H. and Haussman, P. (1977). Genetics of carbon catabolite repression in *Saccharomyces cerevisiae*: genes involved in the derepression process. Molec. Gen. Genet. 151, 95–103.

Protein Adaptations and Signal Transduction. Edited by K.B. Storey and J.M. Storey

CHAPTER 12

Cellular Regulation of Protein Kinase C

Alexandra C. Newton[1]* and Alex Toker[2]
[1]*Department of Pharmacology, University of California at San Diego, La Jolla, CA 92093-0640, U.S.A.;*
[2]*Department of Pathology, Beth Israel Deaconess Medical Center, Harvard Medical School, Boston, MA 02215, U.S.A.*

1. Protein kinase C: a central role in signaling

Members of the protein kinase C (PKC) family of serine/threonine kinases transduce a multitude of signals, operating through diverse receptor mechanisms (Newton, 1997; Nishizuka, 1995). For most family members, a unifying feature of the signals they transduce is the production of the lipid second messenger, diacylglycerol. This second messenger can be produced by activation of G protein-coupled receptors, tyrosine kinase receptors, or even non-receptor tyrosine kinases. The finding that PKC can regulate opposing cellular functions such as cell survival in some cases, and cell death in other cases, underscores the key role of PKC in cellular signaling. It also suggests complex mechanisms to maintain fidelity and specificity in the transduction of signals. This chapter begins by discussing how the function of PKC is regulated in the cell, and then focuses on the role of PKC isozymes in cell survival and cell death.

2. Structure, function, and regulation of protein kinase C

2.1. *Protein kinase C family members*

Molecular cloning of the first PKC family members in the mid 1980s provided the first clue that this enzyme was actually a family of structurally and functionally related proteins (Coussens et al., 1986). To date, there are 10 known mammalian isozymes that fall into three major classes: conventional (α, βI, βII, γ), novel (δ, ε, η, θ), and atypical (ζ, ι) PKC isozymes (Fig. 12.1) (Mellor and Parker, 1998). In addition, PKC μ and ν are considered by some to constitute a fourth class and by others to comprise a distinct family called protein kinase D.

All isozymes have in common a single polypeptide with an amino-terminal regulatory moiety and a carboxyl-terminal kinase core (Newton, 1997). The kinase core is similar to that of protein kinase A, except for protein kinase D isozymes which are more closely related to Ca^{2+}/calmodulin dependent kinases. The regulatory moiety contains two important functional segments: an autoinhibitory pseudosubstrate sequence which allosterically regulates access to the substrate-binding cavity, and a membrane-targeting module. It is the nature of the membrane targeting modules that defines the classes of PKC isozymes. All PKCs have a version of the C1 domain, the diacylglycerol sensor. This domain binds diacylglycerol and the potent functional analogues, phorbol esters, in all isozymes except atypical PKC. For these isozymes, an impaired ligand binding pocket does not support the binding of diacylglycerol or phorbol esters and, as a consequence, the hallmark of atypical PKCs is their complete lack of response to phorbol esters/diacylglycerol. Conventional and novel PKCs

*Corresponding author.

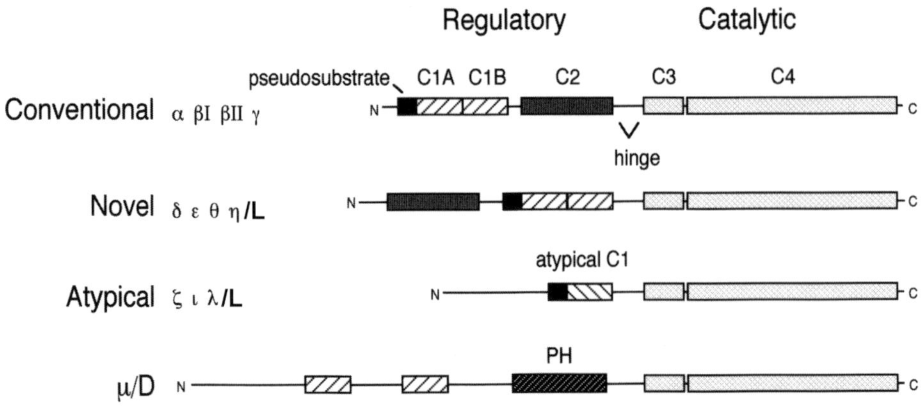

Fig. 12.1. Schematic of primary structure of PKC family members. Conventional, novel, and atypical subclasses are shown, in addition to the closely related protein kinase D; members of each subclass are noted. The amino-terminal moiety contains regulatory elements. These are the autoinhibitory pseudosubstrate sequence (black box) and membrane-targeting modules: the C1A and C1B domains which bind diacylglycerol or phorbol esters in all but the atypical PKCs (hatched box); the C2 domain which binds anionic lipids and, for conventional PKCs, Ca^{2+} (dark grey box); the PH domain which binds phosphoinositides (dark hatched box). The carboxyl-terminal moiety contains the kinase domain (light gray boxes) consisting of the ATP binding lobe (C3) and substrate binding lobe (C4). Adapted from Newton and Johnson (1998).

have a C2 domain; this domain binds anionic lipid in a Ca^{2+}-dependent manner for conventional PKCs. However, an impaired Ca^{2+} binding pocket in the novel PKCs makes them unresponsive to Ca^{2+}. PKC μ and ν have a PH domain which presumably binds phosphoinositides. Interestingly, PKC μ has been shown to interact with both phosphoinositide 4-kinase and phosphoinositide 4-phosphate 5-kinase activities through a region between the N-terminus and the PH domain, although the physiological significance of this interaction is unknown (Nishikawa et al., 1998).

2.2. *Membrane binding modules regulate the function of protein kinase C*

PKC is maintained in an inactive state by the pseudosubstrate, which occupies the substrate-binding cavity and blocks substrate access. Engagement of the enzyme's membrane-binding modules on the membrane provides the energy to release the pseudosubstrate and thus allow substrate phosphorylation (Johnson et al., 2000; Newton and Johnson, 1998). In the case of conventional PKCs, elevation of intracellular Ca^{2+} promotes the membrane association of the C2 domain and generation of diacylglycerol promotes the membrane association of the C1 domain. The binding energy from engaging one domain alone is not sufficient to allow significant activation of PKC, but engagement of both domains results in a high affinity membrane interaction that effectively releases the pseudosubstrate and maximally activates PKC. Novel PKCs respond only to elevations in intracellular diacylglycerol and it is not yet clear what the precise role of their C2 domain is in membrane translocation or enzyme function.

2.3. *Phosphorylation*

Before PKC is competent to respond to second messengers, it must first be processed by a series of ordered phosphorylations (Keranen et al., 1995; Newton, 1997; Parekh et al., 2000). The first phosphorylation is catalyzed by the recently discovered phosphoinositide-dependent kinase-1, PDK-1 (Chou et al., 1998; Dutil et al., 1998; Le Good et al., 1998) (Fig. 12.2). This kinase plays a pivotal role in cellular signaling by providing the 'on' switch to the catalytic function of diverse members of the AGC family of protein kinases (Toker and Newton, 2000). This switch is located on a loop near the entrance to the active site, referred to as the activation loop. Phosphorylation at

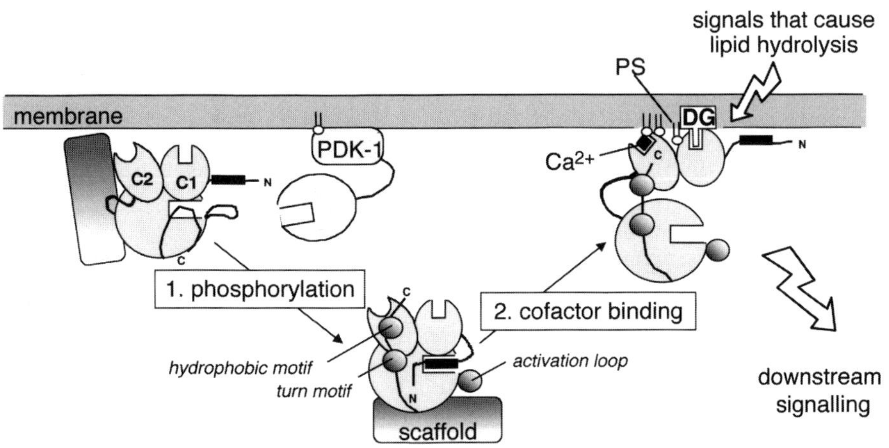

Fig. 12.2. Model summarizing the spatial, structural, and conformational regulation of PKC by phosphorylation, targeting proteins, and cofactors. Newly synthesized PKC (left panel) associates with the membrane in a conformation in which the pseudosubstrate sequence (black rectangle) is out of the active site. The first step in the post-translational modification of PKC is phosphorylation at the activation loop by PDK-1 (gray circle represents phosphate). This phosphorylation correctly aligns residues for catalysis, triggering the autophosphorylation of the two carboxyl-terminal sites (turn motif and hydrophobic motif). The fully phosphorylated species is then released into the cytosol, where it is maintained in an auto-inhibited conformation by the pseudosubstrate (middle panel). Generation of diacylglycerol provides the allosteric switch to activate PKC by engaging the membrane-targeting modules on the membrane (right panel), thus providing the energy to release the pseudosubstrate from the active site, allowing substrate binding and catalysis. In addition to the regulation by phosphorylation and cofactors, scaffold proteins (shaded oblongs) play a key role in PKC function by positioning specific isozymes at particular intracellular locations.

a conserved Thr on this loop correctly aligns residues in the active site for catalysis and causes the activation loop to swing away from the active site to promote substrate binding. Conventional, novel, and atypical PKCs are all regulated by PDK-1.

Following phosphorylation by PDK-1, PKC rapidly autophosphorylates at two conserved positions on the carboxyl-terminus, the turn motif (Thr 641 in PKC βII) and the hydrophobic motif (Ser 660 in PKC βII) (Behn-Krappa and Newton, 1999). These phosphorylation sites are conserved among all PKC family members, except that a Glu is present in the place of the phosphorylatable residue in the atypical PKCs (Keranen et al., 1995). Phosphorylation at these positions locks PKC in a catalytically competent, phosphatase-resistant, and thermally stable conformation (Bornancin and Parker, 1997; Bornancin and Parker, 1996; Edwards et al., 1999; Edwards and Newton, 1997). Indeed, once phosphorylated at the turn motif, phosphate on the activation loop is dispensable for activity.

Studies with conventional PKCs have revealed that newly synthesized PKC associates with the membrane (Dutil and Newton, unpublished data). The conformation of this species of PKC is such that the pseudosubstrate is released from the active site, thus exposing the activation loop phosphorylation site (Dutil and Newton, 2000). Following the phosphorylation events, the pseudosubstrate gains access to the active site, and the mature, fully phosphorylated enzyme is released into the cytosol (Fig. 12.2). This species accounts for most of the PKC in unstimulated cells. Although catalytically competent, it is maintained in an inactive conformation by the bound pseudosubstrate. Signals that cause diacylglycerol production are required to allosterically activate the enzyme, as discussed above.

2.4. Regulation of PDK-1, the upstream kinase for protein kinase C

Similar to PKC, PDK-1 contains a membrane-targeting module (Vanhaesebroeck and Alessi,

2000). In this case, it is a PH domain carboxyl-terminal to the kinase domain. However, unlike PKC, this module does not need to be engaged on the membrane for PDK-1 to phosphorylate substrates. Rather, the majority of evidence to date suggests a model by which PDK-1 is constitutively active in cells, with substrate phosphorylation depending on the conformation and subcellular location of the substrate (Toker and Newton, 2000). Such a mechanism provides an elegant explanation for how one kinase can have so many substrates yet retain specificity in signaling by specific stimuli. For example, phosphorylation of Akt by PDK-1 requires activation of PI3-kinase because 3-phosphoinositides engage the PH domain of Akt on the membrane, unmasking the PDK-1 phosphorylation site (Stokoe et al., 1997). In the case of PKC, the pseudosubstrate must be released in order for PDK-1 to access the activation loop phosphorylation site (Dutil and Newton, 2000).

For PKCs, the phosphorylation by PDK-1 does, indeed, appear to be constitutive (Gao et al., 2000). It occurs equally well in serum-starved cells, where PI 3-kinase activity is off, or under conditions where PI 3-kinase has been activated. Whether stimuli other than 3-phosphoinositides regulate the activity of PDK-1 and hence its processing of PKC is still an open question. In this regard, PDK-1 is regulated by phosphorylation at its own activation loop, although for this kinase the reaction is autophosphorylation (Casamayor et al., 1999). However, the enzyme is multiply phosphorylated at additional sites which may regulate its function (Casamayor et al., 1999; Prasad et al., 2000).

2.5. Protein kinase C anchoring proteins

The subcellular location of PKC is critical for its activation and a battery of binding partners for PKC have been characterized (Jaken and Parker, 2000; Mochly-Rosen, 1995). These proteins localize PKC isozymes at specific intracellular locations, positioning them near their substrates or regulators such as phosphatases, kinases, or second messengers. Proteins that localize unphosphorylated PKC, phosphorylated but inactive PKC, and phosphorylated and activated PKC have been identified. For example, the anchoring protein CG-NAP has recently been shown to localize newly synthesized (unphosphorylated) PKC ε to the Golgi/centrosome (Takahashi et al., 2000). Members of the AKAP family of scaffold proteins (for A Kinase Anchoring Proteins) position phosphorylated but inactive PKC near relevant substrates (Klauck et al., 1996). A family of proteins called RACKs (for Receptors For Activated C Kinase) anchor the active conformation of PKC at specific cellular locations (Mochly-Rosen et al., 1991). Other proteins named STICKs (for Substrates That Interact with C Kinase) tether inactive PKC and release the enzyme following their phosphorylation resulting from activation of PKC (Jaken, 1996). Other PKC adapter proteins negatively regulate signaling. For example, interaction of atypical PKC ζ with the product of the par-4 gene serves to inactivate the protein kinase leading to apoptosis, as discussed below.

The importance of scaffold proteins in signaling by PKC is epitomized by genetic studies focusing on the phototransductive cascade in Drosophila. In this system, the scaffold protein ina D coordinates a number of proteins involved in phototransduction through a series of PDZ domains, each specific for a particular protein. Abolishing its interaction with any one of these proteins, including PKC, results in mislocalization of the relevant signaling protein and disrupts signaling (Tsunoda et al., 1997).

2.6. Summary

PKC family members are regulated by two coordinated mechanisms: phosphorylation which is required to generate catalytically competent and stable enzyme, and cofactors which allosterically regulate enzyme activity by removing the pseudosubstrate from the active site (Fig. 12.2). The phosphorylation mechanism is initiated by PDK-1, a kinase that plays a central role in signaling by modulating a conserved phosphorylation switch in many members of the AGC family of kinases. The allosteric regulation is initiated by generation of diacylglycerol and Ca^{2+}, which engage the C1 and

C2 domains of PKC on membranes, providing the energy to expel the pseudosubstrate from the active site. In addition, protein:protein interactions provide an important mechanism to localize PKC at specific cellular locations.

3. Protein kinase C in cell survival and programmed cell death

Programmed cell death (or apoptosis) is a complex cellular process subject to multiple mechanisms of regulation. Protection from apoptosis (or cellular survival) is typically associated with interference with one of these regulatory events. The current model for apoptosis in higher eukaryotic cells involves an initial disruption of mitochondrial integrity leading to cytochrome c release. Cytochrome c then binds to and activates the apoptotic protease-activating factor (Apaf-1). Apaf-1 in turn binds to and activates the initiator cysteine protease, caspase-9. This triggers a cascade of activation of additional caspases, terminating in the activation of the executioner caspases 3 and 7 (reviewed in Thornberry and Lazebnik, 1998). A large number of proteins have been shown to regulate these series of events, most notably members of the Bcl-2 family of proteins. Protein kinases such as the Akt/PKB proto-oncogene, and the chaperone 14-3-3 proteins are implicated in cellular survival by regulating the function of Bcl-2 family members.

It has long been known that PKC isozymes regulate apoptosis. Paradoxically, however, initial studies suggested that activation of PKC can have pro- as well as anti-apoptotic effects in cells. These studies depended on the use of phorbol esters such as PMA, which in some cell types causes apoptosis, whereas in others can protect cells from a variety of stresses which lead to death (Deacon et al., 1997). Elucidating the role of PKC was further confounded by the fact that the ubiquitously expressed PKC ζ isozyme is an important mediator of cellular survival but it is not PMA responsive. Recent studies using molecular genetic (active and inactive mutants of PKC) and pharmacological (specific inhibitors of distinct isozymes) approaches have begun to shed light on the role of protein kinase C in stress responses. The emerging picture is that conventional and atypical PKCs appear to regulate cellular survival, whereas novel isozymes such as PKC δ mediate pro-apoptotic functions (Fig. 12.3).

3.1. Conventional protein kinase Cs

Considerable evidence suggests that the conventional PKC α and β isozymes are involved in cellular survival, although the precise mechanism by which these isozymes mediate survival is largely unexplored. A loss of PKC α protein is often associated with apoptosis in a variety of cell types, and this loss can be mimicked with chronic phorbol ester treatment or through the use of antisense oligonucleotides (Haimovitz-Friedman et al., 1994; Whelan and Parker, 1998). In promyelocytic U937 cells, a loss of PKC α function has been shown to correlate with dephosphorylation of the enzyme and with increased incidence of apoptosis (Whelan and Parker, 1998). These data suggest that fully phosphorylated PKC α is required for the survival responses in cells, consistent with the discussion above that only fully phosphorylated PKC is competent to become activated by diacylglycerol. PKC α also rescues cells from etoposide-induced apoptosis. Survival responses attributed to PKC α have also been reported in other cells. Further support for a role of PKC α in apoptosis comes from the observation that expression of this isozyme is elevated in a large number of cancers, where a strong survival response is necessary to support tumor growth (Deacon et al., 1997). Elevated levels of ceramide result in induction of apoptosis in many cell types, an event that correlates with dephosphorylation and inactivation of PKC α (Lee et al., 1996; Lee et al., 2000). Activation of the Fas antigen, a member of the tumor necrosis factor (TNF) receptor family, induces a strong apoptotic response and concomitant inhibition of PKC α activity (Chen and Faller, 1999); this Fas-induced apoptotic response can be rescued by pretreating cells with PMA. It has also been proposed that PKC exerts its protective effect upstream of caspases-3 and-8 in the Fas-mediated apoptotic pathway (Gomez-Angelats et al., 2000).

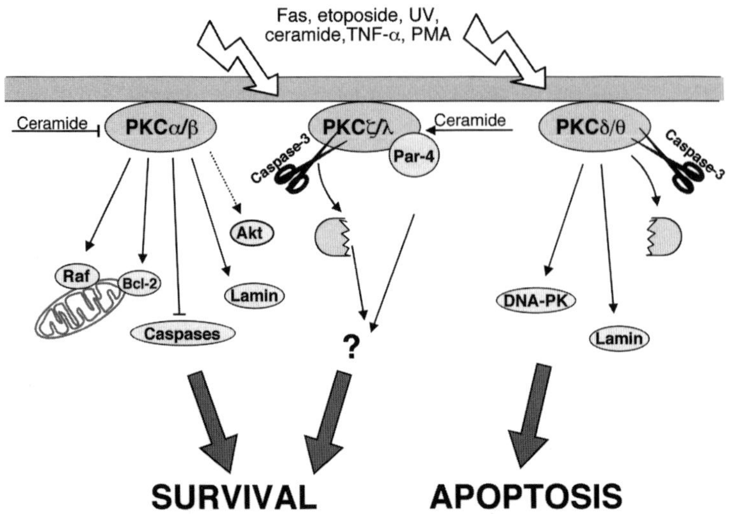

Fig. 12.3. Model depicting the role of PKC family members in cell stress. A variety of cellular stresses induce activation of PKC, either directly by stimulating the generation of diacylglycerol or by membrane recruitment (e.g., PMA). Alternatively, cellular stresses such as etoposide, UV light and ceramide indirectly interfere with PKC activation, in some cases by activating protein phosphatases which dephosphorylate PKC (e.g., ceramide). Activation of conventional PKC isozymes leads to protection from cell death by mechanisms which include phosphorylation of anti-apoptotic proteins such as Bcl-2 and mitochondrial c-Raf-1. Similarly, atypical PKC isozymes such as ζ and λ also mediate anti-apoptotic mechanisms, by interacting with proteins such as par-4, and by serving as substrates for caspases, where the resulting PKC fragment is catalytically inactive. Conversely, novel isozymes, in particular, PKC δ are directly implicated in apoptosis, possibly by regulating the DNA-dependent protein kinase and phosphorylation of the nuclear protein lamin. Activation of caspases also leads to cleavage of PKC δ, although in this case the resulting fragment is catalytically active and mediates the apoptotic signal.

PKC β activity has also been shown to correlate with Fas-mediated apoptosis in human myeloid leukemia cells (Laouar et al., 1999). Specifically, PKC βII has been linked to cellular survival by directly phosphorylating mitotic lamin, an event that regulates breakdown of the nuclear envelope in cells (Goss et al., 1994) (Fig. 12.3). Thus, much evidence points to conventional PKC isozymes, in particular PKC α, in protection from apoptosis.

Although there is ample evidence that conventional PKC isozymes provide a survival signal in response to a variety of stresses, the precise nature by which PKC transduces this survival signal is not clear. A number of observations are noteworthy. Firstly, PKC α has been shown to directly phosphorylate Bcl-2 leading to an increase in its anti-apoptotic function (Ruvolo et al., 1998). Mitochondrial c-Raf-1 is also anti-apoptotic and this has been linked to its phosphorylation by Akt/PKB as well as PKC α (Majewski et al., 1999). In this regard, it is possible that at least in some instances, the protective effects of PKC α and PKC β may be operating through Akt/PKB which is known to provide a strong survival signal in many cells. Overexpression of PKC α stimulates Akt/PKB activity and this suppresses apoptosis induced by growth factor removal (Li et al., 1999). Interestingly, these effects could not be recapitulated with other PKCs such as PKC δ or PKC ε. Elucidation of the precise mechanism by which PKC α/β mediate survival responses, including identification of specific substrates, awaits further research.

3.2. Novel protein kinase Cs

In contrast to the conventional PKCs, novel PKC isozymes have been linked to pro-apoptotic signaling as opposed to survival. The most extensively studied of these isozymes is PKC δ, which is activated in response to a wide variety of cellular stresses, including Fas, cis-platinum and etoposide (Mizuno et al., 1997; Reyland et al., 1999). Unlike conventional PKCs, there is a loss of PKC δ protein in tumors (Lu et al., 1997). Similarly, expression of active PKC δ or activation of this isozyme causes cell cycle arrest, specifically at the G2/M transition (Watanabe et al., 1992). One of the first indications that PKC δ participates in pro-apoptotic responses was the finding that it is proteolytically cleaved by caspase-3 in irradiated U937 cells undergoing apoptosis (Emoto et al., 1995). This cleavage generates a catalytically active fragment which is lipid-independent and thus is constitutively active. Indeed expression of this fragment alone induces an apoptotic phenotype, whereas a catalytically inactive variant does not (Ghayur et al., 1996). Cleavage of PKC δ also correlates with spontaneous neutrophil apoptosis, which can be reversed with PKC δ-specific inhibitors (Pongracz et al., 1999). Nuclear translocation of PKC δ has also been reported in T cells undergoing apoptosis (Scheel-Toellner et al., 1999). PKC δ has also been shown to lead to caspase-3 activation, suggesting a positive feedback loop for further cleavage and activation of the kinase (Basu and Akkaraju, 1999). A number of potential PKC δ substrates have been identified which may mediate this isozyme's pro-apoptotic function; these include the DNA-dependent protein kinase (DNA-PK) and lamin B (Bharti et al., 1998; Cross et al., 2000) (Fig. 12.3). In addition, the tyrosine kinase c-Abl interacts with, and is phosphorylated by, PKC δ in response to cellular oxidative stress, though it is not clear if this is part of the pro-apoptotic response (Sun et al., 2000).

The novel PKC isozyme PKC θ is also implicated in pro-apoptotic functions. PKC θ is unusual in that it has a very restricted tissue distribution, being exclusively found in cells derived from the hematopoietic lineage. As with PKC δ, PKC θ is selectively cleaved by caspase-3 and over-expression of the resulting catalytic fragment induces an apoptotic phenotype (Datta et al., 1997). It is well established that PKC θ participates in the activation of the transcription factor NF-κB in T cells, and several reports suggested that PKC θ was upstream of the stress kinase JNK/SAPK (c-Jun N-terminal kinase/stress-activated protein kinase) (Ghaffari-Tabrizi et al., 1999; Werlen et al., 1998). However, a recent study which made use of PKC θ –/– T lymphocytes demonstrated that this PKC is required for T-cell receptor-mediated activation of NF-κB, but in a JNK/SAPK-independent manner (Sun et al., 2000). Finally, pharmacological studies using phorbol esters and PKC θ inhibitors revealed that induction of thymocyte apoptosis requires activation of PKC θ (Asada et al., 1998).

A number of reports have provided evidence that, unlike PKC δ and PKC θ, PKC ε is implicated in protection from apoptosis. For example, PKC ε antagonists inhibit the protection of cardiac myocytes from hypoxia-induced apoptosis (Gray et al., 1997). In these cells, the same peptide antagonists prevent PKC ε translocation and activation. Similarly, PKC ε inhibitors as well as dominant negative PKC ε mutants inhibit UV-induced apoptosis *in vivo* (Chen et al., 1999), and inhibition of PKC ε activity also blocks the inhibitory effect of PMA on TNFα-induced apoptosis (Mayne and Murray, 1998). Thus, in a number of isolated cases, PKC ε has been linked to cellular survival. Whether this will turn out to be a more general function of PKC ε remains to be determined.

3.3. Atypical protein kinase Cs

Both of the atypical PKC isozymes, PKC ζ and PKC ι/λ control survival signals in cells (Fig. 12.3). As discussed above, atypical PKCs are acutely regulated by the PI 3-K/PDK-1 pathways which lead to their activation in mitogen-stimulated cells. Activation of PI 3-K is required for survival signals in response to cellular stress, and although a major effector of PI 3-K in this pathway is the Akt/PKB kinase, atypical PKCs are likely to also play an important role. For example, PKC ι/λ has been shown to protect cells from drug- and UV-induced apoptosis (Murray and Fields, 1997). Similarly, PKC ζ has been shown to specifically interact with the product of the par-4 gene which is induced during apoptosis (Diaz-Meco et al., 1996). Par-4 interacts with PKC ζ and inhibits its protein kinase activity, an event that is required for the ability of par-4 to induce apoptosis. PKC ζ, but apparently not PKC ι/λ, is also cleaved by caspase-3 during UV-induced apoptosis, as reported for PKC δ (Frutos et al., 1999). However, unlike PKC δ, the resulting PKC ζ fragment is catalytically inactive. Consistent with this observation, a mutant caspase-3-resistant PKC ζ protects transfected cells from apoptosis more efficiently than the wild-type counterpart. These data support a role for atypical PKCs, in particular PKC ζ, in transducing survival signals. To date, no downstream targets of PKC ζ or PKC ι/λ have been described in the survival pathway.

As with PKC θ, there is evidence that links PKC ζ to the activation of the stress-activated kinase pathway. Ceramide activation of JNK/SAPK has been shown to require PKC ζ, and a complex of PKC ζ and SAPK has been detected in cells (Bourbon et al., 2000). Activation of PKC ζ by TNFα has also been linked to transcriptional regulation of NF-κB, and this appears to occur through the direct phosphorylation of an IκB-kinase by PKC ζ (Diaz-Meco et al., 1999; Lallena et al., 1999). However, it is not clear how this pathway affects cellular survival. In summary, atypical PKCs play an important role in anti-apoptotic signaling, though the precise mechanism by which they do so is not known. This will undoubtedly be clarified by the identification of PKC ζ and PKC ι/λ substrates which mediate the survival signal.

4. Perspectives

The function of PKC is exquisitely sensitive to its phosphorylation state, subcellular location, and cofactor interactions. This family of enzymes has been implicated in a multitude of cellular responses, with recent studies pointing to its key involvement in signaling pathways which are activated in response to cellular stress. Initial findings implicated PKC in both cellular survival and death, which seemed contradictory. This was further complicated by the fact that any one cell type expresses multiple isozymes which are activated by similar mechanisms. Molecular genetic approaches have provided an explanation for these observations and have demonstrated that different PKC isozymes play distinct roles in cell survival. The emerging theme is that conventional and atypical PKC family members transduce signals which ultimately result in cell survival, whereas novel isoforms, most notably PKC δ are involved in pro-apoptotic signaling. Despite these opposing functions, PKCs share some similarities in their response to cellular stress; for example, several isoforms are cleaved by caspases and the resulting fragments can positively or negatively regulate apoptotic signals. It is also worth noting that there are some exceptions to the above rule, for example novel PKC ε has been linked to cellular survival, although whether this is an isolated, cell type-specific function as opposed to a more general function for this enzyme remains to be seen. These exciting observations linking PKC to apoptosis or survival immediately raise the question of what are the substrates of the kinases which transduce their signals? Although there are some clues (e.g., the DNA-PK substrate of PKC δ) these questions remain largely unanswered. Identification of precise substrates will provide a key piece to the puzzle of understanding the biology of this remarkably versatile family of protein kinases.

References

Asada, A., Zhao, Y., Kondo, S. and Iwata, M. (1998). Induction of thymocyte apoptosis by Ca^{2+}-independent protein kinase C (nPKC) activation and its regulation by calcineurin activation. J. Biol. Chem. 273, 28392– 28398.

Basu, A. and Akkaraju, G.R. (1999). Regulation of caspase activation and cis-diamminedichloroplatinum(II)-induced cell death by protein kinase C. Biochemistry 38, 4245–4251.

Behn-Krappa, A. and Newton, A.C. (1999). The hydrophobic phosphorylation motif of conventional protein kinase C is regulated by autophosphorylation. Curr. Biol. 9, 728–737.

Bharti, A., Kraeft, S.K., Gounder, M., Pandey, P., Jin, S., Yuan, Z.M., Lees-Miller, S.P., Weichselbaum, R., Weaver, D., Chen, L.B., Kufe, D. and Kharbanda, S. (1998). Inactivation of DNA-dependent protein kinase by protein kinase Cδ: implications for apoptosis. Mol. Cell. Biol. 18, 6719–6728.

Bornancin, F. and Parker, P.J. (1997). Phosphorylation of protein kinase C-α on serine 657 controls the accumulation of active enzyme and contributes to its phosphatase-resistant state. J. Biol. Chem. 272, 3544–3549.

Bornancin, F. and Parker, P.J. (1996). Phosphorylation of

threonine 638 critically controls the dephosphorylation and inactivation of protein kinase C α. Curr. Biol. 6, 1114–1123.

Bourbon, N.A., Yun, J. and Kester, M. (2000). Ceramide directly activates protein kinase C ζ to regulate a stress-activated protein kinase signaling complex. J. Biol. Chem. 275, 35617–35623.

Casamayor, A., Morrice, N.A. and Alessi, D.R. (1999). Phosphorylation of Ser-241 is essential for the activity of 3-phosphoinositide-dependent protein kinase-1: identification of five sites of phosphorylation in vivo. Biochem. J. 342, 287–292.

Chen, C.Y. and Faller, D.V. (1999). Selective inhibition of protein kinase C isozymes by Fas ligation. J. Biol. Chem. 274, 15320–15328.

Chen, N., Ma, W., Huang, C. and Dong, Z. (1999). Translocation of protein kinase C ε and protein kinase C δ to membrane is required for ultraviolet B-induced activation of mitogen-activated protein kinases and apoptosis. J. Biol. Chem. 274, 15389–15394.

Chou, M.M., Hou, W., Johnson, J., Graham, L.K., Lee, M. H., Chen, C. S., Newton, A.C., Schaffhausen, B.S. and Toker, A. (1998). Regulation of protein kinase C ζ by PI 3-kinase and PDK-1. Curr. Biol. 8, 1069–1077.

Coussens, L., Parker, P. J., Rhee, L., Yang-Feng, T.L., Chen, E., Waterfield, M.D., Francke, U. and Ullrich, A. (1986). Multiple, distinct forms of bovine and human protein kinase C suggest diversity in cellular signaling pathways. Science 233, 859–866.

Cross, T., Griffiths, G., Deacon, E., Sallis, R., Gough, M., Watters, D. and Lord, J.M. (2000). PKC-δ is an apoptotic lamin kinase. Oncogene 19, 2331–2337.

Datta, R., Kojima, H., Yoshida, K. and Kufe, D. (1997). Caspase-3-mediated cleavage of protein kinase C θ in induction of apoptosis. J. Biol. Chem. 272, 20317–20320.

Deacon, E.M., Pongracz, J., Griffiths, G. and Lord, J.M. (1997). Isoenzymes of protein kinase C: differential involvement in apoptosis and pathogenesis. Mol. Pathol. 50, 124–131.

Diaz-Meco, M.T., Lallena, M.J., Monjas, A., Frutos, S. and Moscat, J. (1999). Inactivation of the inhibitory κB protein kinase/nuclear factor κB pathway by Par-4 expression potentiates tumor necrosis factor α-induced apoptosis. J. Biol. Chem. 274, 19606–19612.

Diaz-Meco, M.T., Municio, M.M., Frutos, S., Sanchez, P., Lozano, J., Sanz, L. and Moscat, J. (1996). The product of par-4, a gene induced during apoptosis, interacts selectively with the atypical isoforms of protein kinase C. Cell 86, 777–786.

Dutil, E.M. and Newton, A.C. (2000). Dual role of pseudosubstrate in the coordinated regulation of protein kinase C by phosphorylation and diacylglycerol. J. Biol. Chem. 275, 10697–10701.

Dutil, E.M., Toker, A. and Newton, A.C. (1998). Regulation of conventional protein kinase C isozymes by phosphoinositide-dependent kinase 1 (PDK-1). Curr. Biol. 8, 1366–1375.

Edwards, A.S., Faux, M.C., Scott, J.D. and Newton, A.C. (1999). Carboxyl-terminal phosphorylation regulates the function and subcellular localization of protein kinase C βII. J. Biol. Chem. 274, 6461–6468.

Edwards, A.S. and Newton, A.C. (1997). Phosphorylation at conserved carboxyl-terminal hydrophobic motif regulates the catalytic and regulatory domains of protein kinase C. J. Biol. Chem. 272, 18382–18390.

Emoto, Y., Manome, Y., Meinhardt, G., Kisaki, H., Kharbanda, S., Robertson, M., Ghayur, T., Wong, W.W., Kamen, R., Weichselbaum, R. and et al. (1995). Proteolytic activation of protein kinase C δ by an ICE-like protease in apoptotic cells. EMBO J. 14, 6148– 6156.

Frutos, S., Moscat, J. and Diaz-Meco, M.T. (1999). Cleavage of ζPKC but not λ/ιPKC by caspase-3 during UV-induced apoptosis. J. Biol. Chem. 274, 10765–10770.

Gao, T., Toker, A. and Newton, A.C. (2001). The carboxyl-terminus of protein kinase C provides a switch to regulate its interaction with the phosphoinositide-dependent kinase, PDK-1. J. Biol. Chem. J. Biol. chem. 276, 19588–19596.

Ghaffari-Tabrizi, N., Bauer, B., Villunger, A., Baier-Bitterlich, G., Altman, A., Utermann, G., Uberall, F. and Baier, G. (1999). Protein kinase Cθ, a selective upstream regulator of JNK/SAPK and IL-2 promoter activation in Jurkat T cells. Eur. J. Immunol. 29, 132–142.

Ghayur, T., Hugunin, M., Talanian, R. V., Ratnofsky, S., Quinlan, C., Emoto, Y., Pandey, P., Datta, R., Huang, Y., Kharbanda, S., Allen, H., Kamen, R., Wong, W. and Kufe, D. (1996). Proteolytic activation of protein kinase C δ by an ICE/CED 3-like protease induces characteristics of apoptosis. J. Exp. Med. 184, 2399–2404.

Gomez-Angelats, M., Bortner, C.D. and Cidlowski, J.A. (2000). Protein kinase C (PKC) inhibits fas receptor-induced apoptosis through modulation of the loss of K+ and cell shrinkage. A role for PKC upstream of caspases. J. Biol. Chem. 275, 19609–19619.

Goss, V.L., Hocevar, B.A., Thompson, L.J., Stratton, C.A., Burns, D.J. and Fields, A.P. (1994). Identification of nuclear β II protein kinase C as a mitotic lamin kinase. J. Biol. Chem. 269, 19074–19080.

Gray, M.O., Karliner, J.S. and Mochly-Rosen, D. (1997). A selective ε-protein kinase C antagonist inhibits protection of cardiac myocytes from hypoxia-induced cell death. J. Biol. Chem. 272, 30945–30951.

Haimovitz-Friedman, A., Balaban, N., McLoughlin, M., Ehleiter, D., Michaeli, J., Vlodavsky, I. and Fuks, Z. (1994). Protein kinase C mediates basic fibroblast growth factor protection of endothelial cells against radiation-induced apoptosis. Cancer Res. 54, 2591–2597.

Jaken, S. (1996). Protein kinase C isozymes and substrates. Curr. Opin. Cell Biol. 8, 168–173.

Jaken, S. and Parker, P.J. (2000). Protein kinase C binding partners. Bioessays 22, 245–254.

Johnson, J.E., Giorgione, J. and Newton, A.C. (2000). The C1 and C2 domains of protein kinase C are independent membrane targeting modules, with specificity for phosphatidylserine conferred by the C1 domain. Biochemistry 39, 11360–11369.

Keranen, L.M., Dutil, E.M. and Newton, A.C. (1995). Protein kinase C is regulated in vivo by three functionally distinct phosphorylations. Curr. Biol. 5, 1394–1403.

Klauck, T.M., Faux, M.C., Labudda, K., Langeberg, L.K., Jaken, S. and Scott, J.D. (1996). Coordination of three signalling enzymes by AKAP 79, a mammalian scaffold protein. Science 271, 1589–1592.

Lallena, M.J., Diaz-Meco, M.T., Bren, G., Paya, C.V. and Moscat, J. (1999). Activation of IκB kinase β by protein kinase C isoforms. Mol. Cell. Biol. 19, 2180–2188.

Laouar, A., Glesne, D. and Huberman, E. (1999). Involvement of protein kinase C-β and ceramide in tumor necrosis factor-α-induced but not Fas-induced apoptosis of human myeloid leukemia cells. J. Biol. Chem. 274, 23526–23534.

Le Good, J.A., Ziegler, W.H., Parekh, D.B., Alessi, D.R., Cohen, P. and Parker, P.J. (1998). Protein kinase C isotypes controlled by phosphoinositide 3-kinase through the protein kinase PDK1. Science 281, 2042–2045.

Lee, J.Y., Hannun, Y.A. and Obeid, L.M. (1996). Ceramide inactivates cellular protein kinase Cα. J. Biol. Chem. 271, 13169–13174.

Lee, J.Y., Hannun, Y.A. and Obeid, L.M. (2000). Functional dichotomy of protein kinase C (PKC) in tumor necrosis factor-α (TNF-α) signal transduction in L929 cells. Translocation and inactivation of PKC by TNF-α. J. Biol. Chem. 275, 29290–29298.

Li, L., Lorenzo, P.S., Bogi, K., Blumberg, P.M. and Yuspa, S.H. (1999). Protein kinase Cδ targets mitochondria, alters mitochondrial membrane potential and induces apoptosis in normal and neoplastic keratinocytes when overexpressed by an adenoviral vector. Mol. Cell. Biol. 19, 8547–8558.

Lu, Z., Hornia, A., Jiang, Y.W., Zang, Q., Ohno, S. and Foster, D.A. (1997). Tumor promotion by depleting cells of protein kinase C δ. Mol. Cell. Biol. 17, 3418–3428.

Majewski, M., Nieborowska-Skorska, M., Salomoni, P., Slupianek, A., Reiss, K., Trotta, R., Calabretta, B. and Skorski, T. (1999). Activation of mitochondrial Raf-1 is involved in the antiapoptotic effects of Akt. Cancer Res. 59, 2815–2819.

Mayne, G.C. and Murray, A.W. (1998). Evidence that protein kinase Cε mediates phorbol ester inhibition of calphostin C- and tumor necrosis factor-α-induced apoptosis in U937 histiocytic lymphoma cells. J. Biol. Chem. 273, 24115–24121.

Mellor, H. and Parker, P.J. (1998). The extended protein kinase C superfamily. Biochem. J. 332, 281–292.

Mizuno, K., Noda, K., Araki, T., Imaoka, T., Kobayashi, Y., Akita, Y., Shimonaka, M., Kishi, S. and Ohno, S. (1997). The proteolytic cleavage of protein kinase C isotypes, which generates kinase and regulatory fragments, correlates with Fas-mediated and 12-O-tetradecanoyl-phorbol-13-acetate-induced apoptosis. Eur. J. Biochem. 250, 7–18.

Mochly-Rosen, D. (1995). Localization of protein kinases by anchoring proteins: a theme in signal transduction. Science 268, 247–251.

Mochly-Rosen, D., Khaner, H., Lopez, J. and Smith, B.L. (1991). Intracellular receptors for activated protein kinase C. J. Biol. Chem. 266, 14866–14868.

Murray, N. R. and Fields, A. P. (1997). Atypical protein kinase C ι protects human leukemia cells against drug-induced apoptosis. J. Biol. Chem. 272, 27521–27524.

Newton, A.C. (1997). Regulation of protein kinase C. Curr. Opin. Cell Biol. 9, 161–167.

Newton, A.C. and Johnson, J.E. (1998). Protein kinase C: a paradigm for regulation of protein function by two membrane-targeting modules. Biochim. Biophys. Acta 1376, 155–172.

Nishikawa, K., Toker, A., Wong, K., Marignani, P.A., Johannes, F.J. and Cantley, L.C. (1998). Association of protein kinase Cμ with type II phosphatidylinositol 4-kinase and type I phosphatidylinositol-4-phosphate 5-kinase. J. Biol. Chem. 273, 23126–23133.

Nishizuka, Y. (1995). Protein kinase C and lipid signaling for sustained cellular responses. FASEB J. 9, 484–496.

Parekh, D.B., Ziegler, W. and Parker, P.J. (2000). Multiple pathways control protein kinase C phosphorylation. EMBO J. 19, 496–503.

Pongracz, J., Webb, P., Wang, K., Deacon, E., Lunn, O.J. and Lord, J.M. (1999). Spontaneous neutrophil apoptosis involves caspase 3-mediated activation of protein kinase C-δ. J. Biol. Chem. 274, 37329–37334.

Prasad, N., Topping, R.S., Zhou, D. and Decker, S.J. (2000). Oxidative stress and vanadate induce tyrosine phosphorylation of phosphoinositide-dependent kinase 1 (PDK1). Biochemistry 39, 6929–6935.

Reyland, M.E. Anderson, S.M., Matassa, A.A., Barzen, K.A. and Quissell, D.O. (1999). Protein kinase C δ is essential for etoposide-induced apoptosis in salivary gland acinar cells. J. Biol. Chem. 274, 19115–19123.

Ruvolo, P.P., Deng, X., Carr, B.K. and May, W.S. (1998). A functional role for mitochondrial protein kinase Cα in Bcl2 phosphorylation and suppression of apoptosis. J. Biol. Chem. 273, 25436–25442.

Scheel-Toellner, D., Pilling, D., Akbar, A.N., Hardie, D., Lombardi, G., Salmon, M. and Lord, J.M. (1999). Inhibition of T cell apoptosis by IFN-β rapidly reverses nuclear translocation of protein kinase C-δ. Eur. J. Immunol. 29, 2603–2612.

Stokoe, D., Stephens, L.R., Copeland, T., Gaffney, P.R., Reese, C.B., Painter, G.F., Holmes, A.B., McCormick,

F. and Hawkins, P.T. (1997). Dual role of phosphatidylinositol-3,4,5-trisphosphate in the activation of protein kinase B. Science 277, 567–570.

Sun, X., Wu, F., Datta, R., Kharbanda, S. and Kufe, D. (2000). Interaction between protein kinase C δ and the c-Abl tyrosine kinase in the cellular response to oxidative stress. J. Biol. Chem. 275, 7470–7473.

Sun, Z., Arendt, C.W., Ellmeier, W., Schaeffer, E.M., Sunshine, M.J., Gandhi, L., Annes, J., Petrzilka, D., Kupfer, A., Schwartzberg, P.L. and Littman, D.R. (2000). PKC-θ is required for TCR-induced NF-κB activation in mature but not immature T lymphocytes. Nature 404, 402–407.

Takahashi, M., Mukai, H., Oishi, K., Isagawa, T. and Ono, Y. (2000). Association of immature hypophosphorylated protein kinase Cε with an anchoring protein CG-NAP. J. Biol. Chem. 275, 34592–34596.

Thornberry, N.A. and Lazebnik, Y. (1998). Caspases: enemies within. Science 281, 1312–1316.

Toker, A. and Newton, A. (2000). Cellular signalling: pivoting around PDK-1. Cell 103, 185–188.

Tsunoda, S., Sierralta, J., Sun, Y., Bodner, R., Suzuki, E., Becker, A., Socolich, M. and Zuker, C.S. (1997). A multivalent PDZ-domain protein assembles signalling complexes in a G-protein-coupled cascade. Nature 388, 243–249.

Vanhaesebroeck, B. and Alessi, D.R. (2000). The PI3K-PDK1 connection: more than just a road to PKB. Biochem J 346 Pt 3, 561–576.

Watanabe, T., Ono, Y., Taniyama, Y., Hazama, K., Igarashi, K., Ogita, K., Kikkawa, U. and Nishizuka, Y. (1992). Cell division arrest induced by phorbol ester in CHO cells overexpressing protein kinase C-δ subspecies. Proc. Natl. Acad. Sci. USA 89, 10159–10163.

Werlen, G., Jacinto, E., Xia, Y. and Karin, M. (1998). Calcineurin preferentially synergizes with PKC-θ to activate JNK and IL-2 promoter in T lymphocytes. EMBO J. 17, 3101–3111.

Whelan, R.D. and Parker, P.J. (1998). Loss of protein kinase C function induces an apoptotic response. Oncogene 16, 1939–1944.

Protein Adaptations and Signal Transduction. Edited by K.B. Storey and J.M. Storey

CHAPTER 13

Mitogen-activated Protein Kinases and Stress

Klaus P. Hoeflich and James R. Woodgett*
Department of Medical Biophysics, Ontario Cancer Institute, University of Toronto, Toronto, Ontario M5G 2M9, Canada

1. Introduction

Exposure of cells to environmental stress or strong deviations from normal conditions initiates complex cascades of stress-inducible transductory enzymes that impact processes such as gene transcription in an attempt to adapt the cell to its new situation. The mitogen-activated protein kinase (MAPK) superfamily plays an important role in transducing signals from the cell surface to the nucleus, effecting both the cell's ability to cope with outside changes as well as cellular coordination in the case of multicellular organisms. The term MAPK is most widely used as a general denominator of this family of protein kinases. The MAPK acronym originally described the "microtubule-associated protein-2 kinase" but evolved into mitogen-activated protein kinase when it was discovered that the enzyme was induced by a variety of hormones and mitogens. Upon the molecular cloning of these enzymes, it was realized that they existed in several classes that were structurally related but distinctly regulated. The term MAPK is now commonly used to denote the entire class of protein-serine kinases that share the following features: the core functional unit of a MAPK module consists of a triad of three kinases that act sequentially, where MAPKs are activated via phosphorylation on both a threonine and tyrosine residue by selective upstream regulatory kinases, MAPK kinases (MAPKKs or MAP2Ks). MAPKKs are in turn phosphorylated and activated by a group of structurally related kinases termed MAPK kinase kinases (MAPKKKs or MAP3Ks).

The first MAPK to be molecularly identified was isolated in a genetic screen of the budding yeast *Saccharomyces cerevisiae* (Courchesne et al., 1989; Elion et al., 1990). In yeast cells, mating-specific processes are initiated by the binding of mating type-specific peptides, known as α factor and **a** factor, to a G protein-coupled pheromone receptor on the cell surface (Herskowitz, 1995; Madhani and Fink, 1998). Subsequent signal transduction culminates in a set of physiological responses that prepare cells for mating, such as arrest of the cell cycle, changes in gene expression and altered cell polarity and morphology. Genetic screens identified a group of "sterile" mutants defective in mating which were initially grouped into two categories: deficient in pheromone response or in pheromone production. Additional components of the pheromone response pathway were identified by a variety of other approaches, including screens for genetic interactions with the original sterile alleles. The approaches that revealed the first MAPKs employed screens for suppressors of supersensitivity to mating pheromone-induced growth arrest (yielding KSS1, kinase suppressor of *sst2*; Courchesne et al., 1989) and mutations which prevented yeast from proceeding through mating-induced cell fusion (yielding FUS3, fusion-3; Elion et al., 1990). Further genetic analysis identified components for five distinct MAPK pathways in *S. cerevisiae* (Herskowitz, 1995). These MAPK

*Corresponding author.

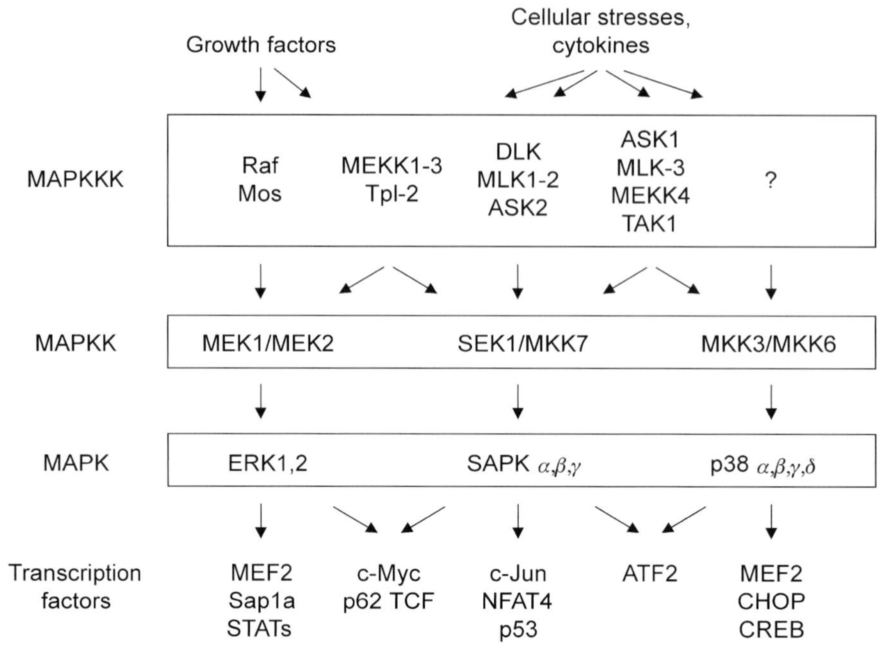

Fig. 13.1. Mammalian MAPK modules. The MAPK module comprises a MAPKKK, MAPKK and a MAPK. These pathways respond to extracellular signals, including growth and differentiation factors, cellular stress and cytokines. Once activated, MAPKs can phosphorylate a wide variety of proteins, including transcription factors and other kinases. See text for full details.

pathways are essential for processes including mating, sporulation, osmoregulation, cell wall integrity, starvation and filamentous growth (Madhani and Fink, 1998; Schaeffer and Weber, 1999).

Full-length sequences of more than a hundred MAPKs from numerous species have since been reported. Several MAPK modules have been identified in mammals including the extracellular signal-regulating kinase (ERK), stress-activated protein kinase (SAPK; or c-Jun NH_2-terminal kinase, JNK), and the p38 group of kinases (Fig. 13.1). A number of extensive reviews on MAPK signal transduction have recently been published (Tibbles and Woodgett, 1999; Widmann et al., 1999). Hence, the goal of this commentary is not to provide another comprehensive review of the literature, but rather to focus on recent developments and to present some perspectives. Furthermore, since the ERK subfamily of MAPKs is, in general, much less sensitive to typical stress agonists, it will not be a focus of this review.

2. The SAPK family

Mammalian MAPKs have been classified on the basis of two criteria: sequence homology and differential activation by agonists (Tibbles and Woodgett, 1999; Widmann et al., 1999). Firstly, the activity of MAPKs is controlled by dual phosphorylation within an amino acid sequence known as the activation loop (Canagarajah et al., 1997). Phosphorylation of the signature motif threonine-X-tyrosine in the activation loop (where X is glutamic acid, proline or glycine for the ERK, SAPK and p38 MAPKs, respectively) is catalyzed by specific MAPKKs and results in a conformational change and a >1000-fold increase in specific activity of the MAPK. In essence, the enzymes are inactive unless phosphorylated by their upstream enzymes. While the ERK class of MAPKs is primarily activated by growth factors and mitogens, SAPKs and p38 MAPKs are preferentially induced by a variety of stress signals (Kyriakis et al., 1994). These stimuli include genotoxic agents (irradiation and carcinogens), pathogenic signals (LPS and dsRNA), proinflammatory cytokines (tumor necrosis factor (TNF)-α and interleukin (IL)-1β), homeostatic perturbations (in temperature, osmolarity and pH), oxygen tension, intracellular calcium, and other chemical insults (e.g. exposure to arsenite or anisomycin).

As expected, a major point of regulation occurs at the level of the MAPK. Since phosphorylation of both threonine and tyrosine residues is required for MAPK activity, dephosphorylation of either is sufficient for inactivation. This can be achieved through complex regulation by tyrosine-specific phosphatases, serine/threonine-specific phosphatases or by dual specificity (threonine/tyrosine) protein phosphatases and is discussed in detail elsewhere (reviewed in Keyse, 2000). It is clear, however, that the duration and magnitude of MAPK activation reflects a balance between the activities of the upstream activating kinases and protein phosphatases.

Three SAPK genes (termed α, β and γ; or JNK2, JNK3, and JNK1, respectively) have been cloned (Hibi et al., 1993; Kyriakis et al., 1994; Gupta et al., 1996). Overall, the family members share 85–92% identity and are 42–45% identical within the catalytic domain to the ERK family. The SAPK genes are further diversified by alternative mRNA splicing into as many as ten isoforms. Each gene generates 54 kDa and 46 kDa polypeptides, the latter variants arising through the introduction of a 5 bp sequence into the carboxy-terminal region which introduces a premature stop codon. To date, clear functional differences between those 46 kDa and 55 kDa isoforms have not been reported. SAPKα and SAPKγ are widely expressed, while SAPKβ is selectively expressed in the brain, heart and testis.

The SAPKs were originally identified as the major serine/threonine kinases responsible for the phosphorylation of the c-Jun transcription factor (Gupta et al., 1996; Hibi et al., 1993; Kyriakis et al., 1994). c-Jun dimerizes with members of the Fos, Jun or activating transcription factor (ATF) family of transcription factors to form the activator protein-1 (AP-1) transcription factor complex. AP-1 activity is induced by a number of stressful stimuli and part of this activation can be attributed to phosphorylation of serines 63 and 73 in the c-Jun transactivation domain, catalyzed by the SAPKs. Through these effects on AP-1, SAPKs influence cell proliferation and oncogenic transformation. Additional SAPK targets include: other Jun proteins (JunB and JunD; Kallunki et al., 1996) and the related activating transcription factor-2 (ATF2; Gupta et al., 1995); the ternary complex factor (TCF) subfamily of ETS-domain transcription factors (Whitmarsh et al., 1995); tumor suppressor p53 (Fuchs et al., 1998); Smad3 (Engel et al., 1999); nuclear factor of activated T cells (NFAT4; Chow et al., 1997); and the basic-helix–loop-helix transcription factor, Myc (Noguchi et al., 1999). While some of the aforementioned SAPK targets still await validation, to date, SAPK targets are exclusively transcription factors. This is in contrast to the ERK and p38 families that phosphorylate substrates outside of the nucleus as well as within. SAPK isoforms have varying substrate affinities and may therefore selectively target transcription factors for distinct biological functions *in vivo* (Gupta et al., 1996). To address this question, mutant mice lacking each member of the SAPK family have been generated and their role in embryonic development assessed (Table 13.2).

Although mutant mice with a single deletion of SAPKα/JNK2 (Yang et al., 1998), SAPKβ/JNK3 (Yang et al., 1997) or SAPKγ/JNK1 (Dong et al., 1998) are viable without overt structural abnormalities, compound deficiencies of the SAPK family have developmental consequences (Kuan et al., 1999; Sabapathy et al., 1999). Mice with SAPKα/JNK2 and SAPKγ/JNK1 dual deficiencies, but not other SAPK double mutations, exhibit aberrant brain apoptosis and early embryonic lethality. The most conspicuous feature of E10.5 SAPKα/SAPKγ (JNK1/JNK2) double mutants is failed closure of the neural folds in the hindbrain region. These embryos display decreased apoptosis in the hindbrain at E9.25 and increased apoptosis in both the hindbrain and forebrain regions at E10.5. Loss of three out of four SAPK alleles also affects embryonic development as 25% of $SAPK\alpha^{+/-}SAPK\gamma^{-/-}$ ($JNK1^{-/-}JNK2^{+/-}$) fetuses exhibited exencephaly similar to the double mutant phenotype. By contrast, mice that completely lack SAPKα/JNK2 and have only one *SAPKγ*/JNK1 allele were obtained in Mendelian ratio and did not display developmental abnormalities. The reason for this difference is not known, although detection of SAPK proteins in brain extracts indicates that gene dosage plays a critical role in controlling SAPK protein levels. It is possible that a certain threshold of SAPK expression is essential whereby both SAPKα/JNK1 and SAPKγ/JNK2 either phosphorylate a common target, or act in parallel to phosphorylate multiple targets, to induce cell death during neural tube closure. These results reveal a functional diversification of the SAPK family *in vivo* and a role for SAPKs in mammalian morphogenesis, analogous to the requirement of the SAPK signaling pathway in *Drosophila* embryos during dorsal closure (Riesgo-Escovar et al., 1996; Sluss et al., 1996), a morphogenetic process that occurs during mid-embryogenesis.

The preferential expression of SAPKβ/JNK3 in neural tissue suggests a unique function. Indeed, gene-targeting studies demonstrate that SAPKβ/JNK3 deficiency, but not SAPKα/JNK2 or SAPKγ/JNK1 null mutations, results in increased resistance to kainic acid-induced seizures and apoptosis of hippocampal neurons (Yang et al., 1997). Kainate elicits epileptic seizures by direct stimulation of the AMPA/kainate class of glutamate receptors and indirectly by increasing the release of excitatory amino acids from nerve terminals. Administration of kainate to wild-type mice induces severe seizures that lasted for 1-2 hours, whereas *SAPKβ*$^{-/-}$ (*JNK3*$^{-/-}$) mice show milder symptoms and faster recovery. These results are phenocopied by mice with a "knock-in" c-Jun mutation that eliminates the SAPK phosphorylation sites (Behrens et al., 1999). While c-Jun appears to be the essential substrate for SAPKβ/JNK3 in stress-induced neuronal apoptosis, the mechanism by which these molecules function in excitotoxicity remains to be defined. Together with the observation that SAPK-deficiency causes defects in thymocyte apoptosis (Rincon et al., 1998; Sabapathy et al., 1999), the data obtained with the *SAPKβ*$^{-/-}$ (*JNK3*$^{-/-}$) and *SAPKα*$^{-/-}$*SAPKγ*$^{-/-}$ (*JNK1*$^{-/-}$*JNK2*$^{-/-}$) mutant mice provides compelling evidence for roles of SAPKs in apoptotic responses. It would therefore be of great interest to use these systems to identify physiologically relevant targets of the SAPK apoptotic pathway.

Recent insight into the mechanism of SAPK in apoptosis has been provided by Tournier et al. (2000). Murine embryonic fibroblasts (MEFs) were derived from *SAPKα*$^{-/-}$*SAPKγ*$^{-/-}$ (*JNK1*$^{-/-}$*JNK2*$^{-/-}$) embryos and, as the neuronal-specific SAPKβ (JNK3) isoform cannot be detected, these MEF lack a functional SAPK and represent a useful model for studying the SAPK signal transduction pathway. To further define the requirement for SAPK in apoptosis, wild-type and SAPK-null MEFs were exposed to a variety of cell killing agents. SAPK-null MEFs were nearly completely protected from apoptosis induced by UV irradiation, methyl methanesulfonate and anisomycin, while normal apoptosis was observed by activation of the Fas death-signaling pathway. Increased survival signaling by the transcription factor NF-κB and the protein kinase PKB/Akt did not account for the resistance of SAPK-null MEFs to UV-induced apoptosis. SAPK-null MEFs express slightly more p53 than their wild-type counterparts but the potential contribution of p53 to the UV resistance of SAPK-null MEFs is difficult to understand, mechanistically. Importantly, SAPK is not required for the death receptor-signaling pathway mediated by caspase-8, but is essential for stress-induced apoptosis utilizing the Apaf1, initiator caspase-9, and effector caspase-3 genetic pathway. Accordingly, mitochondrial membrane permeability and subsequent cytochrome c release is also blocked in SAPK-null cells in response to UV but not in response to Fas.

Clearly, the molecular mechanism by which SAPKs function in apoptotic signal transduction and mitochondrial depolarization is an important, as yet unresolved, question. An important clue comes from the observation that inhibitors of protein and mRNA synthesis (cycloheximide and actinomycin D, respectively) do not inhibit UV-induced apoptosis (Tournier et al., 2000). This implies that SAPKs can promote stress-induced killing in a transcription/translation-independent mechanism. This could occur by affecting members of the Bcl-2 family of apoptotic regulatory proteins, for example. It is possible to reconcile these findings with the data obtained using the $SAPK\alpha^{-/-}SAPK\gamma^{-/-}$ ($JNK1^{-/-}JNK2^{-/-}$) *in vivo* apoptosis model in which these SAPK isoforms temporally mediate both cell survival and apoptosis during brain development. Possible scenarios include both separate and cooperative mechanisms where the cellular outcome is determined by the varying kinetics, isoform selectivity, feedback loops, and autocrine secretions promoted by transcription-dependent and transcription-independent SAPK signaling events. In this way, we could envision a balancing act between SAPK survival and apoptotic signaling analogous to that described for TNF-mediated NF-κB transactivation and caspase processing (Baker and Reddy, 1998).

3. Dual-specificity protein kinases of the SAPK pathway

Two MAPKKs have been identified as upstream activators of the SAPKs, SEK1 (SAPK/ERK kinase-1, also known as MKK4 or JNKK1; Derijard et al., 1995; Lin et al., 1995; Sanchez et al., 1994) and MKK7 (MAPK kinase-7, also known as SEK2 or JNKK2; Moriguchi et al., 1997; Tournier et al., 1997; Yao et al., 1997). While the existence of SAPK activators distinct from SEK1 was suggested by early studies using chromatographically fractionated cell extracts (Moriguchi et al., 1995) and SEK1-deficient murine cell lines (Nishina et al., 1997; Yang et al., 1997), until recently only SEK1 had been molecularly cloned. Thus, information regarding the biological functions of MKK7 is just beginning to become available. The MKK7 gene consists of 14 exons and alternative splicing leads to the inclusion or exclusion of exons located in the 5′ and 3′ regions of the gene, resulting in the expression of six MKK7 isoforms that differ in their amino and carboxy termini (Tournier et al., 1999). Comparison of the activities of the MKK7 isoforms demonstrates that the MKK7α isoforms exhibit lower activity, but a higher level of inducible fold activation, than the corresponding MKK7β and MKK7γ isoforms in response to different upstream components of the SAPK signaling pathway (Tournier et al., 1999). Although the mouse SEK1 and MKK7 genes reside on the same chromosome (Tournier et al., 1999; White et al., 1996), it does not appear that this linkage is evolutionarily conserved. For instance, in *Drosophila* the *hep* (MKK7 homologue; Glise et al., 1995) and *D-MKK4* (Han et al., 1998) genes are located on different chromosomes.

The physiological role of SEK1 has been extensively studied in mice and $SEK1^{-/-}$ embryos display defective liver organization and massive hepatocyte apoptosis (Ganiatsas et al., 1998; Nishina et al., 1999; Yang et al., 1997). These embryos die between E11.5–12.5, later in development than the SAPKα/SAPKγ double knockout (Kuan et al., 1999; Sabapathy et al., 1999). This phenotype can be partially understood by considering the tissue expression of SEK1 and MKK7. Although the genes are fairly ubiquitously expressed in mice, SEK1 expression is highest (and MKK7 is the lowest) in the liver (Nishina et al., 1999; Yao et al., 1997). Thus, while MKK7 can perhaps compensate for some SEK1-related functions in embryogenesis there is insufficient MKK7 present in hepatocytes to rescue SAPK signaling in the liver. It will be intriguing to know which receptors are responsible for triggering these SEK1-dependent survival signals in hepatocytes during embryogenesis. Of note, c-Jun-deficient mice exhibit a similar phenotype although the liver defects are less severe than in $SEK1^{-/-}$ embryos and livers from E12.5 $c\text{-}Jun^{-/-}$ embryos still contain residual hepatocytes (Hilberg et al., 1993; Johnson et al., 1993). These data led to the belief that the SEK1-SAPK-c-Jun pathway was required for the anti-apoptotic function of c-Jun during liver organogenesis.

Hence, to further investigate the physiological relevance of the amino-terminal phosphorylation of c-Jun by SAPKs, mice harboring an allele of c-Jun with serines 63 and 73 mutated to alanine (referred to as JunAA) were generated (Behrens et al., 1999). Surprisingly, JunAA homozygotes were obtained at Mendelian frequency, although a slight but significant reduction in body weight was observed in comparison to wild-type adult animals.

Histological examination of several organs, including the liver, revealed no obvious abnormalities. This implies that SAPK phosphorylation of c-Jun is not essential during hepatogenesis or during developmental regulation of cell differentiation and apoptosis. Of note, *SEK1*$^{-/-}$*c-Jun*$^{-/-}$ double mutants die very early in embryogenesis (between E7.5–8.5; the cause of lethality is not known; Nishina et al., 1999). The additive severity of the double mutant phenotype further supports the notion that the SEK1/SAPK module and c-Jun can function in parallel during development. However, SAPK is clearly required for many other functions of c-Jun as JunAA fibroblasts exhibit a defect in proliferation and reduced transformation by components of the Ras pathway and by oncogenic *fos* (Behrens et al., 2000). It is possible that c-Jun phosphorylation by SAPK can act as a molecular switch that increases the spectrum of functions of c-Jun (possibly by regulating the recruitment of distinct co-activator complexes).

Since SEK1-deficient mice are inviable, its function has been further studied by employing *SEK1*$^{-/-}$ ES cells to complement recombination-activating gene (RAG)-2–deficient blastocysts. Recent genetic evidence suggests that *SEK1*$^{-/-}$ thymocytes and peripheral T cells exhibit increased sensitivity to Fas (CD95) and CD3-mediated apoptosis and are defective in CD28-mediated costimulation for proliferation and IL-2 production (Nishina et al., 1997a; Nishina et al., 1997b). B lymphocyte development is also partially impaired (Nishina et al., 1997a). However, these results and conclusions contrast with those of another study that employed RAG-2-blastocyst complementation with a different line of SEK1-targeted ES cells and reported that SEK1 is dispensable for the development of both the B and T lineages (Swat et al., 1998). Furthermore, it was reported that these cells are phenotypically indistinguishable from those of wild-type mice (Swat et al., 1998). Aging *SEK1*$^{-/-}$*RAG-2*$^{-/-}$ chimeric mice frequently developed lymphadenopathy and polyclonal B and T cell expansions, indicating that SEK1 may be required for maintaining peripheral lymphoid homeostasis (Swat et al., 1998). Further investigation is required to resolve the differences between these two reports.

Extensive characterization of the signal transduction in *SEK1*$^{-/-}$ cells has been performed. Cell lines lacking SEK1 exhibit defective SAPK activation and c-Jun activation in response to some (anisomycin, heat shock, TNF-α and IL-1β), but not all (UV irradiation and sorbitol), cellular stresses (Ganiatsas et al., 1998; Nishina et al., 1997b; Yang et al., 1997). Currently, some discrepancy exists amongst scientists in this field as to the precise levels of SAPK activation by these agonists in the absence of SEK1. There also appear to be cell type-specific effects in as far as deficits in responses are distinct between ES cells and fibroblasts, for example.

SAPK and p38 are activated with both quantitative and qualitative differences after a variety of stress stimuli, which must reflect a divergence in activating pathways immediately upstream of these kinases (Zanke et al., 1996). However, SEK1 has been shown to phosphorylate p38 *in vitro* and this promiscuity has raised the possibility that SAPK and p38 are co-regulated by this kinase (Lin et al., 1995). In support of evidence that dominant-negative SEK1 specifically acts as an inhibitor of the SAPK signal transduction pathway (Zanke et al., 1996), biochemical studies with the homozygous knockout SEK1 cells indicates that despite defective SAPK signaling, the activation of p38 by a variety of agonists was unaltered (Ganiatsas et al., 1998; Nishina et al., 1997). While it still remains theoretically possible that the effect of SEK1 gene disruption on p38 is complemented by the p38 upstream activators MKK3 and MKK6, these data suggest that SEK1 functions as a specific activator of SAPK, and not p38, *in vivo*. A MKK6 knockout mouse has not yet been published, but assaying for any SEK1-dependent p38 activity in a *MKK3*$^{-/-}$*MKK6*$^{-/-}$ background will finally resolve this issue. There is agreement that SAPK does not associate with either MKK3 and MKK6, and it was recently shown that MKK3 disruption has no effect on SAPK activation by UV radiation, osmotic shock, IL-1β and TNF-α (Wysk et al., 1999).

4. Regulation of SAPK by MAPKKKs

A large and diverse array of MAPKKKs has been shown to activate SAPKs when overexpressed in cells (reviewed in Widmann et al., 1999; Table 13.1). These include the MEK/ERK kinase (MEKK) subgroup, the mixed-lineage kinase (MLK) group, tumor progression locus-2 (TPL-2, the product of the Cot oncogene), and TGFβ-activated protein kinase (TAK1). The MEKK group of MAPKKKs includes MEKK1-4 and apoptosis signal-regulated kinase-1 and -2 (ASK1/2 or MAPKKK5/6) which are mammalian homologues of *S. cerevisiae* STE11. The MLK group of MAPKKKs, which share significant sequence identity with both serine/threonine and tyrosine kinases, includes MLK1-3, dual-leucine zipper-bearing kinase (DLK or MUK), and leucine zipper-bearing kinase (LZK). Of these, MEKK1-3 and Tpl-2 can also activate the ERK pathway, while only TAK1, ASK1, MLK3 and MEKK4 have been shown to strongly activate p38s as well. There are no known MAPKKKs that activate only p38 or p38/ERK MAPK pathways.

The importance of Ras family GTPases in mammalian MAPK signal transduction was first appreciated with the discovery that oncogenic Ras could activate the ERK pathway. GTPases of the Rho family (Rho, Rac, Cdc42) were originally thought only to regulate the actin cytoskeleton (Bishop and Hall, 2000). More recently, however, these GTPases have been implicated in MAPK signal transduction since constitutively-active mutants of Rac1 and Cdc42 can activate SAPK and p38 (Bishop and Hall, 2000; Coso et al., 1995). Downstream targets of Rac1 and Cdc42 possess a common motif that is critical for G-protein binding (Burbelo et al., 1995). This site, the CDC42/Rac1 interaction and binding domain (CRIB domain), is present on several protein kinases and has been described for MLK1-3 (but not DLK or LZK; Nagata et al., 1998). It has also been reported that MEKK1 and MEKK4 (but not MEKK2 or MEKK3) bind directly to Cdc42 and Rac1 (Fanger et al., 1997). These interactions between Rho GTPases and MAPKKKs may contribute to the effects of Rho

Table 13.1. Components of mammalian stress-regulated MAPK signaling pathways

MAPKs	
SAPKα/β/γ	stress-activated protein kinase (JNK2/3/1 respectively)
p38α/β/γ/δ	p38 MAPK, p38/HOG1, MPK2, Mxi2, CSBP1/2
MAPKKs	
MKK3	MAPK/ERK kinase 3
MKK6	MAPK/ERK kinase 6
SEK1	SAPK/ERK kinase 1 (MKK4, JNKK1)
MKK7	MAPK/ERK kinase 7 (SEK2, JNKK2)
MAPKKKs	
ASK1/2	apoptosis signal-regulating kinase (ASK1 = MAPKKK5)
DLK	dual leucine-zipper bearing kinase (MUK, ZPK)
MEKK1-4	MAPK/ERK kinase kinase (MEKK4 = MTK1)
MLK2	mixed-lineage kinase (MLK2 = MST; MLK3 = SPRK)
PAK	p21-activated kinase
TAK1	TGF-activated protein kinase
Tpl2	tumor progression locus 2 (Cot)
STE20s	
GCK	germinal center kinase
GCKR	GCK-related
GLK	GCK-like kinase
HGK	HPK/GCK-like kinase
HPK1	hematopoietic progenitor kinase 1
MST1	mammalian Ste20-like protein kinase
NESK	NIK-like embryo specific kinase
NIK	Nck-interacting kinase
TAO1/2	one thousand and one amino acid protein kinase 1
Scaffold proteins	
IB1	Islet-Brain 1
JIP1	JNK-interacting protein 1

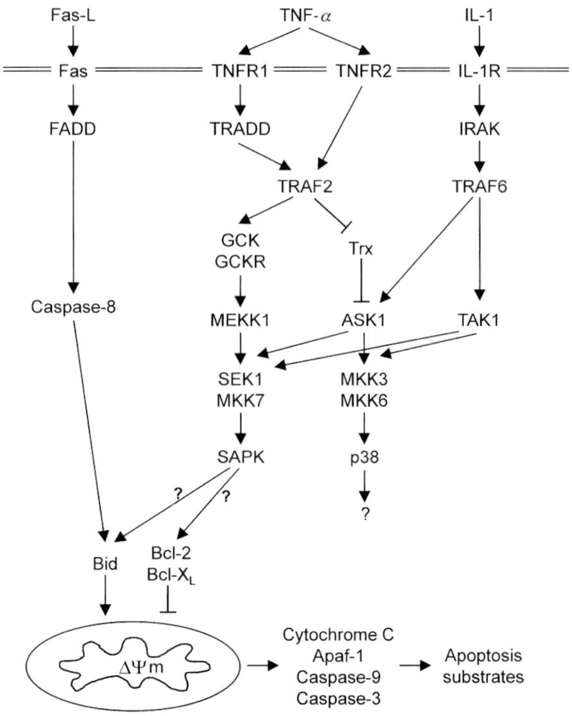

Fig. 13.2. Overview of TNF/SAPK signaling. Inflammation is an important biological response to the exposure of tissues to stress. SAPK and p38 mediate inflammatory signals from the tumor necrosis factor (TNF) family of cytokines. Central to this, the MAPKKK ASK1 associates with TRAF2 and can be negatively regulated by thioredoxin (Trx). As determined by targeted gene disruptions in mice, SAPK is required for caspase-9 activation by the mitochondrial pathway. Potential targets of SAPK include members of the Bcl2 group of apoptotic regulatory proteins.

GTPases on SAPK and p38 activation, however their regulation is not well defined.

The best characterized mechanism of SAPK activation is by proinflammatory cytokines of the tumor necrosis factor (TNF) family (Fig. 13.2). TNF is involved in a variety of biological activities through its binding to two distinct cell surface receptors, p55 TNFR1 and p75 TNFR2 (Baker and Reddy, 1998). Both TNF receptors are part of a receptor superfamily consisting of more than 20 structurally related type I transmembrane proteins that can be divided into two subgroups, depending on whether their intracellular region contains an 80 amino acid motif, termed the “death domain”. The most intensively studied death domain-containing receptors are TNFR1 and Fas: while TNFR1 only induces cell death under certain circumstances and more often induces transcriptional gene activation, Fas is very efficient in cell death induction. TNF receptor family members that do not contain death domains are represented by TNFR2, CD40, CD30, CD27, among others, and are involved primarily in gene transcription for cell survival, growth, and differentiation. Neither TNFR possess intrinsic enzymatic activity and upon ligand binding and receptor aggregation these receptors trigger downstream signaling pathways by recruiting receptor-associated effector molecules. For instance, the 34 kDa protein TNFR-associated death domain protein (TRADD), one of the first identified TNFR1 adapter molecules, is recruited to TNFR1 in a TNF-dependent manner (Hsu et al., 1995) and interacts with another adaptor, TRAF2 (Rothe et al., 1994).

TRAF2 is one of six known mammalian members of the TNF receptor-associated factor (TRAF) family, each of which consists of carboxy-terminal TRAF domains, central zinc finger repeats, and with the exception of TRAF1, an amino-terminal RING finger domain (Baker and Reddy, 1998; Cao et al., 1996). The RING finger is critical for TRAF2 signaling to downstream effectors and the TRAF domains mediate the binding of TRAF proteins to their upstream activators and downstream targets. TRAFs are also genetically conserved across other multicellular organisms including *Drosophila*, *Caenorhabditis elegans*, and *Dictyostelium discoideum*. Transient overexpression of TRAF2, -5, and -6 can activate SAPK and transcription factors in the AP-1 and NF-κB families (Song et al., 1997), while expression of a mutant TRAF construct (TRAF2$^{87\text{-}501}$), in which the zinc RING finger is deleted, blocks TNF activation of SAPK (Liu et al., 1996; Natoli et al., 1997). The TRAF2$^{87\text{-}501}$ construct presumably exerts its effects by binding TRADD and sequestering it from endogenous TRAF2. This result was confirmed by gene targeting studies in which TRAF2 was deleted in mice (Yeh et al., 1997). TRAF2-null embryonic fibroblasts could not activate SAPK, suggesting that TRAF2 is necessary component for coupling TNFR1 to the SAPKs.

Recent studies have identified several MAPKKKs involved in TNF signal transduction. For instance, NF-κB-inducing kinase (NIK) associates with TRAF2 and other members of the TRAF family and mediates activation of NF-κB, but not SAPK (Malinin et al., 1997). TAK1 is also activated by TNF, and was demonstrated to associate with TRAF6, but not TRAF2, in an IL-1-dependent manner (Ninomiya-Tsuji et al., 1999). On the other hand, TRAF2-mediated activation of SAPK is inhibited by catalytically inactive mutants of MEKK1 (Yuasa et al., 1998) and ASK1 (Nishitoh et al., 1998; Hoeflich et al., 1999). ASK1 is responsive to TNF treatment in many cell types (Ichijo et al., 1997) and TNF signaling to SAPK is mediated by ASK1 association with members of the TRAF family (Nishitoh et al., 1998; Hoeflich et al., 1999; Liu et al., 2000). These interactions require the conserved amino-terminal zinc RING and TRAF-N motifs typical to TRAF family members. While overexpression of a wild-type or activated allele of *ASK1* induces apoptosis in various cell types through mitochondria-dependent caspase activation (Hatai et al., 2000), catalytically-inactive ASK1 rescues cells from TNF-mediated killing (Ichijo et al., 1997). This rescue by dominant-negative ASK1 is dependent on the presence of endogenous TRAF2, as determined by comparing TNF-induced apoptosis in TRAF2-deficient and wild-type control fibroblasts (Hoeflich et al., 1999). In addition, through genetic screening for ASK1-binding proteins, the redox-sensing enzyme thioredoxin (Trx) was recently identified as a physiological inhibitor of ASK1 (Liu et al., 2000; Saitoh et al., 1998). Upon treatment of cells with TNF or reactive oxygen species generators such as hydrogen peroxide, Trx appears to be oxidized and ASK1 dissociates from Trx and is bound and activated by TRAF2.

Serine/threonine kinases homologous to yeast STE20, such as germinal center kinase (GCK), GCK-related (GCKR, also referred to as KHS1) and GCK-like kinase (GLK), have also been shown to be important effectors for TNF signaling of SAPK activation (Kyriakis, 1999; Shi and Kehrl, 1997; Yuasa et al., 1998). Antisense constructs of GCKR can block TNF and TRAF2 activation of SAPK. In addition, expression of full-length TRAF2, but not TRAF2 mutants wherein the RING domain has been deleted, activates GCKR and the SAPKs *in vivo*. Via their carboxy terminal domains, GCK and GCKR can both associate *in vivo* with TRAF2 and GCK can also associate *in vivo* with TRAF6. It is noteworthy that the carboxy region of GCK is also required for binding to MEKK1, thereby constituting an ASK1 independent signaling pathway from TRAF2 to SAPK. However, the physiological effect of this interaction is likely to differ somewhat from that of TRAF2-mediated ASK1 activation. For instance, MEKK1 and GCK/GCKR can only activate SAPK, while TNF-α, TRAF2 and ASK1 can activate both the SEK1/MKK7-SAPK and MKK3/MKK6-p38 pathways. Thus, ASK1 is likely to be a physiological target of TRAF2 in recruiting p38. Moreover, coimmunoprecipitation experiments indicate that GCK and ASK1 do not reliably interact *in vivo*. Also, no apoptotic function has been assigned to GCK. MEKK1 has been implicated in the activation of IκB-kinase (IKK), a component in the anti-apoptotic NF-κB pathway (Lee et al., 1997), but antisense constructs of GCKR have no effect on the transactivation of NF-κB. Despite a great deal of interest in the GCK group of kinases and the TNF/SAPK pathway, there has not been a direct demonstration of the activation of a MAPKKK by a GCK homologue. Although these kinases may activate SAPK when over-expressed in cells, their regulation of different MAPKKKs has not been demonstrated biochemically or genetically. Taken together, however, one function of TRAFs may be to regulate the interactions between cytokine-activated GCKs and their effectors, and TNF-induced ternary complex formation of TRAF2-GCK-MEKK1 will be of interest to study.

Efforts to elucidate the mechanisms of MAPKKK regulation have been hampered by the fact that all mammalian SAPK-activating MAPKKKs identified thus far are constitutively active upon overexpression. Correspondingly, while ASK1 can associate with components of the TNFR1 complex, mere overexpression of ASK1 results in its potent activation, overwhelming any endogenous inhibitors present in limiting concentrations. While we still await data from ASK1-deficient cells to determine if ASK1 is selectively required for TNF-induced sustained activation of SAPK and p38, there are already two reports with conflicting results regarding the regulation of SAPK by TNF-α and IL-1β in *MEKK1*$^{+/+}$ and *MEKK1*$^{-/-}$ macrophages and fibroblasts (Xia et al., 2000; Yujiri et al., 2000). It remains possible that as yet unidentified proteins can also bind TRAF2 and mediate SAPK activation.

New evidence indicates that signaling specificity may be mediated through formation of multi-protein complexes held together by "scaffold" proteins. The first example of such a framework molecule is from the yeast mating pheromone pathway where the MAPK FUS3 binds to the scaffold protein STE5 together with the MAPKK STE7 and MAPKKK STE11 (Herskowitz, 1995). Scaffold proteins have now also been identified in mammalian cells. These include MEK partner-1 (MP1) which interacts with the MAPK ERK1 and MAPKK MEK1 (Schaeffer et al., 1998), and the JNK-interacting protein (JIP) group of proteins that bind to SAPK, MKK7 and mixed-lineage protein kinases (Dickens et al., 1997; Yasuda et al., 1999). JIPs have yet to be shown to play a role in TNF receptor-signaling.

5. The p38 MAPK family

Mammalian p38 MAPK was first identified as an LPS-inducible activity in murine peritoneal macrophages (Han et al., 1994). Activation of p38 has traditionally been associated with the stress response and some forms of apoptosis, however, recent studies indicate that a larger variety of cellular processes are regulated by p38 (Nebreda and Porras, 2000). For instance, p38 MAPKs have been proposed to play a physiological role in inflammation and the immune response; inducing differentiation in adipocytes, myoblasts, neurons, chrondrocytes, cardiomyocytes and erythroid cells; and promoting or inhibiting cell proliferation and survival in a cell-type specific manner.

Four p38 MAPKs have been cloned that are 60–70% identical in their amino acid sequences: p38α/Mpk2/CSBP, p38β, p38γ/ERK6, and p38δ. Substrates of p38 MAPKs include protein kinases such as MAPKAPK2 (MAPK-activated protein kinase-2) and several transcription factors including MEF2 (myocyte enhancer factor-2), CHOP/GADD153 (C/EBP homology protein/growth arrest and DNA damage-153), CREB (cAMP-response-element-binding protein) and ATF2 (Gupta et al., 1995; Han et al., 1997; Iordanov et al., 1997; Stokoe et al., 1992; Wang and Ron, 1996). Although p38s have overlapping substrate specificity, some targets appear to be preferentially phosphorylated by one or more isoforms (Cohen, 1997). This, together with the observation that the isoforms have distinct tissue expression patterns (Wang et al., 1997), suggests that p38 MAPKs may have both redundant and specific functions. However, the precise biochemical role that each isoform serves *in vivo* remains unclear.

Studies aimed at understanding the function of p38 have been greatly facilitated by a novel class of pyridinylimidazoles known as cytokine-suppressing anti-inflammatory drugs (CSAIDs; i.e. SB203580) that achieve their effect, at least in part, by inhibition of the α and β isoforms of this kinase (Cuenda et al., 1995). Since p38γ, p38δ and many other protein kinases tested appear to be insensitive to CSAIDs, these drugs have been used extensively to identify substrates and physiological roles of p38α and p38β (Cohen, 1997). However, whether all reported CSAID effects are attributable to p38 inhibition remains to be clarified. Structural and site-directed mutagenesis studies have recently provided a basis for the selectivity of CSAIDs: the drugs are inserted into the ATP-binding pocket of p38α and bind competitively with ATP (Tong et al., 1997; Wilson et al., 1997). CSAIDs, however, do not make contact with residues of the ATP-binding pocket that actually interact with ATP and recent studies have established that threonine-106 of p38α interacts with the 4-fluorophenyl moiety of CSAIDs and plays a critical role in determining drug sensitivity. Mutation of this residue to methionine or glutamine, amino acids present at the equivalent position in other MAPKs, or to other residues with bulky side chains, makes p38α and p38β insensitive to CSAIDs (Wilson et al., 1997).

Table 13.2. Mutant phenotypes of SAPK and p38 pathway components.

Gene	Mouse homozygous phenotype	Reference
ATF2	Viable; hypochondroplasia	Reimold et al. Nature 379, 262–5 (1996)
c-Jun	Lethal E13.5–14.5; impaired hepatogenesis	Hilberg et al. Nature 365, 179–181 (1993); Johnson et al. Genes Dev. 7, 1309–1317 (1993)
JunD	Viable; impaired spermatogenesis	Thepot et al. Development 127, 143–53 (2000)
MEKK1	Viable; no detectable phenotype	Yujiri et al. Science 282, 1911–1914 (1998)
MEKK3	Lethal E11; defective cardiovascular development	Yang et al. Nat Genet. 24, 309–313 (2000)
MKK3	Viable; impaired IL-12 production	Lu et al. EMBO J. 18, 1845–1857 (1999)
MKK7	Lethal	Unpublished (Josef Penninger, personal communication)
p38α	Lethal E13.5; defective placental development and erythropoiesis	Adams et al. Mol Cell 6, 109–116 (2000); Mudgett et al. Proc Natl Acad Sci USA 97, 10454–10459 (2000); Tamura et al Cell 102, 221–231 (2000)
SAPKα (JNK2)	Viable; defective T cell differentiation	Yang et al. Immunity 9, 575–585 (1998)
SAPKβ (JNK3)	Viable; reduction in neuronal apoptosis	Yang et al. Nature 389, 865–870 (1997)
SAPKγ (JNK1)	Viable; defective T cell differentiation	Dong et al. Science 282, 2092–2095 (1998)
SAPKα/SAPKγ	Lethal E10.5; dysregulation of apoptosis in brain	Kuan et al. Neuron 22, 667–676 (1999); Sabapathy et al., Mech. Dev. 89, 115–124 (1999)
SEK1 (MKK4)	Lethal E11.5–12.5; abnormal hepatogenesis	Nishina et al. Development 126, 505–516 (1999); Ganiatsas et al. Proc Natl Acad Sci USA 95, 6881–6886 (1998); Yang et al. Proc Natl Acad Sci USA 94, 3004–3009 (1997)

Conversely, mutation of this residue to threonine in other MAPK family members (SAPKs, p38γ and p38δ) confers sensitivity to CSAIDs (Eyers et al., 1998; Gum et al., 1998). Examination of the sequences of protein kinases in the databases reveals that a bulky residue is almost always found at the position equivalent to threonine-106. However, a small number of proteins do have threonine at this position and recent work demonstrating that CSAIDs can also modulate the activity of type-II TGF-β receptor, Lck (lymphoid cell kinase), Raf-1 MAPKKK, PDK1, and cyclooxygenase (Borsch-Haubold et al., 1998; Eyers et al., 1998; Hall-Jackson et al., 1999; Lali et al., 2000) highlight the potential for CSAID-mediated cellular effects that are independent of p38. Thus, to better understand the biological function of p38 and to what extent the various p38 isoforms participate in separate physiological processes, four independent groups have recently used homologous recombination to disrupt p38α in mice (Table 13.2).

6. Genetic analysis of p38α in mice

In all cases, deletion of the p38α gene resulted in embryonic lethality commencing at E10.5 (Adams et al., 2000; Allen et al., 2000; Mudgett et al., 2000; Tamura et al., 2000). Biochemical assays determining the stimulus-induced phosphorylation of p38α-dependent targets (MAPKAPK2, ATF2) confirmed that the p38α targeting strategy successfully and specifically abolished signaling by p38α. Since mice express multiple p38 MAPK family members, the developmental arrest demonstrates that the different enzymes do not perform entirely redundant activities, at least during embryonic development. Although Allen et al. (2000) noted that mice null for the p38α allele die during embryonic development, no specific phenotype was described. Adams et al (2000) and Mudgett et al. (2000) observe similar phenotypes. These groups indicate that with time, the *p38α*$^{-/-}$ embryos become pale and anemic, have deficiencies in vascularization of the embryo and yolk sac, and show varying degrees of growth retardation. The challenge was to identify the defects that are directly due to loss of p38 function and to distinguish these from secondary defects associated with the loss of viability.

Since a β-galactosidase cassette was inserted into the p38α locus, the first clue came from studying the expression of p38α during embryonic development (Adams et al., 2000). At E10.5, abundant labeling was seen in many regions of the embryo including the heart, branchial arches, limb buds, and somites. High levels of p38α were also found in the extraembryonic tissues, such as the endoderm, mesoderm, and the vasculature of the yolk sac and the placenta. Closer examination revealed that while the maternal part of the placenta was normal, there was a striking reduction in the labyrinthine layer and embryonic blood vessels seemed to be trapped in the superficial layers of the placenta and could not intermingle with maternal blood vessels. Histological analysis by one study (Mudgett et al., 2000) described a greatly reduced spongiotrophoblast layer while the other study (Adams et al., 2000) reported this structure to be normal. However, from the phenotype both groups concluded that p38α plays an essential role in placental organogenesis.

To determine if the primary site of the defect caused by the p38α mutation was indeed the fetal placenta, Adams et al. (2000) generated chimeras between and diploid homozygous mutant embryos and wild-type embryos made tetraploid by electrofusion at the two-cell stage. Tetraploid embryos, with rare exceptions, are incapable of forming the embryo proper, but are capable of forming the extraembryonic tissues (primitive endoderm and the trophoblast lineages). ES cells, by comparison, are unable to form the trophoblast compartment but do form the extra-embryonic mesoderm and the embryo proper. Therefore, if a defect occurs in the function of the extraembryonic cell population, fusion of the mutant embryos to wild-type tetraploid cells rescues the defect and allows the development of the embryo. In this way, tetraploid/mutant embryo chimeras were used to assess whether the primary site of the *p38α*$^{-/-}$ defect was indeed in the placental trophoblast cell population. After aggregation, the *p38α*$^{-/-}$ animals developed to term. In addition, a total of 39 E18.5 embryos were recovered of which seven were *p38α*$^{-/-}$ and appeared completely viable. Histological analysis revealed that heart structures were normal in homozygous mutant embryos after tetraploid rescue, demonstrating that the cardiac and vascular developmental defects originally observed in *p38α*$^{-/-}$ embryos are strongly dependent on placental function. Taken together, this indicates that the cardiovascular malformation and massive reduction of the myocardium at E10.5 was secondary to impaired placental morphogenesis and the primary cause for growth retardation and lethality of homozygous mutants is insufficient oxygen and nutrient transfer across the placenta. These findings are consistent with previous evidence for the relevance of placental function for normal cardiac development (Ihle, 2000). Surprisingly, in spite of being broadly expressed in the embryo, p38α appears to be critical only for placenta organogenesis.

Mammals have developed precise and well-regulated mechanisms required to establish fetal–maternal contact. One of the first differentiation events in a mammalian embryo leads to the generation of trophoblast cells, specialized epithelial cells that form the embryonic component of the fetal–maternal interface during the implantation and placentation processes. The trophoblast giant cells together with the parietal endoderm comprise the earliest placental structure (parietal yolk sac). Trophoblast giant cells produce hormones, proteinases and other molecules which facilitate the breakdown and invasion of the decidua. Placental development is characterized by extensive angiogenesis to establish the vascular structures involved in transplacental exchange, massive proliferation and differentiation of multiple cell types. Failures in implantation and placental development are a significant source of embryonic lethality and therefore there is a need for a tight regulation of these processes.

What might be the upstream activators of p38α in this process? A number of proteins has been shown to be vital for chorioallantoic placental development including the basic-helix–loop-helix transcription factors MASH-2 and TFEB, nuclear hormone receptor peroxisome proliferator-activated receptor-γ (PPARγ), estrogen-receptor–related receptor-β (ERR-β), the von Hippel-Lindau tumor suppressor protein (VHL), heat shock protein-90β (Hsp90β), and retinoid X receptor α or β (Barak et al., 1999; Gnarra et al., 1997; Guillemot et al., 1994; Luo et al., 1997; Sapin et al., 1997; Steingrimsson et al., 1998; Voss et al., 2000; Wendling et al., 1999). An interaction between p38α and some of these known players in placental development is conceivable, as p38s previously have been shown to be activated by heat shock, oxidative stress and hormone signaling. Early response proto-oncogenes, such as c-Jun and JunB (Dungy et al., 1991; Schorpp-Kistner et al., 1999), have been associated with both proliferation and differentiation events of extra-embryonic tissues. These genes are also expressed in the placenta of human and rodents throughout gestation, suggesting an additional mechanism for p38α in placental development. Lastly, the p38α-null mice phenotype is highly reminiscent of the defective labyrinthine layer of the placenta and cardiovascular malformation that has been recently described for mice lacking MEKK3, an upstream activator of p38 MAPK signaling (Yang et al., 2000).

The fourth study by Tamura et al. (2000) reported a different analysis and interpretation of the *p38α*$^{-/-}$ phenotype. The authors indicate that the p38α-deficiency results in two distinct developmental defects. As described by the other groups, the p38α null embryos are initially challenged by placental insufficiency, but a significant portion (6/31 embryos) remain viable until much later in development (E16.5) with normal morphology but highly anemic appearance. The basis for the anemic phenotype was traced to a block in erythroid differentiation in fetal liver cells and deficiency in Epo gene expression. The reasons for the differences between the various examples of p38α mutant animals is currently unclear as all appear to be true nulls (rather than, for example, hypomorphs).

Tamura at al. (2000) injected C57Bl/6 blastocysts with targeted 129/SvEv ES cells and suggested genetic background and strain variation as a possible explanation for their different phenotype. However, whether this could explain all of the differences among the p38α animals is unclear since between Mudgett et al. (2000) and Adams et al. (2000) multiple ES cell lines were used and the progeny were crossed onto 129/SvEv, C57Bl/6, CD1 or Balb/c mice with comparable phenotypes. In addition, in the aggregation of tetraploid *p38α*$^{+/+}$ embryos with diploid *p38α*$^{-/-}$ embryos performed by Adams et al. (2000), animals in mixed 129/SvEv x C57Bl/6 and 129/SvEv × C57Bl/6 × CD1 backgrounds were used with very similar results. Furthermore, Tamura et al. (2000) indicated that the actual time at which the *p38α*$^{-/-}$ mice died varied between the animal facilities used for this study: mice housed in San Diego did not survive beyond E13.5, while those kept in Saitama, Japan were reported viable at E16.5. The basis for this difference was not reported but presumably reflects immunological/infective factors.

Despite some discrepancies, these studies demonstrate several important conclusions. Firstly, the phenotypes associated with the p38α null mutations suggest that the function of p38α is, at least partially, nonredundant with other p38 MAPK family members. This is quite different than what has been revealed from knockouts of the SAPK or ERK family members. None of these mutants revealed a critical role for any of the individual MAPKs in development, and therefore, it has been assumed that this family of protein kinases has extensively overlapping functions within each subtype. In addition, although previous studies have established that MAPKAPK2 is a substrate of p38α, the extent to which other kinases may participate *in vivo* in the activation of MAPKAPK2 remained unclear. Biochemical analysis in the p38α knockout studies found that UV, anisomycin and sodium arsenite-induced activation of MAPKAPK2 was completely impaired in *p38α*$^{-/-}$ cells (Adams et al., 2000; Allen et al., 2000). Thus, while the lack of a comparable phenotype in MKK3-deficient embryos suggests that this kinase is not uniquely required for p38α activation, it can be concluded that MAPKAPK2 is a nonredundant component of the pathway.

7. Concluding remarks

The ultimate merit of basic research in the so-called “life sciences” is largely gauged by its impact on medicine and human health. This prompts the question/concern: has what we have learned mechanistically about MAPK pathways, their stress-inducing agonists, and their physiological consequences given us any better insight into human disease?

For instance, it was previously demonstrated that some human cancer tissues and cell lines have SEK1 genetic changes and lose SEK1 protein expression or activity. Homozygous deletions or missense mutations were detected in pancreatic (2/92; 2%), biliary (1/16; 6%) and breast (3/22; 15%) carcinomas, as well as in cancer cell lines originating from pancreas and lung cancers (in total, 6/213; 3%; Su et al., 1998; Teng et al., 1997). This indicates that SEK1 may have a role as a tumor suppressor. However, a third group, employing a different method of tumor sample preparation, examined the correlation between the SEK1 protein expression and clinicopathological features of tumors and reported that patients with SEK1 protein expressed in gastric cancer tissues have significantly poorer survival rates than patients lacking SEK1 (Wu et al., 2000). Taken together, although we can speculate that SEK1 may be a significant prognostic factor and player in cancer progression, further work and clarification is clearly required. This example shows that along with more analysis of clinical samples, genomic and proteomic-based approaches will be needed to make significant strides in understanding events downstream of MAPK activation in disease states. In the six years since the initial identification of SAPKs, less than ten candidate substrates have been identified (only half of which have been confirmed by more than one group).

With the new large-scale approaches to studying protein function that are available, the role of these signaling pathways in disease (including the aforementioned suggestion that control of SEK1 and MKK7 activity and expression may provide novel approaches to cancer therapy) can be better understood in the near future.

Although much research is focussed on understanding the action of disease-related/causing genes, these gene products are often not amenable to pharmaceutical drug intervention. It is therefore important to also concentrate efforts on the enzymes, receptors or channels (which do make good drug targets) upstream of these disease genes. Significant success has already been achieved with the development of small molecule inhibitors to a number of protein-tyrosine kinases (with several inhibitors in clinical trials) and similar efforts for serine/threonine protein kinases are already under way. Most of these inhibitors target the ATP binding pocket. This raises the spectre of low specificity given the predicted existence of 2000 protein kinases in the human genome. However, practice has proven that remarkable selectivity can be achieved. An important side-product of this pharmaceutical work is the release of valuable tools to the research community for examining the consequences of inhibition of these stress-sensing pathways. That said, it is unlikely that small molecules will effectively discriminate between the highly related splice isoforms of the proteins that populate the stress-kinase response pathways. Use of inhibitors has important caveats. Although our understanding of the physiological role of p38α and p38β has been significantly increased through utilization of the CSAID class of inhibitors, CSAIDs can also affect other enzymes (although usually with lower potency) and important control experiments must be used to validate *in vivo* results. The best way to evaluate new inhibitors to MAPKs will be to use a combination of approaches allowing comparison of results obtained through genetic inactivation of multiple MAPK enzymes or isoforms, dominant-negative and inhibitor-insensitive mutants and small molecule antagonists. The increasing use of conditional inactivation alleles for gene targeted mice (e.g. Cre-loxP recombinase technology) allows examination of the consequences of deletion within an adult tissue of a gene that is needed for embryonic development. Indeed, perhaps the most unpredicted result of analysis of the physiological roles of stress-activated signaling proteins is their important function in normal developmental processes. Clearly, these systems evolved first as stress-responsive machines. They were later co-opted into less volatile responses in embryogenesis and differentiation without shedding their primordial raison d'etre. That said, perhaps their largest "stress-inducing effect" has been on investigators trying to understand their complexity and roles. Although progress in the field has been significant, it is safe to say that the scientists working in the area are unlikely to adapt as rapidly to new information as the systems they are perturbing.

Acknowledgements

We apologize to the authors whose original work is not included in the references owing to space limitations. K.P.H. is supported by a Medical Research Council Studentship. J.R.W. is supported by grants from the Medical Research Council and Howard Hughes Medical Institute and is a Medical Research Council Senior Scientist.

References

Adams, R.H., Porras, A., Alonso, G., Jones, M., Vintersten, K., Panelli, S., Valladares, A., Perez, L., Klein, R. and Nebreda, A.R. (2000). Essential role of p38α MAP kinase in placental but not embryonic cardiovascular development. Mol. Cell. 6, 109–116.

Allen, M., Svensson, L., Roach, M., Hambor, J., McNeish, J. and Gabel, C.A. (2000). Deficiency of the stress kinase p38α results in embryonic lethality: characterization of the kinase dependence of stress responses of enzyme-deficient embryonic stem cells. J. Exp. Med. 191, 859–870.

Baker, S.J., and Reddy, E.P. (1998). Modulation of life and death by the TNF receptor superfamily. Oncogene 17, 3261–3270.

Barak, Y., Nelson, M.C., Ong, E.S., Jones, Y.Z., Ruiz-Lozano, P., Chien, K.R., Koder, A. and Evans, R.M. (1999). PPAR gamma is required for placental, cardiac, and adipose tissue development. Mol. Cell. 4, 585–595.

Behrens, A., Jochum, W., Sibilia, M. and Wagner, E.F. (2000). Oncogenic transformation by ras and fos is mediated by c-Jun N-terminal phosphorylation. Oncogene 19, 2657–2663.

Behrens, A., Sibilia, M. and Wagner, E.F. (1999). Amino-terminal phosphorylation of c-Jun regulates stress-induced apoptosis and cellular proliferation. Nat. Genet. 21, 326–329.

Bishop, A.L. and Hall, A. (2000). Rho GTPases and their effector proteins. Biochem. J. 348 Pt 2, 241–255.

Borsch-Haubold, A.G., Pasquet, S. and Watson, S.P. (1998). Direct inhibition of cyclooxygenase-1 and -2 by the kinase inhibitors SB 203580 and PD 98059. SB 203580 also inhibits thromboxane synthase. J. Biol. Chem. 273, 28766–28772.

Burbelo, P.D., Drechsel, D. and Hall, A. (1995). A conserved binding motif defines numerous candidate target proteins for both Cdc42 and Rac GTPases. J. Biol. Chem. 270, 29071–29074.

Canagarajah, B.J., Khokhlatchev, A., Cobb, M.H. and Goldsmith, E.J. (1997). Activation mechanism of the MAP kinase ERK2 by dual phosphorylation. Cell 90, 859–869.

Cao, Z., Xiong, J., Takeuchi, M., Kurama, T. and Goeddel, D.V. (1996). TRAF6 is a signal transducer for interleukin-1. Nature 383, 443–446.

Chow, C.W., Rincon, M., Cavanagh, J., Dickens, M. and Davis, R.J. (1997). Nuclear accumulation of NFAT4 opposed by the JNK signal transduction pathway. Science 278, 1638–1641.

Cohen, P. (1997). The search for physiological substrates of MAP and SAP kinases in mammalian cells. Trends Cell. Biol. 7, 353–361.

Coso, O.A., Chiariello, M., Yu, J.C., Teramoto, H., Crespo, P., Xu, N., Miki, T. and Gutkind, J.S. (1995). The small GTP-binding proteins Rac1 and Cdc42 regulate the activity of the JNK/SAPK signaling pathway. Cell 81, 1137–1146.

Courchesne, W.E., Kunisawa, R. and Thorner, J. (1989). A putative protein kinase overcomes pheromone-induced arrest of cell cycling in *S. cerevisiae*. Cell 58, 1107–1119.

Cuenda, A., Rouse, J., Doza, Y.N., Meier, R., Cohen, P., Gallagher, T.F., Young, P.R. and Lee, J.C. (1995). SB 203580 is a specific inhibitor of a MAP kinase homologue which is stimulated by cellular stresses and interleukin-1. FEBS Lett 364, 229–233.

Derijard, B., Raingeaud, J., Barrett, T., Wu, I.H., Han, J., Ulevitch, R.J. and Davis, R.J. (1995). Independent human MAP-kinase signal transduction pathways defined by MEK and MKK isoforms [published erratum appears in Science 1995 Jul 7;269(5220):17]. Science 267, 682–685.

Dickens, M., Rogers, J.S., Cavanagh, J., Raitano, A., Xia, Z., Halpern, J.R., Greenberg, M.E., Sawyers, C.L. and Davis, R.J. (1997). A cytoplasmic inhibitor of the JNK signal transduction pathway. Science 277, 693–696.

Dong, C., Yang, D.D., Wysk, M., Whitmarsh, A.J., Davis, R.J. and Flavell, R.A. (1998). Defective T cell differentiation in the absence of Jnk1. Science 282, 2092–2095.

Dungy, L.J., Siddiqi, T.A. and Khan, S. (1991). C-jun and jun-B oncogene expression during placental development. Am. J. Obstet. Gynecol. 165, 1853–1856.

Elion, E.A., Grisafi, P.L. and Fink, G.R. (1990). FUS3 encodes a cdc2+/CDC28-related kinase required for the transition from mitosis into conjugation. Cell 60, 649–664.

Engel, M.E., McDonnell, M.A., Law, B.K. and Moses, H.L. (1999). Interdependent SMAD and JNK signaling in transforming growth factor-beta-mediated transcription. J. Biol. Chem. 274, 37413–37420.

Eyers, P.A., Craxton, M., Morrice, N., Cohen, P. and Goedert, M. (1998). Conversion of SB 203580-insensitive MAP kinase family members to drug-sensitive forms by a single amino-acid substitution. Chem. Biol. 5, 321–328.

Fanger, G.R., Johnson, N.L. and Johnson, G.L. (1997). MEK kinases are regulated by EGF and selectively interact with Rac/Cdc42. Embo J. 16, 4961–4972.

Fuchs, S.Y., Adler, V., Pincus, M.R. and Ronai, Z. (1998). MEKK1/JNK signaling stabilizes and activates p53. Proc. Natl. Acad. Sci. USA 95, 10541–10546.

Ganiatsas, S., Kwee, L., Fujiwara, Y., Perkins, A., Ikeda, T., Labow, M.A. and Zon, L.I. (1998). SEK1 deficiency reveals mitogen-activated protein kinase cascade cross-regulation and leads to abnormal hepatogenesis. Proc. Natl. Acad. Sci. USA 95, 6881–6886.

Glise, B., Bourbon, H. and Noselli, S. (1995). hemipterous encodes a novel *Drosophila* MAP kinase kinase, required for epithelial cell sheet movement. Cell 83, 451–461.

Gnarra, J.R., Ward, J.M., Porter, F.D., Wagner, J.R., Devor, D.E., Grinberg, A., Emmert-Buck, M.R., Westphal, H., Klausner, R.D. and Linehan, W.M. (1997). Defective placental vasculogenesis causes embryonic lethality in VHL-deficient mice. Proc. Natl. Acad. Sci. USA 94, 9102–9107.

Guillemot, F., Nagy, A., Auerbach, A., Rossant, J. and Joyner, A.L. (1994). Essential role of Mash-2 in extraembryonic development. Nature 371, 333–336.

Gum, R.J., McLaughlin, M.M., Kumar, S., Wang, Z., Bower, M.J., Lee, J.C., Adams, J.L., Livi, G.P., Goldsmith, E.J. and Young, P.R. (1998). Acquisition of sensitivity of stress-activated protein kinases to the p38 inhibitor, SB 203580, by alteration of one or more amino acids within the ATP binding pocket. J. Biol. Chem. 273, 15605–15610.

Gupta, S., Barrett, T., Whitmarsh, A.J., Cavanagh, J., Sluss, H.K., Derijard, B. and Davis, R.J. (1996). Selective interaction of JNK protein kinase isoforms with transcription factors. Embo J. 15, 2760–2770.

Gupta, S., Campbell, D., Derijard, B. and Davis, R.J. (1995). Transcription factor ATF2 regulation by the JNK signal transduction pathway. Science 267, 389–393.

Hall-Jackson, C.A., Goedert, M., Hedge, P. and Cohen, P. (1999). Effect of SB 203580 on the activity of c-Raf in vitro and in vivo. Oncogene 18, 2047–2054.

Han, J., Jiang, Y., Li, Z., Kravchenko, V.V. and Ulevitch, R.J. (1997). Activation of the transcription factor MEF2C by the MAP kinase p38 in inflammation. Nature 386, 296–299.

Han, J., Lee, J.D., Bibbs, L. and Ulevitch, R.J. (1994). A MAP kinase targeted by endotoxin and hyperosmolarity in mammalian cells. Science 265, 808–811.

Han, Z.S., Enslen, H., Hu, X., Meng, X., Wu, I.H., Barrett, T., Davis, R.J. and Ip, Y.T. (1998). A conserved p38 mitogen-activated protein kinase pathway regulates *Drosophila* immunity gene expression. Mol. Cell. Biol. 18, 3527–3539.

Hatai, T., Matsuzawa, A., Inoshita, S., Mochida, Y., Kuroda, T., Sakamaki, K., Kuida, K., Yonehara, S., Ichijo, H. and Takeda, K. (2000). Execution of apoptosis signal-regulating kinase 1 (ASK1)-induced apoptosis by the mitochondria-dependent caspase activation. J. Biol. Chem. 275, 26576–26581.

Herskowitz, I. (1995). MAP kinase pathways in yeast: for mating and more. Cell 80, 187–197.

Hibi, M., Lin, A., Smeal, T., Minden, A. and Karin, M. (1993). Identification of an oncoprotein- and UV-responsive protein kinase that binds and potentiates the c-Jun activation domain. Genes Dev. 7, 2135–2148.

Hilberg, F., Aguzzi, A., Howells, N. and Wagner, E.F. (1993). c-jun is essential for normal mouse development and hepatogenesis.Nature 365, 179–181.

Hoeflich, K.P., Yeh, W.C., Yao, Z., Mak, T.W. and Woodgett, J.R. (1999). Mediation of TNF receptor-associated factor effector functions by apoptosis signal-regulating kinase-1 (ASK1). Oncogene 18, 5814–5820.

Hsu, H., Xiong, J. and Goeddel, D.V. (1995). The TNF receptor 1-associated protein TRADD signals cell death and NF-κ B activation. Cell 81, 495–504.

Ichijo, H., Nishida, E., Irie, K., ten Dijke, P., Saitoh, M., Moriguchi, T., Takagi, M., Matsumoto, K., Miyazono, K. and Gotoh, Y. (1997). Induction of apoptosis by ASK1, a mammalian MAPKKK that activates SAPK/JNK and p38 signaling pathways. Science 275, 90–94.

Ihle, J.N. (2000). The challenges of translating knockout phenotypes into gene function. Cell 102, 131–134.

Iordanov, M., Bender, K., Ade, T., Schmid, W., Sachsenmaier, C., Engel, K., Gaestel, M., Rahmsdorf, H.J. and Herrlich, P. (1997). CREB is activated by UVC through a p38/HOG-1-dependent protein kinase. Embo J. 16, 1009–1022.

Johnson, R.S., van Lingen, B., Papaioannou, V.E. and Spiegelman, B.M. (1993). A null mutation at the c-jun locus causes embryonic lethality and retarded cell growth in culture. Genes Dev. 7, 1309–1317.

Kallunki, T., Deng, T., Hibi, M. and Karin, M. (1996). c-Jun can recruit JNK to phosphorylate dimerization partners via specific docking interactions. Cell 87, 929–939.

Keyse, S.M. (2000). Protein phosphatases and the regulation of mitogen-activated protein kinase signalling.

Curr. Opin. Cell Biol. 12, 186–192.

Kuan, C.Y., Yang, D.D., Samanta Roy, D.R., Davis, R.J., Rakic, P. and Flavell, R.A. (1999). The Jnk1 and Jnk2 protein kinases are required for regional specific apoptosis during early brain development. Neuron 22, 667–676.

Kyriakis, J.M. (1999). Signaling by the germinal center kinase family of protein kinases. J. Biol. Chem. 274, 5259–5262.

Kyriakis, J.M., Banerjee, P., Nikolakaki, E., Dai, T., Rubie, E.A., Ahmad, M.F., Avruch, J. and Woodgett, J.R. (1994). The stress-activated protein kinase subfamily of c-Jun kinases. Nature 369, 156–160.

Lali, F.V., Hunt, A.E., Turner, S.J. and Foxwell, B.M. (2000). The pyridinyl imidazole inhibitor SB203580 blocks phosphoinositide-dependent protein kinase activity, protein kinase B phosphorylation, and retinoblastoma hyperphosphorylation in interleukin-2-stimulated T cells independently of p38 mitogen-activated protein kinase. J. Biol. Chem. 275, 7395–7402.

Lee, F.S., Hagler, J., Chen, Z.J. and Maniatis, T. (1997). Activation of the IκB α kinase complex by MEKK1, a kinase of the JNK pathway. Cell 88, 213–222.

Lin, A., Minden, A., Martinetto, H., Claret, F.X., Lange-Carter, C., Mercurio, F., Johnson, G.L. and Karin, M. (1995). Identification of a dual specificity kinase that activates the Jun kinases and p38-Mpk2. Science 268, 286–290.

Liu, H., Nishitoh, H., Ichijo, H. and Kyriakis, J.M. (2000). Activation of apoptosis signal-regulating kinase 1 (ASK1) by tumor necrosis factor receptor-associated factor 2 requires prior dissociation of the ASK1 inhibitor thioredoxin. Mol. Cell. Biol. 20, 2198–2208.

Liu, Z.G., Hsu, H., Goeddel, D.V. and Karin, M. (1996). Dissection of TNF receptor 1 effector functions: JNK activation is not linked to apoptosis while NF-κB activation prevents cell death. Cell 87, 565–576.

Luo, J., Sladek, R., Bader, J.A., Matthyssen, A., Rossant, J. and Giguere, V. (1997). Placental abnormalities in mouse embryos lacking the orphan nuclear receptor ERR-β. Nature 388, 778–782.

Madhani, H.D. and Fink, G.R. (1998). The riddle of MAP kinase signaling specificity. Trends Genet. 14, 151–155.

Malinin, N.L., Boldin, M.P., Kovalenko, A.V. and Wallach, D. (1997). MAP3K-related kinase involved in NF-κB induction by TNF, CD95 and IL-1. Nature 385, 540–544.

Moriguchi, T., Kawasaki, H., Matsuda, S., Gotoh, Y. and Nishida, E. (1995). Evidence for multiple activators for stress-activated protein kinase/c-Jun amino-terminal kinases. Existence of novel activators. J. Biol. Chem. 270, 12969–12972.

Moriguchi, T., Toyoshima, F., Masuyama, N., Hanafusa, H., Gotoh, Y. and Nishida, E. (1997). A novel SAPK/JNK kinase, MKK7, stimulated by TNFα and cellular stresses. Embo J. 16, 7045–7053.

Mudgett, J.S., Ding, J., Guh-Siesel, L., Chartrain, N.A., Yang, L., Gopal, S. and Shen, M.M. (2000). Essential role for p38α mitogen-activated protein kinase in placental angiogenesis. Proc. Natl. Acad. Sci. USA 97, 10454–10459.

Nagata, K., Puls, A., Futter, C., Aspenstrom, P., Schaefer, E., Nakata, T., Hirokawa, N. and Hall, A. (1998). The MAP kinase kinase kinase MLK2 co-localizes with activated JNK along microtubules and associates with kinesin superfamily motor KIF3. Embo J. 17, 149–158.

Natoli, G., Costanzo, A., Ianni, A., Templeton, D.J., Woodgett, J.R., Balsano, C. and Levrero, M. (1997). Activation of SAPK/JNK by TNF receptor 1 through a noncytotoxic TRAF2-dependent pathway. Science 275, 200–203.

Nebreda, A.R. and Porras, A. (2000). p38 MAP kinases: beyond the stress response. Trends Biochem. Sci. 25, 257–260.

Ninomiya-Tsuji, J., Kishimoto, K., Hiyama, A., Inoue, J., Cao, Z. and Matsumoto, K. (1999). The kinase TAK1 can activate the NIK-I κB as well as the MAP kinase cascade in the IL-1 signaling pathway. Nature 398, 252–256.

Nishina, H., Bachmann, M., Oliveira-dos-Santos, A.J., Kozieradzki, I., Fischer, K.D., Odermatt, B., Wakeham, A., Shahinian, A., Takimoto, H., Bernstein, A., Mak, T.W., Woodgett, J.R., Ohashi, P.S. and Penninger, J.M. (1997a). Impaired CD28-mediated interleukin 2 production and proliferation in stress kinase SAPK/ERK1 kinase (SEK1)/mitogen-activated protein kinase kinase 4 (MKK4)-deficient T lymphocytes. J. Exp. Med. 186, 941–953.

Nishina, H., Fischer, K.D., Radvanyi, L., Shahinian, A., Hakem, R., Rubie, E.A., Bernstein, A., Mak, T.W., Woodgett, J.R. and Penninger, J.M. (1997b). Stress-signalling kinase Sek1 protects thymocytes from apoptosis mediated by CD95 and CD3. Nature 385, 350–353.

Nishina, H., Vaz, C., Billia, P., Nghiem, M., Sasaki, T., De la Pompa, J.L., Furlonger, K., Paige, C., Hui, C., Fischer, K.D., Kishimoto, H., Iwatsubo, T., Katada, T., Woodgett, J.R. and Penninger, J.M. (1999). Defective liver formation and liver cell apoptosis in mice lacking the stress signaling kinase SEK1/MKK4. Development 126, 505–516.

Nishitoh, H., Saitoh, M., Mochida, Y., Takeda, K., Nakano, H., Rothe, M., Miyazono, K. and Ichijo, H. (1998). ASK1 is essential for JNK/SAPK activation by TRAF2. Mol. Cell. 2, 389–395.

Noguchi, K., Kitanaka, C., Yamana, H., Kokubu, A., Mochizuki, T. and Kuchino, Y. (1999). Regulation of c-Myc through phosphorylation at Ser-62 and Ser-71 by c-Jun N-terminal kinase. J. Biol. Chem. 274, 32580–32587.

Riesgo-Escovar, J.R., Jenni, M., Fritz, A. and Hafen, E.

(1996). The *Drosophila* Jun-N-terminal kinase is required for cell morphogenesis but not for DJun-dependent cell fate specification in the eye. Genes Dev. 10, 2759–2768.

Rincon, M., Whitmarsh, A., Yang, D.D., Weiss, L., Derijard, B., Jayaraj, P., Davis, R.J. and Flavell, R.A. (1998). The JNK pathway regulates the In vivo deletion of immature CD4(+)CD8(+) thymocytes. J. Exp. Med. 188, 1817–1830.

Rothe, M., Wong, S.C., Henzel, W.J. and Goeddel, D.V. (1994). A novel family of putative signal transducers associated with the cytoplasmic domain of the 75 kDa tumor necrosis factor receptor. Cell 78, 681–692.

Sabapathy, K., Hu, Y., Kallunki, T., Schreiber, M., David, J.P., Jochum, W., Wagner, E.F. and Karin, M. (1999). JNK2 is required for efficient T-cell activation and apoptosis but not for normal lymphocyte development. Curr. Biol. 9, 116–125.

Sabapathy, K., Jochum, W., Hochedlinger, K., Chang, L., Karin, M. and Wagner, E.F. (1999). Defective neural tube morphogenesis and altered apoptosis in the absence of both JNK1 and JNK2. Mech. Dev. 89, 115–124.

Saitoh, M., Nishitoh, H., Fujii, M., Takeda, K., Tobiume, K., Sawada, Y., Kawabata, M., Miyazono, K. and Ichijo, H. (1998). Mammalian thioredoxin is a direct inhibitor of apoptosis signal-regulating kinase (ASK) 1. Embo J. 17, 2596–2606.

Sanchez, I., Hughes, R.T., Mayer, B.J., Yee, K., Woodgett, J.R., Avruch, J., Kyriakis, J.M. and Zon, L.I. (1994). Role of SAPK/ERK kinase-1 in the stress-activated pathway regulating transcription factor c-Jun. Nature 372, 794–798.

Sapin, V., Dolle, P., Hindelang, C., Kastner, P. and Chambon, P. (1997). Defects of the chorioallantoic placenta in mouse RXRα null fetuses. Dev. Biol. 191, 29–41.

Schaeffer, H.J., Catling, A.D., Eblen, S.T., Collier, L.S., Krauss, A. and Weber, M.J. (1998). MP1: a MEK binding partner that enhances enzymatic activation of the MAP kinase cascade. Science 281, 1668–1671.

Schaeffer, H.J. and Weber, M.J. (1999). Mitogen-activated protein kinases: specific messages from ubiquitous messengers. Mol. Cell. Biol. 19, 2435–2444.

Schorpp-Kistner, M., Wang, Z.Q., Angel, P. and Wagner, E.F. (1999). JunB is essential for mammalian placentation. Embo J. 18, 934–948.

Shi, C.S. and Kehrl, J.H. (1997). Activation of stress-activated protein kinase/c-Jun N-terminal kinase, but not NF-κB, by the tumor necrosis factor (TNF) receptor 1 through a TNF receptor-associated factor 2- and germinal center kinase related-dependent pathway. J. Biol. Chem. 272, 32102–32107.

Sluss, H.K., Han, Z., Barrett, T., Davis, R.J. and Ip, Y.T. (1996). A JNK signal transduction pathway that mediates morphogenesis and an immune response in *Drosophila*. Genes Dev. 10, 2745–2758.

Song, H.Y., Regnier, C.H., Kirschning, C.J., Goeddel, D.V. and Rothe, M. (1997). Tumor necrosis factor (TNF)-mediated kinase cascades: bifurcation of nuclear factor-κB and c-jun N-terminal kinase (JNK/SAPK) pathways at TNF receptor-associated factor 2. Proc. Natl. Acad. Sci. USA 94, 9792–9796.

Steingrimsson, E., Tessarollo, L., Reid, S.W., Jenkins, N.A. and Copeland, N.G. (1998). The bHLH-Zip transcription factor Tfeb is essential for placental vascularization. Development 125, 4607–4616.

Stokoe, D., Campbell, D.G., Nakielny, S., Hidaka, H., Leevers, S.J., Marshall, C. and Cohen, P. (1992). MAPKAP kinase-2; a novel protein kinase activated by mitogen-activated protein kinase. Embo J. 11, 3985–3994.

Su, G.H., Hilgers, W., Shekher, M.C., Tang, D.J., Yeo, C.J., Hruban, R.H. and Kern, S.E. (1998). Alterations in pancreatic, biliary, and breast carcinomas support MKK4 as a genetically targeted tumor suppressor gene. Cancer Res. 58, 2339–2342.

Swat, W., Fujikawa, K., Ganiatsas, S., Yang, D., Xavier, R.J., Harris, N.L., Davidson, L., Ferrini, R., Davis, R.J., Labow, M.A., Flavell, R.A., Zon, L.I. and Alt, F.W. (1998). SEK1/MKK4 is required for maintenance of a normal peripheral lymphoid compartment but not for lymphocyte development. Immunity 8, 625–634.

Tamura, K., Sudo, T., Senftleben, U., Dadak, A.M., Johnson, R. and Karin, M. (2000). Requirement for p38α in erythropoietin expression: a role for stress kinases in erythropoiesis. Cell 102, 221–231.

Teng, D.H., Perry, W.L., Hogan, J.K., Baumgard, M., Bell, R., Berry, S., Davis, T., Frank, D., Frye, C., Hattier, T., Hu, R., Jammulapati, S., Janecki, T., Leavitt, A., Mitchell, J.T., Pero, R., Sexton, D., Schroeder, M., Su, P.H., Swedlund, B., Kyriakis, J.M., Avruch, J., Bartel, P., Wong, A.K., Tavtigian, S.V., et al. (1997). Human mitogen-activated protein kinase kinase 4 as a candidate tumor suppressor. Cancer Res. 57, 4177–4182.

Tibbles, L.A. and Woodgett, J.R. (1999). The stress-activated protein kinase pathways. Cell. Mol. Life Sci. 55, 1230–1254.

Tong, L., Pav, S., White, D.M., Rogers, S., Crane, K.M., Cywin, C.L., Brown, M.L. and Pargellis, C.A. (1997). A highly specific inhibitor of human p38 MAP kinase binds in the ATP pocket. Nat. Struct. Biol. 4, 311–316.

Tournier, C., Hess, P., Yang, D.D., Xu, J., Turner, T.K., Nimnual, A., Bar-Sagi, D., Jones, S.N., Flavell, R.A. and Davis, R.J. (2000). Requirement of JNK for stress-induced activation of the cytochrome c-mediated death pathway. Science 288, 870–874.

Tournier, C., Whitmarsh, A.J., Cavanagh, J., Barrett, T. and Davis, R.J. (1997). Mitogen-activated protein kinase kinase 7 is an activator of the c-Jun NH2-terminal kinase. Proc. Natl. Acad. Sci. USA 94, 7337–7342.

Tournier, C., Whitmarsh, A.J., Cavanagh, J., Barrett, T. and Davis, R.J. (1999). The MKK7 gene encodes a group of c-Jun NH2-terminal kinase kinases. Mol. Cell. Biol. 19, 1569–1581.

Voss, A.K., Thomas, T. and Gruss, P. (2000). Mice lacking HSP90β fail to develop a placental labyrinth. Development 127, 1–11.

Wang, X.S., Diener, K., Manthey, C.L., Wang, S., Rosenzweig, B., Bray, J., Delaney, J., Cole, C.N., Chan-Hui, P.Y., Mantlo, N., Lichenstein, H.S., Zukowski, M. and Yao, Z. (1997). Molecular cloning and characterization of a novel p38 mitogen-activated protein kinase. J. Biol. Chem. 272, 23668–23674.

Wang, X.Z. and Ron, D. (1996). Stress-induced phosphorylation and activation of the transcription factor CHOP (GADD153) by p38 MAP Kinase. Science 272, 1347–1349.

Wendling, O., Chambon, P. and Mark, M. (1999). Retinoid X receptors are essential for early mouse development and placentogenesis. Proc. Natl. Acad. Sci. USA 96, 547–551.

White, R.A., Hughes, R.T., Adkison, L.R., Bruns, G. and Zon, L.I. (1996). The gene encoding protein kinase SEK1 maps to mouse chromosome 11 and human chromosome 17. Genomics 34, 430–432.

Whitmarsh, A.J., Shore, P., Sharrocks, A.D. and Davis, R.J. (1995). Integration of MAP kinase signal transduction pathways at the serum response element. Science 269, 403–407.

Widmann, C., Gibson, S., Jarpe, M.B. and Johnson, G.L. (1999). Mitogen-activated protein kinase: conservation of a three-kinase module from yeast to human. Physiol. Rev. 79, 143–180.

Wilson, K.P., McCaffrey, P.G., Hsiao, K., Pazhanisamy, S., Galullo, V., Bemis, G.W., Fitzgibbon, M.J., Caron, P.R., Murcko, M.A. and Su, M.S. (1997). The structural basis for the specificity of pyridinylimidazole inhibitors of p38 MAP kinase. Chem. Biol. 4, 423–431.

Wu, C.W., Li, A.F., Chi, C.W., Huang, C.L., Shen, K.H., Liu, W.Y. and Lin, W. (2000). Human gastric cancer kinase profile and prognostic significance of MKK4 kinase. Am. J. Pathol. 156, 2007–2015.

Wysk, M., Yang, D.D., Lu, H.T., Flavell, R.A. and Davis, R.J. (1999). Requirement of mitogen-activated protein kinase kinase 3 (MKK3) for tumor necrosis factor-induced cytokine expression. Proc. Natl. Acad. Sci. USA 96, 3763–3768.

Xia, Y., Makris, C., Su, B., Li, E., Yang, J., Nemerow, G.R. and Karin, M. (2000). MEK kinase 1 is critically required for c-Jun N-terminal kinase activation by proinflammatory stimuli and growth factor-induced cell migration. Proc. Natl. Acad. Sci. USA 97, 5243–5248.

Yang, D., Tournier, C., Wysk, M., Lu, H.T., Xu, J., Davis, R.J. and Flavell, R.A. (1997). Targeted disruption of the MKK4 gene causes embryonic death, inhibition of c-Jun NH2-terminal kinase activation, and defects in AP-1 transcriptional activity. Proc. Natl. Acad. Sci. USA 94, 3004–3009.

Yang, D.D., Conze, D., Whitmarsh, A.J., Barrett, T., Davis, R.J., Rincon, M. and Flavell, R.A. (1998). Differentiation of CD4+ T cells to Th1 cells requires MAP kinase JNK2. Immunity 9, 575–585.

Yang, D.D., Kuan, C.Y., Whitmarsh, A.J., Rincon, M., Zheng, T.S., Davis, R.J., Rakic, P. and Flavell, R.A. (1997). Absence of excitotoxicity-induced apoptosis in the hippocampus of mice lacking the Jnk3 gene. Nature 389, 865–870.

Yang, J., Boerm, M., McCarty, M., Bucana, C., Fidler, I.J., Zhuang, Y. and Su, B. (2000). Mekk3 is essential for early embryonic cardiovascular development. Nat. Genet. 24, 309–313.

Yao, Z., Diener, K., Wang, X.S., Zukowski, M., Matsumoto, G., Zhou, G., Mo, R., Sasaki, T., Nishina, H., Hui, C.C., Tan, T.H., Woodgett, J.P. and Penninger, J.M. (1997). Activation of stress-activated protein kinases/c-Jun N-terminal protein kinases (SAPKs/JNKs) by a novel mitogen-activated protein kinase kinase. J. Biol. Chem. 272, 32378–32383.

Yasuda, J., Whitmarsh, A.J., Cavanagh, J., Sharma, M. and Davis, R.J. (1999). The JIP group of mitogen-activated protein kinase scaffold proteins. Mol. Cell. Biol. 19, 7245–7254.

Yeh, W.C., Shahinian, A., Speiser, D., Kraunus, J., Billia, F., Wakeham, A., de la Pompa, J.L., Ferrick, D., Hum, B., Iscove, N., Ohashi, P., Rothe, M., Goeddel, D.V. and Mak, T.W. (1997). Early lethality, functional NF-κB activation and increased sensitivity to TNF-induced cell death in TRAF2-deficient mice. Immunity 7, 715–725.

Yuasa, T., Ohno, S., Kehrl, J.H. and Kyriakis, J.M. (1998). Tumor necrosis factor signaling to stress-activated protein kinase (SAPK)/Jun NH2-terminal kinase (JNK) and p38. Germinal center kinase couples TRAF2 to mitogen-activated protein kinase/ERK kinase kinase 1 and SAPK while receptor interacting protein associates with a mitogen-activated protein kinase kinase kinase upstream of MKK6 and p38. J. Biol. Chem. 273, 22681–22692.

Yujiri, T., Ware, M., Widmann, C., Oyer, R., Russell, D., Chan, E., Zaitsu, Y., Clarke, P., Tyler, K., Oka, Y., Fanger, G.R., Henson, P. and Johnson, G.L. (2000). MEK kinase 1 gene disruption alters cell migration and c-Jun NH2-terminal kinase regulation but does not cause a measurable defect in NF-κ B activation. Proc. Natl. Acad. Sci. USA 97, 7272–7277.

Zanke, B.W., Rubie, E.A., Winnett, E., Chan, J., Randall, S., Parsons, M., Boudreau, K., McInnis, M., Yan, M., Templeton, D.J. and Woodgett, J.R. (1996). Mammalian mitogen-activated protein kinase pathways are regulated through formation of specific kinase-activator complexes. J. Biol. Chem. 271, 29876–29881.

Protein Adaptations and Signal Transduction. Edited by K.B. Storey and J.M. Storey
2001 Elsevier Science B.V.

CHAPTER 14

How to Activate Intrinsic Stress Resistance Mechanisms to Obtain Therapeutic Benefit

Prasanta K. Ray[1]*, Tanya Das[2] and Gaurisankar Sa[2]
[1]*Department of Surgery, Beth Israel Hospital, A.J. Antenucci Medical Research Building, Room 301, 432 W. 58th Street, New York, NY 10019, USA;* [2]*Bose Institute, Animal Physiology Section, Calcutta-700054, India*

1. General introduction

The environment in which we live is always changing, often becoming inhospitable for normal life processes. From the beginning of the evolution of life on the surface of this planet, each living cell has had to struggle for its existence in the otherwise adverse and stressful environment. Stress can originate from diverse sources. Any chemical, biochemical or harmful infection-causing organism, beyond its threshold limit, can be considered as a stressor, i.e. a substance which can induce stress. Moreover, numerous physical, chemical or biological agents to which we are exposed in our day-to-day lives, are being added to the list of stressors almost every day. Mounting effects of these stresses on our body are a common cause of deviations and alterations from the normal homeostatic functioning of organs and systems, which under certain circumstances may be considered pathological. However, every organism alters its cellular physiology in an attempt to counter the imbalance created by stresses. Thus, during the exposure to any stress inducing substance, there is an orchestration of many events that are required to fight against the stress-induced damages. These events include altered hormonal responses to environmental pressures, altered gene activity with the elicitation of new gene products, and alterations in the metabolic profile, etc. Normally, to avoid exposure to various stresses, all organisms have developed a number of anatomical, physiological, biochemical, and immunological barriers. Thus, following exposure to any stressor, the physiological system, with the aid of those intrinsic defense mechanisms, may initiate a large number of biochemical processes as mentioned above. The organism selects the best-suited mechanisms to help it endure the environmental conditions to which it is exposed at any given time. Sometimes, more than one such mechanism may be induced depending upon what will be required to cope with the situation, since very often it is a question of life and death type of situations (Walt, 1971; Ray, 1998, 1999).

2. Body's defense against different forms of stress

2.1. Immune defense

The immune system is one of the major defense mechanisms of the body that fights against stresses caused by insults inflicted by foreign substances. When exposed for the first time to even a minuscule amount of stressor, the immune system is alerted. Its operational machinery is switched on through both cellular and humoral modes to develop the capability (immunity) to resist the

*Corresponding author.

otherwise harmful effects of the stressor. The immediate danger is fought while retaining the memory of that encounter so that a faster attack can be mounted towards the same stressor during secondary exposures.

An exposure of B lymphocytes to the stressor leads to the synthesis of a class of compounds, called immunoglobulins or antibodies. This offers humoral (antibody or immunoglobulin mediated) immunity to the host, and is utilized to destroy (lyse) the foreign substances and to remove them from the circulation (Yang and Glaser, 2000).

A series of immune mechanisms are cell mediated. Cells like macrophages, dendritic cells, monocytes, neutrophils, T and B cells take part in different types of immune reactions, but still maintain a close harmony for a composite whole, like that of an orchestra. Readers are directed to any textbook on immunology for more details of the immune system and its function. We do not have the scope to discuss these mechanisms in detail in this review.

The immune system may be depressed under various stressful conditions. For example, under surgical stress the immune system is depressed, but soon recovers to render a greater resistance power to the host during a subsequent exposure. It is known that the cellular (Th1-type) immune response is centrally involved in the fight against foreign substances, to deal with pathological insults that are inflicted during the course of various diseases, disorders and malfunctions. Within the immunological cascades of Th1-type immunity, cytokines such as tumor necrosis factor-alpha (TNF-α), interleukin (IL)-1α, -6, -8, and the anti-inflammatory cytokine IL-1β, which are mainly produced by mononuclear cells, are known to play an important role in the response to and pathogenesis of surgical stress (Ono and Mochizuki, 2000). Moreover, interferon-gamma (IFN-γ) has been known to trigger a series of immune-relevant reactions mostly directed towards forward regulation of the antigen specific immune response (Widner et al., 2000). Recent studies suggest that dendritic cells are the potent initiators of primary immune responses and hold the key to immune reactions through their ability to sense changes in their local environment and respond appropriately to induce T-cell immunity, or possibly tolerance at times (McLellan et al., 2000). Exposure of macrophages to inflammatory stimuli activates their cytocidal activity as well as potentiates the expression of complement components (Laszlo et al., 1993). All this information demonstrates that the immune system of the host takes part in the fight against stress factors either directly or by modulating the effector systems, i.e. cytokines, immunoglobulins, etc. Thus, as long as the host is immunologically competent, it can adapt to the stressful environment in a better way.

2.2. Detoxification process

In our day-to-day life, we are exposed to stress from a myriad of man-made chemical compounds and products developed by man's synthetic ingenuity. Even naturally occurring compounds already present before prehistoric days came into use through a synthetic process. But it is interesting to note that a small amount of chemicals may not cause serious injury or jeopardize our physiological system. The concentrations of such chemicals causing harm to any organism may vary from one chemical to another, and also from one organism to another organism. In fact, higher organisms appear to have developed a class of genes encoding various proteins and enzymes to detoxify such toxic or harmful compounds, which are mostly of no physiological value. Such detoxification reactions occur in two distinct phases. The Phase I biotransformation process involves a series of catalytic enzymes, for example, cytochrome P450, alcohol dehydrogenase, superoxide dismutase, glutathione peroxidase, monoamine oxidase, etc., whereas the Phase II detoxification system involves enzymes such as glucuronidase, glutathione-S-transferase, methyltransferases, etc. In general, the toxic chemicals are metabolized to water-soluble products by these reactions and, thus, are more easily eliminated from the body than the parent compounds (Ray and Das, 1998). It has been documented that cytochrome P450, which is induced as a fight-back response of the body to stress, can change the chemical nature of different drugs, pesticides, and anesthetics as well as various carcinogens (Coon and Persson, 1980). It has also been reported that induced synthesis of many other Phase I enzymes as well as Phase II enzymes occurs as the intrinsic defense mechanism of the body (Ray and Das, 1998). Thus, if the host is competent to activate the detoxification system in an accelerated manner, it should be able to withstand stressor-induced toxic insults, in order to keep itself fit and functional in an otherwise stressful environment.

2.3. Cell regeneration and replenishment

The pathologies associated with defects in cell death (apoptosis) phenomena include environmental stress, cancer, developmental defects, autoimmune disease, and neurodegenerative disorders due to senescence, etc. (Ishizaki et al., 1995; Badely et al., 1999; Roy et al., 1995). Necrosis or apoptosis are the two main mechanisms of cell death (Deveraux et al., 1999). Programmed cell death or apoptosis is switched on through a precisely orchestrated sequence of events and is one of the physiological processes that is essential to maintain the homeostasis of the body. In addition to the normal process of cell turnover, our exposure to various kinds of stress inducers may also cause damage to cells and the tissues that lead to activation of this intrinsic mechanism of cell death. Regardless of cell type or signal transduction mechanism, a number of morphological changes occur during apoptosis and the manifestation of specific events such as chromatin condensation, DNA fragmentation, and the exposure of phosphatidylserine on the outer leaflet of cells are observed (Yuan, 1997). However, the host has its intrinsic machinery to repair any minor damage in an accelerated manner, and/or regenerate the depleted cells quickly to maintain the homeostatic balance, which is required for its own survival. Also, by altering the availability of the required materials, the intrinsic defense system helps in the regeneration of damaged tissue, e.g., shear stress alters chondrocyte metabolism to limit matrix destruction and stimulate cartilage repair and regeneration. In fact, articular chondrocytes exhibit a dose- and time-dependent response to shear stress that results in the release of soluble mediators and extracellular matrix macromolecules. This chondrocyte response to mechanical stimulation contributes to the maintenance of articular cartilage homeostasis *in vivo* (Lane et al., 2000).

2.4. DNA repair

Once it was recognized that DNA is the informationally active chemical component of essentially all genetic material of all living organisms (with the notable exception of RNA viruses). It was assumed that this macromolecule must be extraordinarily stable in order to maintain the high degree of fidelity required of a master blueprint. In fact, the DNA of living cells reacts very easily with a variety of stressors, e.g., chemical compounds and physical agents, including radiation, electromagnetic radiation, gravitational force, etc. These are present in the environment, such as products of metabolism or decomposition of other living forms, man-made chemicals contributing to the genetic insult, mutagens, teratogens, carcinogens, etc. Thus, DNA damage, either spontaneous or environmental, is an inescapable aspect of life in our biosphere. However, the normal cellular response to such damage is through a DNA repair mechanism that is associated with the restoration of the normal base sequence and chemistry of DNA (Friedberg et al., 1995a).

In response to DNA damage, the cell-cycle checkpoints integrate cell-cycle control with DNA repair (Yonis-Rouach et al., 1993). In a damaged cell, the pro-apoptotic protein p53 arrests the cell cycle progression at G0/G1 phases and gives the cell a chance to repair its damaged DNA (Walworth, 2000). However, if the damage is beyond repair, the cell is directed to undergo apoptosis through the p21-mediated cell death pathway (Waldman et al., 1995). Cells with damaged DNA may also develop "DNA damage tolerance" mechanisms that result in permanent mutation in the genome (Friedberg et al., 1995b). Thus, by either reversing the DNA damage or by developing tolerance to the "bad" DNA, the cell tries to withstand the stress-induced insult in order to maintain its own existence.

2.5. Growth factors/cytokines/hormones/chaperones

Many growth factors, cytokines and hormones mediate their diverse biological responses by binding to and activating cell surface receptors (Darnell, 1994; Geer, 1994). Activation of these receptors by their specific ligands control important physiological processes, such as cell proliferation, survival, differentiation, cell metabolism, etc. (Hunter, 1995) The functions of intratumoral lymphocytes in many human malignant tumors are inhibited by reactive oxygen species (ROS) generated by adjacent monocytes/macrophages (MO). Immunotherapeutic cytokines such as interleukin-2 (IL-2) or interferon-alpha (IFN-α) only weakly activate T cells or natural killer (NK) cells in a reconstituted environment of oxidative stress and inhibitors of the formation of ROS or scavengers of ROS synergize with IL-2 and IFN-alpha to activate T cells and NK cells. Histamine optimizes cytokine-induced activation of several subsets of T cells by affording protection against MO-inflicted oxidative inhibition. The putative clinical benefit of histamine as an adjunct to immunotherapy with IL-2 and/or IFN-α is currently being evaluated in clinical trials in metastatic malignant melanoma and acute myelogenous leukemia (Hellstrand et al., 2000).

As indicated earlier, stress genes can be ascribed to have been generated by organisms during the evolutionary process to serve their intrinsic urge to survive against a changing and challenging environment. Stress may produce a variety of stress responses in mammalian cells with implications for cell integrity, shape, locomotion, metabolism, proliferation and survival. Heat-induced stress at the molecular level shares many features with the heat shock response. This includes the differential sensitivity of the stress signal pathway elements to the magnitude of the stress, stressor-specific activation of the response elements, and the protective role of the heat shock response (Creagh et al., 2000). The signal cascade triggered by stress induces the activity of the mitogen-activated protein (MAP) kinases, c-jun terminal kinase (JNK), and p38 MAPK or stress-activated serine/threonine protein kinase (Obata et al., 2000). Diverse extracellular stimuli including environmental stress, irradiation, heat shock, high osmotic stress, proinflammatory cytokines and certain mitogens trigger a stress-regulated protein kinase cascade. This culminates in activation of p38 MAPK which appears to play a major role in phosphorylation of heat shock proteins, cytokine production, nitric oxide production, transcriptional regulation, and cytoskeletal reorganization (Obata et al., 2000). The availability of all these factors may ultimately offer therapeutic benefit for certain critically ill patients. Heat shock proteins have been implicated as having a role in providing resistance to the host against different types of stressors. A minute amount of stress inducers have been observed to aid expression of stress resistance genes, providing increased capability to the host to protect itself against a myriad of biotic and abiotic stressors (Ray, 1999; Maulik et al., 1995).

3. Failure of the intrinsic defense

It is clear from the above discussion that each animal has its intrinsic defense machinery to ensure proper management of stress. The better an animal can exploit its own defense mechanisms, the greater its chances for survival in a stressful environment. However, each animal or organism can fight against the adverse effects of stress only up to a certain limit. Beyond that limit, the intrinsic system fails to operate properly giving way to stress-induced damage to the body. It is known that due to various kinds of stress (e.g., oxidative stress, toxic insult, etc.), the detoxification system of the body (phase I and phase II biotransformation and detoxification enzymes) may also be depressed. This can compromise the ability of the host to detoxify and eliminate toxic chemicals from the body and prevent stress-induced damage. As a result, accumulated toxic chemicals would induce a second degree of stress response by the body and may cause more damage.

Reports from our laboratory have shown that stress due to toxic chemicals depresses the biotransformation enzymes (Dwivedi, 1989;

Srivastava, 1987; Dohadwala and Ray, 1985). Recently, alteration of Phase I and Phase II detoxification genes by pro-oxidant environmental pollutants has been observed (Maier et al., 2000). Stress beyond a limit induces dysfunction of the host's immunological system, thereby jeopardizing one of the body's major defense systems against the intrusion of bacteria or viruses, etc. Stress due to many diseases (e.g., cancer, AIDS) also suppresses immune function (Subbulakshmi et al., 1997; Ghosh et al., 1999a,b; Mafune and Tanaka, 2000) as does the stress of anesthesia and surgery (Mafune and Tanaka, 2000; Fehder, 1999). Immunodepression caused by stress due to toxic chemicals or pesticides is also known (Raisuddin, 1994; Singh, 1990). Environmental stressors and pollutants that activate cytokine and growth factor receptors can lead to oncogenesis and other disorders associated with excessive cell proliferation. On the other hand, stressors that inactivate these receptors can lead to a variety of developmental disorders, e.g., inhibition of proliferation and replenishment of tissue damage (Ullrich and Schlessinger, 1990; Erlebacher et al., 1995). Stress-induced failure of the DNA repair system leading to apoptosis of the cell has also been documented (Harrouk, 2000). The role of oxidative DNA damage and its effect in carcinogenesis has been well documented by Simon et al. (2000). Failure of the DNA repairing system may even lead to infertility (Shen and Ong, 2000). Thus, due to continuous exposure to stressful conditions or due to high amounts of stressors, the body's defense machinery can become completely shattered. As a result, the animal may lose its fitness to survive.

4. Possible avenues for reversal of stress-injury

All this information makes it clear that when, due to the exposure of the host to stress, all the intrinsic defense capabilities are grossly compromised, the host loses its fitness for survival. Therefore, we hypothesized (Fig. 14.1) that the possible avenues for reversal of stress-injury depend on the success in finding a suitable biological response modifier or

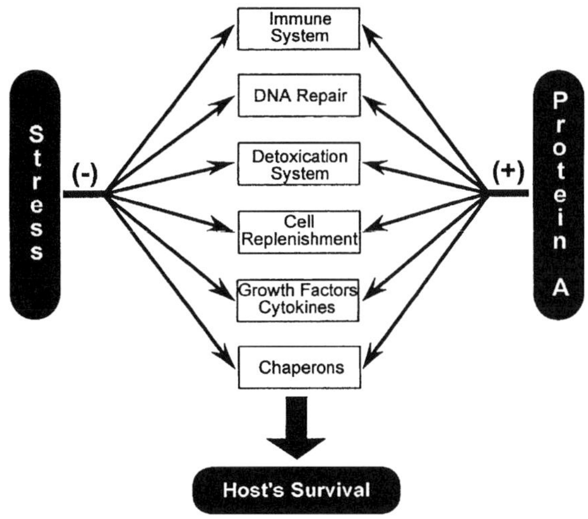

Fig. 14.1. Protection from stress-induced damage of the host's intrinsic defense system by Protein A.

similar such agent that can activate the intrinsic defense systems of the host to fight back the toxic and stressful insults. We have constructed this hypothesis, on the basis of the results from a series of investigations using Protein A (PA) of *Staphylococcus aureus* as a probe.

Protein A is a unique protein having multifarious biomodulatory properties. It has antitumor (Verma et al., 1999; Shukla et al., 1996), anticarcinogenic (Ray et al., 1996; Das et al. 2000), and antitoxic (Ray et al., 1985; Ghosh et al. 1999a) as well as immunopotentiating (Das et al., 1999a,b; Ghosh et al., 1999b) properties. During our search for a biological response modifier that can ameliorate stress-induced damage, we found that PA can revert the depressed phase I and phase II biotransformation and detoxification enzymes (Ray and Das, 1998). Recent studies from our laboratory have shown that PA can protect bone marrow progenitor cells from zidovudin (AZT)-induced apoptosis (Ghosh et al., 1999a,b; Ray et al., 1998) and accelerates the plasma clearance rate of AZT and its toxic metabolites by activating both the biotransformation and detoxification systems of the host (Subbulakshmi et al., 1998). In this way, PA rendered increased resistance to the host against stress.

Interestingly, PA can act as both a mitogenic and an apoptogenic agent in the same cell depending on its concentration (Das et al., 1999b). Moreover, PA regulates the balance between the pro-apoptosis proteins, p53 and Bax, and the pro-proliferative Bcl-2 in favor of cell survival in normal cells (Das et al., 1999b) and thus helps in regeneration of depleted cells from stress-induced insult. Preliminary studies from our laboratory also indicated that PA-induced an increase in Hsps in normal cells to protect them from stress-induced apoptosis (unpublished data). PA has also been found to stimulate immunocyte proliferation through a signaling cascade in which NO is downstream of tyrosine kinase > PLCγ1 > PKC (Goenka et al., 1998, Das et al., 1999b, 2000) and protect the immunocytes from stress-induced damage. Stimulation of the host immune system by PA, results in increased elicitation of various cytokines, e.g., IL1, 1L2, TNF-α, IFN-γ, etc. (Sinha et al., 1999) and possibly some growth factors. These in turn help in replenishment of the host immunocytes and also in regeneration of other cell types. In all these cases, prior inoculation of a very small amount of PA could provide the host with the ability to withstand a larger than normal amount of stressors, e.g., drugs, chemicals, oxidative stress, tumor challenge, etc. The above observations led us to conclude that PA, by its overall bio-regulatory activities, can potentiate the intrinsic defense machineries of the host and protect cells from stress-induced damage as well as help them to proliferate and grow. Thus, the therapeutic implications of substances like PA as a "rescue molecule" is high in so far as its effectiveness in reversal of stress-injury is concerned.

5. Future directions

Intensive investigations are necessary to understand and delineate the details of different mechanisms involved in the repair and reconstruction processes of different organelles, cells and tissues which have suffered toxic injuries. How such signals are transmitted and what regulates them needs to be understood. Various genes involved in the entire reconstruction processes need to be identified and their cooperativity as well as the regulatory activities of different biomolecules at different phases of these reactions need to be elaborated. Then one has to search out, test and develop from natural products or synthetic routes, very effective 'rescue molecules' in order to treat mankind for alleviation of, and/or to reduce the toxic injuries caused by, drugs, environmental pollutants, microbes, and various physical agents.

Acknowledgements

PKR is grateful to the Chairman, Department of Surgery, Montefiore Medical Center, Albert Einstein College of Medicine, New York for offering him a Senior Visiting Fellowship, and to the Director, Chanin Institute for Cancer Research and to Dr. Howard Kaufman for proving space and laboratory facilities during the course of the preparation of this manuscript. The author is very much indebted to Dr. R.S. Chamberlain for his continuous support, interests, serious discussions, and various other help during the period. Thanks are due to Prof. Arun Roy, Chairman, Department of Animal Physiology and to the Director, Bose Institute, Calcutta, India for allowing his colleagues (TD and GS) to collaborate during the course of the preparation of this manuscript.

References

Badly, A.D., Parato, K., Cameron, D.W., Kravcik, S., Phenix, B.N., Ashby, D., Kumar, A., Lynch, D.H., Tschopp, J. and Angel, J.B. (1999). Dynamic correlation of apoptosis and immune activation during treatment of HIV infection. Cell Death Differ. 6, 420–432.

Coon, M.J. and Persson, A.V. (1980). In: Enzymatic Basis of Detoxification. (Jacoby, W.B., Ed.), pp. 117–134. Academic Press, New York.

Creagh, E.M., Sheehan, D. and Cotter, T.G. (2000). Heat shock proteins-modulators of apoptosis in tumor cells. Leukemia 14, 1161–1173.

Darnell, J.E., Kerr, I.M. and Stark, G.R. (1994). Jak-STAT pathways and transcriptional activation in response to IFNs and other extracellular signaling proteins. Science 264, 1415–1421.

Das, T., Sa, G. and Ray, P.K. (1999a). Mechanisms of Protein A superantigen-induced signal transduction for proliferation of mouse B cell. Immunol. Lett. 70, 43–51.

Das, T., Sa, G., Sinha, P. and Ray, P.K. (1999b). Induction of cell proliferation and apoptosis: dependence on the dose of the inducers. Biochem. Biophys. Res. Commun. 260, 105–110.

Das, T., Sa, G., Subbulakshmi, V., Subramaniam, S., Sen, P.C., Biswas, S. and Ray, P.K. (2000). Protein A-activated rat splenic lymphocyte proliferation involves tyrosine kinase–phospholipase C–protein kinase C pathway. Immunopharmacol. Immunotoxicol. 22, 75–90.

Deveraux, Q.L., Stennicke, H.R., Salvesen, G.S. and Reed, J.C. (1999). Endogenous inhibitors of caspases. J. Clin. Immunol. 19, 388–398.

Dohadwala, M. and Ray, P.K. (1985). In vivo protection by protein A of hepatic microsomal mixed function oxygenase system of cyclophosphamide-treated rats. Cancer Chemo. Pharmacol. 4, 59–62.

Dwivedi, P.D., Verma, A.S., Mishra, A., Singh, K.P., Prasad, A.K., Saxena, A.K., Dutta, K.K., Mathur, N. and Ray, P.K. (1989). Protein A protects mice from depletion of biotransformation enzymes and mortality induced by *Salmonella typhimurium* endotoxin. Toxicol Lett. 49, 1–13.

Erlebacher, A., Filvaroff, E.H., Gitelman, S.E. and Derynck, R. (1995). Toward a molecular understanding of skeletal development. Cell 80, 371–378.

Friedberg, E.C., Walker, G.C. and Siede, W. (1995a). DNA Repair and Mutagenesis. pp. 2–58, ASM Press, Washington DC.

Friedberg, E.C., Walker, G.C. and Siede, W. (1995b). DNA Repair and Mutagenesis. pp. 91–133. ASM Press, Washington DC.

Fehder, W.P. (1999). Alterations in immune response associated with anxiety in surgical patients. CRNA 10, 124–129.

Geer, P. van der, Hunter, T. and Linberg, R.A. (1994). Phosphopeptide mapping and phosphoamino acid analysis by electrophoresis and chromatography on thin-layer cellulose plates. Annu. Rev. Cell. Biol. 10, 2251–2337.

Ghosh, A.K., Jana, S., Das, T., Mandal, N., Sa, G. and Ray, P.K. (1999a). Protection by Protein A of apoptotic cell death caused by anti-AIDS drug, zidovudine. Biochem. Biophys. Res. Commun. 264, 601–604.

Ghosh, A.K., Sinha, P., Das, T., Sa, G. and Ray, P.K. (1999b). *S. aureus* superantigen Protein A expand CD4(+)/CD8(+)/CD19(+)/CD34(+) cells in mice: a potential immunorestorer. Biochem. Biophys. Res. Commun. 256, 142–146.

Goenka, S., Das, T., Sa, G. and Ray, P.K. (1998). Protein A induces NO production: involvement of tyrosine kinase, phospholipase C and protein kinase C. Biochem. Biophys. Res. Commun. 250, 425–429.

Harrouk, W., Codrington, A., Vinson, R., Robaire, B. and Hales, B.F. (2000). Paternal exposure to cyclophosphamide induces DNA damage and alters the expression of DNA repair genes in the rat preimplantation embryo. Mutat. Res. 461, 229–41.

Hellstrand, K., Brune, M., Naredi, P., Mellqvist, U.H., Hansson, M., Gehlsen, K.R., Hermodsson, S. (2000). Histamine: a novel approach to cancer immunotherapy. Cancer Invest. 18, 347–355.

Hunter, T. (1995). Protein kinases and phosphatases: the yin and yang of protein phosphorylation and signaling. Cell 80, 225–236.

Ishizaki, Y., Cheng, L., Mudge, A.W. and Raff, M.C. (1995). Programmed cell death by default in embryonic cells, fibroblasts, and cancer cells. Mol. Biol. Cell 6, 1443–1458.

Lane Smith, R., Trindade, M.C., Ikenoue, T., Mohtai, M., Das, P., Carter. D.R., Goodman, S.B. and Schurman, D.J. (2000). Effects of shear stress on articular chondrocyte metabolism. Biorheology 37, 95–107.

Laszlo, D.J., Henson, P.M., Remigio, L.K., Weinstein, L., Sable, C., Noble, P.W. and Riches, D.W. (1993). Development of functional diversity in mouse macrophages. Mutual exclusion of two phenotypic states. Am. J. Pathol. 143, 587–597.

Mafune, K. and Tanaka, Y. (2000). Influence of multimodality therapy on the cellular immunity of patients with esophageal cancer. Ann. Surg. Oncol. 7, 609–616.

Maier, A., Dalton, T.P. and Puga, A. (2000). Disruption of dioxin-inducible phase I and phase II gene expression patterns by cadmium, chromium, and arsenic. Mol. Carcinog. 28, 225–235.

Maulik, N., Watanabe, M., Engelman, D., Engelman, R.D., Kagan, V.E., Kishin, E., Tyurin, V., Cordis, G.A. and Das, D.K. (1995). Myocardial adaptation to ischemia by oxidative stress induced by endotoxin. Am. J. Physiol. 269, 907–916.

McLellan, A.D., Brocker, E.B. and Kampgen, E. (2000). Dendritic cell activation by danger and antigen-specific T-cell signaling. Exp Dermatol. 9, 313–322.

Obata, T., Brown, G.E. and Yaffe, M.B. (2000). MAP kinase pathways activated by stress: the p38 MAPK pathway. Crit. Care Med. 28, N67–77.

Ono, S. and Mochizuki, H. (2000). Cytokine production in surgical stress. Nippon Geka Gakkai Zasshi. 101, 582–587.

Raisuddin, S., Singh, K.P., Zaidi, S.A. and Ray, P.K. (1994). Immunostimulating effects of protein A in immunosuppressed aflatoxin-intoxicated rats. Int. J. Imunopharmacol. 16, 977–984.

Ray, P.K. (1998). Protective role of stress genes. Environ. Health Perspect. 106, A217–218.

Ray, P.K. (1999). Stress genes and species survival. Mol. Cell. Biochem. 196, 117–123.

Ray, P.K. and Das, T. (1998).Molecular adaptation to toxic chemicals and drugs. Adv. Organ Biol. 6, 251–265.

Ray, P.K. and Srivastava, M. (1996). A new concept in cancer chemoprevention. The Cancer J. 9, 221–225.

Ray, P.K., Dohadwala, M., Bandyopadhyay, S.K. et al.

(1985). Rescue of rats from large dose cyclophosphamide toxicity using Protein A. Cancer Chemother. Pharmacol. 14, 59–62.

Ray, P.K., Goenka, S., Das, T., Sa, G. et al. (1998). Role of nitric oxide in immune function and amelioration of toxicity and carcinogenicity of drugs and chemicals. In: Biological Oxidants: Molecular Mechanisms and Health Effects. (Packer, L. and Augustine, S.H., Eds.), pp. 42–53. AOCS Press, USA.

Roy, N., Mahadevan, M.S., McLean, M., Shutler, G. et al. (1995). The gene for neuronal apoptosis inhibitory protein is partially deleted in individuals with spinal muscular atrophy. Cell 80, 167–178.

Shen, H. and Ong, C. (2000). Detection of oxidative DNA damage in human sperm and its association with sperm function and male infertility. Free Rad. Biol. Med. 28, 529–536.

Shukla, Y., Verma, A.S., Mehrotra, N.K. and Ray, P.K. (1996). Antitumor activity of Protein A in mouse skin model of two stage carcinogenesis. Cancer Lett. 103, 41–47.

Simon, M.S., Heilbrun, L.K., Stephens, D., Lababidi, S. and Djuric, Z. (2000). Recruitment for a pilot case control study of oxidative DNA damage and breast cancer risk. Am. J. Clin. Oncol. 23, 283–287.

Singh, K.P., Zaidi, S.A., Raisuddin, S., Saxena, A.K., Dwivedi, P.D., Seth, P.K. and Ray, P.K. (1990). Protection against carbon-tetrachloride-induced lymphoid organotoxicity in rats by protein A. Toxicol. Lett. 51, 339–351.

Sinha, P., Ghosh, A.K., Das, T., Sa, G. and Ray, P.K. (1999). Protein A of *S. aureus* evokes Th1 type response in mice. Immunol. Lett. 67, 157–165.

Srivastava, S.P., Singh, K.P., Saxena, A.K., Seth, P.K. and Ray, P.K. (1987). In vivo protection by protein A of hepatic microsomal mixed function oxidase system of CC14-administered rats. Biochem. Pharmacol. 36, 4055–4058.

Subbulakshmi, V., Ghosh, A.K., Das, T. and Ray, P.K. (1998). Mechanism of Protein A-induced amelioration of toxicity of anti-AIDS drug, zidovudine. Biochem. Biophys. Res. Commun. 250, 15–21.

Ullrich, A. and Schlessinger, J. (1990). Signal transduction by receptors with tyrosine kinase activity. Cell 61, 230–212.

Verma, A.S., Dwivedi, P.D., Mishra, A. and Ray, P.K. (1999). Ehrlich's ascites fluid adsorbed over Protein A containing *Staphylococcus aureus* Cowan I produces inhibition of tumor growth. Immunopharmacol. Immunotoxicol. 21, 89–108.

Walt, R.S. (1971). Biochemical Evolution and the Origin of Life (Schoffeniels, E., Ed.). pp. 14–42. North Holland Publishing, Amsterdam.

Walworth, N.C. (2000). Cell–cycle checkpoint kinases: checking in on the cell cycle. Curr. Opin. Cell. Biol. 12, 697–704.

Waldman, T., Kinzler, K.W. and Vogelstein, B. (1995). p21 is necessary for the p53-mediated G1 arrest in human cancer cells. Cancer Res. 55, 5187–5190.

Widner, B., Wirleitner, B., Baier-Bitterlich, G., Weiss, G. and Fuchs, D. (2000). Cellular immune activation, neopterin production, tryptophan degradation and the development of immunodeficiency. Arch. Immunol. Ther. Exp. (Warsz) 48, 251–258.

Yang, E.V. and Glaser, R. (2000). Stress-induced immunomodulation: impact on immune defenses against infectious disease. Biomed. Pharmacother. 54, 245–250.

Yonis-Rouach, E., Grunwald, D., Wilder, S., Kimchi, A., May, E., Laurence, J.J., May, P. and Oren, M. (1993). p53-mediated cell death: relationship to cell cycle control. Mol. Cell. Biol. 13, 1415–1423.

Yuan, J. (1997). Transducing signals of life and death. Curr. Opin. Cell. Biol. 9, 247–251.

Protein Adaptations and Signal Transduction. Edited by K.B. Storey and J.M. Storey

CHAPTER 15

Regulation of Ion Channel Function and Expression by Hypoxia

Chris Peers
Institute for Cardiovascular Research, University of Leeds, Leeds LS2 9JT, U.K.

1 Cellular responses to acute hypoxia

Virtually all known organisms have a requirement for O_2 in order to obtain the energy required for normal cell function. O_2 delivery in complex, multicellular organisms such as mammals requires co-ordinated systems designed to optimise uptake from the environment and delivery to different tissues. Ventilation–perfusion matching is tightly controlled to meet these demands, and is also highly adaptable to meet the varying O_2 demands required under different conditions, such as exercise and rest. This adaptability and close control is due in large part to the presence of specialized chemoreceptors such as the carotid body arterial chemoreceptors and neuroepithelial body airway chemoreceptors (Fidone and Gonzalez, 1986; Gonzalez et al., 1994; Cutz and Jackson, 1999). Cells within these tissues are known to respond to changes in local O_2 levels in a rapid manner, due to the presence of ion channels that are regulated by local O_2 levels. Such responses are the essence of chemoreception, initiating complex, whole-organism reflexes designed to optimise O_2 uptake. The prototype O_2-sensing organ is the carotid body, and below an account of this tissue and its functioning is given as an introduction to the increasingly widespread phenomenon of O_2 sensitive ion channels.

2. The carotid body

The carotid body is the major arterial chemoreceptor, and is located at the bifurcation of the common carotid artery (Fig. 15.1A). It is a highly vascularized organ, ideally suited to the detection of levels of O_2, CO_2 and pH of arterial blood (Gonzalez et al., 1994). When arterial blood becomes hypoxic, hypercapnic and/or acidic, the carotid body detects this and transduces such stimuli into increased activity of afferent chemosensory fibres running in the carotid sinus nerve (Fig. 15.1B). In this way, information concerning the blood gas and pH status is relayed to the respiratory centre of the CNS, allowing the initiation of corrective changes in breathing pattern, such as altered rate and depth of ventilation, so that blood gas and pH levels can be restored to physiologically acceptable levels.

Although this role of the carotid body has been recognized for decades, the underlying mechanisms have yet to be determined. It had, however, been established that chemoreception required the presence of type I (glomus) cells (Fig. 15.1C), and that these cells released numerous transmitters in response to hypoxia and other stimuli that presumably initiated increased action potential frequency in afferent chemosensory nerves (Fidone and Gonzalez, 1986). The nature of the transmitter(s) involved has long been contested, and for several years catecholamines (particularly dopamine) were considered the primary transmitters (reviewed by Gonzalez et al., 1994). However, most recent evidence suggests that chemoreception relies on release of acetylcholine and ATP in order to excite postsynaptic afferent nerve endings (Nurse and Zhang, 1999; Zhang et al., 2000). This notwithstanding, the fundamental question of how type I cells sense hypoxia and transduce this

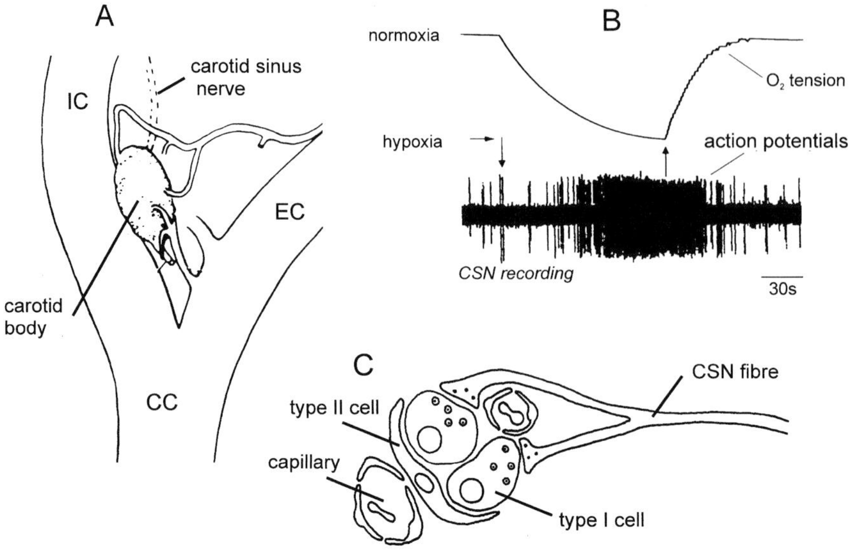

Fig. 15.1. (A) Diagram illustrating the location and vascularization of the carotid body. *CC*, common carotid body; *IC*, internal carotid artery; *EC* external carotid artery. (B) Example recording of an afferent chemosensory nerve fiber from the carotid sinus nerve (CSN), utilizing an *in vitro*, intact carotid body preparation (courtesy of Dr. Prem Kumar, University of Birmingham, U.K.). Note the graded increase in afferent nerve discharge in response to graded hypoxia. (C) Schematic of the major cell types found within the carotid body. Note the synaptic connection between type I cell and afferent nerve ending. Panel (B) adapted from Lopez-Barneo (1996), with permission.

stimulus into a secretory response remained for several years. In 1988 a report was published which went a long way towards addressing this issue. Using type I cells dissociated from the rabbit carotid body, Lopez-Barneo et al. (1988) found that these cells possessed K^+ channels which could be reversibly inhibited by acute hypoxia. This finding, confirmed later by others using type I cells from other species (Delpiano and Hescheler, 1989; Peers, 1990; Stea and Nurse, 1991; Buckler, 1997), initiated the "membrane hypothesis" to account for chemoreception. Put simply, this hypothesis indicates that hypoxia, by causing K^+ channel closure, depolarises type I cells sufficiently to activate voltage-gated Ca^{2+} channels. The subsequent influx of Ca^{2+} then triggered exocytosis of neurotransmitters to excite afferent chemosensory nerve endings. It is pertinent to note at this point that several aspects of this hypothesis have been contested. One such point of contention was the identity of the O_2-sensitive K^+ channel. Originally, a voltage-gated, slowly inactivating K^+ channel was identified in rabbit type I cells, but in the rat, evidence suggested that a high-conductance Ca^{2+}-activated K^+ channel (maxi-K channel) was selectively inhibited by hypoxia (Peers, 1990). These differences now appear to be genuinely species-related, and not age-related (Hatton et al., 1997). However, more recently another K^+ channel has been identified as being O_2 sensitive in rat type I cells. This channel appears to be TASK, an acid-sensitive member of the tandem P-domain family of low conductance, voltage-insensitive K^+ channels (Buckler, 1997; Buckler et al., 2000). The relative importance of TASK versus maxi-K channels has been debated and remains to be fully resolved. However, recent reports have indicated that selective pharmacological inhibition of maxi-K channels mimics the action of hypoxia to evoke neurosecretion from type I cells (Jackson and Nurse, 1997; Pardal et al., 2000). Thus, different K^+ channel types have been identified, even from the same preparation. This brief list has more recently been extended with the identification of other O_2-sensitive K^+ channels that serve specific physiological roles in a diverse range of tissues.

3. O_2-sensitive K^+ channels in other tissues

Since the discovery of O_2 -sensitive K^+ channels in type I carotid body cells, other O_2-sensing tissues have also been shown to possess such channels. Secretory cells of neuroepithelial bodies (NEBs) also have a voltage-gated K^+ channel (Youngson et al., 1993). NEBs appear to be the airway counterpart of the arterial chemoreceptor; they are located at the branching points of the airways, and release 5-HT and other vasoactive agents in response to hypoxia (Cutz and Jackson, 1999). Their exact role has yet to be clearly defined, but their involvement in pathological situations such as bronchopulmonary dysplasia, congenital central hypoventilation syndrome and sudden infant death syndrome (Cutz, 1997) has been implicated. Superficially, hypoxia may act in a similar manner to its effect on the type I cell (K^+ channel inhibition leading to depolarisation, Ca^{2+} influx and hence exocytosis), but the mechanism(s) coupling hypoxia to channel inhibition appear quite different (see later).

The pulmonary circulation is unique within the vasculature in that it contracts, rather than dilates under hypoxic conditions (Weir and Archer, 1995; Ward and Aaronson, 1999). This is an appropriate response for this vascular bed, since hypoxic constriction leads to the diversion of blood away from poorly ventilated regions of the lung, thus contributing to the optimisation of ventilation–perfusion matching. Several groups have demonstrated that vascular smooth muscle cells isolated from pulmonary resistance vessels possess O_2-sensitive K^+ channels (Post et al., 1992; Yuan et al., 1993; Osipenko et al., 1997), leading to the suggestion that hypoxic pulmonary vasoconstriction involves membrane depolarisation arising from hypoxic inhibition of K^+ channels which is sufficient to trigger Ca^{2+} influx through voltage-gated Ca^{2+} channels. This Ca^{2+} influx then initiates constriction. However, there is much debate as to the identity of the O_2-sensitive K^+ channel in this tissue (Post et al., 1992; Patel et al., 1997; Osipenko et al., 2000) and, moreover, the importance of these channels has more recently been contested, with evidence to suggest other endothelium-mediated responses and Ca^{2+} release from intracellular stores might account, at least in part, for hypoxic pulmonary vasoconstriction (Ward and Robertson, 1995; Ward and Aaronson, 1999).

Hypoxic/ischemic conditions have long been known to modulate the firing of central neurons. However, the possibility that central neurones might possess K^+ channels that can be modulated by hypoxia in a manner comparable to that seen in, for example, the carotid body, has only been pursued in recent years. Using neurons isolated from the neocortex and substantia nigra, Jiang and Haddad identified a high conductance, Ca^{2+}-sensitive K^+ channel that is inhibited by ATP that could be reversibly inhibited by hypoxia, thus indicating that the phenomenon of O_2-sensitive K^+ channels extended into the CNS (Jiang and Haddad, 1994; Haddad and Jiang, 1997). Since this discovery, O_2 sensitive K^+ channels have more recently been identified in neonatal (but not adult) chromaffin cells (Rychkov et al., 1998; Thompson and Nurse, 1998), and also in two cell lines which are proving to be of considerable use. Firstly, the PC12 cell line (Zhu et al., 1996), a rat pheochromocytoma line which behaves superficially like the type I cell of the carotid body, having a K^+ channel (identified as Kv1.2; Conforti and Millhorn 1997) that is inhibited by hypoxia, which leads to membrane depolarisation, a rise of $[Ca^{2+}]_i$ and a consequent, quantal release of catecholamines (Zhu et al., 1996; Taylor and Peers, 1998). The second cell line is a small cell carcinoma line, H-146, which is derived from a NEB tumor and shares many similarities with the parent tissue (O'Kelly et al., 1998, 1999). Both PC12 cells and H-146 cells have provided useful insights into the mechanisms coupling hypoxia to K^+ channel inhibition (detailed below) and the ease of their maintenance and use, as compared with native cells which are difficult to isolate, means that their usefulness will continue.

An additional advance in our understanding of the mechanisms coupling hypoxia to K^+ channel inhibition is likely to come from recombinant studies. To date, only a few different classes of K^+ channel have been expressed in recombinant systems to investigate their O_2 sensitivity (e.g. Patel et al., 1997; McKenna et al., 1998; Perez-Garcia et al., 1999), but already it seems that this approach, combined with mutagenesis studies, will rapidly advance our understanding of the structural requirements necessary for O_2 sensing by K^+ and other channels.

4. O_2-sensitive Ca^{2+} channels

Some seven years after the discovery of O_2-sensitive K^+ channels, a report appeared which described the O_2 sensitivity of voltage-gated Ca^{2+} channels (Franco-Obregon et al., 1995). The activity of these channels was recorded in isolated systemic smooth muscle cells, where hypoxia was observed to inhibit L-type (dihydropyridine-sensitive) Ca^{2+} channel activity in a voltage-dependent manner (inhibition being most striking at lower, more physiologically relevant membrane potentials) which was associated with a slight but detectable slowing of activation kinetics. This was an important observation, since it provided a simple and direct explanation for hypoxic dilation of the systemic vasculature. Indeed, such effects were also observed in type I carotid body cells of the rabbit (but not the rat; Lopez-Lopez et al., (1997)) where voltage-dependent Ca^{2+} channel inhibition prevented hypersensitivity of the cells to hypoxia (Montoro et al., 1996). Furthermore, the same group progressed to report a similar effect in large conduit vessels of the pulmonary vasculature, yet found the opposite effect—a voltage-dependent *potentiation* of Ca^{2+} currents—in pulmonary resistance vessels (Franco-Obregon and Lopez-Barneo, 1996). These findings were once more important contributing factors in our understanding of hypoxic pulmonary vasoconstriction as well as hypoxic dilation of the systemic vasculature. However, there is as yet no explanation as to why such diversity of responses exists within the same vascular bed.

Following these reports of hypoxic inhibition of native L-type Ca^{2+} channels, we examined the ability of hypoxia to exert such effects in a recombinant system (Fearon et al., 1997). Using the human cardiac L-type Ca^{2+} channel α_{1C} subunit stably expressed in human embryonic kidney (HEK 293) cells in the absence of auxiliary subunits, we found hypoxia to cause voltage dependent channel inhibition, associated with a slowing of activation kinetics, which was indistinguishable from effects seen in native tissues. Thus, in contrast to the effects of hypoxia on certain recombinant K^+ channels (Perez-Garcia et al., 1999), hypoxic inhibition of L-type channels was not dependent on auxiliary subunits (Fearon et al., 1997). This expression system is therefore potentially very useful for investigating the structural requirements of Ca^{2+} channels necessary for O_2 sensing.

5. Other O_2-sensitive ion channels

Although space does not permit an exhaustive documentation of O_2-sensitive ion channels, it should be noted that the list of channel types responsive to hypoxia continues to grow. In addition to O_2 sensitive K^+ and Ca^{2+} channels, reports have documented hypoxic inhibition of neuronal Na^+ channels via a protein kinase C dependent mechanism (O'Reilly et al., 1997), enhancement of non-inactivating Na^+ channels (Ju et al., 1996) and inhibition of intracellular muscle Cl^- channels (Kourie, 1997). The vast superfamily of ligand-gated ion channels is also largely unexplored in terms of acute O_2 sensitivity.

6. Mechanisms of O_2 sensing

"How does a fall of Po_2 lead to channel inhibition?" and "What is the O_2 sensor?" are two of the major questions frequently asked of researchers in the field. The answers remain elusive, and their determination is, of course, an immediate major aim. Attempts to address these questions are complicated by the fact that different groups use different O_2-sensing systems to address the same question and, as mentioned earlier, there are species and age-related differences even within the same preparation. However, the use of different systems has, paradoxically, allowed the establishment of some comparable observations that lead this author at least to conclude that different O_2 sensing mechanisms exist which can couple to the modulation of ion channel function.

One surprising observation is the wide range of time courses over which hypoxia exerts its inhibitory effects. On the one hand, Lopez-Lopez and Gonzalez (1992) elegantly demonstrated that hypoxia inhibited K^+ channel activity at a faster rate that the block of Na^+ channels by the rapidly acting toxin tetrodotoxin. On the other hand, hypoxic inhibition of K^+ channels in central neurones required several minutes, and could often be preceded by a transient enhancement of activity (Jiang and Haddad, 1994; Haddad and Jiang, 1997).

Several candidate mechanisms have been put forward to account for O_2 sensing by the carotid body, some of which are illustrated in Fig. 15.2. The involvement of cytochrome P-450 (cP-450; Fig. 15.2A) has been implicated in both pulmonary smooth muscle and rat type I carotid body cells (Yuan et al., 1995; Hatton and Peers, 1996). This idea suggests that cP-450, under normoxic conditions, generates reactive oxygen species which increase channel activity via redox modulation. The major criticism of this suggestion is that it is based largely on the dangerous assumption that cP-450 inhibitors employed are selective in their effects on cP-450, and other approaches are required before this idea can be validated. Redox modulation (i.e. the idea that hypoxia alters the reduced:oxidised forms of redox couples such as glutathione (GSSG:GSH)) via other means has long been considered, based on studies such as those illustrated in Fig. 15.2B (Benot et al., 1993). However, it should be noted that hypoxia can inhibit these same channels (recorded in isolated membrane patches from rabbit type I cells) in the absence of GSH, which casts doubt on this effect being the major means by which hypoxia inhibits K^+ channels in this tissue. Thirdly, and perhaps most convincingly, the involvement of a heme-containing protein has been implicated from the

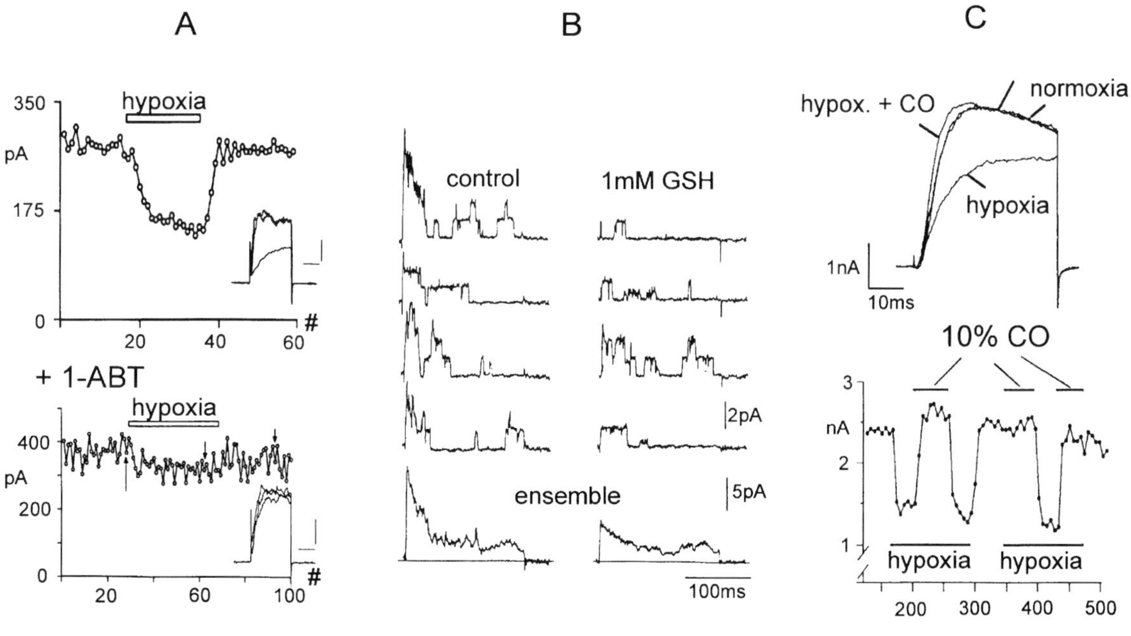

Fig. 15.2. Evidence for different mechanisms coupling hypoxia to K^+ channel inhibition. (A) *Upper trace*: time series plot of whole-cell K^+ current amplitude evoked in a rat type I cell by repeated depolarisations to +20 mV from a holding potential of –70 mV. A period of hypoxia causes reversible current inhibition. Lower trace: as upper, except that the cell was pre-treated with the cP-450 inhibitor 1-aminobenzotriazole before recording – this abolishes hypoxic inhibition. (B) Oxygen-sensitive K^+ channels recorded in an outside-out membrane patch taken from a rabbit type I cell. Application of glutathione to the intracellular face of the channel suppresses its activity. (C) Whole-cell K^+ currents evoked in a rabbit type I cell before and during hypoxia, and under hypoxic conditions in the additional presence of carbon monoxide (CO). From these traces, and the time series below, it is apparent that hypoxic inhibition is reversed by CO. Adapted with permission from Hatton and Peers (1996); Benot et al. (1993); Lopez-Lopez and Gonzalez (1992), with permission.

work of Lopez-Lopez and Gonzalez (1992), who demonstrated that hypoxic inhibition of whole cell K^+ currents could be fully reversed by carbon monoxide (CO): since heme proteins are the only known interactive site for O_2 and CO, this findings would seem compelling, yet further supporting evidence remains to appear.

One earlier suggestion for a candidate O_2 sensor was NADPH oxidase. This work, pioneered by Acker and co-workers, suggested that the oxidase generated H_2O_2, which maintained channel activity through oxidation, and that hypoxic inhibition was a result of lack of substrate (i.e. O_2) for the oxidase (see Acker and Xue, 1995, for review). Much of the idea for the involvement of NADPH as an O_2 sensor was based on the fact that diphenylene iodide (DPI), an inhibitor of NADPH oxidase, inhibited hypoxic excitation of the carotid body sinus nerve afferents. However, evidence has since accumulated to discount this oxidase as an O_2 sensor, at least in the carotid body and in pulmonary vascular smooth muscle. In both tissue types, DPI appears to act as a non-selective (and possibly direct) channel inhibitor (Weir et al., 1994; Wyatt et al., 1994). Furthermore, DPI and other NADPH oxidase inhibitors failed to alter hypoxia evoked transmitter release from intact rabbit carotid bodies (Obeso et al., 1999). Finally, hypoxic pulmonary vasoconstriction was found to be unaltered in mice lacking one of the functional subunits of the oxidase (Archer et al., 1999).

In contrast to the evidence arguing against NADPH oxidase as an O_2 sensor in the carotid body and pulmonary vasculature, its role as an O_2 sensor in airway chemoreceptor (NEB) cells is well supported: most importantly, Cutz and colleagues have recently demonstrated that O_2 sensing in a mouse NADPH oxidase knockout model is completely inhibited in terms of hypoxic inhibition of NEB cell K^+ currents (Fu et al., 2000). Furthermore, using the NEB-derived cell line H146, we have shown that O_2 sensing was modulated by activation of protein kinase C (O'Kelly et al., 2000). Such kinase activation stimulates NADPH oxidase activity, so that in the face of reduced O_2 levels, the enzyme could continue to generate H_2O_2 production and so maintain K^+ channel activity. Thus, the hypoxic sensitivity of K^+ channels in H146 cells was suppressed. Therefore, at present, evidence supports the idea that different mechanisms are in place to permit O_2 sensing in different tissue types. Although this fundamental issue seems resolved, far greater detail of these molecular mechanisms will be required in order to account fully for the diverse, rapid responses of ion channels to altered O_2 levels.

7. Chronic hypoxia

Prolonged periods of hypoxia—at Po_2 values less severe than those required to exert rapid, 'direct' effects on ion channels as described above—have long been known to exert cellular effects at the level of transcription. Such responses can be physiological (e.g. increased production of erythropoeitin at high altitude) or can arise from pathophysiological situations, such as chronic lung disease or congestive heart failure (reviewed by Semenza, 2000). The effects of prolonged hypoxia (physiological or pathophysiological) on ion channel expression are not as thoroughly studied as such effects on other proteins such as erythropoeitin (Jelkmann, 1992) or tyrosine hydroxylase (TH; Czyzyk-Krzeska et al., 1992, 1994). However, a small number of studies are beginning to shed light on this potentially important aspect of ion channel research.

In addition to the rapid responses of ion channels to acute hypoxia in carotid body type I cells, their expression is now known to be altered following prolonged (chronic) hypoxia. Nurse and colleagues found that rat type I cells, cultured under hypoxic conditions over a period of two weeks, increased their expression of voltage-gated Na^+ channels, an effect which could be mimicked by increasing intracellular cAMP levels (Stea et al., 1992, 1995). This observation might account for increased sensitisation of the ventilatory reflex caused by chronic hypoxia. By contrast, we studied the expression of ion channels in carotid body type I cells isolated from rats born and reared under chronically hypoxic conditions for 10 days—conditions which are known to cause blunting of the hypoxic ventilatory response (Wyatt et al., 1995). We found that there was a selective suppression of high conductance Ca^{2+} activated K^+ channels in the chronically hypoxic cells. These channels are the ones that are inhibited by acute hypoxia to cause membrane depolarisation, and so the chronically hypoxic cells were unable to depolarise under acute hypoxia. This observation could account for hypoxic ventilatory blunting, but the underlying mechanisms were not established.

The best studied effect of chronic hypoxia on ion channel expression and activity comes from Millhorn and colleagues, who first noted a selective increase in the O_2 sensitive component of the whole-cell K^+ current in PC12 cells following a period of chronic hypoxia (Conforti and Millhorn, 1997). The underlying channel was identified as Kv1.2, and although the pathways coupling hypoxia to increased Kv1.2 gene expression remain to be determined, it is of interest to note this group's investigation of the regulation of another gene, which encodes TH (reviewed by Millhorn et al., 2000). In contrast to other studies which implicate hypoxia inducible factor (HIF-1) (see Semenza, 2000, and references therein) as a major factor in hypoxic gene regulation, Millhorn et al found that HIF-1 was not induced in PC12 cells, but that TH gene expression was dependent on a rise of $[Ca^{2+}]_i$ and possibly involved Ca^{2+} binding to calmodulin to exert its regulation. Subsequent studies indicated that mitogen-activated protein kinase (MAP kinase) was also involved in TH gene regulation. Clearly, such approaches must be adopted for the study of ion channel expression in order for us to understand electrophysiological adaptation to hypoxic conditions.

Our own recent studies have investigated the release of catecholamines from PC12 cells in response to acute and chronic hypoxia, prompted by the above-described work of Millhorn and colleagues. Perhaps as anticipated, acute hypoxia evoked exocytosis from PC12 cells, as determined amperometrically. This was fully dependent on external Ca^{2+} entering the cells through voltage-gated Ca^{2+} channels, since removal of external Ca^{2+} or blockade of Ca^{2+} channels with Cd^{2+} fully abolished ongoing secretion (Fig. 15.3A, B). The Ca^{2+} channels were activated by depolarisation arising, presumably, from hypoxic inhibition of Kv1.2 (Taylor and Peers, 1998). Following a period of chronic hypoxia, secretion evoked by acute hypoxia was enhanced. Again, this secretion was wholly dependent on external Ca^{2+} but could not be blocked completely by Cd^{2+} (Fig. 15.3 C,D). This indicated that chronic hypoxia induced a Cd^{2+} resistant Ca^{2+} influx pathway coupled to exocytosis. Subsequently, we identified this pathway as being attributable to Ca^{2+} permeable channels formed from amyloid β peptides associated with Alzheimer's disease (Taylor et al., 1999; Taylor and Peers, 1999). It is established that individuals who suffer a hypoxic or ischemic episode are more likely to develop Alzheimer's disease in later life (Kokmen et al., 1996; Moroney et al., 1996), and it is our hope that we have uncovered a simple cell system to exploit, raising the hope that we can identify the mechanisms coupling hypoxia to amyloid β peptide formation with a view to future therapeutic intervention.

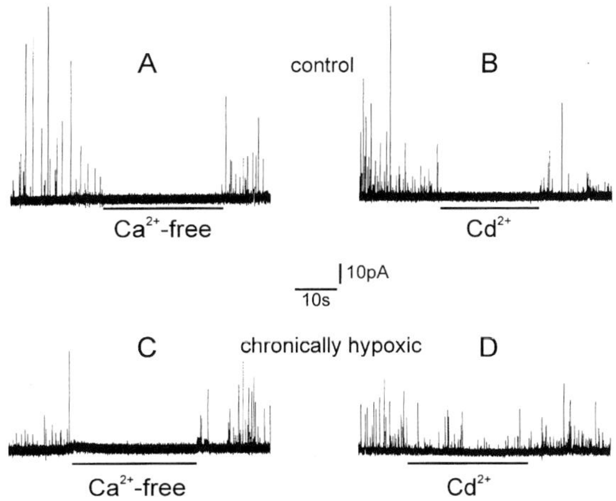

Fig. 15.3. Amperometric recordings of exocytosis from PC12 cells in response to ongoing hypoxia (Po_2 20 mmHg). Cells were cultured normoxically (A,B), or in an atmosphere of 10% O_2 (C,D). For the periods indicated by the horizontal bars, either external Ca^{2+} was removed and replaced by 1 mM EGTA (A,C), or Cd^{2+} (200 μM) was applied. Scale bars apply to all traces. Note the lack of complete inhibition caused by Cd^{2+} in the chronically hypoxic cell (D), but not the control cell (B). Adapted from Taylor et al. (1999), with permission.

8. Conclusions

It is now twelve years since the discovery of O_2 sensitive ion channels. Whilst several key questions remain to be answered concerning the coupling of hypoxia to channel inhibition, several important observations have arisen. Clearly, O_2 sensitive channels are not confined to chemoreceptor cells, but appear to be far more widely distributed. Secondly, O_2 sensing is a property (most likely an indirect property) of a variety of different ion channels. Thirdly, findings to date suggest strongly that different tissues possess different mechanisms for linking a fall of Po_2 to altered channel activity.

Regulation of channel expression by sustained episodes of hypoxia is a topic in its infancy. However, should the mechanisms underlying channel expression mirror those by which hypoxia regulates expression of other genes, then progress may be rapid. This subject matter is of immediate clinical importance, and may allow future beneficial intervention for a variety of conditions which arise as a consequence of prolonged hypoxia.

Acknowledgements

I would like to thank colleagues in the field who have given permission to reproduce their findings in this chapter. I am also grateful to past and present graduate students Chris Wyatt, Chris Hatton, Ita O'Kelly Tony Lewis and Shafeena Taylor, postdoctoral workers Liz Carpenter, Ian Fearon and Matt Hartness, and my collaborator Paul Kemp for their expertise and contributions to this field. Finally, the support of the Wellcome Trust and the British Heart Foundation is greatly appreciated.

References

Acker, H. and Xue, D. (1995). Mechanisms of O_2 sensing in the carotid body in comparison with other O_2-sensing cells. News Physiol. Sci. 10, 211–216.

Archer, S.L., Reeve, H.L., Michelakis, E., Puttagunta, L., Waite, R., Nelson, D.P., Dinauer, M.C. and Weir, E.K. (1999). O_2 sensing is preserved in mice lacking the gp91 phox subunit of NADPH oxidase. Proc. Natl. Acad. Sci. USA 96, 7944–7949.

Benot, A.R., Ganfornina, M.D. and Lopez-Barneo, J. (1993). Potassium channel modulated by hypoxia and the redox status in glomus cells of the carotid body. In: Ion Flux in Pulmonary Vascular Control (Weir, E.K., Hume, J.R. and Reeves, J.T., Eds.) pp. 177–187. Plenum Press, New York.

Buckler, K.J. (1997). A novel oxygen-sensitive potassium current in rat carotid body type I cells. J. Physiol. 498, 649–662.

Buckler, K.J., Williams, B.A. and Honore, E. (2000). An oxygen-, acid- and anaesthetic-sensitive TASK-like background potassium channel in rat arterial chemoreceptor cells. J. Physiol. 525, 135–142.

Conforti, L. and Millhorn, D.E. (1997). Selective inhibition of a slow-inactivating voltage-dependent K^+ channel in rat PC12 cells by hypoxia. J. Physiol. 502, 293–305.

Cutz, E. (1997). Studies on neuroepithelial bodies under experimental and disease conditions. In: Cellular and Molecular Biology of Airway Chemoreceptors (Cutz, E., Ed.) pp. 109–129. Landes Bioscience, Texas.

Cutz, E. and Jackson, A. (1999). Neuroepithelial bodies as airway oxygen sensors. Resp. Physiol. 115, 201–214.

Czyzyk-Krzeska, M.F., Bayliss, D.A., Lawson, E.E. and Millhorn, D.E. (1992). Regulation of tyrosine hydroxylase gene expression in the rat carotid body by hypoxia. J. Neurochem. 58, 1538–1546.

Czyzyk-Krzeska, M.F., Furnari, B.A., Lawson, E.E. and Millhorn, D.E. (1994). Hypoxia increases rate of transcription and stability of tyrosine hydroxylase mRNA in pheochromocytoma (PC12) cells. J. Biol. Chem. 269, 760–764.

Delpiano, M.A. and Hescheler, J. (1989). Evidence for a Po_2-sensitive K^+ channel in the type–I cell of the rabbit carotid body. FEBS Lett.. 249, 195–198.

Fearon, I.M., Palmer, A.C.V., Balmforth, A.J., Ball, S.G., Mikala, G., Schwartz, A. and Peers, C. (1997). Hypoxia inhibits the recombinant α_{1C} subunit of the human cardiac L-type Ca^{2+} channel. J. Physiol. 500, 551–556.

Fidone, S. and Gonzalez, C. (1986). Initiation and control of chemoreceptor activity in the carotid body. In: Handbook of Physiology. The Respiratory System. Control of Breathing (Cherniack, N.S. and Widdicombe, J.G., Eds.) pp. 247–312. American Physiological Society, Bethesda, MD, USA.

Franco-Obregon, A. and Lopez-Barneo, J. (1996). Differential oxygen sensitivity of calcium channels in rabbit smooth muscle cells of conduit and resistance pulmonary arteries. J Physiol. 491, 511–518.

Franco-Obregon, A., Urena, J. and Lopez-Barneo, J. (1995). Oxygen-sensitive calcium channels in vascular smooth muscle and their possible role in hypoxic arterial relaxation. Proc. Natl. Acad. Sci. USA 92, 4715–4719.

Fu, X.W., Wang, D., Nurse, C.A., Dinauer, M.C. and Cutz, E. (2000). NADPH oxidase is an O_2 sensor in airway chemoreceptors: evidence from K^+ current modulation in wild-type and oxidase deficient mice. Proc. Natl. Acad. Sci. USA 97, 4374–4379.

Gonzalez, C., Almarez, L., Obeso, A. and Rigual, R. (1994). Carotid body chemoreceptors: from natural stimuli to sensory discharges. Physiol. Rev. 74, 829–898.

Haddad, G.G. and Jiang, C. (1997). O_2-sensing mechanisms in excitable cells: Role of plasma membrane K^+ channels. Ann. Rev. Physiol. 59, 23–42.

Hatton, C.J., Carpenter, E., Pepper, D.R., Kumar, P. and Peers, C. (1997). Developmental changes in isolated rat type I carotid body cell K^+ currents and their modulation by hypoxia. J. Physiol. 501, 49–58.

Hatton, C.J. and Peers, C. (1996). Inhibition of K^+ and Ca^{2+} currents in isolated rat type I carotid body cells by cytochrome P-450 inhibitors Am. J. Physiol. 271, C85–C92.

Jackson, A. and Nurse, C.A. (1997). Dopaminergic properties of cultured rat carotid body chemoreceptors grown in normoxic and hypoxic environments. J. Neurochem. 69, 645–654.

Jelkmann, W. (1992). Erythropoeitin: Structure, control of production and function. Physiol. Rev. 72, 449–489.

Jiang, C. and Haddad, G.G. (1994). A direct mechanism for sensing low–oxygen levels by central neurons. Proc. Natl. Acad. Sci. USA 91, 7198–7201.

Ju, Y.K., Saint, D.A. and Gage, P.W. (1996). Hypoxia increases persistent sodium current in rat ventricular myocytes. J. Physiol. 497, 337–347.

Kokmen, E., Whisnant, J.P., O'Fallon, W.M., Chu, C.P. and Beard, C.M. (1996). Dementia after ischemic stroke: a population–based study in Rochester, Minnesota (1960–1984). Neurology 46, 154–159.

Kourie, J.I. (1997). A redox O_2 sensor modulates the SR Ca^{2+} countercurrent through voltage- and Ca^{2+}-dependent Cl^- channels. Am. J. Physiol. 272, C324–C332.

Lopez–Barneo, J. (1996). Oxygen-sensing by ion channels and the regulation of cellular functions. Trends Neurosci. 19, 435–440.

Lopez–Barneo, J., Lopez–Lopez, J.R., Urena, J. and Gonzalez, C. (1988). Chemotransduction in the carotid body: K^+ current modulated by Po_2 in type I chemoreceptor cells. Science 241, 580–582.

Lopez-Lopez, J.R. and Gonzalez, C. (1992). K^+ current inhibition by low oxygen in chemoreceptor cells of adult rabbit carotid body. Effects of carbon monoxide. FEBS Lett. 299, 251–254.

Lopez-Lopez, J.R., Gonzalez, C., and Perez-Garcia, M.T. (1997). Properties of ionic currents from isolated adult rat carotid body chemoreceptor cells: effect of hypoxia J. Physiol. 499, 429–441.

McKenna, F., Ashford, M.L.J. and Peers, C. (1998). Hypoxia reversibly inhibits the activity of cloned human brain BK_{Ca} channels stably expressed in HEK 293 cells. J. Physiol. 509, 188P.

Millhorn, D.E., Beitner-Johnson, D., Conforti, L., Conrad, P.W., Kobyashi, S., Yuan, Y., and Rust, R. (2000). Gene regulation during hypoxia in excitable oxygen-sensing cells: depolarisation–transcriptional coupling. Adv. Exp. Med. Biol. 475, 131–142.

Montoro, R.J., Urena, J., Fernandez-Chacon, R., Alvarez de Toledo, G. and Lopez-Barneo, J. (1996). Oxygen sensing by ion channels and chemotransduction in single glomus cells. J. Gen. Physiol. 107, 133–143.

Moroney, J.T., Bagiella, E., Desmond, D.W., Paik, M.C., Stern, Y. and Tatemichi, T.K. (1996). Risk factors for incident dementia after stroke. Role of hypoxic and ischemic disorders. Stroke 27, 1283–1289.

Nurse, C.A. and Zhang, M. (1999). Acetylcholine contributes to hypoxic chemotransmission in co-cultures of rat type I cells and petrosal neurons. Resp. Physiol. 115, 189–200.

Obeso, A., Gomez-Nino, A. and Gonzalez, C. (1999). NADPH oxidase inhibition does not interfere with low Po_2 transduction in rat and rabbit CB chemoreceptor cells. Am. J. Physiol. 276, C593–C601.

O'Kelly, I., Lewis, A., Peers, C. and Kemp, P.J. (2000). O_2 sensing by airway chemoreceptor-derived cells. Protein kinase c activation reveals functional evidence for involvement of NADPH oxidase. J. Biol. Chem. 275, 7684–7692.

O'Kelly, I., Peers, C. and Kemp, P.J. (1998). Oxygen-sensitive K^+ channels in neuroepithelial body-derived small cell carcinoma cells of the human lung. Am. J. Physiol. 275, L709–L716.

O'Kelly, I., Stephens, R.H., Peers, C. and Kemp, P.J. (1999). Potential identification of the O_2-sensitive K^+ channel in a human neuroepithelial body-derived cell line. Am. J. Physiol. 276, L96–L104.

O'Reilly, J.P., Cummins, T.R. and Haddad, G.G. (1997). Oxygen deprivation inhibits Na+ current in rat hippocampal neurones via protein kinase C. J. Physiol. 503, 479–488.

Osipenko, O.N., Evans, A.M., and Gurney, A.M. (1997). Regulation of the resting potential of rabbit pulmonary artery myocytes by a low threshold, O_2-sensing potassium current. Br. J. Pharmacol. 120, 1461–1470.

Osipenko, O.N., Tate, R.J. and Gurney, A.M. (2000). Potential role for Kv3.1b channels as oxygen sensors. Circ. Res. 86, 534–540.

Pardal, R., Ludewig, U., Garcia-Hirschfeld, J. and Lopez-Barneo, J. (2000). Secretory responses to hypoxia and tetraethylammonium of intact glomus cells in thin slices of rat carotid body. Proc. Natl. Acad. Sci. USA 97, 2361–2366.

Patel, A.J., Lazdunski, M. and Honore, E. (1997). Kv2.1/Kv9.3, a novel ATP-dependent delayed rectifier K^+ channel in oxygen-sensitive pulmonary artery myocytes.

EMBO J. 16, 6615–6625.

Peers, C. (1990). Hypoxic suppression of K^+ currents in type I carotid body cells: selective effect on the Ca^{2+}-activated K^+ current. Neurosci. Lett. 119, 253–256.

Perez-Garcia, M.T., Lopez-Lopez, J.R. and Gonzalez, C. (1999). Kvβ1.2 subunit coexpression in HEK293 cells confers O_2 sensitivity to Kv4.2 but not to *Shaker* channels. J. Gen. Physiol. 113, 897–907.

Post, J.M., Hume, J.R., Archer, S.L. and Weir, E.K. (1992). Direct role for potassium channel inhibition in hypoxic pulmonary vasoconstriction. Am. J. Physiol. 262, C882–C890.

Rychkov, G.Y., Adams, M.B., McMillen, I.C. and Roberts, M.L. (1998). Oxygen-sensing mechanisms are present in the chromaffin cells of the sheep adrenal medulla before birth. J. Physiol. 509, 887–893.

Semenza, G.L. (2000). Chairman's summary: mechanisms of oxygen homeostasis, circa 1999. Adv. Exp. Med. Biol. 475, 303–310.

Stea, A., Jackson, A. and Nurse, C.A. (1992). Hypoxia and $N^6,O^{2'}$–dibutyryladenosine 3′,5′-cyclic monophosphate, but not nerve growth factor, induce Na+ channels and hypertrophy in chromaffin-like arterial chemoreceptors. Proc. Natl. Acad. Sci. USA 89, 9469–9473.

Stea, A., Jackson, A., MacIntyre, L. and Nurse, C.A. (1995). Long-term modulation of inward currents in O_2 chemoreceptors by chronic hypoxia and cyclic AMP *in vitro*. J. Neurosci. 15, 2192–2202.

Stea, A. and Nurse, C.A. (1991). Whole-cell and perforated-patch recordings from O_2-sensitive rat carotid body cells grown in short- and long-term culture. Pflugers Archiv. 418, 93–101.

Taylor, S.C., Batten, T.F.C. and Peers, C. (1999). Hypoxic enhancement of quantal catecholamine secretion: evidence for the involvement of amyloid β-peptides. J. Biol. Chem. 274, 31217–31223.

Taylor, S.C. and Peers, C. (1998). Hypoxia evokes catecholamine secretion from rat pheochromocytoma PC12 cells. Biochem. Biophys. Res. Comm. 248, 13–17.

Taylor, S.C. and Peers, C. (1999). Chronic hypoxia enhances the secretory response of rat pheochromocytoma (PC-12) cells to acute hypoxia. J. Physiol. 514, 483–491.

Thompson, R.J. and Nurse, C.A. (1998). Anoxia differentially modulates multiple K^+ currents and depolarizes neonatal rat adrenal chromaffin cells. J. Physiol. 512, 421–434.

Ward, J.P.T. and Aaronson, P.I. (1999). Mechanisms of hypoxic pulmonary vasoconstriction: can anyone be right? Resp. Physiol. 115, 261–271.

Ward, J.P.T. and Robertson, T.P. (1995). The role of the endothelium in hypoxic pulmonary vasoconstriction. Exp. Physiol. 80, 793–801.

Weir, E.K. and Archer, S.L. (1995). The mechanism of acute hypoxic pulmonary vasoconstriction: the tale of two channels. FASEB J. 9, 183–189.

Weir, E.K., Wyatt, C.N., Reeve, H.L., Huang, J., Archer, S.L. and Peers, C. (1994). Diphenylene iodonium inhibits both potassium and calcium currents in isolated pulmonary artery smooth muscle cells. J. Appl. Physiol. 76, 2611–2615.

Wyatt, C.N., Wright, C., Bee, D. and Peers, C. (1995). O_2-Sensitive K^+ currents in carotid body chemoreceptor cells from normoxic and chronically hypoxic rats and their roles in hypoxic chemotransduction. Proc. Natl. Acad. Sci. USA 92, 295–299.

Wyatt, C.N., Weir, E.K. and Peers, C. (1994). Diphenylene iodonium blocks K^+ and Ca^{2+} currents in type I cells isolated from the neonatal rat carotid body. Neurosci. Lett. 172, 63–66.

Youngson, C., Nurse, C., Yeger, H. and Cutz, E. (1993). Oxygen sensing in airway chemoreceptors. Nature 365, 153–155.

Yuan, X.J., Goldman, W.F., Tod, M.L., Rubin, L.J. and Blaustein, M.P. (1993). Hypoxia reduces potassium currents in cultured rat pulmonary but not mesenteric arterial myocytes. Am. J. Physiol. 264, L116–L123.

Yuan, X.J., Tod, M.L., Rubin, L.J. and Blaustein, M.P. (1995). Inhibition of cytochrome P-450 reduces voltage-gated K^+ currents in pulmonary arterial myocytes. Am. J. Physiol. 268, C259–C270.

Zhang, M., Zhong, H., Vollmer, C. and Nurse, C.A. (2000). Co-release of ATP and ACh mediates hypoxic signalling at rat carotid body chemoreceptors. J. Physiol. 525, 143–158.

Zhu, W.H., Conforti, L., Czyzyk-Krzeska, M.F. and Millhorn, D.E. (1996). Membrane depolarization in PC12 cells during hypoxia is regulated by an O_2-sensitive K^+ current. Am. J.Physiol. 271, C658–C665.

CHAPTER 16

Ca^{2+} Dynamics Under Oxidant Stress in the Cardiovascular System

Tapati Chakraborti[1], Sudip Das[2], Malay Mandal[2], Amritlal Mandal[2] and Sajal Chakraborti[2]*
[1]Department of Neurosciences, Brain Institute, University of Florida, Gainesville, Florida 32610, U.S.A.;
[2]Department of Biochemistry and Biophysics, University of Kalyani, Kalyani 741235, West Bengal, India

1. Introduction

Oxidative stress causes cellular injuries that are mediated, at least in part, by an increase in cytosolic Ca^{2+} concentration $[Ca^{2+}]_i$ (Yoshida et al., 2000). Disturbances in a variety of mechanisms that normally maintain intracellular Ca^{2+} homeostasis have been found to occur during oxidant stress (Coetzee et al., 1994; Chakraborti et al., 1998). For example, oxidants such as hypochlorous acid (HOCl) cause an increase in $[Ca^{2+}]_i$, which can be abolished by pretreatment with caffeine (Tani, 1990). This observation indicated that oxidants can modify the activity of internal Ca^{2+} stores. An increase in Ca^{2+} permeability, as a consequence of oxidant stress, also occurred in mitochondria and sarco(endo)plasmic reticular vesicles. Furthermore, a depressed Ca^{2+} uptake and an inhibition of sarco(endo)plasmic reticular Ca^{2+}ATPase activity have also been found to occur under oxidant stress (Huang et al., 1992; Kaneko et al., 1994). Oxidant stress, therefore, not only promotes Ca^{2+} release but also impairs its uptake mechanisms into internal stores with a consequent increase in $[Ca^{2+}]_i$.

Cytosolic Ca^{2+} overload can occur either by an increase in Ca^{2+} influx from the extra-cellular space to the cytosol or as a result of insufficient Ca^{2+} extrusion from the cytosol. Cytosolic Ca^{2+} concentration is also affected by subcellular Ca^{2+} store sites such as the sarco(endo)plasmic reticulum and mitochondria (Kaneko et al., 1994; Chakraborti et al., 1999a).

*Corresponding author.

2. Ca^{2+} influx from extracellular to intracellular space

In addition to an initial transient increase in $[Ca^{2+}]_i$, Roveri et al. (1992) observed a sustained elevation of cytosolic Ca^{2+} when coronary artery smooth muscle cells were exposed to H_2O_2. This sustained elevation was not observed in Ca^{2+} free media, suggesting that H_2O_2 also affects Ca^{2+} transport mechanisms that are associated with the plasma membrane. It is well accepted that excitation–contraction coupling in mammalian heart includes two critical Ca^{2+} components: (i) Ca^{2+} influx across the sarcolemmal membrane and (ii) Ca^{2+} derived from the sarcoplasmic reticulum via the process of Ca^{2+}-induced Ca^{2+} release (Fabiato and Fabiato, 1978; Fabiato, 1983; Kaneko et al., 1994). The Ca^{2+} channel and the Na^+/Ca^{2+} exchanger are recognized as responsible for influx of Ca^{2+}, although the Na^+/Ca^{2+} exchanger also operates in the net efflux of Ca^{2+} (Fig. 16.1) (Mullins, 1979; Langer et al., 1982; Philipson and Ward, 1986; Kaneko et al., 1994).

2.1. ATP independent Ca^{2+} binding

The sources of the Ca^{2+} that enters the cell across the sarcolemma appear to be the extracellular

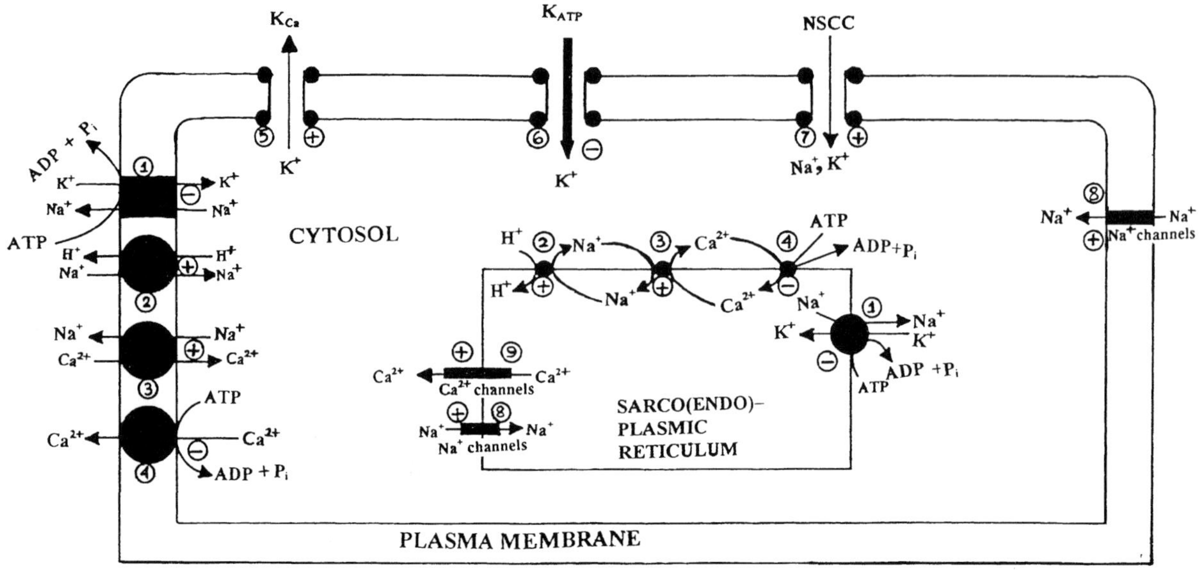

Fig. 16.1. Proposed scheme of the effects of oxidant stress on transport processes in myocardial cells. Several transport systems which are affected by oxidants are summarized. Details are given in the text. The net effects of oxidant stress are indicated by + and – symbols. 1: Na^+/K^+ ATPase; 2: Na^+/H^+ exchanger; 3: Na^+/Ca^{2+} exchanger; 4: Ca^{2+} ATPase; 5: Ca^{2+} activated K^+ channel; 6: ATP regulated K^+ channel; 7: nonselective cation channel (NSCC); 8: Na^+ channel; 9: Ca^{2+} channel.

space and the Ca^{2+} binding sites are present within the sarcolemmal membrane (Bers and Langer, 1979; Bers et al., 1981; Langer, 1986; Kaneko et al., 1994). In the presence of H_2O_2 or the O_2^- generating system, xanthine plus xanthine oxidase (X+XO), both low and high affinity Ca^{2+} activities, for example, Na^+/Ca^{2+} and Ca^{2+} ATPase (ATP dependent Ca^{2+} uptake), respectively, were found to be increased at the initial period of incubation. Philipson et al. (1980) showed that a large amount of Ca^{2+} was bound to membrane phospholipids at physiological levels of extracellular Ca^{2+}. Since reactive oxygen species (ROS) are known to promote the peroxidation of membrane phospholipids (Freeman and Crapo, 1982), it is likely that the changes in ATP-independent Ca^{2+} entry by ROS are due to alterations in the phospholipid composition of the membrane.

2.2. Ca^{2+} channels

Dihydropyridine antagonists inhibit smooth muscle L-type Ca^{2+} currents thereby decreasing intracellular Ca^{2+} concentration and concomitantly inducing smooth muscle relaxation (Tani, 1990). It has been shown recently that dihydropyridines can induce the release of nitric oxide (NO) from the coronary vascular endothelium (Dhein et al., 1999).These findings of the dual mode of action (i.e., a direct relaxing effect by inhibition of the smooth muscle L-type Ca^{2+} current and an indirect relaxing effect by release of NO from coronary vascular endothelium) may help us to understand the beneficial antihypertensive effect of dihydropyridine Ca^{2+} antagonists. Additionally, NO release from both the vascular endothelium and the platelets may contribute to the anti-atherosclerotic and antithrombotic effects of dihydropyridines (Dhein et al., 1999).

Adenosine has been shown to exert significant cardioprotective effects through the activation of adenosine A1 receptors (Thornton et al., 1992). Activation of the A1 receptors attenuate H_2O_2 induced cardiodepression (Karmazyn and Cook, 1992). In addition, ischemia preconditioning confers protection against H_2O_2 via adenosine dependent mechanisms (Gan et al., 1996). Adenosine A1 receptor agonists have been shown in various studies to protect ischemic-reperfused (I-R) myocardium (Thornton et al., 1992), although the precise mechanism(s) involved for this protection is not known. One of the possible mechanisms of action of adenosine A1 receptor agonists could be due to its effect on L-type Ca^{2+} channels in the myocardium. This was evident from studies with cyclopiazonic acid (CPA), a selective adenosine A1 receptor agonist. Treatment of rat ventricular myocardium with CPA has been shown to inhibit H_2O_2 induced stimulation of L-type Ca^{2+} current (Thomas et al., 1998). Although the A1 receptor is the dominant adenosine receptor subtype present in the ventricular myocardium, A2 and A3 receptors also exist in the ventricular myocytes. A2 receptor agonists have been shown to produce a minimum protective influence on the heart, whereas the anti-neutrophil and anti-platelet actions of A2 receptor activation can induce protection of the heart (Zucchi et al., 1992; Schlack et al., 1993; Thomas et al., 1998). The role of the A3 receptors in this context is currently unknown.

2.3. β-Adrenergic receptors

Although β-adrenergic receptor (β-AR) blockers are used for the treatment of ischemic heart disease, the mechanisms of their beneficial actions have not been fully elucidated. In view of the role of sarcoplasmic reticulum (SR) abnormalities in cardiac dysfunction due to I-R, Temsah et al. (2000) examined the effects of β-AR blockers on the I-R induced changes in SR Ca^{2+} uptake and release as well as gene expression of ryanodine receptors, SR Ca^{2+} pump, phospholamban and calsequestrin. I-R in isolated rat hearts was induced by stopping perfusion for 30 minutes and then reperfusing for 60 min. Hearts were treated with or without the β1-AR blockers, atenolol and propranolol, before inducing ischemia as well as during the reperfusion period. I-R significantly depressed the cardiac performance, SR Ca^{2+} uptake, and Ca^{2+} release activities as well as Ca^{2+}/calmodulin dependent protein kinase and cAMP dependent protein kinase mediated phosphorylation of proteins. The mRNA levels for the SR Ca^{2+} pump, ryanodine receptor, phospholamban and calsequestrin were also reduced by I-R. All of these changes that occurred due to I-R were ameliorated by treatment with β-AR blockers (Temsah et al., 2000). This study suggested beneficial effects of β-AR blockers on cardiac performance in the I-R hearts. The beneficial effects produced by β-AR blockers may be due to prevention of changes in SR Ca^{2+} uptake and release activities as well as Ca^{2+}/calmodulin dependent and cAMP dependent protein phosphorylations of SR proteins. On the other hand, the protection of I-R induced alterations in mRNA levels of SR proteins by β-AR blockers suggest that cardiac SR gene expression is also a site of their cardioprotective action (Temsah et al., 2000).

The β-adrenergic-adenylyl cyclase system has been reported to play a role in intracellular Ca^{2+} overload (Chakraborti et al., 2000). At low concentrations, the oxidant HOCl promotes intracellular Ca^{2+} overload by increasing the entry of Ca^{2+} into cardiomyocytes whereas at high concentrations, this agent depresses Ca^{2+} influx. Therefore, both an increase and a decrease in Ca^{2+} influx were found upon exposing cardiomyocytes to HOCl (Kaminishi et al., 1989; Fukui et al., 1994). In view of the biphasic effects of H_2O_2 and the oxyradical generating system, X+XO, on the isoproterenol stimulated adenylyl cyclase activity in cardiac membranes (Persad et al., 1997, 1998), it seems that the biphasic action of HOCl may represent a phenomenon associated with oxidative stress. It was suggested that the observed changes by HOCl at high concentrations may be due to the loss of -SH groups or degradation of the proteins by HOCl. This view was supported by the observation that treatment of cardiac membranes with relatively high concentrations of HOCl (0.1 mM HOCl) decreases the -SH group content as well as the level of Gs protein (Persad et al., 1999). Further investigations are necessary to determine the exact mechanisms for the effects of HOCl on the β-adrenoceptor mediated signal transduction pathways in the myocardium.

ROS such as O_2^- have gained acceptance as modulators of receptor-mediated signal transduction in a variety of cell types. Ligand–receptor binding has been demonstrated to induce production of ROS (Finkel, 1998, Fukui et al., 1999a). The peptide hormone angiotensin II (Ang II) which acts on G-protein coupled AT1 receptors induces a rapid increase in intracellular H_2O_2 that is involved in its hypertropic response via an increase in $[Ca^{2+}]_i$ (Zafari et al., 1998). AngII directly activates an NADH/NADPH oxidase in coronary vascular smooth muscle cells (Griendling et al., 1994; Fukai et al., 1998). AngII induced hypertension is also associated with an increase in vascular superoxide production (Fukai et al., 1999b). Interestingly, in a species of rat that lacks vascular superoxide dismutase (SOD) activity, angII infusion produces hypertension which appeared to be substantially more severe than that observed in rats having SOD activity. In addition, AngII-induced hypertension in rats can be prevented by liposome-encapsulated SOD which is similar to the vascular SOD. Thus, in species that express vascular SOD, such as the mouse, upregulation of endogenous vascular SOD may represent an important compensatory mechanism that blunts the blood pressure response under conditions when Ang II is chronically elevated (Carlsson et al., 1996; Laursen et al., 1997; Fukui et al., 1997, Fukai et al 1999b).

2.4. Na^+/Ca^{2+} and Na^+/H^+ exchange, and Na^+/K^+ ATPase activities

The Na^+/Ca^{2+} exchange system can move Ca^{2+} either into or out of the cytosol across the plasma membrane in exchange for Na^+. The Na^+ /Ca^{2+} exchange inhibitor KB-R7943 has been shown to ameliorate reoxygenation-induced arrhythmias which suggests that Ca^{2+} influx by the Na^+/Ca^{2+} exchanger may play a key role in reoxygenation injury (Mukai et al., 2000).

The Na^+/Ca^{2+} exchanger is also influenced by changes in the activities of the Na^+/K^+ATPase and Na^+/H^+ exchanger (Kaneko et al., 1994). Inhibition of Na^+/K^+ ATPase results in an increase in the intracellular level of Na^+ (Deitmer and Ellis, 1978a,b). An elevation of intracellular Na^+ has been shown to cause a rise in intracellular Ca^{2+} that could occur by a decrease in Ca^{2+} efflux via Na^+/Ca^{2+} exchange (Philipson and Ward, 1986). The inhibition of Na^+/K^+ ATPase also results in an intracellular acidification that is thought to be a consequence of a rise in intracellular Ca^{2+} produced via Na^+/Ca^{2+} exchange (Vaughan-Jones et al., 1983; Philipson and Ward, 1986; Kim et al., 1987; Kim and Smith, 1988). Thus, these three ion transport systems can affect each other and play important roles in Ca^{2+} handling in cardiac cells.

Kramer et al. (1984) demonstrated that Na^+/K^+ATPase activity was reduced by ROS generating systems in canine cardiac sarcolemmal membranes. Kukreja et al (1990) also showed inhibition of Na^+/K^+ ATPase activity in the presence of ROS generating agents such as H_2O_2, H_2O_2 plus Fe^{2+}, and HOCl in the system. Bhatnagar et al. (1990) observed that perfusion of frog ventricular single cells with *tert*-butylhydroperoxide (*t*-buOOH) increases intracellular Na^+ which could result from a decrease in Na^+ efflux via the Na^+/K^+ pump. On the other hand, Xie et al. (1990) reported that both Na^+/K^+ ATPase and Na^+/H^+ exchange activities were reduced by X+XO and that Na^+/H^+ exchange appeared to be more sensitive to oxidative stress than did Na^+/K^+ ATPase.

Accumulation of toxic metabolites may play an important role in regulating sarcolemmal Na^+/K^+ATPase activity in ischemia-reperfusion injury. For example, accumulation of palmitoyl carnitine and related compounds during ischemia has been suggested to cause Na^+/K^+ATPase inhibition (Pitts and Okhuysen, 1984; Kim and Akera, 1987). These compounds may account for Na^+/K^+ATPase injury that occurs during prolonged ischemia but cannot explain Na^+/K^+ATPase inhibition that develops during reperfusion (Kim et al., 1983). Acidosis due to lactate accumulation is another potential mechanism by which Na^+/K^+ATPase might be inactivated because exposure of Na^+/K^+ATPase to acidic medium may cause a permanent inactivation of the enzyme. Accumulation of lactate, however, is also rapidly reversed during reperfusion, and, therefore, cannot explain the injury to Na^+/K^+ATPase that occurs during reperfusion (Kim and Akera, 1987).

Myocardial ischemia results in an increase in $[Na^+]_i$ which may lead to intracellular Ca^{2+} overload that appears to be mediated by membrane transporters (Fligel and Frohlich, 1993). Because protein kinase C (PKC) has been shown to reduce Na^+/K^+ATPase activity, it has been postulated that pharmacological inhibition of PKC would directly increase Na^+/K^+ATPase activity, reduce $[Na^+]_i$ during ischemia and provide protection from ischemic injury (Lundmark et al., 1999). Treatment with chelerythrine, a PKC inhibitor, in rat hearts 30 min before global ischemia increased Na^+/K^+ATPase activity, reduced PKC activity in both the membrane and cytosolic fractions, and also decreased creatine kinase release upon reperfusion. The rise in $[Na^+]_i$ during ischemia was significantly reduced in the hearts treated with chelerythrine (Lundmark et al., 1999; Chien, 1999). Thus, it appears that pharmacological inhibition of PKC before ischemia may provide cardioprotection by reducing intracellular Na^+ overload, at least partly, via an increase in Na^+/K^+ATPase activity.

2.5. Ischemic preconditioning and K^+ channels

Several studies have reported effects of H_2O_2 on various ion channels and exchangers. H_2O_2 has been shown to alter the function of the delayed rectifier K^+ current, inward rectifier K^+ current and ATP sensitive K^+ current (Pignac et al., 1996). The hyperpolarization and depolarization of membrane potential by H_2O_2 apparently occur via two different mechanisms. Low H_2O_2 concentrations inhibited inward rectifying K^+ currents whereas higher H_2O_2 concentrations increased the amplitude of the outward K^+ current (Bychkov et al., 1999).

ATP dependent potassium channels (K_{ATP}) may serve an important pathophysiological function in ischemia-reperfused myocardium. This is apparent from studies with ischemia preconditioned rat hearts treated with low levels of hydrogen peroxide. In ischemic preconditioning, the preconditioned hearts develop ultrastructural damage more slowly than non-preconditioned hearts. The rate of ATP depletion in the preconditioned hearts during the initial phases of prolonged occlusion is reduced as a result of reduced ATP utilization. The rates of glycogen breakdown and anaerobic glycolysis have also been shown to decrease in preconditioned myocardium. Therefore, preservation of ATP levels or a reduction of the cellular content of toxic metabolites may be responsible for preconditioning. Activation of K_{ATP} channels could result in these favorable metabolic effects by shortening the action potential duration and attenuating membrane depolarization. By shortening the action potential duration, activation of K_{ATP} channels would decrease Ca^{2+} levels during ischemia by indirectly regulating voltage regulated Ca^{2+} channels, modulating SR Ca^{2+}ATPase, and possibly by maintaining the Na^+/Ca^{2+} exchanger, Na^+/H^+ exchanger and Na^+/K^+ ATPase in a productive mode of extruding Ca^{2+}. These effects would lead to a decrease in $[Ca^{2+}]_i$, a rapid loss of contractile activity and a decrease in ATP utilization as well as a decrease in the accumulation of metabolites that could delay cell death (Yamashita et al., 1994; Tamargo et al., 1995; Rahman et al., 1996; Miyawaki et al., 1998). Thus, enhancing or accelerating myocardial K_{ATP} channel activation with the channel openers may be a useful therapeutic strategy for the treatment of ischemic heart disease.

3. Ca^{2+} extrusion from intracellular space to extracellular space

To maintain the appropriate levels of Ca^{2+} ions in the cytosol and subcellular Ca^{2+} stores, Ca^{2+} which entered the cells from the extracellular space during the action potential has to be pumped back out again into the extracellular space. It is believed that heart sarcolemma contains two important mechanisms for extruding Ca^{2+}, the Na^+/Ca^{2+} exchanger and Ca^{2+}ATPase (Fig. 16.1) (Grover and Samson, 1997; Azma et al., 1999).

3.1. Na^+/Ca^{2+} exchange

In the heart, Na^+/Ca^{2+} exchange is thought to function primarily as a mechanism for pumping Ca^{2+} out of the cell. It is, however, possible that Na^+/Ca^{2+} exchange can promote the net entry of Ca^{2+} into the cell under certain circumstances, for example, membrane depolarization (Kaneko et al., 1994). Kutryk and Pierce (1988) showed that Na^+/Ca^{2+}

exchange activity in canine heart sarcolemmal vesicles was depressed by oxidants, for instance, the $O_2^{\cdot -}$ generating system X+XO. Furthermore, Kaneko et al. (1991) showed that various types of ROS generating systems inhibited the Na^+/Ca^{2+} exchange activity in sarcolemmal membrane vesicles isolated from rat, bovine, canine and porcine hearts.

3.2. Ca^{2+}ATPase of sarcolemmal membrane

Both the Ca^{2+} ATPase activity and ATP-dependent Ca^{2+} accumulation in rat heart sarcolemmal inside-out vesicles were reduced by X+XO, H_2O_2 or H_2O_2 plus Fe^{2+} in a dose and time dependent manner. Since Ca^{2+} pump ATPase in cardiac sarcolemma is intimately involved in the extrusion of Ca^{2+} across the cell membrane, the inhibition of sarcolemmal Ca^{2+} pump ATPase by ROS could lead to a decrease in Ca^{2+} extrusion from the cytosol resulting in an increase in cytosolic Ca^{2+} concentration (Hess et al., 1981, 1983).

3.3. Effect of ROS on sulfhydryl groups

The inhibition of sarcolemmal Ca^{2+} pump activity by ROS was prevented by dithiothreitol (DTT) and cysteine. N-ethylmaleimide (NEM) inhibition of Ca^{2+} pump activity was prevented upon pretreatment with dithiothreitol (DTT) and cysteine. Heart sarcolemmal sulfhydryl groups were reduced by various types of ROS both in a dose and time dependent manner (Kaneko et al., 1989, 1994). Free radical scavengers showed protective effects on sulfhydryl group depression by ROS. Furthermore, there was a significant correlation between changes in sarcolemmal Ca^{2+} pump ATPase activity and sarcolemmal sulfhydryl groups (Kaneko et al., 1989, 1994). These results indicate that ROS depress the heart sarcolemmal Ca^{2+} pump activity by modifying the sulfhydryl groups in the sarcolemmal membrane. In addition, because sulfhydryl groups are known to regulate other membrane-bound ion transporting systems such as Na^+/K^+ ATPase, Na^+/Ca^{2+} exchange and voltage dependent Ca^{2+} channels in sarcolemma, as well as Ca^{2+} release protein(s) and Ca^{2+} pump ATPase in the sarcoplasmic reticulum, it seems possible that the oxidation of sulfhydryl groups in these membrane-bound ion transport systems may lead to a general depression of all these activities by ROS (Kaneko et al., 1994; Coetzee et al., 1994; Ingbar and Wendt, 1997).

Not only the direct reactions of ROS with membrane-bound enzyme proteins but also the peroxidation of membrane lipids by ROS can affect the enzyme activities (Chakraborti et al., 1989; Chakraborti and Chakraborti, 1995; Chakraborti et al., 1996, 2000). Accumulation of hydroperoxides produced by peroxidation of membrane phospholipids can inactivate enzymes by oxidizing amino acid residues or by mediating polypeptide chain polymerization reactions (Ghosh et al., 1996a,b; Chakraborti et al., 1995).

Under physiological conditions, mammalian cells exist in a redox environment which reflects a balance between the activities of oxidant and reductant chemical species. Oxidant stress may be operationally defined as the condition in which one or more oxidant moieties elicit a measurable biological response. This is especially pertinent to the vasculature in which blood borne compounds can shift the redox balance within the endothelial cells to a more oxidized state and alter cellular function (Ghosh et al., 1996a,b). Exposure of vascular endothelial cells to the model oxidant *t*-buOOH results in an oxidant stress which is characterized by inhibition of Ca^{2+} signaling and that occurs in three temporal phases. Initially, *t*-buOOH inhibits agonist-stimulated Ca^{2+} influx. Subsequently, *t*-buOOH inhibits agonist-stimulated release of Ca^{2+} from internal stores, an effect related to a decrease in the production of D-myoinositol 1,4,5 trisphosphate/(IP_3) rather than depletion of intracellular Ca^{2+} stores. Finally, basal $[Ca^{2+}]_i$ is significantly elevated, likely due to inhibition of plasma membrane Ca^{2+}ATPase (Chakraborti et al., 1996, Chakraborti and Chakraborti, 1998).

Several components of Ca^{2+} signaling machinery, including the plasmalemmal Ca^{2+} pump and the IP3 receptor, possesses functionally important sulfhydryl groups that may be influenced by intracellular thiol status (Bowie and O'Neill, 2000). Availability of reduced glutathione is a major determinant of the rate and extent by which oxidants alter Ca^{2+} dependent signal transduction in myocardial cells (Elliot et al., 1995).

3.4. Effect of ROS on protein fragmentation

The alteration in function observed upon oxidation of sarcoplasmic reticulum vesicles could be due to direct effects on the Ca^{2+} ATPase. Electrophoretic analysis of purified Ca^{2+} ATPase after *in vitro* oxidation indicated a decrease in intensity of the 110 kDa Ca^{2+}ATPase protein and the appearance of low molecular weight peptides. Based on this observation, Castilho et al. (1996) proposed that impairment of function of the Ca^{2+} pump may be related to amino acid oxidation and fragmentation of Ca^{2+} ATPase. However, this *in vitro* study needs to be verified in appropriate *in vivo* systems before any conclusion can be made.

4. Ca^{2+} translocating processes of sarcoplasmic reticulum

Hess et al. (1981, 1983) studied the effects of various types ROS on Ca^{2+} transporting systems in cardiac sarcoplasmic reticulum. They observed that Ca^{2+} pump ATPase activity and steady state Ca^{2+} uptake were depressed by ROS. Direct measurement of the number and turnover of the pump units indicated that the number of the units was unchanged, but the turnover rate was decreased by ROS. Furthermore, exposure to ROS increased the passive permeability of the sarcoplasmic reticular vesicles to Ca^{2+} but the increased permeability *per se* appears insufficient to explain the effects of ROS on Ca^{2+} pump ATPase (Kass and Orrenius, 1999). Holmberg et al. (1991) have studied the effects of free radicals (produced by illumination of rose Bengal) on sheep cardiac sarcoplasmic reticular Ca^{2+} release channels that were inserted into synthetic lipid bilayers. They found that Ca^{2+} release channels open probability was increased initially, and that continued illumination resulted in a reversible loss of channel function and subsequent bilayer disruption. They also showed that ryanodine binding in isolated cardiac membrane was reduced by ROS, with associated degradation of a 340 kD protein which is thought to be the ryanodine receptor and sarcoplasmic reticular Ca^{2+} release channel complex Holmberg et al (1991). Previous studies indicated that oxidizing agents induced rapid Ca^{2+} efflux from actively loaded sarcoplasmic reticular vesicles isolated from rabbit skeletal muscle. These studies suggested that when sarcoplasmic reticulum is exposed to ROS, Ca^{2+} release from sarcoplasmic reticulum to the cytosol is promoted and Ca^{2+} sequestration from cytosol into the lumen of the sarcoplasmic reticulum is inhibited (Aviv, 1996; Natarajan et al., 1998).

Two principal pathways of Ca^{2+} release from the sarcoplasmic reticulum of myocardial cells have been described. One pathway is dependent on IP_3 and the other pathway is sensitive to Ca^{2+} and regulated by caffeine and ryanodine. It was suggested that the pathway of IP_3-dependent Ca^{2+} release from the sarcoplasmic reticulum may be sensitive to ROS such a O_2^- (Elliot et al., 1995; Gibson et al., 1998).

Treatment with linoleic acid hydroperoxide (LOOH) has been shown to cause a rapid but transient increase in intracellular free Ca^{2+} concentration in the presence of extracellular Ca^{2+}. In the absence of extracellular Ca^{2+}, the increase in intracellular free Ca^{2+} caused by LOOH was of lesser magnitude and immediately returned to basal levels. The LOOH evoked rise in intracellular free Ca^{2+} concentration was not mediated through IP_3 sensitive pool(s) via stimulation of IP_3 formation. However, pretreatment with LOOH strongly inhibited the rise in intracellular free Ca^{2+} concentration that occurred upon the subsequent addition of agents that mediate Ca^{2+} release from IP_3 sensitive pool (s) (Suzuki et al., 1997). These findings suggested that reuptake of Ca^{2+} into intracellular membrane pool(s) may be reduced in the presence of LOOH and/or the availability of Ca^{2+} from agonist-sensitive sites may be inhibited. Additionally, like *t*-buOOH, the ability of LOOH to oxidize cellular GSH and increase formation of mixed disulfides may release Ca^{2+} from IP_3 sensitive pool(s) even in the absence of IP_3 formation (Suzuki et al., 1997; Castro and Bhatnager, 1993).

5. Protein bound Ca^{2+}

Recently the role of protein bound Ca^{2+} has been recognized as a source of elevated $[Ca^{2+}]_i$ under oxidant stimulation in myocardial cells. Exposure to *t*-buOOH resulted in an increase in $[Ca^{2+}]_i$ and also translocation of membrane-associated cytoskeletal annexins from the membrane to the cytosol (Hoyal and Forman, 1995).The annexins are a family of Ca^{2+} binding proteins that reversibly bind Ca^{2+} in the presence of phospholipid. A study using confocal microscopy demonstrated a similar pattern of Ca^{2+} movement from plasma membrane to the cytosolic space in response to *t*-buOOH (Suzuki et al., 1997). These studies suggested that protein bound Ca^{2+} plays an important role in oxidant-mediated intracellular signaling.

The signaling function of Ca^{2+} is controlled by reversible complexation to specific proteins. Most of the soluble proteins belong to the exchange factor "E-F hand" family and act as decoders of the Ca^{2+} information. They do so by changing conformation twice, once upon complexing Ca^{2+} and later upon interacting with target enzymes (Suzuki et al., 1997). The most important among the "E-F hand" proteins known to date is calmodulin which plays an important role in Ca^{2+} signaling phenomena under oxidant stimulation in cardiovascular systems (Pereira et al., 1992; Carafoli, 1994).

Thus, the fine and rapid tuning of cellular Ca^{2+} is performed essentially by pumps, although plasma membrane Na^+/Ca^{2+} exchanger and Ca^{2+} pumps are also important (Fig. 16.1). Long-term, low affinity Ca^{2+} regulation, particularly in the presence of pathological increases in Ca^{2+} entry, is probably performed by mitochondrial uptake/release systems (Luft and Landau, 1995; Chakraborti et al., 1999b).

6. Mitochondrial Ca^{2+} dynamics

Mitochondria are active in the continuous generation of reactive oxygen species (ROS). An alteration in mitochondrial Ca^{2+} concentration has been suggested to be an important event in the induction of oxidative stress (Luft and Landau, 1995; Chakraborti et al., 1999b). Maintenance of low cytosolic Ca^{2+} is necessary for proper functioning of cells. Mitochondria transport Ca^{2+} (i) to regulate cytosolic Ca^{2+}; (ii) to serve as a store of Ca^{2+} when its concentration in the cytosol is excessive; (iii) to serve as a releasable source of activator Ca^{2+}; and (iv) to regulate mitochondrial matrix Ca^{2+} and thereby control the activity state of Ca^{2+} sensitive metabolic enzymes. Thus, mitochondria play an important role in controlling cellular Ca^{2+} dynamics (Luft and Landau, 1995; Chakraborti et al., 1999b; Jornot et al., 1999).

The spontaneous discharge of Ca^{2+} from mitochondria is associated with the following sequence of events: (i) increased nonspecific permeability of the mitochondrial inner membrane; (ii) swelling of the mitochondria; (iii) loss of K^+ from the matrix; (iv) loss of matrix adenine nucleotides; (v) oxidation, hydrolysis or leakage of matrix nicotinamide adenine nucleotides; (vi) stimulation of the inner membrane phospholipase A2 (PLA2) activity and accumulation of unsaturated fatty acids; and (vii) collapse of membrane potential ($\Delta\psi$) (Chakraborti et al., 1999b).

In principle, Ca^{2+} can leave mitochondria by different ways: by nonspecific leakage through the inner membrane by "pore formation", by changes in membrane lipid phase, by reversal of the uniport influx carrier, by the specific Ca^{2+}/H^+ (or Na^+) antiport system, by channel-mediated release pathways or by a combination of two or more of these pathways (Chakraborti et al., 1999b). The mitochondrial Ca^{2+} cycle is schematically represented in Fig. 16.2. Additionally, the release of Ca^{2+} from mitochondria can also occur either by oxidation of internal nicotinamide adenine nucleotides to ADP ribose and nicotinamide or by oxidation of thiols in membrane proteins (Richter et al., 1995).

In cardiac injury by ischemic insult, mitochondrial damage is observed early in the sequence of pathological events, as evidenced by an increase in mitochondrial swelling and a decrease in respiratory rate (Roychoudhury et al., 1996a). Several classes of ion channel activities are associated with the outer and inner membranes of the mitochondrion. In guinea pig working hearts subjected to global ischemia, pretreatment with Ca^{2+} channel

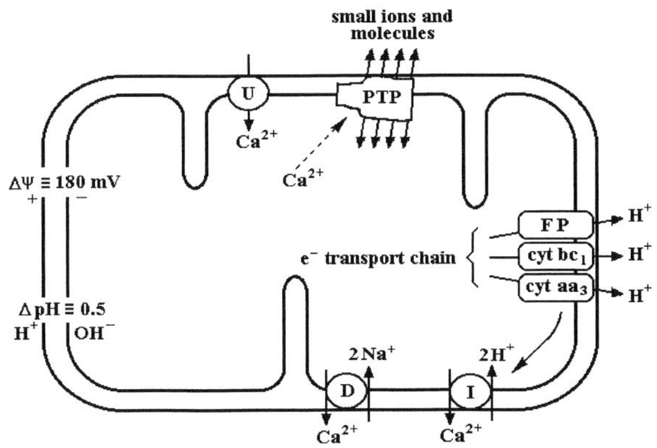

Fig. 16.2. The Ca^{2+} transport system in mitochondria. The mitochondrial Ca^{2+} uniporter (U) facilitates the transport of Ca^{2+} in an inward direction down the electrochemical gradient of this ion. The Na+ independent influx mechanism (I) is depicted here as an active $Ca^{2+}/2H^+$ exchanger, receiving energy from the electron transport chain (ETC). The Na^+-dependent efflux mechanism (D) is depicted here as a $Ca^{2+}/2Na^+$ exchanger. The Ca^{2+} activated permeability transition pore (PTP) is also shown [Taken from Gunter and Gunter (1994) with permission].

antagonists of different subclasses markedly improved the recovery of myocardial function during reperfusion. Nifedepine reduces the left ventricular stiffness, improves the left ventricular compliance and decreases the Ca^{2+} content of the left ventricular myocardial mitochondria (Uceda et al., 1995). In agreement with these observations, other investigators also demonstrated that the administration of Ca^{2+} channel blockers such as nifedepine, verapamil and diltiazem prior to a period of ischemia decreased the deleterious effects evoked by myocardial ischemia and reperfusion in isolated hearts of rats and rabbits (Chakraborti et al., 1999b; Roychoudhury et al., 1996b). A megachannel in the inner mitochondrial membrane known as the "permeability transition pore", may be opened by high concentrations of inorganic phosphate due to the hydrolysis of creatine phosphate under oxidative stress (e.g. reoxygenation injury to vascular cells) (Chakraborti et al., 1999b).

Ca^{2+} release from mitochondria by the oxidized pyridine nucleotides was postulated some time ago. Lotscher et al. (1979) demonstrated that *t*-buOOH-induced release of accumulated Ca^{2+} is an electroneutral process dependent on the oxidation of NAD(P)H. Reversal of *t*-buOOH-induced Ca^{2+} efflux can be achieved by the addition of $NAD(P)^+$ reducing agents, which suggested the existence of a metabolic link between pyridine nucleotides and Ca^{2+} fluxes (Lotscher et al., 1979; Chakraborti et al., 1999b). Whether the role of oxidant-induced Ca^{2+} release from mitochondria is limited to the mechanism of oxygen toxicity or may serve as a component of cell signaling is currently unknown.

Previous research indicated that oxidative stress caused a permeability transition of the inner mitochondrial membrane, which involves increased permeability to ions, mitochondrial swelling, uncoupling of oxidative phosphorylation and collapse of $\Delta\psi$ (Luft and Landau, 1995; Richter et al., 1995; Chakraborti et al., 1999b). Evidence suggested that this transition may be due to some change in the deacylation–reacylation cycle of the inner mitochondrial membrane phospholipids. Oxidative stress inhibited reacylation of lysophosphatides, and the deacylation was stimulated by enhanced phospholipase A2 (PLA2) activity in response to mobilizable Ca^{2+} in mitochondria. The accumulation of lysophospholipids and free fatty acids resulted in an alteration in the permeability of the inner mitochondrial membrane. This permeability transition can be stimulated by exogenous lysophospholipids and ameliorated by PLA2 inhibitors such as br-phenacylbromide and mepacrine (Gunter and Pfeiffer, 1990). Therefore, it is possible that oxidants stimulate an increase in mobilizable free Ca^{2+} in the mitochondria and that plays an important role in activating the PLA_2 activity in the inner mitochondrial membrane (Chakraborti et al 1999b). Additionally Roychoudhury et al. (1996a) have demonstrated that redox regulation of pyridine nucleotides, but not glutathione, regulates Ca^{2+} release from mitochondria under oxidant triggered conditions. It is conceivable that both the permeability transition and redox regulation of pyridine nucleotides theories of Ca^{2+} homeostasis could prevail and the extent of oxidant stress may dictate the predominance of one pathway over the other under different pathophysiological conditions.

7. Consequences of oxidant induced increase in $[Ca^{2+}]_i$

Over the past few years, a number of reports have appeared in which components of the Ca^{2+} homeostasis and signaling machinery have been identified as important regulators of apoptosis. For example, expression of the Ca^{2+} binding protein calbindin provides protection against a number of apoptotic stimuli. Similarly, the expression of the inositol trisphosphate receptor ($InsP_3R$) is increased many-fold during apoptosis. Apoptosis was found to be stimulated by TNF-α and ultraviolet light, which at least partly, act via Ca^{2+}/calmodulin dependent kinase and inhibition of kinase activity was found to prevent apoptosis (Suzuki et al., 1997; Chakraborti et al., 1999). In agreement with this, a recent discovery indicated that Ca^{2+}/calmodulin dependent kinase induces apoptosis when overexpressed (Richter et al., 1995; Suzuki et al., 1997). It is now apparent that Ca^{2+} signaling machinery plays a crucial role in apoptosis. However, the exact molecular mechanisms(s) by which Ca^{2+} participates in regulating apoptosis is currently unknown.

Many recent studies of apoptosis have focused on caspases (Vissers et al., 1999). However, there is considerable evidence that additional proteases including serine proteases and members of the calpain family of Ca^{2+}-activated proteases participate in TNFα-mediated apoptosis (Chakraborti et al., 1999b). The cellular targets for calpains in apoptosis are not well known. Fodrin and lamins are cleaved by calpains under oxidant stress elicited by TNFα (Suzuki et al., 1997). Ca^{2+} has recently been shown to increase caspase 3 activity in a cell free system when added to nonapoptotic cell cytosol and to elicit nuclear morphologic changes and DNA fragmentation (Chakraborti et al., 1999b). These findings indicated a role of an increase in $[Ca^{2+}]_i$ in caspase-mediated apoptosis.

A connection between alteration of endoplasmic reticular (ER) Ca^{2+} and apoptosis has recently been suggested. Inhibition of sarcoendoplasmic reticulum Ca^{2+}ATPase (SERCA) by thapsigargin (Tg) directly leads to depletion of the intracellular Ca^{2+} store and induction of apoptosis. Since oxidants deplete Tg sensitive Ca^{2+} stores, it may be suggested that a decrease in Ca^{2+} efflux from the ER can delay the onset of apoptosis. How intracellular Ca^{2+} fluxes signal apoptosis is presently unknown. One possible mechanism involves activation of oxidative stress responsive nuclear transcription factors, such as NF-κB or AP-1. Furthermore, C-fos, the early response gene that has been implicated in signaling apoptosis, is transcriptionally regulated by ER Ca^{2+} release in response to ROS produced during I-R injury to myocardium (Distelhorst et al., 1996, Kuo et al., 1998).

7.1. AP-1 transcription factors

The AP-1 transcription factor is comprised of a heterodimer of fos and jun proteins which are products of c-fos and c-jun protooncogenes. Oxidants induce expression of these genes and modulate signal transduction mechanisms associated with apoptosis and oncogenesis through the involvement of an increase in $(Ca^{2+})_i$. C-fos mRNA was found to be induced by ROS generated, for example, by X+XO (Chakraborti et al., 1999a, 1998b).

7.2. NF-κB transcription factors

The predominant inactive form of NF-κB exists as a trimer that consists of p65, p50 and IκBα subunits which dissociates to form activated NF-κB (p^{65}/p^{50} dimer) and that migrates to the nucleus for binding with the relevant DNA (Chakraborti and Chakraborti, 1998; Chakraborti et al., 1998). NF-κB binding sites were also found in genes for a variety of chemokines and cell adhesion molecules including murine colony stimulating factor (CSF-1), human monocyte chemotactic protein-1 (MCP-1), and human leukocyte adhesion molecule-1. NF-κB, therefore, gained considerable attention as a regulator of atherosclerotic lesions which may have resulted by Ca^{2+} signaling phenomena under oxidant-triggered conditions (Jordan et al., 1999). Advanced glycation end products (AGE) which may be responsible for the pathogenesis of diabetes-induced atherosclerosis, were found to activate NF-κB, presumably by Ca^{2+} generated through ROS (Chappey et al., 1997). Oxidants were shown to cause endothelial adhesion by Ca^{2+} via induced intracellular cell adhesion moleculae-1 (ICAM-1) gene expression (Jordan et al., 1999).

Mitochondria take up and buffer cytosolic Ca^{2+} when its concentration increases to levels that allow the operation of the mitochondrial low-affinity uptake system. Because the influx of Ca^{2+} across a damaged plasma membrane is a common pathway by which cells are killed, mitochondria act as a safety device against a toxic increase in cytosolic Ca^{2+} (Chakraborti et al., 1999b). When this ability to retain Ca^{2+} is compromised or lost, the injured cells will die.

Intramitochondrial free Ca^{2+} has been considered a mechanism that controls the rate of ATP output by mitochondria independently of substrate stimulation or product inhibition. The dehydrogenases such as pyruvate dehydrogenase, isocitrate dehydrogenase and oxoglutarate dehydrogenase were activated by Ca^{2+}. In the heart, mitochondria are able to control matrix Ca^{2+} so that it is roughly the same as an average of cytosolic Ca^{2+} under physiological conditions. A rapid and frequent Ca^{2+} pulse would engender a higher average level of mitochondrial matrix Ca^{2+}, a greater activation of dehydrogenases and faster ATP production. In contrast, a decrease in matrix Ca^{2+} due to its release from mitochondria causes inhibition of the dehydrogenases and reduces ATP production (Chakraborti et al., 1999b).

An oxidant-induced hydrolytic product of NAD, the cADP ribose produced by the mitochondrial enzyme NAD^+-glycohydrolase, stimulates Ca^{2+} release from the ryanodine sensitive Ca^{2+} release pool(s) of the endoplasmic reticulum which further contributes to the Ca^{2+} overload in the cell. The increase in cytosolic Ca^{2+} may play an important role in initiating the final common pathway(s) for toxic cell death (Chakraborti et al., 1999b). Thus, the effects of oxidants appear to be, at least partly, due to an alteration of mitochondrial Ca^{2+} dynamics.

8. Future prospects

The available data suggest that mitochondrial Ca^{2+} efflux pathway(s) is a central coordinating event of the apoptotic effector phase. The hypothesis predicts that various pathways of apoptosis converge at the level of Ca^{2+} efflux. When Ca^{2+} efflux is triggered, a series of common pathways of apoptosis are initiated, each of which may be sufficient to destroy the cell. Further studies are needed to explore the nature of the apoptosis-inducing pathway(s), the precise mechanism(s) of Ca^{2+} efflux, the molecular biology of apoptosis inducing factor formation and release and the essential molecular targets of apoptosis triggering protease(s). Clarification of these issues are important for better understanding of the effects produced by oxidant stress and associated molecular mechanisms.

Ca^{2+} overload occurs when hearts are exposed to an excess amount of ROS. Although ATP independent Ca^{2+} binding is increased, Ca^{2+} influx through Ca^{2+} channels marginally increase in the presence of ROS. Another possible pathway through which Ca^{2+} can enter the myocytes is the Na^+/H^+ exchanger. Although the activities of the Na^+/K^+ ATPase and the Na^+/H^+ exchanger are inhibited by ROS, it is unknown whether oxidant stress raises $[Na^+]_i$. This point may be illuminated upon understanding the importance of the Na^+/Ca^{2+} exchanger in the Ca^{2+} influx process from extracellular spaces. Another question is which way does the Na^+/Ca^{2+} exchanger work under oxidative stress? Net influx or efflux? Answers to these questions will have important implications for elucidating the mechanisms that regulate Ca^{2+} dynamics in the cardiovascular system under oxidant stress.

Phosphotyrosine phosphatase (PTPs) serve as important regulators of cellular signal transduction pathways. PTPs are sensitive targets of oxidative stress and may be inhibited by treatments that induce intracellular oxidation. The effects of PTP inactivation by ROS are amplified by the redox-linked activation of key protein tyrosine kinases (PTKs) thereby leading to the initiation of phosphotyrosine signaling cascades (Krejsa and Schieven, 1998; Natarajan et al., 1998). This results in the accumulation of protein phosphotyrosine, the generation of second messengers, the activation of downstream kinases and the nuclear translocation of nuclear factor kappa-B (NF-κB) through an increase in $[Ca^{2+}]_i$ (Krejsa and Schieven, 1998; Natarajan et al., 1998). It would be important to undertake studies in myocardial cells using PTP inhibitors with the goal of identifying crucial phosphotyrosine signaling pathways leading to changes in cellular function and to distinguish between effects of oxidative stress on PTK activation versus PTP inhibition.

Oxidants such as H_2O_2 have also recently been reported to stimulate the activity of mitogen activated protein kinases (MAPKs) and the expression of nuclear transcription factors such as c-fos and c-jun (Chakraborti and Chakraborti, 1998). Recent research suggests that arachidonic acid metabolites play a role in stimulating the activities of MAPKs and the nuclear transcription factors which may profoundly modulate cell growth and differentiation, endothelial-monocyte adhesion and atherosclerosis. Since a rise in $[Ca^{2+}]_i$ under oxidant stress plays an important role in the stimulation of phospholipase A2 activity resulting in an increase in the production of arachidonic acid metabolites and the subsequent activation of NF-κB (Chakraborti and Chakraborti, 1998), one important aspect of future work will be to ascertain the mechanism(s) by which oxidant-mediated increases in $[Ca^{2+}]_i$ trigger the action of MAPKs and nuclear transcription factors in the myocardium.

Acknowledgements

This work was supported partly by the Indian Council of Medical Research (ICMR), New Delhi, and The Department of Biotechnology (DBT), New Delhi and the Defence Research and Development Organization (DRDO), New Delhi.

References

Aviv, A. (1996). Recent advances in cellular Ca^{2+} homeostasis: Implications to altered regulation of cellular Ca^{2+} and Na^+/H^+ exchange in hypertension. Curr. Opin. Cardiol. 11, 477–482.

Azma, T., Saeki, N. and Yuge, O. (1999). Cytosolic Ca^{2+} movements of endothelial cells exposed to reactive oxygen intermediates: role of hydroxyl radical-mediated redox alteration of cell membrane Ca^{2+} channels. Br. J. Pharmacol. 126, 1462–1470.

Bers, D.M. and Langer D.A. (1979). Uncoupling cation effects on cardiac contractility and sarcolemmal Ca^{2+} binding. Am. J. Physiol. 237, H332–H341.

Bers, D.M., Philipson, K.D. and Langer, G.A. (1981). Cardiac contractility and sarcolemmal calcium binding in several cardiac muscle preparations. Am. J. Physiol. 240, H576–H583.

Bhatnagar, A., Srivastava, S.K. and Szabo, G. (1990). Oxidative stress alters specific membrane currents in isolated cardiac myocytes. Circ. Res. 67, 535–549.

Bowie, A. and O'Neill, L.A. (2000). Oxidative stress and nuclear factor kappaβ activation: a reassessment of the evidence in the light of recent discoveries. Biochem. Pharmacol. 59, 13–23.

Bychkov, R., Pieper, K., Reid, C., Milosheva, M., Bychkov, E., Luft, F.C. and Haller, H. (1999). Hydrogen peroxide, potassium current, and membrane potential in human endothelial cells. Circulation 99, 1719–1725.

Carafoli, E. (1994). The signaling function of calcium and its regulation. J. Hypertension 12, S47–S56.

Carlsson, L.M., Marklund, S.L. and Edlund, T. (1996). The rat extracellular superoxide dismutase dimer is converted to a tetramer by the exchange of a single amino acid. Proc. Natl. Acad. Sci. USA 93, 5219–5222.

Castilho, R.F., Alves, P.C.C., Vercesi, A.E. and Ferreira, S.T. (1996). Oxidative damage to sarcoplasmic reticulum Ca^{2+} pump induced by Fe^{2+}/H_2O_2/ascorbate is not mediated by lipid peroxidation of thiol oxidation and leads to protein fragmentation. Mol. Cell. Biochem. 159, 105–114.

Castro, G.J. and Bhatnagar, A. (1993). Effects of extracellular ions and modulators of Ca^{2+} transport on survival of tertiary butyl hydroperoxide exposed cardiac myocytes. Cardiovasc. Res. 27, 1873–1881.

Chakraborti, S. and Chakraborti, T. (1998). Oxidant-mediated activation of mitogen activated protein kinases and nuclear transcription factors in the cardiovascular system: an overview. Cell Signal. 10, 675–683.

Chakraborti, S., Batabyal, S.K. and Chakraborti., T. (1995). Role of hydroxyl radical in the stimulation of arachidonic acid release caused by H_2O_2 in pulmonary smooth muscle cells. Mol. Cell. Biochem. 146, 91–98.

Chakraborti, S., Chakraborti, T. and Shaw, G. (2000). β-Adrenergic mechanisms in cardiac diseases: a perspective. Cell. Signal.12, 499–513.

Chakraborti, S., Michael, J.R. and Gurtner, G.H. (1989). Oxidant-mediated activation of PLA_2 in pulmonary endothelium. Am. J. Physiol. 257, L430–L437.

Chakraborti, T. and Chakraborti, S. (1995). Down regulation of protein kinase C attenuates the oxidant H_2O_2-mediated activation of phospholipase A_2 in pulmonary smooth muscle cells. Cell. Signal. 7, 75–83.

Chakraborti, T., Das, S., Mandal, M., Roychoudhury, S. and Chakraborti, S. (1999) Oxidant, mitochondria and calcium: an overview. Cell Signal. 11, 77–85.

Chakraborti, T., Ghosh, S.K., Michael, J.R. and Chakraborti, S. (1996). Role of an aprotinin sensitive protease in the stimulation of Ca^{2+} ATPase by superoxide radical ($O_2^{\cdot-}$) in microsomes of pulmonary smooth muscle. Biochem. J. 317, 885–890.

Chakraborti, T., Ghosh, S.K., Michael, J.R., Batabyal, S.K. and Chakraborti, S. (1998). Targets of oxidative stress in cardiovascular system; an overview. Mol. Cell. Biochem. 187, 1–10.

Chappey, O., Dosquet, C., Wautier, M.P. and Wautier, J.L. (1997). Advanced glycation and products, oxidative stress and vascular lesions. Eur. J. Clin. Invest. 27, 97–108.

Chien, K.R. (1999). Stress pathways and heart failure. Cell 98, 555–558.

Coetzee, W.A., Ichikawa, H. and Hearse, D.J. (1994). Oxidant stress inhibits Na^+/Ca^{2+} exchange current in cardiac myocytes: mediation by sulfhydryl groups? Am. J. Physiol. 266, H909–H919.

Deitmer, J.W. and Ellis, D. (1978a). Change in intracellular sodium activity of sheep heart Purkinje fibres produced by calcium and other divalent ions. J. Physiol. 277, 437–453.

Deitmer, J.W. and Ellis, D. (1978b). The intracellular sodium activity of cardiac Purkinje fibres during inhibition and reactivation of the Na^+/K^+ pump. J. Physiol. 284, 241–259.

Dhein, S., Salameh, A., Berkels, R. and Klaus, W. (1999). Dual mode of action of dihydropyridine calcium antagonists: a role for nitric oxide. Drugs 58, 397–404.

Distelhorst, C.W., Lam, M. and McLormick, T.S. (1996). Bcl-2 inhibits hydrogen peroxide induced ER Ca^{2+} pool depletion. Oncogene 16, 2051–2055.

Fabiato, A. and Fabiato, F. (1978). Calcium induced release of calcium from the sarcoplasmic reticulum of skinned cells from adult human, dog, cat, rabbit, rat and frog hearts and from fetal and newborn rat ventricles. Ann. N.Y. Acad. Sci. 307, 491–522.

Fabiato, A. (1983). Calcium induced release of calcium from the cardiac sarcoplasmic reticulum. Am. J. Physiol. 245, C1–C14.

Finkel, T. (1998). Oxygen radicals and signaling. Curr. Opin. Cell. Biol. 10, 248–253.

Fliegel, L. and Frohlich, O. (1993). The Na^+/H^+ exchanger: an update on structure, regulation and cardiac physiology. Biochem. J. 296, 273–285.

Freeman, B.A. and Crapo, J.D. (1982). Biology of disease: free radicals and tissue injury. Lab. Invest. 47, 412–426.

Fukai, U.M., Alexander, R.W., Akers, M. and Griendling, K.K. (1998). p^{38} mitogen activated protein kinase is a critical component of the redox sensitive signaling pathways activated by angiotensin II. Role in vascular smooth muscle cell hypertrophy. J. Biol. Chem. 273, 15022–15029.

Fukai, M.U., Alexander, R.W., Akers, M., Yin, Q.Q., Fujio, Y., Walsh, K. and Griendling, K.K. (1999). Reactive oxygen species mediate the activation of Akt/protein kinase B by angiotensin II in vascular smooth muscle cells. J. Biol. Chem. 274, 22699–22704.

Fukai, T., Siegfried, M.L., Fukai, M.U., Griendling, K.K. and Harrison, D.G (1999). Modulation of extracellular superoxide dismutase expression by angiotensin II and hypertension. Circ. Res. 85, 23–28.

Fukui, K., Kanedo, M., Takahashi, E., Washio, M. and Doi, K. (1994). Protective effects of sulfhydryl compounds on HOCl induced intracellular Ca^{2+} increase in single rat ventricular myocytes. J. Mol. Cell. Cardiol. 26, 455–461.

Fukui, T., Ishizaka, N., Rajagopalan, S., Laursen, J.B., Capers, Q., Taylor, W.R., Harrison, D.G., de Leon, H., Wilcox, J.N. and Griendling, K.K. (1997). p^{22}phox mRNA expression and NADPH oxidase activity are increased in aortas from hypertensive rats. Circ. Res. 80, 45–51.

Gan, X.T., Cook, M.A., Moffat, M.P. and Karmazyn, M. (1996). Transient ischemia in the presence of an adenosine deaminase plus a nucleotide transport inhibitor confers protection against contractile depression produced by hydrogen peroxide. Possible role of glycogen. J. Mol. Cell. Cardiol. 28, 1165–1176.

Ghosh, S.K., Chakraborti, T., Michael, J.R. and Chakraborti, S. (1996). Oxidant-mediated proteolytic activation of Ca^{2+} ATPase in microsomes of pulmonary smooth muscle. FEBS Lett. 387, 171–174.

Ghosh, S.K., Chakraborti, T., Banerjee, A.B., Roychoudhury, S. and Chakraborti, S. (1996). Role of hydroxyl radical in superoxide caused microsomal lipid peroxidation: protective effect of anion channel blocker J. Biosci. 21, 35–43.

Gibson, A., McFadzean I., Wallace, P. and Wayman, C.P. (1998). Capacitative Ca^{2+} entry and the regulation of smooth muscle tone. Trends Physiol. Sci. 19, 266–269.

Griendling, K.K., Minieri, C.A., Ollerenshaw, J.D. and Alexander, R.W. (1994). Angiotensin II stimulates NADH and NADPH oxidase activity in cultured vascular smooth muscle cells. Circ. Res. 74, 1141–1148.

Grover, A.K. and Samson, S.E. (1997). Peroxide resistance of ER Ca^{2+}-pump in endothelium: implications to coronary artery function. Am. J. Physiol. 273, C1250–C1258.

Gunter, K.K. and Gunter, T.E. (1994). Transport of calcium by mitochondria. J. Bioenerg. Biomembr. 26, 471–85.

Gunter, T.E. and Pfeiffer, D.R. (1990). Mechanisms by which mitochondria transport calcium. Am. J. Physiol.

258, C755–C786.

Hess, M.L., Krause, S. and Kontos, H.A. (1983). Mediation of sarcoplasmic reticulum disruption in the ischemic myocardium: proposed mechanism by the interaction of hydrogen ions and oxygen free radicals. Adv. Exp. Biol. Med. 161, 377–389.

Hess, M.L., Okabe, E. and Kontos, H.A. (1981). Proton and free oxygen radical interaction with the calcium transport system of cardiac sarcoplasmic reticulum. J. Mol. Cell. Cardiol. 13, 767–772.

Holmberg, S.R., Cumming, D.V., Kusama, Y., Hearse, D.J., Poole-Wilson, P.A., Shattock, M.J. and Williams, A.J. (1991). Reactive oxygen species modify the structure and function of the cardiac sarcoplasmic reticulum calcium-release channel. Cardioscience 2, 19–25.

Hoyal, C.R. and Forman, H.J. (1995). Calcium binding proteins as a potential source of elevated $[Ca^{2+}]_i$ during exposure to oxidative stress. Presented at the Annual Meeting of the Oxygen Society, Pasadena, CA, USA.

Huang, W.H., Wang, Y. and Askari, A. (1992). (Na^+–K^+)-ATPase: inactivation and degradation induced by oxygen radicals. Int. J. Biochem. 24, 621–626.

Ingbar, D.H. and Wendt, C.H. (1997). The sodium pump and oxidant stress: if only it were too simple. J. Lab. Clin. Med. 130, 119–122.

Jordan, J.E., Zhao, Z.Q. and Vinten-Johansen, J. (1999). The role of neutrophils in myocardial ischemia-reperfusion injury. Cardiovasc. Res. 43, 860–878.

Jornot, L., Macchler, P., Wollheim, C.B. and Junod, A.F. (1999). Reactive oxygen metabolites increase mitochondrial calcium in endothelial cells: implication of the Ca^{2+}/Na^+ exchanger. J. Cell Sci. 112, 1013–1022.

Kaminishi, T., Matsuoka, T., Yanagishita, T. and Kako, K.J. (1989). Increase vs. decrease of calcium uptake by isolated heart cells induced by H_2O_2 vs. HOCl. Am. J. Physiol. 256, C598–C607.

Kaneko, M., Chapman, D.C., Ganguly, P.K., Beamish, R.E. and Dhalla, N.S. (1991). Modifications of cardiac adrenergic receptors by oxygen free radicals. Am. J. Physiol. 260, H821–H826.

Kaneko, M., Elimban, V. and Dhalla, N.S. (1989). Mechanism for depression of heart sarcolemmal Ca^{2+} pump by oxygen free radicals. Am. J. Physiol. 257, H804–H811.

Kaneko, M., Matsumoto, Y., Hayashi, H., Kobayashi, A. and Yamazaki, N. (1994). Oxygen free radicals and calcium homeostasis in the heart. Mol. Cell. Biochem. 139, 91–100.

Karmazyn, M. and Cook, M.A. (1992). Adenosine A1 receptor activation attenuates cardiac injury produced by hydrogen peroxide. Circ. Res. 71, 1101–1110.

Kass, G.E. and Orrenius, S. (1999). Calcium signaling and cytotoxicity. Environ. Health Perspect. 107, 25–35.

Kim, D. and Smith, T.W. (1988). Cellular mechanisms underlying calcium–proton interactions in cultured chick ventricular cells. J. Physiol. 398, 391–410.

Kim, D., Cragoe, E.J. and Smith, T.W. (1987). Relations among sodium pump inhibition, Na^+–Ca^{2+} and Na^+–H^+ exchange activities, and Ca^{2+}–H^+ interaction in cultured chick heart cells. Circ. Res. 60, 185–193.

Kim, D.H., Akera, T. and Kennedy, R.H. (1983). Ischemia-induced enhancement of digitalis sensitivity in isolated guinea-pig heart. J. Pharmacol. Exp. Ther. 226, 335–342.

Kim, M.S. and Akera, T. (1987). O_2 free radicals: cause of ischemia-reperfusion injury to cardiac Na^+–K^+-ATPase. Am. J. Physiol. 252, H252–H257.

Kramer, J.H., Mak, I.T. and Weglicki, W.B. (1984). Differential sensitivity of canine cardiac sarcolemmal and microsomal enzymes to inhibited by the free radical-induced lipid peroxidation. Circ. Res. 55, 120–124.

Krejsa, C.M. and Schieven, G.L. (1998). Impact of oxidative stress on signal transduction control by phosphotyrosine phosphatases. Environ. Health Perspect. 106, 1179–1184.

Kukreja, R.C., Weaver, A.B. and Hess, M.L. (1990). Sarcolemmal Na^+–K^+-ATPase: inactivation by neutrophil-derived free radicals and oxidants. Am. J. Physiol.259, H1330–H1336.

Kuo, T.H., Kim, H.R., Zhu, L., Yu, Y., Lin, H.M. and Tsang, W. (1998). Modulation of endoplasmic reticulum calcium pump by Bcl-2. Oncogene 17, 1903–1910.

Kutryk, M.J. and Pierce, G.N. (1988). Stimulation of sodium calcium exchange by cholesterol incorporation into isolated cardiac sarcolemmal vesicles. J. Biol. Chem. 263, 13167–13172.

Langer, G.A., Frank, J.S. and Philipson, K.D. (1982). Ultrastructure and calcium exchange of the sarcolemma, sarcoplasmic reticulum and mitochondria of the myocardium. Pharmacol. Ther. 16, 331–376.

Langer, G.A. (1986) In: Cardiac Muscle: the Regulation of Excitation and Contraction. (Nathan, R.D., Ed.), pp. 269–281. Academic Press, Orlando.

Laursen, J.B., Rajagopalan, S., Galis, Z., Tarpey, M., Freeman, B.A. and Harrison, D.G. (1997). Role of superoxide in angiotensin II-induced but not catecholamine-induced hypertension. Circulation 95, 588–593.

Lotscher, H.R., Winterhalter, K.H., Carafoli, E. and Richter, C. (1979). Hydroperoxides can modulate the redox state of pyridine nucleotide and the calcium balance in rat liver mitochondria. Proc. Natl. Acad. Sci USA 76, 4340–4344.

Luft, R. and Landau, B.R. (1995). Mitochondrial medicine. J Intern. Med. 238, 405–21.

Lundmark, J.L., Ramasamy, R., Vulliet, P.R. and Schaefer, S. (1999). Chelerythrine increases Na^+/K^+ATPase activity and limits ischemic injury in isolated rat hearts. Am. J. Physiol. 277, H999–H1006.

Miyawaki, H., Wang, Y. and Ashraf, M. (1998). Oxidant stress with hydrogen peroxide attenuates calcium paradox injury: role of protein kinase C and ATP-sensitive potassium channel. Cardiovasc Res. 37, 691–699.

Mukai, M., Terada, H., Sugiyama, S., Satoh, H. and Hayashi, H. (2000). Effects of a selective inhibitor of Na^+/Ca^{2+} exchange, KB-R7943, on reoxygenation-induced injuries in guinea pig papillary muscles. J. Cardiovasc. Pharmacol. 35, 121–128.

Mullins, L.J. (1979). The generation of electric currents in cardiac fibers by Na^+/Ca^{2+} exchange. Am. J. Physiol. 236, C103–110.

Natarajan, V., Scribner, W.M., Al-Hassani, M. and Vepa, S. (1998). Reactive oxygen species signaling through regulation of protein tyrosine phosphorylation in endothelial cells. Environ. Health Perspect. 106, 1205–1212.

Pereira, R.S., Bertocchi, A.P. and Vercesi, A.E. (1992). Protective effect of trifluoperazine on the mitochondrial damage induced by Ca^{2+} plus prooxidants. Biochem. Pharmacol. 44, 1795–1801.

Persad, S., Rupp, H., Jindal, R., Arneja, J. and Dhalla, N.S. (1998) Modification of cardiac beta-adrenoceptor mechanisms by H_2O_2. Am. J. Physiol. 274, H416–H423.

Persad, S., Elimban, V., Siddiqui, F. and Dhalla, N.S. (1999). Alterations in cardiac membrane beta-adrenoceptors and adenylyl cyclase due to hypochlorous acid. J. Mol. Cell. Cardiol. 31, 101–111.

Persad, S., Elimban, V., Kaila, J. and Dhalla, N.S. (1997). Biphasic alterations in cardiac beta-adrenoceptor signal transduction mechanism due to oxyradicals. J. Pharmacol. Exp. Ther. 282, 1623–1631.

Philipson, K.D. and Ward, R. (1986). Ca^{2+} transport capacity of sarcolemmal Na^+–Ca^{2+} exchange. Extrapolation of vesicle data to in vivo conditions. J. Mol. Cell. Cardiol. 18, 943–951.

Philipson, K.D., Bers, D.M. and Nishimoto, A.Y. (1980). The role of phospholipids in the Ca^{2+} binding of isolated cardiac sarcolemma. J. Mol. Cell. Cardiol. 12, 1159–1173.

Pignac, J., Lacaille, C. and Dumont, L. (1996). Protective effects of the K^+ ATP channel opener, aprikalim, against free radicals in isolated rabbit hearts. Free Radic. Biol. Med. 20, 383–389.

Pitts, B.J. and Okhuysen, C.H. (1984). Effects of palmitoyl carnitine and LPC on cardiac sarcolemmal Na^+K^+-ATPase. Am. J. Physiol. 247, H840–H846.

Rahman, F., Kato, A., Kawahara, K. and Nakajima, T. (1996) Y-26763 protects the working rat myocardium from ischemia/reperfusion injury through opening of KATP channels. Eur. J. Pharmacol. 317, 293–299.

Richter, C., Gogvadze, V., Laffranchi, R., Schlapbach, R., Schweizer, M., Suter, M., Walter, P. and Yaffee, M. (1995). Oxidants in mitochondria: from physiology to diseases. Biochim. Biophys. Acta. 1271, 67–74.

Roveri, A., Coassin, M., Maiorino, M., Zamburlini, A., Van Amsterdam, F.T., Ratti, E. and Ursini, F. (1992). Effect of hydrogen peroxide on calcium homeostasis in smooth muscle cells. Arch. Biochem. Biophys. 297, 265–270.

Roychoudhury, S., Chakraborti, T., Ghosh, S.K. and Chakraborti, S. (1996). Redox state of pyridine nucleotides, but not glutathiones, regulate Ca^{2+} release by H_2O_2 from mitochondria of pulmonary smooth muscle. Ind. J. Biochem. Biophys. 33, 218–222.

Roychoudhury, S., Ghosh, S.K., Chakraborti, T. and Chakraborti, S. (1996). Role of hydroxyl radical in the oxidant H_2O_2-mediated Ca^{2+} release from pulmonary smooth muscle mitochondria. Mol. Cell. Biochem. 159, 95–103.

Schlack, W., Schafer, M., Uebing, A., Schafer, S., Borchard, U. and Thamer, V. (1993). Adenosine A2–receptor activation at reperfusion reduces infarct size and improves myocardial wall function in dog heart. J. Cardiovasc. Pharmacol. 22, 89–96.

Suzuki, Y.J., Forman, H.J. and Sevanian, A. (1997). Oxidants as stimulators of signal transduction. Free Radic. Biol. Med. 22, 269–285.

Tamargo, J., Perez, O., Delpon, E., Garcia-Rafanell, J., Gomez, L. and Cavalcanti, F. (1995). Cardiovascular effects of the novel potassium channel opener UR-8225. J. Cardiovasc. Pharmacol. 26, 295–305.

Tani, M. (1990). Mechanisms of Ca^{2+} overload in reperfused ischemic myocardium. Annu. Rev. Physiol. 52, 543–559.

Temsah, R.M., Dyck, C., Netticadan, T., Chapman, D., Elimban, V. and Dhalla, N.S. (2000). Effect of beta-adrenoceptor blockers on sarcoplasmic reticular function and gene expression in the ischemic-reperfused heart. J. Pharmacol. Exp. Ther. 293, 15–23.

Thomas, G.P., Sims, S.M., Cook, M.A. and Karmazyn, M. (1998). Hydrogen peroxide-induced stimulation of L-type calcium current in guinea pig ventricular myocytes and its inhibition by adenosine A1 receptor activation. J. Pharmacol. Exp. Ther. 286, 1208–1214.

Thornton, J.D., Liu, G.S., Olsson, R.A. and Downey, J.M. (1992). Intravenous pretreatment with A1-selective adenosine analogues protects the heart against infarction. Circulation. 85, 659–665.

Uceda, G., Garcia, A.G., Guantes, J.M., Michelena, P. and Montiel, C. (1995). Effects of Ca^{2+} channel antagonist subtypes on mitochondrial Ca^{2+} transport. Eur. J. Pharmacol. 289, 73–80.

Vaughan-Jones, R.D., Lederer, W. and Eisner, D.A. (1983). Ca^{2+} ions can affect intracellular pH in mammalian cardiac muscle. Nature 301, 522–524.

Vissers, M.C., Pullar, J.M. and Hampton, M.B. (1999). Hypochlorous acid causes caspase activation and apoptosis or growth arrest in human endothelial cells. Biochem J. 344, 443–449.

Xie, Z.J,. Wang, Y.H., Askari, A, Huang, W.H., Klaunig, J.E. and Askari A. (1990). Studies on the specificity of the effects of oxygen metabolites on cardiac sodium pump. J. Mol. Cell. Cardiol. 22, 911–920.

Yamashita, T., Masuda, Y. and Tanaka, S. (1994). Potassium channel openers relax A23187-induced nifedipine-resistant contraction of rat aorta. J. Cardiovasc. Pharmacol. 24, 914–920.

Yoshida, Y., Hirai, M., Yamada, T., Tsuji, Y., Kondo, T., Inden, Y., Akahoshi, M., Murakami, Y., Tsuda, M., Tsuboi, N., Hirayama, H., Okamoto, M., Ito, T., Saito, H. and Toyama, J. (2000). Antiarrhythmic efficacy of dipyridamole in treatment of reperfusion arrhythmias : evidence for cAMP-mediated triggered activity as a mechanism responsible for reperfusion arrhythmias. Circulation 101, 624–630.

Zafari, A.M., Ushio-Fukai, M., Akers, M., Yin, Q., Shah, A., Harrison, D.G., Taylor, W.R. and Griendling, K.K. (1998). Role of NADH/NADPH oxidase-derived H_2O_2 in angiotensin II-induced vascular hypertrophy. Hypertension 32, 488–495.

Zucchi, R,. Ronca-Testoni, S., Galbani, P., Yu, G., Mariani, M. and Ronca, G. (1992). Cardiac A2 adenosine receptors—influence of ischaemia. Cardiovasc. Res. 26, 549–554.

Protein Adaptations and Signal Transduction. Edited by K.B. Storey and J.M. Storey

CHAPTER 17

Role of NF-E2 Related Factors in Oxidative Stress

David Bloom, Saravanakumar Dhakshinamoorthy, Wei Wang, Claudia M. Celli and Anil K. Jaiswal*
Department of Pharmacology, Baylor College of Medicine, Houston, TX 77030, U.S.A.

1. Oxidative stress

Reactive oxygen species (ROS) cause oxidative stress and have a profound impact on the survival and evolution of all living organisms (Breimer, 1990; Meneghini, 1997). ROS include both free radicals, such as the superoxide anion and the hydroxyl radical, and oxidants such as hydrogen peroxide (H_2O_2). ROS are produced endogenously by a variety of oxidases, as well as being by-products of cellular respiration (Grisham and McCord, 1986). Physiological responses, such as the activation of neutrophils during inflammation and infection (Thelen et al., 1993), protective cytotoxic processes (Kerr et al., 1996), oxidative phosphorylation (Wei, 1998), and auto-oxidation reactions catalyzed by transition metal ions (Kasprzak, 1995), produce ROS as byproducts. Activation of other intrinsic factors, such as IL-1, IL-6, IL-8, TNF-α, TNF-β, PDGF, also results in the production of ROS (Suzuki et al., 1997), as do a range of environmental phenomena, including ionizing radiation (Breen and Murphy, 1995). Cytochrome P450 and Cytochrome P450 reductase, activated by environmental exposure, generate ROS during oxidative metabolism of specific compounds. Long wavelength UV light (UVA and UVB) as well as photochemical oxidant pollutants, such as ozone and nitrogen oxides (NO_2, N_2O, O=NOO–, NO, O=NOOH), generate singlet oxygen species (Last et al., 1994). Radiolysis of water, caused by exposure to environmental radiation, leads to the formation of hydroxyl radicals, which are the most reactive species known *in vivo* (Ward, 1994). Therefore, it is clear that all cells must continuously strive to keep the levels of ROS in check.

ROS attack DNA and other cellular macromolecules, causing oxidative stress and many other physiological and pathological conditions. These conditions include aging, neurodegenerative diseases, arthritis, arteriosclerosis, inflammatory responses, and tumor induction and promotion (Grisham and McCord, 1986; Ward, 1994; Breen and Murphy, 1995; Rosen et al., 1995). Many of these conditions, including cancer, are often preceded by damage or mutation of genomic DNA (Ames et al., 1995). Much of the research on ROS has been centered on the damaging effects of oxidative stress. However, it is now apparent that ROS and H_2O_2 activate a battery of cellular enzymes that detoxify ROS and thereby protect the cell against damage caused by oxidative stress. The mechanism by which ROS and H_2O_2 activate these genes is unknown. They may act as or mimic endogenous second messengers. It should be noted that the concentration at which ROS provide protection to a cell might be only slightly lower than the concentration at which damage occurs.

2. Oxidative stress-activated defensive mechanisms

Much of what we know about the mechanisms of protection against oxidative stress has come from the study of prokaryotic cells (Fig. 17.1) (Bauer et al., 1999; Zheng and Storz, 2000). Prokaryotic

*Corresponding author.

cells utilize transcription factors to sense the redox state of the cell, and in times of oxidative stress these transcription factors induce the expression of about eighty antioxidant/defensive genes (Zheng and Storz, 2000). OxyR and SoxR are two of the best studied of these transcription factors, and are found in *Escherichia coli*. OxyR contains a thio-disulfide switch that is sensitive to H_2O_2. SoxR contains a 2Fe-2S cluster that is sensitive to superoxide and nitric oxide. Both of these sensors can be turned on and off very quickly, which allows the cell to respond promptly to subtle changes in the concentrations of ROS.

Similar mechanisms of protection against oxidative stress have been found in eukaryotic cells (Fig. 17.1) (Venugopal et al., 1997). Over one hundred genes have been found to be involved in the response to oxidative stress. Their products regulate a wide variety of cellular activities including signal transduction, proliferation, and immunologic defense reactions. The signal transduction pathways responsible for sensing oxidative stress and activating these responses is still not well understood in eukaryotes, and is an area of intense study. There are only a few transcription factors, that include NF-κB and the NF-E2 related factors (Nrfs), known at this time that are activated by ROS (Venugopal and Jaiswal, 1996; Schulze-Osthoff et al., 1997). This review is focused on the role of the NF-E2 related factors: Nrf1, Nrf2, and Nrf3. However, the mechanism of NF-κB activation is better studied than that of the Nrfs, so it is helpful to briefly describe the NF-κB pathway first.

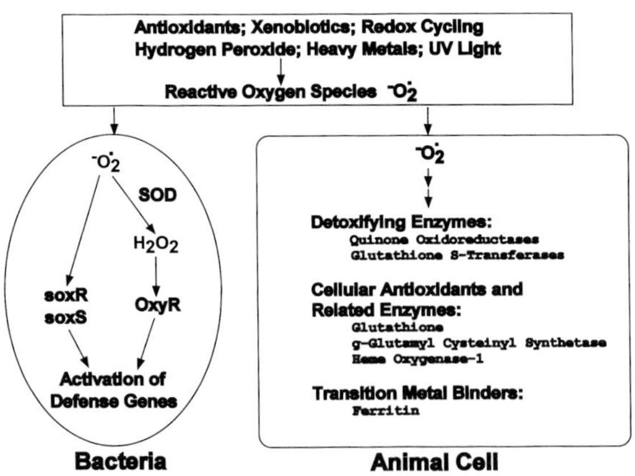

Fig. 17.1. Comparison of oxidative stress regulation in bacterial and mammalian cells. OxyR and soxRS are redox regulated transcription factors in bacteria.

3. Transcription factor NF-κB

NF-κB was the first eukaryotic transcription factor reported to respond directly to oxidative stress (Schulze-Osthoff et al., 1997). NF-κB also plays an important role in the regulation of many genes controlling immune and inflammatory responses (Liou and Baltimore, 1993; Baeuerle and Henkel, 1994). NF-κB can be composed of homodimers or heterodimers of several subunits including p50, p65 (RelA), p52, c-Rel and RelB. The subunits can form a variety of heterodimers, each controlling a specific subset of genes (Sen and Baltimore, 1986; Nolan and Baltimore, 1992; Liou and Baltimore, 1993). Interestingly, activation of NF-κB does not require the synthesis of new protein. A family of inhibitors, IκBs, retain NF-κB in the cytoplasm, thereby inhibiting DNA binding and transcriptional activity (Baeuerle and Baltimore, 1988; Beg and Baldwin, 1993). When cells are stimulated, NF-κB is released from IκB and it is free to translocate to the nucleus where it is transcriptionally active (Zabel et al., 1990). However, it has recently been reported that IκB is also present in the nucleus, probably to rapidly turn off the NF-κB signal when the stimulus is terminated (Renard et al., 2000).

IκB-α, the best characterized of the IκB family, binds p65, c-Rel, and RelB. As with all of the IκB family members, IκB-α contains SW16 ankyrin repeats (Blank et al., 1992). Mutational analysis has shown that these repeats are necessary, but not solely responsible, for inhibition of NF-κB by IκB (Inoue et al., 1992).

Stimuli of NF-κB include the cytokines tumor necrosis factor (TNF) and interleukin-1 (IL-1), phorbol esters, lipopolysaccharide (LPS), double-stranded RNA, inhibitors of protein synthesis, UV and ionizing irradiation, and viral transactivator proteins. These all share the characteristic of being proinflammatory, and many increase the intracellular concentration of ROS. Data suggest that H_2O_2 activates NF-κB in response to pro-oxidants, while antioxidants might elicit a response through other ROS. Yet, other reports suggest that antioxidants block NF-κB activation (Schreck et al., 1992a and b). In contrast, the Nrfs are clearly activated by antioxidants (Dhakshinamoorthy et al., 2000).

4. NF-E2 Related factors

NF-E2 was first cloned in 1993 (Andrews et al., 1993). The mouse NF-E2 protein contains 373 amino acids and has an apparent molecular mass of 45 kDa; it is therefore also referred to as p45. NF-E2 is expressed only in erythroid cells, megakaryocytes, and mast cells. It binds to an Activator Protein 1 (AP1)-like, NF-E2 recognition site (GCTGAGTCA), and regulates tissue specific expression of the globin genes (Ney et al., 1990; Moi and Kan, 1990; Liu et al., 1992; Mignotte et al., 1989). NF-E2 functions as a heterodimer with the ubiquitously expressed small Maf proteins (Igarashi et al., 1994). NF-E2 –/– mice have no circulating platelets, and most die due to hemorrhage (Shivdasani and Orkin, 1995). The loss of NF-E2 has only a mild effect on erythroid cells, so mice that survive to adulthood are normal except for exhibiting conditions consistent with a minor decrease in hemoglobin.

NF-E2 related factors Nrf1 and Nrf2, both 66–68 kDa proteins, were cloned using a yeast complementation assay (Chan et al., 1993; Moi et al., 1994). Both display a significant amount of homology to NF-E2, but unlike NF-E2 both are ubiquitously expressed. More recently, a third family member of the Nrfs, Nrf3, was cloned and sequenced (Kobayashi et al., 1999). The Nrfs belong to the family of basic leucine zipper proteins (bZIP) (Fig. 17.2). The basic region, just upstream of the leucine zipper region, is responsible for DNA binding. The acidic region is required for transcriptional activation. The cap'n'collar region, so called because of its homology to the Drosophila cap'n'collar protein, is highly conserved among the Nrfs, but the function of this region remains unknown.

Nrf1 and Nrf2 have been shown to regulate β-globin gene expression by binding to the AP1

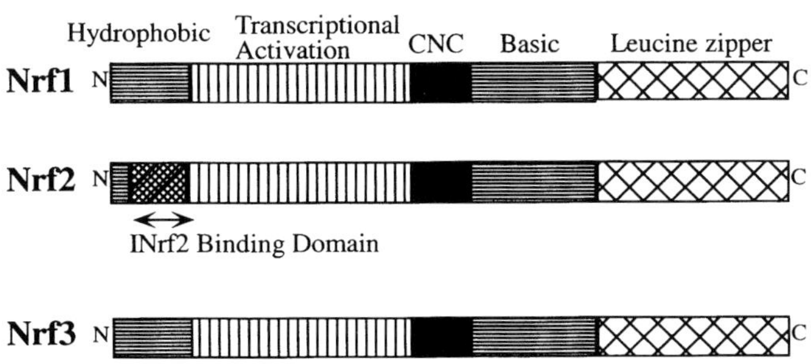


```
Nrf1 RDIRRRGKNKMAAQN C RKRKLDTILN L ERDVED L QRDKAR L LREKVE F LRSLRQ M KQKVQS L Y
Nrf2 RDIRRRGKNKMAAQN C RKRKLENIVE L EQDLDH L KDEKEK L LKEKGE N DKSLHL L KKQLST L Y
Nrf3 RDIRRRGKNKVAAQN C RKRKLDIILN L EDDICN L QAKKEA L KNEQTQ C SKAIDI M RQKLHG L H
     Basic region                 Leucine zipper region
```

Fig. 17.2. NF-E2 related factors. The various domains of NF-E2 related factors Nrf1, Nrf2 and Nrf3 are shown. The domain of Nrf2 that interacts with cytosolic inhibitor INrf2 is shown. The amino acids of the basic and leucine zipper regions of Nrf1, Nrf2 and Nrf3 are aligned to demonstrate sequence conservation. The cysteine in the basic (DNA binding) region, and the leucines in the leucine zipper region are shown in bold letters. Nrf2 contains one mutated leucine, as compared to two mutated leucines in Nrf1 and Nrf3. CNC, cap'n'collar domain.

like NF-E2 recognition site (Chan et al., 1993; Moi et al., 1994). Nrf1 –/– mice die *in utero* due to a decreased number of enucleated red blood cells and severe anemia (Chan et al., 1998). Nrf2 –/– mice are viable and live to adulthood. Nrf2 is therefore not required for erythropoeisis, development, or growth (Chan et al., 1996).

5. Role of NF-E2 related factors in protection against oxidative stress

The first evidence demonstrating the role of Nrf1 and Nrf2 in protection against oxidative stress came from studies of the regulation of NAD(P)H: quinone oxidoreductase1 (NQO1) gene expression (Venugopal and Jaiswal, 1996). NQO1 is a flavoprotein that competes with cytochrome P450 reductase and catalyzes two-electron reduction and detoxification of quinones and other redox cycling compounds (Joseph and Jaiswal, 1994). This function prevents the generation of ROS and provides protection to the cells against oxidative stress. Overexpression of Nrf1 and Nrf2 cDNA has been shown to upregulate the expression and induction of the NQO1 gene in response to antioxidants and xenobiotics (Venugopal and Jaiswal, 1996). The role of Nrf2 in transcriptional activation of NQO1 was further confirmed by results from studies on Nrf2$^{-/-}$ mice (Itoh et al., 1997). Mice lacking the Nrf2 gene exhibited a marked decrease in the expression and induction of NQO1, indicating that Nrf2 plays an essential role in the *in vivo* regulation of NQO1 in response to oxidative stress. Further studies have shown that Nrf2 is also a prevailing factor in the regulation of other defensive genes including glutathione S-transferase Ya (GST Ya), γ-glutamylcysteine synthetase (γ-GCS) and heme oxigenase-1 (HO-1) (Alam et al., 1999; Wild et al., 1999; Nguyen et al., 2000). GST Ya conjugates hydrophobic electrophiles and ROS with glutathione, aiding in their excretion (Pickett and Lu, 1989; Tsuchida and Sato, 1992). γ-GCS plays a role in the metabolism of glutathione (Mulkahy et al., 1997). And HO-1 catalyzes the first and rate-limiting step in heme catabolism (Choi and Alam, 1996). Recent studies have shown that the presence of Nrf1 is critical for cells to cope with oxidative stress. Nrf1 deficient fibroblasts have lower levels of glutathione and are more sensitive to oxidative stress producing compounds (Kwong et al., 1999).

These data demonstrate that Nrf1 and Nrf2 have a significant role in the regulation of defensive/antioxidant genes, and their induction by oxidative stress inducing agents (e.g. xenobiotics, antioxidants, heavy metals, UV light and ionizing radiations). The coordinated induction of this battery of defensive genes provides the cell with the necessary protection against oxidative stress. Nrf3 is also expected to play a part in the response to oxidative stress, though no results have been reported yet.

Nrf1 and Nrf2 induce the antioxidant response element (ARE)-mediated expression of defensive genes in response to β-naphthoflavone (β-NF), 3-(2)-*tert*-butyl-4-hydroxyanisole (BHA), *tert*-butyl hydroquinone (tBHQ) and hydrogen peroxide (H_2O_2) (Venugopal and Jaiswal, 1996; Alam et al., 1999; Wild et al., 1999; Nguyen et al., 2000). An alignment of AREs from several genes is shown in Fig. 17.3. Nucleotide sequence analysis of the human NQO1 gene ARE revealed that it contains one perfect and one imperfect AP1 (TPA response element) element. These elements are arranged as inverse repeats that are separated by three base pairs and are followed by a ‘GC’ box (Jaiswal, 1994). The collagenase and metallothionein genes contain the seven base pair AP1 element in their promoter regions. This AP1 element is responsible for their induction in response to TPA (Angel and Karin, 1991). It has been shown that the ARE is a unique element, responding independently from the AP1 element. It is the ARE, and not AP1, that is utilized for the induction of human NQO1 and other detoxifying genes in response to xenobiotics and antioxidants (Xie et al., 1995). Mutational analysis revealed GTGACA***GC to be the core sequence of the ARE (Rushmore et al., 1991; Rushmore et al., 1993; Jaiswal, 1994; Xie et al., 1995). However, other neighboring sequences and elements also affect the ARE-mediated expression and induction of detoxifying genes (Li and Jaiswal, 1992; Prestera et al., 1993; Wasserman and Fahl, 1997).

		AP1-Like		AP1/ AP1-Like	GC Box	
hNQO1	AAATC	GCAGTCA	CAG	TGACTCA	GCA	GAATC
rNQO1	AGTCT	AGAGTCA	CAG	TGACTTG	GCA	AAATC
rNQO2	AGG	TGACTGC	AAA	TGAGGTG	GCA	GAAGC
rGSTP	CAAAAGTAG	TCAGTCA	CTA	TGATTCA	GCA	ACAAA
mFL		GAGCTCA	GCG	TGACTCA	GCA	GAACT
rGST Ya	GAGCTTGGAAA	TGGCATT	GCTAATGG	TGACAAA	GCA	ACTTT
rGST Ya	TAGCTTGGAAA	TGACATT	GCTAATGG	TGACAAA	GCA	ACTTT
hγGCS			CTCCCCG	TGACTCA	GCT	TTG
ARE CORE SEQUENCE			**G**	**TGACNNN**	**GCN**	

Fig. 17.3. Alignment of AREs from the various detoxifying enzyme genes. hNQO1, human NAD(P)H:quinone oxidoreductase1 gene ARE; rNQO1, rat NAD(P)H:quinone oxidoreductase1 gene ARE; hNQO2, human NAD(P)H:quinone oxidoreductase2 gene ARE; rGSTP, rat glutathione S-transferase P subunit gene ARE; mFL, mouse ferritin L subunit gene ARE; rGST Ya, rat glutathione S-transferase Ya gene ARE; mGST Ya, mouse glutathione S-transferase Ya gene ARE; hγGCS, human γ-glutamyl cysteine synthetase gene ARE

6. Nrf1 and Nrf2 associated factors

Although the cellular sensors of oxidative stress remain unknown, Nrf1 and Nrf2 are possible candidates because of their role in ARE-mediated induction of defensive genes. However, it is likely that there are other factors involved in signaling between antioxidant and xenobiotics and the Nrfs. Nrf1 and Nrf2, bZIP proteins, do not form homodimers or heterodimerize with each other, but they do require heterodimerization with another bZIP protein to be active (Venugopal and Jaiswal, 1996; Venugopal and Jaiswal, 1998; Chan et al., 1993; Moi et al., 1994). c-Jun, Jun-B, and Jun-D have all been shown to heterodimerize with both Nrf1 and Nrf2, and these complexes are capable of inducing ARE-mediated expression of NQO1 and GST Ya in response to antioxidants and xenobiotics (Venugopal and Jaiswal, 1998). Interestingly, heterodimerization of Nrf2 and c-jun requires unknown cytosolic factors (Venugopal and Jaiswal, 1998). Recently a cytosolic inhibitor of Nrf2, INrf2/KEAP1 (Kelch-like ECH-associated protein1), was discovered (Fig. 17.4) (Itoh et al., 1999; Dhakshinamoorthy and Jaiswal, 2001). Under normal conditions INrf2 retains Nrf2 in the cytoplasm. Upon treatment of cells with oxidative stress inducing factors Nrf2 is released from INrf2. Nrf2 is then free to translocate into the nucleus and induce the expression of defensive genes.

Analysis of the INrf2 amino acid sequence revealed a BTB/POZ BTB (broad complex, tramtrack, bric-a-brac)/POZ (poxvirus, zinc finger) domain and a Kelch domain (Fig. 17.4). The role of the BTB/POZ domain remains unknown in INrf2, but in other proteins it has been shown to be a protein-protein interaction domain. In the Drosophila Kelch protein and in PIP, the Kelch domain binds to actin (Albagli et al., 1995; Kim et al., 1999). Therefore, it is expected that INrf2 binds to actin in the cytoskeleton. If this is the case, does INrf2 really release Nrf2? Or does the INrf2–Nrf2 complex dissociate from the cytoskeleton and move to the nucleus together? These questions have not yet been answered. There is a single report that shows the C-terminal portion of INrf2 interacts with Nrf2 (Itoh et al., 1999). This portion contains the Kelch domain, but it is unlikely that the Kelch domain alone binds Nrf2. INrf2 appears to be a specific cytosolic inhibitor for Nrf2, because it does not interact with Nrf1 or Nrf3 (Dhakshinamoorthy and Jaiswal, unpublished). This begs the question, are there cytosolic inhibitors of Nrf1 and Nrf3? If so, they remain unknown at this time.

The small Maf proteins (MafG, MafK, and MafF) are a family of nuclear transcription factors. They act as both activators and repressors of a number of eukaryotic genes (Fujiwara et al., 1993; Kataoka et al., 1993; Kataoka et al., 1994; Kim and

A.

MQPEPKPSGAPRSSQFLPLWSK**C**PEGAGDAVMYASTE**C**KAEVTPSQDGNRTFSYTLEDHTKQAFGIMNELRLSQQL**C**DVTLQVK
YEDIPAAQFMAHKVVLASSSPVFKAMFTNGLREQGMEVVSIEGIHPKVMERLIEFAYTASISVGEK**C**VLHVMNGAVMYQIDSVVR
A**C**SDFLVQQLDPSNAIGIANFAEQIG**C**TELHQRAREYIYMHFGEVAKQEEFFNLSH**C**QLATLISRDDLNVR**C**ESEVFHA**C**IDWV
KYD**C**PQRRFYVQALLRAVR**C**HALTPRFLQTQLQK**C**EILQADAR**C**KDYLVQIFQELTLHKPTQAVP**C**RAPKVGRLIYTAGGYFRQ
SLSYLEAYNPSNGSWLRLADLQVPRSGLAG**C**VVGGLLYAVGGRNNSPDGNTDSSALD**C**YNPMTNQWSP**C**ASLSVPRNRSGGGVID
GHIYAVGGSHG**C**IHHSSVERYEPDRDEWHLVAPMLTRRIGVGVAVLNRLLYAVGGFDGTNRLNSAE**C**YYPERNEWRMITPMNTIR
SGAGV**C**VLHS**C**IYAAGGYDGQDQLNSVERYDVETETWTFVASMKHRRSALGIAVHQGRIYVLGGYDGHTFLDSVE**C**YDPDTDTWS
EVTRLTSGRSGVGVAVTMEP**C**RKQIDQQN**C**T**C**

B.

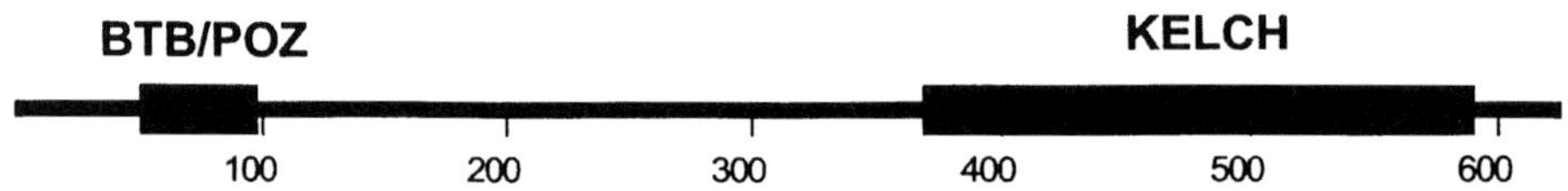

Fig. 17.4. Rat INrf2 protein. The BTB/POZ and KELCH binding domains and the cysteine residues are highlighted.

Andrews, 1997). The small Mafs are homologous to viral-Maf (v-Maf), especially in the DNA binding domain and the leucine zipper domain. However, the small Mafs lack the transactivation domain that is present in v-Maf (Kim and Andrews, 1997; Fujiwara et al., 1993; Kataoka et al., 1993). Small Mafs form homodimers, as well as heterodimers with Nrf2. The homodimers repress, while small Maf–Nrf2 heterodimers activate β-globin expression (Marini et al., 1997). The small Mafs have also been visualized in the ARE-nuclear protein complexes, which contain Nrf1, Nrf2, Jun, Fos, and Fra (Itoh et al., 1997; Wild et al., 1999). MafG and MafK overexpression represses the γ-GCS gene (Wild et al., 1999). Recently, Maf homodimers and Maf–Nrf2 heterodimers have been shown to repress the ARE-mediated expression and antioxidant induction of NQO1 and GST Ya genes (Dhakshinamoorthy and Jaiswal, 2000; Nguyen et al., 2000). In addition to Maf proteins, overexpression of c-Fos or Fra1 also represses ARE-mediated gene expression (Venugopal and Jaiswal, 1996). Mice lacking c-Fos have increased expression of NQO1 and GST Ya, substantiating a negative role for c-Fos in ARE-mediated gene expression *in vivo* (Wilkinson et al., 1998). As with many response elements, it appears that ARE-mediated gene expression is a balance between positive and negative regulatory factors. One theory is that it is always necessary to have a small amount of ROS present to keep the cellular defenses active. Since activation of detoxifying enzymes and other defensive proteins leads to significant reduction in the levels of superoxide and other free radicals, the cell may require negative regulatory factors like the small Mafs and c-Fos to keep the expression of defensive proteins "in check".

7. Mechanism of Nrf signaling and activation of ARE-mediated expression and coordinated induction of defensive genes

The signal transduction pathway that leads from antioxidants and xenobiotics to Nrf1, Nrf2, Jun, Fos, and Mafs, the proteins which regulate ARE-mediated expression of detoxifying genes, remains unknown. Because the metabolism of both xenobiotics and antioxidants results in the creation of superoxides and electrophiles (De Long et al., 1987), it is thought that these molecules might act as second messengers, activating ARE-mediated expression of a host of defensive genes including NQO1, GST Ya, γ-GCS and HO-1. These genes protect the cell against the damaging effects of oxidative stress. For example, expression of the mouse GST gene is induced by hydroxyl radicals, generated by quinones (Pinkus et al., 1996). However, this effect has not yet been demonstrated using the AREs of other defensive genes. H_2O_2, on the other hand, does induce ARE-mediated expression of rat GST Ya, rat NQO1, and human NQO1 (Favreau and Pickett, 1991; Li and Jaiswal, 1994). Therefore, H_2O_2 might play an important part in the induction of defensive genes by xenobiotics and antioxidants. Even though electrophiles are thought to be possible messengers in the oxidative stress pathway, their role, if any, has not been demonstrated.

A hypothetical model illustrating the activation of defensive genes by xenobiotics and antioxidants is shown in Fig. 17.5. The superoxide signal presumably passes through an unknown cytosolic factor(s). This factor(s) then catalyzes the modification of either Nrf2, INrf2, or both. This modification is expected because treatment of cells with antioxidants and xenobiotics does not alter the protein levels of Nrf2 or INrf2 (Venugopal et al., unpublished). As a result of the modification of the Nrf2–INrf2 complex, Nrf2 is released. Nrf2 then translocates to the nucleus where it heterodimerizes with c-Jun and induces the expression of NQO1 and other ARE-regulated genes. The levels of c-Jun increase significantly in response to oxidative stress (Venugopal and Jaiswal, 2000), perhaps to increase expression of ARE-mediated genes at first, and then later, at higher levels or when the stimulus is terminated, to repress their expression. In fact, c-Jun has an ARE in its 5′ untranslated region, and this has been shown to respond to oxidative stress (Venugopal and Jaiswal, 1999). Nrf1 is expected to function in a manner similar to Nrf2. Likewise, Jun-B and Jun-D are expected to function in a manner similar to c-Jun. Their participation in ARE-mediated expression of NQO1 has already been demonstrated (Venugopal and Jaiswal, 1998).

The identification of the unknown cytosolic factor(s) is crucial for our understanding of how the oxidative stress signal is passed from endogenous substances, xenobiotics, and antioxidants to the Nrf and Jun proteins. However, even the nature of this factor(s) remains unknown. It could be a kinase, phosphatase, or oxidation-reduction (redox) protein. The modification could be of Nrf2, INrf2, or even possibly c-Jun. Most likely, the cytosolic factor(s) acts as the cellular sensor of oxidative stress, receiving the signal from superoxides and related ROS, and activating Nrf2. This sequence of events results in the increased expression of ARE-regulated defensive genes that protect the cell against the damaging effects of oxidative stress.

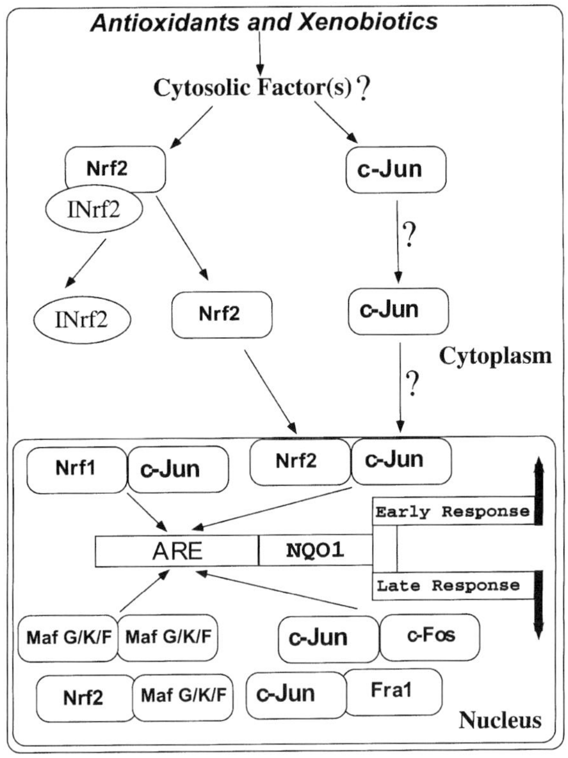

Fig. 17.5. Model to illustrate the role of NF-E2 related factors, Nrf1 and Nrf2, in ARE-mediated expression and induction of defensive genes in response to oxidative stress-inducing agents (xenobiotics and antioxidants).

Acknowledgements

We are thankful to our colleagues from Baylor College of Medicine, Houston, Texas for valuable suggestions. This work was supported by NIH grants RO1 GM47644, RO1 ES07943 and RO1 CA81057.

References

Andrews, N.C., Erdjument-Bromage, H., Davidson, M.B., Tempst, P. and Orikin, S.H. (1993). Erythroid transcription factor NF-E2 is a haematopoietic-specific basic-leucine zipper protein. Nature 339, 722–727.

Alam, J., Stewart, D., Touchard, C., Boinapally, S., Choi, M.K. and Cook, J.L. (1999). Nrf2, a Cap'n'Collar transcription factor, regulates induction of the heme oxygenase-1 gene. J. Biol. Chem. 274, 26071–26078.

Ames, B.N., Gold, L.S. and Willett, W.C. (1995). The causes and prevention of cancer. Proc. Natl. Acad. Sci. USA 92, 5258–5265.

Albagli, O., Dhordain, P., Deweindt, C., Lecocq, G. and Leprince, D. (1995). The BTB/POZ domain: a new protein–protein interaction motif common to DNA- and actin-binding proteins. Cell Growth & Differentiation 6, 1193–1198.

Angel, P. and Karin, M. (1991). The role of Jun, Fos and AP-1 complex in cell proliferation and transformation. Biochem. Biophys. Acta, 1072, 129–157.

Baeuerle, P.A. and Baltimore, D. (1988). A specific inhibitor of the NF-κB transcription factor. Science 242, 540–546.

Baeuerle, P.A. and Henkel, T. (1994). Function and activation of NF-κB in the immune system. Annu. Rev. Immunol. 12, 141–179.

Bauer, C.E., Elsen, S. and Bird, T.H. (1999). Mechanisms for redox control of gene expression. Annu. Rev. Microbiol. 53, 495–523.

Beg, A.A. and Baldwin Jr., A.S. (1993). The IkB proteins: multifunctional regulators of Rel/NF-kB transcription factors. Genes Dev. 7, 2064–2070.

Blank, V.P., Kourilsky, P. and Israel, A. (1992). NF-kB and related proteins: Rel/ dorsal homologies meet ankyrin-like repeats. Trends Biochem. Sci. 17, 135–140.

Breen, A.P. and Murphy, J. A. (1995). Reactions of oxyl radicals with DNA. Free Rad. Biol. Med. 18, 1033–1077.

Breimer, L.H. (1990). Molecular mechanisms of oxygen radical carcinogenesis and mutagenesis: the role of DNA base damage. Mol. Carcinogenesis 3, 188–197.

Chan, J.Y., Han, X. and Kan, Y.W. (1993). Cloning of Nrf1, an NF-E2-related transcription factor, by genetic selection in yeast. Proc. Natl. Acad. Sci. USA 90, 11371–11375.

Chan, K., Lu, R., Chang, J.C. and Kan, Y.W. (1996). NRF2, a member of the NFE2 family of transcription factors, is not essential for murine erythropoiesis, growth, and development. Proc. Natl. Acad. Sci. U.S.A. 93, 13943–13948.

Chan, J.Y., Kwong, M., Lu, R., Chang, J., Wang, B., Yen, T.S. and Kan, Y.W. (1998). Targeted disruption of the ubiquitous CNC-bZIP transcription factor, Nrf-1, results in anemia and embryonic lethality in mice. EMBO J. 17, 1779–1787.

Choi, A.M. and Alam, J. (1996). Heme oxygenase-1: function, regulation, and implication of a novel stress-inducible protein in oxidant-induced lung injury. Am. J. Respir. Cell. Mol. Biol. 15, 9–19.

De Long, M.J., Santamaria, A.B. and Talalay, P. (1987). Role of cytochrome P_1-450 in the induction of NAD(P)H:quinone reductase in a murine hepatoma cell line and its mutants. Carcinogenesis 8, 1549–1553.

Dhakshinamoorthy.S., Long II, D.J. and Jaiswal, A.K. (2000). Antioxidant regulation of genes encoding enzymes that detoxify xenobiotics and carcinogens. Curr. Top. Cell. Regul. 36, 201–216.

Dhakshinamoorthy, S. and Jaiswal, A.K. (2000). Small Maf (MafG and MafK) proteins negatively regulate ARE-mediated expression and antioxidant induction of the NAD(P)H: quinone oxidoreductasel gene. J. Biol. Chem. 275, 40134–40141.

Dhakshinamoorthy, S. and Jaiswal, A.K. (2001). Functional characterization and role of INrf2 in antioxidant response element-mediated expression and antioxidant induction of NAD(P)H:quinone oxidoreductase 1 gene. Oncogene. In press.

Favreau, L.V. and Pickett, C.B. (1991). Transcriptional regulation of the rat NAD(P)H:quinone reductase gene. Identification of regulatory elements controlling basal level expression and inducible expression by planar aromatic compounds and phenolic antioxidants. J. Biol. Chem. 266, 4556–4561.

Fujiwara, K.T., Ashida, K., Nishina, H., Iba, H., Miyajima, N., Nishizawa, M. and Kawai, S. (1993). Two new members of the maf oncogene family, mafK and mafF, encode nuclear b-zip proteins lacking putative trans-activator domain. Oncogene 8, 2371–2380.

Grisham, M.B. and McCord, J.M. (1986). Chemistry and cytotoxicity of reactive oxygen metabolites. In: Physiology of Oxygen Radicals, pp. 1–18. Waverly Press, Baltimore.

Igarashi, K., Kataoka, K., Nishizawa, M. and Yamamoto, M. (1994). Regulation of transcription of erythroid factor NF–E2 p45 with small Maf proteins. Nature 367, 568–572.

Inoue, J.I., Kerr, L.K., Rashid, D., Davis, N., Bose, H.R. and Verma, I.M. (1992). Direct association of pp40/IkB β with rel/NF-kB transcription factors-role of ankyrin repeats in the inhibition of DNA binding activity. Proc. Natl. Acad. Sci. USA 89, 4333–4337.

Itoh, K., Chiba, T., Takahashi, S., Ishii, T., Igarashi, K., Katoh, K., Oyake, T., Hayashi, N., Satoh, K., Hatayama, I., Yamamoto, M. and Nabeshima, Y. (1997). An Nrf2/ small maf heterodimer mediates the induction of phase II detoxifying enzyme genes through antioxidant response elements. Biochem. Biophys. Res. Commun. 236, 313–322.

Itoh, K., Wakabayashi, N., Katoh, Y., ishii, T., Igarashi, K., Engel, J.D. and Yamamoto, M. (1999). Keapl represses nuclear activation of antioxidant responsive elements by Nrf2 through binding to the amino-terminal Neh2 domain. Genes Dev. 13, 76–86.

Jaiswal, A.K. (1994). Antioxidant response element.

Biochem. Pharmacol. 48, 439–444.

Joseph, P. and Jaiswal, A.K. (1994). NAD(P)H:Quinone oxidoreductase1 (DT diaphorase) specifically prevents the formation of benzo(a)pyrene quinone-DNA adducts generated by cytochrome P4501A1 and P450 reductase. Proc. Natl. Acad. Sci. USA 91, 8413–8417.

Kasprzak, K.S. (1995). Possible role of oxidative damage in metal-induced carcinogenesis. Cancer Invest. 13, 411–430.

Kataoka, K., Nishizawa, M. and Kawai, S. (1993). Structure-function analysis of the maf oncogene product, a member of the β-zip family. J. Virol. 67, 2133–2141.

Kataoka, K., Noda, M. and Nishizawa, M. (1994). Maf nuclear oncoprotein recognizes sequences related to an AP-1 site and forms heterodimers with both Fos and Jun. Mol. Cell. Biol. 14, 700–712.

Kerr, M.E., Bender, C.M. and Monti, E.J. (1996). An introduction to oxygen free radicals. Heart Lung 25, 200–209.

Kim, M.J. and Andrews, N.C. (1997). Human MafG is a functional partner for p45 NF–E2 in activating globin gene expression. Blood 89, 3925–3935.

Kim, I.F., Mohammadi, E. and Huang, R.C.C. (1999). Isolation and characterization of IPP, a novel human gene coding an actin-binding, KELCH-like protein. Gene 228, 73–83.

Kobayashi, A., Ito, E., Toki, T., Kogame, K., Takahashi, S., Igarashi, K., Hayashi, N. and Yamamoto, M. (1999). Molecular cloning and functional characterization of a new cap'n'collar family transcription factor Nrf3. J. Biol. Chem. 274, 6443–6452.

Kwong, M., Kan, Y.W. and Chan, J.Y. (1999). The CNC basic leucine zipper factor, Nrf1, is essential for cell survival in response to oxidative stress-inducing agents. J. Biol. Chem. 274, 37491–37498.

Last, J.A., Sun, W.M. and Witschi, H. (1994). Ozone, NO, and NO_2: oxidant air pollutants and more. Environ. Health Perspect. 102 Suppl 10, 179–184.

Li, Y. and Jaiswal, A.K. (1992). Regulation of human NAD(P)H:quinone oxidoreductase gene: role of AP1 binding site contained within human antioxidant response element. J. Biol. Chem. 267, 15097–15104.

Li, Y. and Jaiswal, A.K. (1994). Human antioxidant response element regulation of type 1 NAD(P)H:quinone oxidoreductase gene expression: Effect of sulfhydryl modifying agents. Eur. J. Biochem. 226, 31–39.

Liou, H.C. and Baltimore, D. (1993). Regulation of the NF-kB/Rel transcription factor and IkB inhibitor system. Curr. Opin. Cell Biol. 5, 477–487.

Liu, D., Chang, J.C., Moi, P., Liu, W., Kan, Y.W. and Curtin, P.T. (1992). Dissection of the enhancer activity of β-globin 5' DNaseI-hypersensitive site 2 in transgenic mice. Proc. Natl. Acad. Sci. USA 89, 3899–3903.

Marini, M.G., Chan, K., Casula, L., Kan, Y.W., Cao, A. and Moi, P. (1997). hMAF, a small human transcription factor that heterodimerizes specifically with Nrf1 and Nrf2. J. Biol. Chem. 272, 16490–16497.

Meneghini, R. (1997). Iron homeostasis, oxidative stress, and DNA damage. Free Rad. Biol. Med. 23, 783–792.

Mignotte, V., Eleouet, J.F., Raich, N. and Romeo, P.H. (1989). Cis- and trans-acting elements involved in the regulation of the erythroid promoter of the human porphobilinogen deaminase gene. Proc. Natl. Acad. Sci. USA 86, 6548–6522.

Moi, P. and Kan, Y.W. (1990). Synergistic enhancement of globin gene expression by activator protein-1-like proteins. Proc. Natl. Acad. Sci. USA 87, 9000–9004.

Moi, P., Chan, K., Asunis, I., Cao, A. and Kan, Y.W. (1994). Isolation of NF-E2-related factor 2 (Nrf2), a NF-E2 like basic leucine zipper transcriptional activator that binds to the tandem NF–E2/AP1 repeat of β-globin locus control region. Proc. Natl. Acad. Sci. USA 91, 9926–9930.

Mulkahy, R.T., Wartman, M.A., Bailey, H.H. and Gipp, J.J. (1997). Constitutive and β-naphthoflavone-induced expression of the human γ-glutamylcysteine synthetase heavy subunit gene is regulated by a distal antioxidant response element/TRE sequence. J. Biol. Chem. 272, 7445–7454.

Ney, P.A., Sorrentino, B.P., McDonaugh, K.T. and Nienhuis, A.W. (1990). Tandem AP-1-binding sites within the human beta-globin dominant control region function as an inducible enhancer in erythroid cells. Genes Dev. 4, 993–1006.

Nolan, G.P. and Baltimore, D. (1992). The inhibitory ankyrin and activator Rel proteins. Curr. Opin. Genet. Dev. 2, 211–220.

Nguyen, T., Huang, H.C. and Pickett, C.B. (2000). Transcriptional regulation of the antioxidant response element -activation by Nrf2 and repression by MafK. J. Biol. Chem. 275, 15466–15473.

Pickett, C.B. and Lu, A.Y.H. (1989). Glutathione S-transferases: gene structure, regulation and biological function. Ann. Rev. Biochem. 58, 743–764.

Pinkus, R., Weiner, L.M. and Daniel, V. (1996). Role of oxidants and antioxidants in the induction of AP-1, NF-kB, and glutathione S-transferase gene expression. J. Biol. Chem. 271, 13422–13429.

Prestera, T., Holtzclaw, W.D., Zhang, Y. and Talalay, P. (1993). Chemical and molecular regulation of enzymes that detoxify carcinogens. Proc. Natl. Acad. Sci. USA 90, 2965–2969.

Renard, P., Percherancier, Y., Kroll, M., Thomas, D., Virelizier, J. Arenzana, F. and Bachelerie, F. (2000). Inducible NF-κB activation is permitted by simultaneous degradation of nuclear IκBα. J. Biol. Chem. 275, 15193–15199.

Rosen, G.M., Pou, S., Ramos, C.L., Cohen, M.S. and Britigan, B.E. (1995). Free radicals and phagocytic cells. FASEB J. 9, 200–209.

Rushmore, T.H., Morton, M.R. and Pickett, C.B. (1991).

The antioxidant responsive element. Activation by oxidative stress and identification of the DNA consensus sequence required for functional activity. J. Biol. Chem. 266, 11632–11639.

Rushmore, T.H. and Pickett, C.B. (1993). Glutathione-S-transferases, structure, regulation and therapeutic implications. J. Biol. Chem. 268, 11475–11478.

Schreck, R., Meier, B., Mannel, D.N., Droge, W. and Baeuerle, P.A. (1992). Dithiocarbamates as potent inhibitors of NF-kB activation in intact cells. J. Exp. Med. 175, 1181–1194.

Schreck, R., Albermann, K. and Baeuerle, P.A. (1992). Nuclear factor kB: an oxidative stress-responsive transcription factor of eukaryotic cells (a review). Free Rad. Res. Commun. 17, 221–237.

Schulze-Osthoff, K., Bauer, M., Vogt, M., Weesselborg, S. and Baeuerle, P.A. (1997). Reactive oxygen intermediates as primary signals and second messengers in the activation of transcription factors. (Forman, H.J. & Cadenas, E., Eds.), pp. 239–259. Chapman & Hall, New York.

Sen, R. and Baltimore, D. (1986). Multiple nuclear factors interact with the immuno-globulin enhancer sequence. Cell 46, 705–716.

Shivdasani, R. and Orkin, S.H. (1995). Erythropoiesis and globin gene expression in mice lacking the transcription factor NF-E2. Proc. Natl. Acad. Sci. 92, 8690–8694.

Suzuki, Y.J., Forman, H.J. and Sevanian, A. (1997). Oxidants as stimulators of signal transduction. Free Rad. Biol. Med. 22, 269–285.

Thelen, M., Dewald, B. and Baggiolini, M. (1993). Neutrophil signal transduction and activation of the respiratory burst. Physiol. Rev. 73, 797–821.

Tsuchida, S. and Sato, K. (1992). Glutathione transferases and cancer. Crit. Rev. Biochem. Mol. Biol. 27, 337–384.

Venugopal R and Jaiswal A.K. (1996). Nrf1 and Nrf2 positively and c-Fos and Fra1 negatively regulate the human antioxidant response element-mediated expression of NAD(P)H:quinone oxidoreductase1 gene. Proc. Natl. Acad. Sci. USA 93, 14960–14965.

Venugopal, R., Joseph, P. and Jaiswal, A.K. (1997). Gene expression of DT-diaphorase in cancer cells. In: Oxidative stress and signal transduction. (Forman, H.J. and Cadenas, E., Eds.), pp. 441–475. Chapman & Hall, New York.

Venugopal, R. and Jaiswal, A.K. (1998). Nrf2 and Nrf1 in association with Jun proteins regulate antioxidant response element-mediated expression and coordinated induction of genes encoding detoxifying enzymes. Oncogene 17, 3145–3156.

Venugopal, R. and Jaiswal, A.K. (1999). Coordinated induction of c-Jun gene with genes encoding quinone oxidoreductases in response to xenobiotics and antioxidants. Biochem. Pharm. 58, 597–603.

Ward, J.F. (1994). The complexity of DNA damage: relevance to biological consequences. Int. J. Rad. Biol. 66, 427–432.

Wasserman, W. and Fahl, W.E. (1997). Functional antioxidant responsive elements. Proc. Natl. Acad. Sci. USA. 94, 5361–5366.

Wei, Y.H. (1998). Oxidative stress and mitochondrial DNA mutations in human aging. Proc. Soc. Exp. Biol. Med. 217: 53–63.

Wild, A.C., Moinova, H.R. and Mulcahy, R.T. (1999). Regulation of γ-glutamylcysteine synthetase subunit gene expression by transcription factor Nrf2. J. Biol. Chem. 274, 33627–33636.

Wilkinson IV, J., Radjendirane, V., Pfeiffer, G.R., Jaiswal, A.K. and Clapper, M.L. (1998). Disruption of c-Fos leads to increased expression of NAD(P)H:quinone oxidoreductase1 and glutathione S-transferase. Biochem. Biophys. Res. Commun. 253, 855–858.

Xie, T., Belinsky, M., Xu, Y. and Jaiswal, A.K. (1995). ARE- and TRE- mediated regulation of gene expression: Response to xenobiotics and antioxidants. J. Biol. Chem. 270, 6894–6900.

Zabel, U. and Baeuerle, P.A. (1990). Purified human IkB can rapidly dissociate the complex of the NF–kB transcription factor with its cognate DNA. Cell 61, 255–265.

Zheng, M. and Storz, G. (2000). Redox sensing by prokaryotic transcription factors. Biochem. Pharm. 59, 1–6.

Protein Adaptations and Signal Transduction. Edited by K.B. Storey and J.M. Storey

CHAPTER 18

Signal Transduction Cascades Responsive to Oxidative Stress in the Vasculature

Zheng-Gen Jin and Bradford C. Berk
Center for Cardiovascular Research, University of Rochester Medical Center, 601 Elmwood Avenue, Box 679, Rochester, New York 14642, U.S.A.

1. Introduction: Oxidative stress is implicated in the pathogenesis of vascular diseases

Oxidative stress and the production of intracellular reactive oxygen species (ROS), such as superoxide ($O_2\cdot^-$), hydrogen peroxide (H_2O_2) and hydroxyl radical (OH·), have been implicated in the pathogenesis of a variety of diseases. Prominent roles for ROS have been proposed in Alzheimer's disease, cancer, and vascular diseases such as atherosclerosis, hypertension and restenosis. Risk factors for atherosclerosis such as hypercholesterolemia, diabetes mellitus, cigarette smoking, and hypertension generate ROS and have been demonstrated to change vessel redox state (Fig. 18.1). Furthermore, it has become clear that intracellular ROS are generated in response to a variety of stimuli including growth factors (Sundaresan et al., 1995), hormones (Griendling et al., 1994), and inflammatory cytokines (Rahman et al., 1998) and hemodynamic forces (Laurindo et al., 1994) (Fig. 18.1). Vascular injury, which occurs during percutaneous coronary interventions for patients with coronary disease, also produces a large amount of ROS (Nunes et al., 1995; Tardif et al., 1997) (Fig. 18.1). Several enzyme systems, including mitochondrial oxidative phosphorylation, arachidonic acid (AA) pathway enzymes (e.g. lipoxygenase, cyclooxygenase, and cytochrome P450 monooxygenase), xanthine oxidase, NADH/NADPH oxidase, uncoupled endothelial nitric oxide synthase (eNOS), and peroxidases, are likely enzymatic sources contributing to increased production of ROS in the vasculature. When produced in excess, ROS and their byproducts could be cytotoxic to cells. However, it is now well established that moderate levels of ROS play a role as second messengers to regulate signal transduction processes that ultimately control gene expression and posttranslational modification of proteins. In the vasculature, redox-sensitive signaling pathways regulate multiple functions, including increased intracellular calcium concentration, cytokine production, adhesion molecule expression (De Keulenaer et al., 2000), vascular smooth muscle cell (VSMC) growth (Baas and Berk, 1995), endothelial cell (EC) apoptosis (Lin et al., 2000), and decreased nitric oxide bioactivity (Peterson et al., 1999). Eventually, these effects of ROS lead to an increase in vessel tone (Wilson et al., 1999), up-regulation of vascular inflammatory gene expression (De Keulenaer et al., 2000; Marui et al., 1993), and to alteration of vascular remodeling (Nunes et al., 1995; Tardif et al., 1997), which eventually contribute to the development of vascular diseases (Fig. 18.1).

Several key findings from our laboratories support the concept that ROS contribute to vascular diseases. Initial studies in our lab demonstrated that ROS stimulate cultured VSMC proliferation and activate intracellular kinases such as mitogen-activated protein kinases (MAPKs) which are associated with proto-oncogene expression and cell

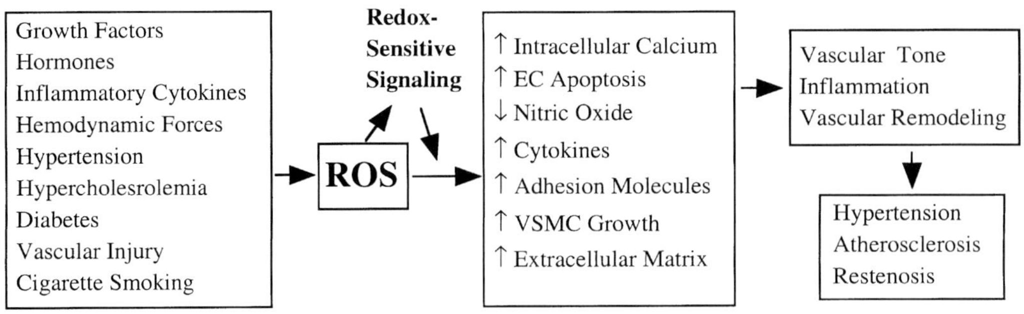

Fig. 18.1. Oxidative stress is involved in the pathogenesis of vascular diseases. The scheme illustrates the potential stimuli of ROS production and the physiological and pathophysiological effects of ROS on VSMC and EC as well as their implication in vascular diseases.

growth (Baas and Berk, 1995; Rao and Berk, 1992; Rao et al., 1993). Furthermore, we observed that in the injured vessel wall there is increased ROS production. Antioxidants decrease neointimal VSMC proliferation and promote outward vessel remodeling in the pig coronary injury model (Nunes et al., 1997; Nunes et al., 1995). We showed that many protein kinases are involved in ROS-mediated signaling and gene expression (Abe and Berk, 1999; Abe et al., 1996; Abe et al., 2000; Baas and Berk, 1995; Yoshizumi et al., 2000). Recently we have demonstrated the novel finding that ROS-stimulated VSMC synthesize and secrete proteins which may activate intracellular events in an autocrine and paracrine fashion (Jin et al., 2000; Liao et al., 2000). In this chapter, we discuss several important mechanisms by which ROS modulate the function of VSMCs and ECs. We will focus on signal transduction pathways that are involved in the regulation of gene expression and cell growth by oxidative stress, and discuss their physiological and pathological relevance.

2. Cellular sensors of oxidative stress

The functions of cells are regulated by a variety of extracellular signals (ligands), including growth factors, hormones, cytokines, extracellular matrix (ECM), mechanical stress and ROS. The transmission of extracellular signals to their intracellular targets is mediated by a network of interacting proteins that governs a large number of cellular functions, such as growth, migration, differentiation and death (Takahashi and Berk, 1998). These signaling pathways are normally under the control of ligands binding to their cognate receptors. However, ROS are unique in their ability to either bypass this normal receptor control or directly activate these pathways as well as co-opting these receptors by a process termed transactivation. There are a number of potential sensors of oxidative stress in cells, such as protein tyrosine kinases (PTKs), GTP-binding protein (G-protein) and ion channels, to transduce the redox signals to intracellular mediators present in both the cytoplasm and the nucleus (Fig. 18.2).

2.1. Receptor tyrosine kinases (RTKs)

Growth factor receptors with intrinsic tyrosine kinase (RTK) activity can transmit growth signals from the cell exterior to interior. Upon the binding of growth factors, RTKs are activated and autophosphorylated on intracellular tyrosine residues, leading to recruitment of Src homology 2 (SH2) domain-containing adaptor proteins such as Shc and Grb2. These proteins, in turn, activate Ras by Src homology 3 (SH3)-mediated interactions with guanine nucleotide exchange proteins such as Sos. This triggers protein phosphorylation events involving sequential activation of MAPK cascades, which play a pivotal role in cell proliferation and differentiation. In VSMC, H_2O_2 rapidly induced tyrosine phosphorylation of epidermal growth factor (EGF) receptor (a process we will term

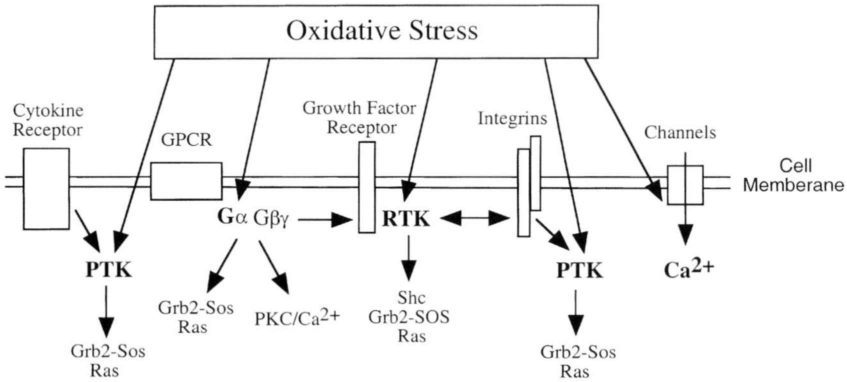

Fig. 18.2. Schematic diagram of proposed cellular sensors responsive to oxidative stress. GPCR: G protein-coupled receptor; RTK: Receptor tyrosine kinases; PTK: Protein tyrosine kinases.

transactivation) and then induced the association of the Shc–Grb2–Sos complex with the EGF receptor, in a ligand-independent manner, resulting in activation of downstream signaling (Rao, 1996). H_2O_2 also induces tyrosine phosphorylation of the platelet-derived growth factor (PDGF) receptor in VSMC (Jin et al., 2000). We have also found that antioxidants inhibited angiotensin II (Ang II)-induced phosphorylation of both the PDGF-receptor and Shc, suggesting ROS is involved in transactivation of the PDGF receptor by Ang II (Heeneman et al., 2000). In EC, oxidized LDL and 4-hydroxynonenal induce tyrosine phosphorylation of the EGF receptor (Suc et al., 1998). We have observed that H_2O_2 stimulated tyrosine phosphorylation of the vascular endothelial growth factor (VEGF) receptor, FIK-1, resulting in downstream signals PI_3K and ERK1/2 activation (Ueba and Berk, unpublished data). These results suggest that RTKs, act as redox sensors in a ligand-independent manner, and sequentially transduce the signals of oxidative stress to downstream targets.

2.2. *G protein-coupled receptors and G proteins*

A family of heterotrimeric G proteins, consisting of α, β, and γ subunits, plays a pivotal role in signal transduction by G protein-coupled receptors (GPCR). In the cardiovascular system, G proteins are involved in hormonal signaling (e.g. norepinephrine, acetylcholine, Ang II and endothelin) that regulate adenylyl cyclases, phospholipases, and ion channels as well as signal pathways that activate protein kinases including MAPKs (Berk, 1999). Recently, it has been shown that ROS-induced activation of extracellular-regulated protein kinases 1 and 2 (ERK1/2, a member in MAPK family) is attenuated through inhibition by the βγ-subunits of G protein ($G_{\beta\gamma}$) in cardiomyocytes (Nishida et al., 2000). G_i, G_o and G_s are three subfamilies of the heterotrimeric G proteins, and H_2O_2 directly activates G_i and G_o (but not G_s) without receptor activation. Further analysis indicates that H_2O_2 modifies the α-subunit of heterotrimeric G_i and G_o (termed $G_{\alpha i}$ and $G_{\alpha o}$, respectively) but not $G_{\beta\gamma}$, which leads to liberation of $G_{\beta\gamma}$ and activation of downstream signal transduction, such as activation of ERK1/2. These results demonstrated that $G_{\alpha i}$ and $G_{\alpha o}$ are target proteins of ROS for activation of ERK1/2 in cardiomyocytes (Nishida et al., 2000). The ability of ROS to activate G proteins directly in other cell types, such as VSMC and EC remains to be determined.

2.3. *Non-receptor protein tyrosine kinases (PTKs)*

Many receptors that lack intrinsic tyrosine kinase activity are nonetheless capable of activating tyrosine kinase signal pathways by associating with non-receptor cytoplasmic protein tyrosine kinases

(PTKs). PTKs of the Src family, including c-Src, Lck, Fyn and ZAP-70, are the first tyrosine kinases activated by many receptors that lack intrinsic tyrosine kinase activity. It has been reported that H_2O_2, pervanadate and UV light induce tyrosine phosphorylation and activation of Lck, Fyn and ZAP-70 in T cells in a receptor-independent manner (Griffith et al., 1998). We have observed that H_2O_2 stimulates c-Src activation in fibroblasts and VSMC, which is required for oxidative stress-mediated activation of big mitogen-activated protein kinase 1 (BMK1) (Abe et al., 1997). We also demonstrated that c-Src mediated JNK activation and Fyn mediated ERK1/2 activation by ROS in VSMC (Abe and Berk, 1999; Yoshizumi et al., 2000). Other non-receptor PTKs, such as focal adhesion kinases (FAK) and FAK-related kinase Pyk2 (proline-rich tyrosine kinase 2), are also stimulated by oxidative stress (see below).

Although oxidative stress stimulates PTK activity in cells, there is to date no evidence that has shown that oxidants directly activate PTKs. Three mechanisms have been proposed by which ROS may activate tyrosine kinases. First, ROS may directly activate kinases by altering protein-protein interactions depending on sulfhydryl groups. Second, inhibition of phosphotyrosine phosphatases (PTPases) may be important since the level of tyrosine phosphorylation of cellular proteins is determined by the balance of PTKs and PTPases. Third, protein oxidation may stimulate proteolysis of regulatory proteins that normally inhibit tyrosine kinase activity.

2.4. Integrin signaling

Vascular cells are surrounded by basement membranes that consist of a large number of extracellular matrix (ECM) proteins. Changes in ECM composition are closely associated with the development of many vascular diseases. The best studied signal transduction system for ECM–cell interaction are the integrin-matrix ligand pathways (Boudreau and Jones, 1999). Ligation of the heterodimeric ($\alpha\beta$) integrin receptor to ECM proteins such as fibronectin, collagen and vitronectin leads to elevation of intracellular calcium and the tyrosine phosphorylation of diverse PTKs, such as Src kinases, FAK and PyK2, as well as adaptor proteins such as Shc, Grb2, p130Cas, Crk (Schlaepfer et al., 1999). These signaling events trigger a number of downstream signals, including not only cytoskeleton reorganization and cell migration, but also activation of MAPK signaling pathways, resulting in a mitogenic response (Boudreau and Jones, 1999; Ishida et al., 1996; Takahashi and Berk, 1996). So far, it is not clear how ROS directly stimulate integrins. However, important effects of ROS on integrin downstream signaling molecules have been reported recently. H_2O_2 induced a time- and dose-dependent tyrosine phosphorylation of FAK, p130Cas and paxillin, three major cytoskeleton proteins, in EC. We have also demonstrated that shear stress stimulated activation of FAK (Ishida et al., 1996), p130Cas (Okuda et al., 1999), and Pyk2 (Okuda and Berk, unpublished results) in EC. Shear stress-induced p130Cas tyrosine phosphorylation requires calcium-dependent c-Src activation (Okuda et al., 1999), and activation of Pyk2 is inhibited by antioxidants (Okuda and Berk, unpublished data), suggesting that shear stress-induced ROS production may mediate these downstream signaling of integrins. In addition, focal adhesion kinase-related tyrosine kinase Pyk2 is activated by ROS in VSMC (Frank et al., 2000). However, the exact role of oxidative stress in integrin signaling remains to be determined.

2.5. Ion channels

Within cells of the arterial wall, membrane potential and cell–cell signaling are governed by an array of ion channel types, which are functionally modulated by extracellular and intracellular factors. Many ion channels possess thiol residues in regulation domains that are susceptible to oxidation (Kourie, 1998). A glutathione-operated cation channel was recently discovered in vascular EC. H_2O_2 activates this channel and causes membrane depolarization. Ca^{2+} entry via this channel may contribute to the increase in resting cytosolic Ca^{2+} concentration during prolonged oxidative stress. The large-conductance Ca^{2+} activated K^+ channels in VSMC are also modulated by glutathione, which is likely to represent a key mechanism by which blood flow and tissue perfusion are modulated under conditions of oxidative stress (Elliott et al., 1998). Like K^+ channels, voltage-sensitive Ca^{2+} channels also contain thiol residues that are targets for oxidation. Thiol oxidants trigger the opening of the ryanodine receptor, a Ca^{2+}-selected ion channel, and the release Ca^{2+} from sarcoplasmic reticulum. Thus, the redox regulation of ion channels is likely involved in the physiological and pathological roles of oxidative stress in vascular system.

A key role for ROS-mediated receptor transactivation and cross-talk among signal pathways is suggested by recent studies. Cells can not only transmit and integrate multiple signals from a vast array of extracellular stimuli into intracellular mediators by shared, convergent intracellular signaling molecules, but also convert a single extracellular stimulus to a variety of signal pathways by cross-talk between heterologous signaling systems of the cell to achieve the regulation of diverge functions. This cross-talk represents a new dimension of complexity in the molecular communication network that governs physiological processes. There is accumulating evidence that GPCR, such as the Ang II type 1 receptor, transactivate growth factor receptors, such as EGF receptors (Wang et al., 2000) and PDGF receptors (Heeneman et al., 2000) to mediate GPCR-induced cell functions. Furthermore, ligated integrins recruit growth factor receptors to the focal adhesion site and result in enhanced receptor signaling (Borges et al., 2000). Alternatively, ligated growth factor receptors, such as vascular endothelial growth factor (VEGF), activate integrins which are involved in VEGF-induced angiogenesis (Byzova et al., 2000). In addition, both RTKs and PTKs modulate ion channels (Hu et al., 1998). Considering the fact that these cellular sensors cross-talk with each other, it is possible that oxidative stress-induced multiple signaling might be explained by signaling cross-talk. Alternatively, ROS may themselves mediate signaling cross-talk, since activation of growth factor receptors, hormonal receptors and cytokine receptors produce ROS. Indeed, we have demonstrated that Ang II-induced tyrosine phosphorylation of PDGF receptor was inhibited by antioxidants (Heeneman et al., 2000). Similar results were observed for Ang II-induced transactivation of the EGF receptor (Wang et al., 2000). These results suggest that ROS play a role in receptor transactivation.

3. Redox regulation of phospholipid-dependent signaling

Phospholipase A_2, phospholipase C, and phospholipase D (PLA_2, PLC and PLD) have been recognized as important effector enzymes involved in transducing signals from the exterior through specific receptor binding to the generation of second messages. Activation of PLC, causes hydrolysis of PIP_2 to generate IP_3 and DAG (Fig. 18.3). IP_3 acts as an intracellular second messenger, stimulating

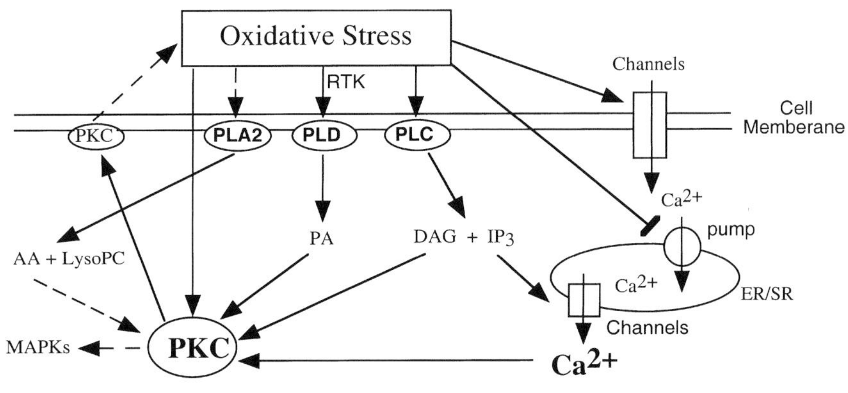

Fig. 18.3. Schematic representation of oxidative stress-mediated phospholipid-dependent signaling.

the release of Ca^{2+} from endoplasmic reticulum (ER), while DAG activates protein kinases C (PKC). Several studies suggest that these phospholipid-dependent signaling are important targets for oxidative stress in the vasculature.

3.1. Phospholipase and oxidative stress

At least two important phospholipases, PLC and PLD, have been shown to be activated by ROS. Exposure of EC to oxidants causes a release of DAG and IP_3, suggesting activation of PLC. H_2O_2 and 4-hydroxynonenal stimulated PLD in EC (Min et al., 1998). Both H_2O_2 and endogenous ROS generator, menadione, stimulated accumulation of both IP_3 (PLC activation) and phosphatidic acid (PA) (PLD activation) (Servitja et al., 2000). The mechanisms of ROS mediated activation of PLC and PLD are not clear. As ROS can directly activate G proteins, ROS-stimulated activation of phospholipases may result from G-protein(s) activation. In addition, the effects of vanadate (an inhibitor of PTPases) and genistein (a PTK inhibitor) on H_2O_2-induced activation of PLC and PLD (Servitja et al., 2000), suggests that PTKs/PTPases may be involved.

3.2. Calcium signaling and oxidative stress

Both EC and VSMC are highly dependent on changes in intracellular calcium for normal biological functions. For example, eNOS is a calcium-calmodulin dependent enzyme and activation of myosin light chain kinase is dependent on a rise in calcium. ROS have been shown to stimulate an increase in intracellular calcium by several pathways (Fig. 18.3). First, ROS increase IP_3 production by stimulating PLC activity. IP_3 binds to an IP_3-sensitive calcium channel, and causes intracellular calcium release from endoplasmic reticulum (ER) and sarcoplasmic reticulum (SR). ROS also increase the sensitivity of intracellular Ca^{2+} stores to IP_3. It has been proposed that ROS stimulate formation of mixed disulfides which oxidize glutathione (GSSG) that interact with the IP_3 receptor/channel to stimulate calcium release. Second, a calcium-ATPase (pump) is required to maintain calcium within the IP_3 sensitive pool and inhibiting this ATPase would increase intracellular calcium. Additionally, ROS-induced change in permeability of calcium channels and exchangers in cell membrane may dramatically alter intracellular calcium (Kourie, 1998).

3.3. Protein kinase C (PKC) and oxidative stress

PKC represents a family of phospholipid-dependent serine/threonine kinases that are involved in a variety of pathways that regulate cell growth, death, and stress responsiveness. For example, we have shown that H_2O_2 stimulated expression of the proto-oncogenes, c-myc and c-fos, was primarily PKC dependent (Rao and Berk, 1992), PKC-ζ mediated Ang II activation of ERK1/2 in VSMC (Liao et al., 1997) and PKC-ε is required for shear stress-induced ERK1/2 activation in EC (Traub et al., 1997). Oxidative stress regulates PKC activation not only by mediating the activity of phospholipases (Fig. 18.3), but also by modifying PKC itself. Both the regulatory and catalytic domains of PKC contain cysteine-rich regions that are targets for redox regulation. H_2O_2 selectively modifies the regulatory domain and leads to constitutive activation of PKC. Like phorbol esters, sustained H_2O_2 treatment caused PKC down-regulation. In EC, H_2O_2 induced membrane translocation of PKC. The catalytic domain is also a target for inactivation by antioxidants. In addition, PKC activation may increase ROS production by regulation of NADPH oxidase activity.

4. Mitogen activated protein kinases as the primary redox-sensitive signal mediators

MAPKs are among the key downstream signaling components of ROS-activated PTKs. Small G proteins serve as intermediaries from PTKs to MAPK signal cascades in response to oxidative stress. Subsequently, the stimulation of MAPKs plays a role in ROS-induced activation of nuclear transcription factors.

4.1. Small G proteins as intermediates from PTKs to MAPK signaling in response to oxidative stress

Small G proteins such as Ras, Rac/CDC42 and Rho, have been proposed to be important mediators of ROS signaling. Activation of ERK1/2 by H_2O_2 is prevented in cells in which Ras activity is blocked (Lander et al., 1995). Using recombinant Ras *in vitro* it was found that ROS directly promote guanine nucleotide exchange on Ras. These results suggest that Ras may act as a signal mediator for the redox signal in cells. We demonstrated that H_2O_2 stimulated Ras activation as well as its downstream ERK1/2, and activation of Ras and ERK1/2 by ROS is mediated by Fyn but not c-Src (Abe and Berk, 1999; Abe et al., 2000). Dominant negative Ras completely blocked H_2O_2-induced activation of ERK1/2 and its substrate p90RSK (Abe et al., 2000). Our findings further support the concept that small G-proteins act as intermediates from PTKs to MAPK signaling cascades by oxidative stress. In addition it also appears that another small G protein, Rac, may regulate ROS production (Sundaresan et al., 1996).

4.2. MAPK pathways in redox sensitive signal transduction

MAPK family members have been proposed to mediate the many changes in gene expression observed in response to ROS. MAPKs are serine and threonine protein kinases. Three classes of dual-specificity MAPKs are defined based on their activation motifs (TEY, TPY, and TGY), which we term ERK1/2 and BMK1 (also called ERK5), c-jun N-terminal protein kinase (JNK), and p38, respectively. ERK6 (also called SAPK3 and p38γ) and ERK7 are recently identified as members of MAPK family. The specificity for MAPK activation is determined, in part, by members of the MEK family, which exhibit unique pairing with downstream MAPKs (Fig. 18.4). The specificity of activation of MAPKs by individual stimuli is reiterated by specific substrates for each class. Common substrates for the MAPKs are transcription factors that, upon phosphorylation, induce changes in gene expression. ERK1/2 phosphorylate the ternary complex factor (TCF)/Elk-1 on sites essential for transactivation. JNK phosphorylates c-jun and increases its transcriptional

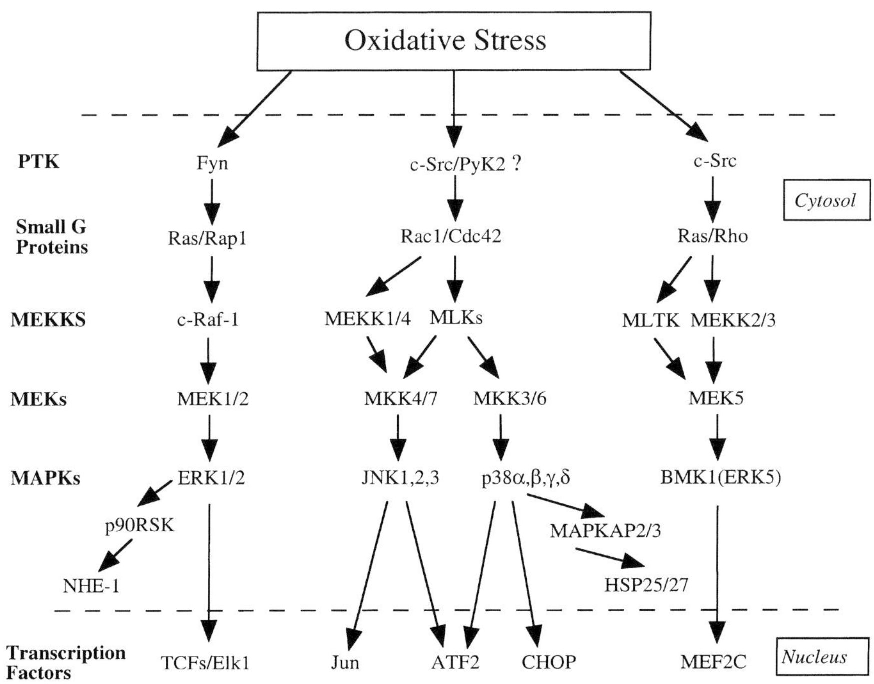

Fig. 18.4. Model for signal transduction cascades for activation of MAPKs by oxidative stress.

activating potential. ATF2 is phosphorylated and activated by both JNK and p38. BMK1 (ERK5) activates the myocyte enhancer factor-2 (MEF2). BMK1 contains a unique long C-terminal domain, which has MEF2-interacting domain and potent transcriptional activation domain (Kasler et al., 2000). The target substrates for ERK7 remain to be determined.

Extracellular signal regulated kinases 1 and 2 (ERK1/2)

Several reports demonstrate the involvement of the ERK1/2 signaling pathway in redox-mediated signaling in the vasculature. Initially, our group demonstrated that addition of the superoxide-generating agent LY83583 to VSMC resulted in a concentration-dependent increase in ERK1/2 activity and cell proliferation (Baas and Berk, 1995). Various physiological agents including oxLDL (Kusuhara et al., 1997), PDGF (Sundaresan et al., 1995), linoleic acid and its metabolites (Rao et al., 1995), also activate ERK1/2 activity in VSMC via the generation of intracellular ROS. The mechanisms underlying activation of ERK1/2 by ROS are not fully understood. Recently, we found that Fyn mediated ROS-induced Ras activation and ERK1/2 activation (Abe and Berk, 1999). In EC, oxLDL increased intracellular ROS and stimulated ERK1/2 activation, and shear stress produced ROS and mediated ERK1/2 activation. The ERK1/2 activity stimulated by ROS in EC may be involved in the activation of AP-1 and afford cellular protection against oxidative stress. ERK1/2 phosphorylates TCF/ Elk-1 on sites essential for transactivation. We found that H_2O_2 stimulated c-fos transcription and ERK1/2 activation in VSMC (Baas and Berk, 1995; Rao and Berk, 1992). Also, bradykinin-induced ERK1/2 activation and expression of c-fos mRNA in VSMC are mediated via the generation of ROS (Greene et al., 2000). p90RSK kinase is an additional substrate for ERK1/2 (Abe et al., 2000; Takahashi et al., 1999), which in turn activates sodium hydrogen exchanger 1 (NHE1) (Takahashi et al., 1999).We also found that H_2O_2 stimulated p90RSK in VSMC and cardiac myocytes (Abe et al., 2000; Takeishi et al., 1999).

Stress activated protein kinases (SAPK): c-Jun N-terminal kinase (JNK) and p38

ROS are involved in activation of JNK and p38 in the vasculature. We have shown that H_2O_2 stimulated both JNK and p38 in a concentration dependent manner in VSMC (Yoshizumi et al., 2000). c-Src and p130Cas are involved in activation of JNK but not ERK1/2 and p38 by ROS, as only JNK activation was blocked in c-Src deficient cells and Cas dominant negative mutant-transfected cells (Yoshizumi et al., 2000). Ang II stimulated an increase of intracellular H_2O_2 which mediated Ang II-induced activation of JNK and p38 but not ERK1/2 (Ushio-Fukai et al., 1998; Zafari et al., 1998). Also we have demonstrated that Ang II rapidly stimulated p21-activated kinase 1 (PAK1), an upstream regulation of both p38 and JNK in VSMC (Schmitz et al., 1998). In EC, H_2O_2 activated p38 activity with an associated reorganization of the stress-fiber network. ROS are involved in leptin-induced signaling in EC, including JNK activation, activator protein 1 (AP-1) DNA binding and monocyte chemoattractant protein-1 (MCP-1) expression (Bouloumie et al., 1999). JNK is tightly bound to and phosphorylates its substrates, transcription factors including c-Jun, ATF2 and Elk1, whereas p38 kinase phosphorylates its substrates, transcription factors ATF2, CHOP and MEF2C, increasing their transactivating activity. In addition, p38 activates the MAPK-activated protein kinases 2 and 3 (MAPKAP2/3), which in turn phosphorylate the small heat shock proteins (Hsp25/27) to mediate reorganization of the stress-fiber network.

Big MAP kinase 1 (BMK1)

In our laboratory we have observed that, in response to several different agonists, BMK1 was stimulated to the greatest extent by H_2O_2, with a relative potency of H_2O_2 > PDGF > PMA = TNF in VSMC (Abe et al., 1996). H_2O_2 activation of BMK1 was calcium and tyrosine kinase dependent but independent of PKC action (Abe et al., 1996). Stimulation of BMK1 by H_2O_2 appeared ubiquitous as shown by increases in BMK1 activity in human fibroblasts, VSMC and EC (Abe et al., 1996). An essential role for c-Src in H_2O_2-mediated

BMK1 activation in VSMC is suggested as shown by inhibition with specific Src family kinase inhibitors and in c-Src deficient cells (Abe et al., 1996). MEK5-BMK1 activation results in the phosphorylation of MEF2C, a transcription factor belonging to MEF2 family, and BMK1 is required for cell proliferation induced by EGF (Kato et al., 1998).

5. Regulation of gene expression and protein secretion by oxidative stress

5.1. Redox regulation of transcription factor activity and gene expression

As consequences of oxidative stress-mediated signal transduction, a large number of genes are induced in vascular cells undergoing oxidative stress or in cellular redox status. Transcription factors are central to gene regulation, as they are the nuclear components that are regulated by upstream signaling events. In the vasculature, oxidative stress has been shown to regulate many transcription factors that are implicated in transcriptional regulation of a wide range of genes involved in inflammation and proliferation in the vasculature, which was implicated in the pathogenesis of restenosis and atherosclerosis. AP-1 and NF-κB are the two best characterized transcription factors subject to redox regulation in vascular cells.

Activator protein (AP-1)

AP-1 is a heterodimeric complex of the Jun and Fos proteins. AP-1 proteins bind to TPA-response elements (TRE) or AP-1 binding sites to transcriptionally activate effector genes. AP-1 behaves as a redox-sensitive transcription factor. Our group was the first to report that H_2O_2 induced AP-1 expression or DNA binding activity in VSMC (Rao and Berk, 1992). Similar results have been observed with oxLDL and lipid peroxidation product, 4-hydroxynonenal, in VSMC. In EC, H_2O_2 and oxLDL activate AP-1 DNA binding activity, resulting in expression of vascular inflammatory gene MCP-1 and intracellular adhesion molecule-1 (ICAM-1) (Wung et al., 1997). Posttranslational modifications by oxidative stress may modulate AP-1 activity. We have shown that PKC and arachidonic acid (AA) are involved in induction of c-fos and c-jun expression in VSMC (Rao et al., 1993; Rao et al., 1993). In addition, both tyrosine and serine/threonine phosphorylation are required for H_2O_2-induced AP-1 activation (Barchowsky et al., 1995).

Nuclear factor κB (NFκB)

Functional NF-κB is composed of heterodimers formed from the p50 and p65 subunits. In unstimulated cells, the nuclear localization sequence of NF-κB is bound to inhibitory proteins (IκB), including IκBα, IκBβ, and IκBγ. Agents that activate NF-κB induce specific phosphorylation of IκB via IκB kinase activity, which directs IκB to a degradation pathway. Dissociation of IκB and NF-κB proteins permits translocation of the active NF-κB dimer to the nucleus, where it can bind to DNA and initiate transcription. H_2O_2 activates NF-κB and increases its DNA binding activity, and the antioxidant PDTC inhibits PMA-induced NF-κB activation in EC (Marui et al., 1993). Both TNFα and oxLDL-induced NF-κB activation and the resultant expression of adhesion molecules (such as E-selectin) are mediated by ROS (Cominacini et al., 2000). In VSMC, human cytomegalovirus infection generates ROS and thereby activates NF-κB, which causes gene expression involved in the pathogenesis of atherosclerosis (Speir et al., 1998). Recently, it was demonstrated that parallel ROS-dependent and ERK1/2-dependent pathways converge to regulate TNFα-induced MCP-1 gene expression in VSMC (De Keulenaer et al., 2000). Flow-adapted endothelial cells generate ROS with ischemia that results in activation of NF-κB and AP-1 and an increase of DNA synthesis (Wei et al., 1999). The precise mechanisms through which oxidants and reductants influence activation of NF-κB is presently unknown; however, there is evidence that antioxidants inhibit IκB kinase activity and prevent the phosphorylation and subsequent degradation of IκB (Jin et al., 1997).

5.2. *Protein secretion in response to oxidative stress*

Increasing evidence suggests secretion of growth factors in response to VSMC agonists mediate their mitogenic activity. For example, Gas6 is secreted from rat VSMC after stimulation by Ang II and thrombin, and modulates VSMC survival (Melaragno et al., 1998). Epiregulin, an EGF-related growth factor, is a potent VSMC-secreted mitogen whose expression is regulated by Ang II, endothelin-1 and thrombin (Taylor et al., 1999). Expression and secretion of heparin-binding EGF-like growth factor (HB-EGF) was also found in growth factor stimulated VSMC (Dluz et al., 1993). However, no factors have been identified as mediators of VSMC proliferation in response to ROS.

Recently we have observed that brief exposure of VSMC to ROS generators such as LY83583, menadione, and xanthine/xanthine oxidase, stimulated ERK1/2 with two peaks, an early peak at 10 min and a late peak at 120 min, with return to baseline by 360 min (Liao et al., 1997; Liao et al., 2000). In contrast, LY83583-induced $O_2^{\cdot-}$ generation peaked at 15 min and returned to baseline by 2 h (Liao et al., 2000). These results suggest the late activation of ERK1/2 by ROS may be due to secretion of factors into the medium. Purification of potential secreted oxidative stress-induced factors (SOXF) from conditioned medium from ROS-stimulated VSMC by sequential chromatography suggested that heat shock protein 90 (HSP90) and proteins of the cyclophilin family might act as a SOXF (Liao et al., 2000). Most recently, we have demonstrated that oxidative stress regulated not only expression of HSP90 and cyclophilin A (CyPA) but also secretion of HSP90 and CyPA (Jin et al., 2000; Liao et al., 2000). Secreted HSP90 and CyPA contribute to ROS-induced late-phase ERK1/2 activation and growth of VSMC (Fig. 18.5). This was the first study to identify factors involved in ROS-mediated signaling and physiological consequence in VSMC. The peptidyl–prolyl isomerase activity (PPIase) of CyPA is required for the function of secreted CyPA since CyPA-induced ERK1/2 activation was inhibited by the immunosuppressive drug cyclosporine A (Jin et al., 2000). To study the role of intracellular ROS in CyPA function, we studied cells transfected with *nox1* (homolog of the NADPH oxidase catalytic subunit) to stimulate ROS production (Suh et al., 1999). CyPA expression and secretion were increased, and antioxidants blocked the secretion of CyPA from *nox1*-transfected cells (Jin et al., 2000). In the balloon-injured rat carotid model, CyPA expression was dramatically increased with a time-course that paralleled neointima formation (Jin et al., 2000). These novel findings suggest that CyPA and HSP90 are secreted redox-sensitive mediators, which may play important roles in the pathogenesis of vascular diseases. Understanding the mechanisms by which ROS stimulate gene expression and protein secretion should provide important insights into the cellular response to oxidative stress.

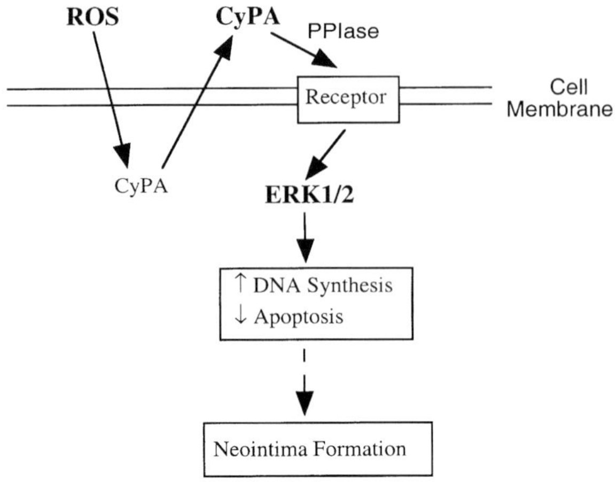

Fig. 18.5. Model for cyclophilin A (CyPA) expression and secretion in response to oxidative stress.

6. Conclusion

Our current knowledge of the oxidative stress-induced signal transduction pathways in the vasculature has been briefly reviewed. The events of oxidative stress-induced signal transduction occur at different levels. The first events include sensing and converting oxidative stress mediated by RTKs, G proteins, PTKs and ion channels. Second events include integrating intracellular signals mediated by calcium, PKC, and MAPKs. The final events are the cellular biological responses, including transcription factor activation, and gene expression and protein secretion, leading to cell proliferation or apoptosis. It is noted that the regulation of cellular signaling by oxidative stress does not have to follow the above signaling sequence. In fact, the specific response elicited by ROS is determined by their specific intracellular target(s) at different levels. This response is dependent on the species of ROS produced, the source and subcellular localization of ROS, the kinetics of production, and the quantities produced. It is imperative that future efforts be directed toward better defining and characterizing the signaling pathways regulated by ROS in vascular cells. Such efforts will likely yield new molecular targets and ultimately more effective therapies for preventing or ameliorating vascular diseases such as atherosclerosis, restenosis, and hypertension, through fine modulation of ROS-regulated signaling.

References

Abe, J. and Berk, B.C. (1999). Fyn and JAK2 mediate Ras activation by reactive oxygen species. J. Biol. Chem. 274, 21003–21010.

Abe, J., Kusuhara, M., Ulevitch, R.J., Lee, J.-D. and Berk, B. C. (1996). The role of big MAP kinase 1 (BMK1) as a redox sensitive kinase in vascular smooth muscle cells. Circulation (abstract) 94, I–280.

Abe, J., Kusuhara, M., Ulevitch, R.J., Berk, B.C. and Lee, J.D. (1996). Big mitogen-activated protein kinase 1 (BMK1) is a redox-sensitive kinase. J. Biol. Chem. 271, 16586–16590.

Abe, J., Okuda, M., Huang, Q., Yoshizumi, M. and Berk, B.C. (2000). Reactive oxygen species activate p90 ribosomal S6 kinase via fyn and ras. J. Biol. Chem. 275, 1739–1748.

Abe, J., Takahashi, M., Ishida, M., Lee, J.D. and Berk, B.C. (1997). c-Src is required for oxidative stress-mediated activation of big mitogen-activated protein kinase 1. J. Biol. Chem. 272, 20389–20394.

Abe, M.K., Kuo, W.L., Hershenson, M.B. and Rosner, M.R. (1999). Extracellular signal-regulated kinase 7 (ERK7), a novel ERK with a C- terminal domain that regulates its activity, its cellular localization and cell growth. Mol. Cell. Biol. 19, 1301–1312.

Baas, A.S. and Berk, B.C. (1995). Differential activation of mitogen-activated protein kinases by H_2O_2 and O_2^- in vascular smooth muscle cells. Circ. Res. 77, 29–36.

Barchowsky, A., Munro, S.R., Morana, S.J., Vincenti, M.P. and Treadwell, M. (1995). Oxidant-sensitive and phosphorylation-dependent activation of NF-κ B and AP-1 in endothelial cells. Am. J. Physiol. 269, L829–36.

Berk, B.C. (1999). Angiotensin II signal transduction in vascular smooth muscle: pathways activated by specific tyrosine kinases. J. Am. Soc. Nephrol. 10 Suppl 11, S62–68.

Borges, E., Jan, Y. and Ruoslahti, E. (2000). PDGF-receptor-β and VEGF-receptor-2 bind to the β3 integrin through its extracellular domain. J. Biol. Chem. 275, 39867–39873.

Boudreau, N.J. and Jones, P.L. (1999). Extracellular matrix and integrin signaling: the shape of things to come. Biochem. J. 339, 481–488.

Bouloumie, A., Marumo, T., Lafontan, M. and Busse, R. (1999). Leptin induces oxidative stress in human endothelial cells. FASEB J. 13, 1231–1238.

Byzova, T.V., Kim, W., Midura, R.J. and Plow, E.F. (2000). Activation of integrin α(V)β(3) regulates cell adhesion and migration to bone sialoprotein. Exp. Cell. Res. 254, 299–308.

Cominacini, L., Pasini, A.F., Garbin, U., Davoli, A., Tosetti, M.L., Campagnola, M., Rigoni, A., Pastorino, A.M., Lo Cascio, V. and Sawamura, T. (2000). Oxidized low density lipoprotein (ox-LDL) binding to ox-LDL receptor-1 in endothelial cells induces the activation of NF-κB through an increased production of intracellular reactive oxygen species. J. Biol. Chem. 275, 12633–12638.

De Keulenaer, G.W., Ushio-Fukai, M., Yin, Q., Chung, A.B., Lyons, P.R., Ishizaka, N., Rengarajan, K., Taylor, W.R., Alexander, R.W. and Griendling, K.K. (2000). Convergence of redox-sensitive and mitogen-activated protein kinase signaling pathways in tumor necrosis factor-α-mediated monocyte chemoattractant protein-1 induction in vascular smooth muscle cells. Arterioscler. Thromb. Vasc. Biol. 20, 385–391.

Dluz, S. M., Higashiyama, S., Damm, D., Abraham, J. A. and Klagsbrun, M. (1993). Heparin-binding epidermal growth factor-like growth factor expression in cultured fetal human vascular smooth muscle cells. Induction of mRNA levels and secretion of active mitogen. J. Biol. Chem. 268, 18330–18334.

Elliott, S.J., Lacey, D.J., Chilian, W.M. and Brzezinska, A.K. (1998). Peroxynitrite is a contractile agonist of cerebral artery smooth muscle cells. Am. J. Physiol. 275, H1585–H1591.

Frank, G.D., Motley, E.D., Inagami, T. and Eguchi, S. (2000). PYK2/CAKβ represents a redox-sensitive tyrosine kinase in vascular smooth muscle cells. Biochem. Biophys. Res. Commun. 270, 761–765.

Greene, E.L., Velarde, V. and Jaffa, A.A. (2000). Role of reactive oxygen species in bradykinin-induced mitogen-activated protein kinase and c-fos induction in vascular cells. Hypertension 35, 942–947.

Griendling, K.K., Minieri, C.A., Ollerenshaw, J.D. and Alexander, R.W. (1994). Angiotensin II stimulates NADH and NADPH oxidase activation in cultured vascular smooth muscle cells. Circ. Res. 74, 1141–1148.

Griffith, C.E., Zhang, W. and Wange, R.L. (1998). ZAP-70-dependent and -independent activation of Erk in Jurkat T cells. Differences in signaling induced by H_2O_2 and Cd3 cross-linking. J. Biol. Chem. 273, 10771–10776.

Heeneman, S., Haendeler, J., Saito, Y., Ishida, M. and Berk, B.C. (2000). Angiotensin II induces transactivation of two different populations of the PDGFβ-receptor: Key role for the adaptor protein Shc. J. Biol. Chem. 275, 15926–15932.

Hu, X.Q., Singh, N., Mukhopadhyay, D. and Akbarali, H.I. (1998). Modulation of voltage-dependent Ca^{2+} channels in rabbit colonic smooth muscle cells by c-Src and focal adhesion kinase. J. Biol. Chem. 273, 5337–5342.

Ishida, T., Peterson, T.E., Kovach, N.L. and Berk, B.C. (1996). MAP kinase activation by flow in endothelial cells. Role of beta 1 integrins and tyrosine kinases. Circ. Res. 79, 310–316.

Jin, D.Y., Chae, H.Z., Rhee, S.G. and Jeang, K.T. (1997). Regulatory role for a novel human thioredoxin peroxidase in NF-κB activation. J. Biol. Chem. 272, 30952–30961.

Jin, N., Hatton, N.D., Harrington, M.A., Xia, X., Larsen, S.H. and Rhoades, R.A. (2000). H_2O_2-induced egr-1, fra-1 and c-jun gene expression is mediated by tyrosine kinase in aortic smooth muscle cells. Free Rad. Biol. Med. 29, 736–746.

Jin, Z.G., Melaragno, M.G., Liao, D.F., Yan, C., Haendeler, J., Suh, Y.A., Lambeth, J.D. and Berk, B.C. (2000). Cyclophilin A is a secreted growth factor induced by oxidative stress. Circ. Res. 87, 789–796.

Kasler, H.G., Victoria, J., Duramad, O. and Winoto, A. (2000). ERK5 is a novel type of mitogen-activated protein kinase containing a transcriptional activation domain. Mol. Cell. Biol. 20, 8382–8389.

Kato, Y., Tapping, R.I., Huang, S., Watson, M.H., Ulevitch, R.J. and Lee, J.D. (1998). Bmk1/Erk5 is required for cell proliferation induced by epidermal growth factor. Nature 395, 713–716.

Kourie, J.I. (1998). Interaction of reactive oxygen species with ion transport mechanisms. Am. J. Physiol. 275, C1–C24.

Kusuhara, M., Chait, A., Cader, A. and Berk, B.C. (1997). Oxidized LDL stimulates mitogen-activated protein kinases in smooth muscle cells and macrophages. Arterioscler. Thromb. Vasc. Biol. 17, 141–148.

Lander, H.M., Ogiste, J.S., Teng, K.K. and Novogrodsky, A. (1995). p21ras as a common signaling target of reactive free radicals and cellular redox stress. J. Biol. Chem. 270, 21195–21198.

Laurindo, F.R., Pedro, M. d. A., Barbeiro, H.V., Pileggi, F., Carvalho, M.H., Augusto, O. and da Luz, P.L. (1994). Vascular free radical release. Ex vivo and in vivo evidence for a flow-dependent endothelial mechanism. Circ. Res. 74, 700–709.

Liao, D.-F., Baas, A.S., Daum, G. and Berk, B.C. (1997). Purification of a secreted protein factor induced by reactive oxygen species (ROS) in vascular smooth muscle cells (VSMC). Circulation 96, I–901.

Liao, D.-F., Jin, Z.-G., Baas, A.S., Daum, G., Gygi, S.P., Aebersold, R. and Berk, B.C. (2000). Purification and identification of secreted oxidative stress-induced factors from vascular smooth muscle cells. J. Biol. Chem. 275, 189–196.

Liao, D.F., Monia, B., Dean, N. and Berk, B.C. (1997). Protein kinase C-ζ mediates angiotensin II activation of ERK1/2 in vascular smooth muscle cells. J. Biol. Chem. 272, 6146–6150.

Lin, C.P., Lynch, M.C. and Kochevar, I.E. (2000). Reactive oxidizing species produced near the plasma membrane induce apoptosis in bovine aorta endothelial cells. Exp. Cell. Res. 259, 351–359.

Marui, N., Offermann, M.K., Swerlick, R., Kunsch, C., Rosen, C.A., Ahmad, M., Alexander, R.W. and Medford, R.M. (1993). Vascular cell adhesion molecule-1 (VCAM-1) gene transcription and expression are regulated through an antioxidant-sensitive mechanism in human vascular endothelial cells. J. Clin. Invest. 92, 1866–1874.

Melaragno, M.G., Wuthrich, D.A., Poppa, V., Gill, D., Lindner, V., Berk, B.C. and Corson, M.A. (1998). Increased expression of Axl tyrosine kinase after vascular injury and regulation by G protein-coupled receptor agonists in rats. Circ. Res. 83, 697–704.

Min, D.S., Kim, E.G. and Exton, J.H. (1998). Involvement of tyrosine phosphorylation and protein kinase C in the activation of phospholipase D by H_2O_2 in Swiss 3T3 fibroblasts. J. Biol. Chem. 273, 29986–29994.

Nishida, M., Maruyama, Y., Kontani, K., Nagao, T. and Kurose, H. (2000). G-α(i) and G-aα(o) are target proteins of reactive oxygen species. Nature 408, 492–495.

Nunes, G.L., Robinson, K., Kalynych, A., King III, S.B., Sgoutas, D.S. and Berk, B.C. (1997). Vitamins C and E inhibit O_2-production in the pig coronary artery. Circulation 96, 3593–3601.

Nunes, G.L., Sgoutas, D.S., Redden, R.A., Sigman, S.R., Gravanis, M.B., King, S.B.R. and Berk, B.C. (1995). Combination of vitamins C and E alters the response to coronary balloon injury in the pig. Arterioscler. Thromb. Vasc. Biol. 15, 156–165.

Okuda, M., Takahashi, M., Suero, J., Murry, C. E., Traub, O., Kawakatsu, H. and Berk, B.C. (1999). Shear stress stimulation of p130(cas) tyrosine phosphorylation requires calcium-dependent c-Src activation. J. Biol. Chem. 274, 26803–26809.

Peterson, T., Poppa, V., Ueba, H., Wu, A., Yan, C. and Berk, B.C. (1999). Opposing effects of reactive oxygen species and cholesterol on endothelial nitric oxide synthase and endothelial cell caveolae. Circ. Res. 85, 29–37.

Rahman, A., Kefer, J., Bando, M., Niles, W.D. and Malik, A.B. (1998). E-selectin expression in human endothelial cells by TNF-α-induced oxidant generation and NF-κB activation. Am. J. Physiol. 275, L533–L544.

Rao, G.N. (1996). Hydrogen peroxide induces complex formation of SHC-Grb2-SOS with receptor tyrosine kinase and activates Ras and extracellular signal-regulated protein kinases group of mitogen-activated protein kinases. Oncogene 13, 713–719.

Rao, G.N., Alexander, R.W. and Runge, M.S. (1995). Linoleic acid and its metabolites, hydroperoxyoctadecadienoic acids, stimulate c-Fos, c-Jun and c-Myc mRNA expression, mitogen-activated protein kinase activation and growth in rat aortic smooth muscle cells. J. Clin. Invest. 96, 842–847.

Rao, G.N. and Berk, B.C. (1992). Active oxygen species stimulate vascular smooth muscle cell growth and proto-oncogene expression. Circ. Res. 70, 593–599.

Rao, G.N., Lassegue, B., Griendling, K.K. and Alexander, R.W. (1993). Hydrogen peroxide stimulates transcription of c-jun in vascular smooth muscle cells: role of arachidonic acid. Oncogene 8, 2759–2764.

Rao, G.N., Lassegue, B., Griendling, K.K., Alexander, R.W. and Berk, B.C. (1993). Hydrogen peroxide-induced c-fos expression is mediated by arachidonic acid release: Role of protein kinase C. Nucl. Acids Res. 21, 1259–1263.

Schlaepfer, D.D., Hauck, C.R. and Sieg, D.J. (1999). Signaling through focal adhesion kinase. Prog. Biophys. Mol. Biol. 71, 435–478.

Schmitz, U., Ishida, T., Ishida, M., Surapisitchat, J., Hasham, M.I., Pelech, S. and Berk, B.C. (1998). Angiotensin II stimulates p21-activated kinase in vascular smooth muscle cells: Role in activation of JNK. Circ. Res. 82, 1272–1278.

Servitja, J.M., Masgrau, R., Pardo, R., Sarri, E. and Picatoste, F. (2000). Effects of oxidative stress on phospholipid signaling in rat cultured astrocytes and brain slices. J. Neurochem. 75, 788–794.

Speir, E., Yu, Z.X., Ferrans, V.J., Huang, E.S. and Epstein, S.E. (1998). Aspirin attenuates cytomegalovirus infectivity and gene expression mediated by cyclooxygenase-2 in coronary artery smooth muscle cells. Circ. Res. 83, 210–216.

Suc, I., Meilhac, O., Lajoie-Mazenc, I., Vandaele, J., Jurgens, G., Salvayre, R. and Negre-Salvayre, A. (1998). Activation of EGF receptor by oxidized LDL. FASEB J. 12, 665–671.

Suh, Y.A., Arnold, R.S., Lassegue, B., Shi, J., Xu, X., Sorescu, D., Chung, A.B., Griendling, K.K. and Lambeth, J.D. (1999). Cell transformation by the superoxide-generating oxidase Mox1. Nature 401, 79–82.

Sundaresan, M., Yu, Z.X., Ferrans, V.J., Irani, K. and Finkel, T. (1995). Requirement for generation of H_2O_2 for platelet-derived growth factor signal transduction. Science 270, 296–299.

Sundaresan, M., Yu, Z.X., Ferrans, V.J., Sulciner, D.J., Gutkind, J.S., Irani, K., Goldschmidt, C.-P.J. and Finkel, T. (1996). Regulation of reactive-oxygen-species generation in fibroblasts by Rac1. Biochem. J. 318, 379–382.

Takahashi, E., Abe, J., Gallis, B., Aebersold, R., Spring, D.J., Krebs, E. G. and Berk, B.C. (1999). p90RSK is a serum-stimulated NHE1 kinase: regulatory phosphorylation of serine 703 of Na^+/H^+ exchanger isoform-1. J. Biol. Chem. 274, 20206–20214.

Takahashi, E. and Berk, B.C. (1998). MAP kinases and vascular smooth muscle function. Acta Physiol. Scand. 164, 611–621.

Takahashi, M. and Berk, B.C. (1996). Mitogen-activated protein kinase (ERK1/2) activation by shear stress and adhesion in endothelial cells. Essential role for a herbimycin-sensitive kinase. J. Clin. Invest. 98, 2623–2631.

Takeishi, Y., Abe, J., Lee, J.D., Kawakatsu, H., Walsh, R.A. and Berk, B.C. (1999). Differential regulation of p90 ribosomal S6 kinase and big mitogen-activated protein kinase 1 by ischemia/reperfusion and oxidative stress in perfused guinea pig hearts. Circ. Res. 85, 1164–1172.

Tardif, J.C., Coté, G., Lespérance, J., Bourassa, M., Lambert, J., Doucet, S., Bilodeau, L., Nattel, S. and de Guise, P. (1997). Probucol and multivitamins in the prevention of restenosis after coronary angioplasty. Multivitamins and Probucol Study Group. N. Engl. J. Med. 337, 365–372.

Taylor, D.S., Cheng, X., Pawlowski, J.E., Wallace, A.R., Ferrer, P. and Molloy, C.J. (1999). Epiregulin is a potent vascular smooth muscle cell-derived mitogen induced by angiotensin II, endothelin-1 and thrombin. Proc. Natl. Acad. Sci. USA 96, 1633–1638.

Traub, O., Monia, B.P., Dean, N.M. and Berk, B.C. (1997). PKC-ε is required for mechano-sensitive activation of ERK1/2 in endothelial cells. J. Biol. Chem. 272, 31251–31257.

Ushio-Fukai, M., Alexander, R.W., Akers, M. and Griendling, K.K. (1998). p38 Mitogen-activated protein kinase is a critical component of the redox-sensitive signaling pathways activated by angiotensin II. Role in vascular smooth muscle cell hypertrophy. J. Biol. Chem. 273, 15022–15029.

Wang, D., Yu, X., Cohen, R.A. and Brecher, P. (2000). Distinct effects of N-acetylcysteine and nitric oxide on angiotensin II-induced epidermal growth factor receptor phosphorylation and intracellular Ca^{2+} levels. J. Biol. Chem. 275, 12223–12230.

Wei, Z., Costa, K., Al-Mehdi, A.B., Dodia, C., Muzykantov, V. and Fisher, A.B. (1999). Simulated ischemia in flow-adapted endothelial cells leads to generation of reactive oxygen species and cell signaling. Circ. Res. 85, 682–689.

Wilson, S.H., Best, P.J., Lerman, L.O., Holmes, D.R., Jr., Richardson, D.M. and Lerman, A. (1999). Enhanced coronary vasoconstriction to oxidative stress product, 8-epi-prostaglandinF2 α, in experimental hypercholesterolemia. Cardiovasc. Res. 44, 601–617.

Wung, B.S., Cheng, J.J., Hsieh, H.J., Shyy, Y.J. and Wang, D.L. (1997). Cyclic strain-induced monocyte chemotactic protein-1 gene expression in endothelial cells involves reactive oxygen species activation of activator protein 1. Circ. Res. 81, 1–7.

Yoshizumi, M., Abe, J., Haendeler, J., Huang, Q. and Berk, B.C. (2000). Src and cas mediate JNK activation but not ERK1/2 and p38 kinases by reactive oxygen species. J. Biol. Chem. 275, 11706–11712.

Zafari, A.M., Ushio-Fukai, M., Akers, M., Yin, Q., Shah, A., Harrison, D.G., Taylor, W.R. and Griendling, K.K. (1998). Role of NADH/NADPH oxidase-derived H_2O_2 in angiotensin II-induced vascular hypertrophy. Hypertension 32, 488–495.

Protein Adaptations and Signal Transduction. Edited by K.B. Storey and J.M. Storey

CHAPTER 19

Oxidative Stress Signaling

Hasem Habelhah and Ze'ev Ronai*
Ruttenberg Cancer Center, Mount Sinai School of Medicine, New York, NY 10029, U.S.A.

1. Introduction

Oxygen is an essential molecule for metabolism and energy generation in aerobic organisms, but it is also the primary source for the generation of reactive oxygen species (ROS) (Winrow et al., 1993; Ames et al., 1993). ROS include superoxide (O_2^-), singlet oxygen ($'O_2$), hydrogen peroxide (H_2O_2) and highly reactive hydroxyl radical (HO·) (Winrow et al., 1993; Ames et al., 1993). ROS are generated in the mitochondria as a by-product of the electron transport chain, as well as by non-mitochondrial enzymes such as oxidases or peroxidases (Lander, 1997). In eukaryotic organisms, O_2^- is a primary source of ROS and is converted efficiently by superoxide dismutase (SOD) to H_2O_2, which is subsequently processed to H_2O and O_2 by peroxidases and catalase (Winrow et al.,1993). Excess ROS is harmful because of its potent ability to interact with, and modify, a wide range of cellular macromolecules implicated in cytotoxicity and mutagenic damage (Winrow et al., 1993; Ames et al., 1993). Conversely, a low level of ROS is necessary in order to maintain cell proliferation and serve as intracellular signaling molecules (Finkel, 1998; Irani, 2000). Accordingly, the level of ROS is tightly controlled by several enzymatic (e.g., superoxide dismutases, glutathione peroxidase, glutathione reductase and catalase) and non-enzymatic (glutathione and vitamins C and E) antioxidants (Sen and Packer, 1996). To offset the oxidant effects and to restore a state of redox balance, cells often activate or silence genes encoding regulatory transcription factors (reviewed in Adler et al., 1999b), antioxidant defense enzymes and structural proteins. ROS, which is secreted by specialized cells including macrophages and neutrophils, serves as a key player in host defense against pathogens (Bogdan et al., 2000; Rosen et al., 1995). Control of ROS levels is one of the key determinants in maintaining cell growth (Irani, 2000), senescence (Powis et al., 1995) and death (Jacobson, 1996; Tan et al., 1998b). In this review, we summarize current understanding of mechanisms underlying the cellular response to ROS at the level of signal transduction and gene expression and their implications for cell growth, senescence and death.

*Corresponding author.

2. Key sources of ROS generation

In mammalian cells, five endogenous sources appear to account for most of the ROS produced by cells (Fig. 19.1): (1) as part of normal aerobic respiration mitochondria consume O_2, which is then reduced by sequential steps to produce H_2O; (2) production of an oxidative burst by phagocytic cells results in the generation of nitric oxide (NO), O_2^-, H_2O_2 and OCl^-, which efficiently destroy bacteria or virus-infected cells; (3) H_2O_2 is generated as a by-product in the process of fatty acid degradation by peroxisomes; (4) cytochrome P450 enzymes, which constitute one of the primary defense systems against natural toxic chemicals, generate oxidant by-products; (5) NAD(P)H

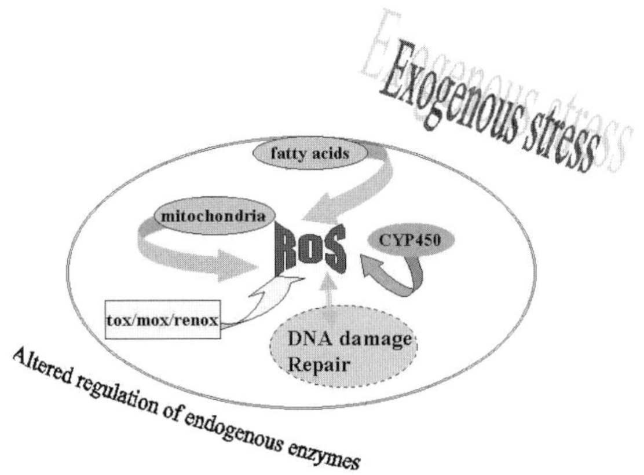

Fig. 19.1. Sources of ROS. Major sources for reactive oxygen species are illustrated; the balance between exogenous exposure, which results in the generation of ROS, and altered expression of endogenous ROS producing proteins is key determinant in the overall ROS present at any given time in the cell, and the concomitant cellular response.

oxidase homologues, discovered most recently in kidney (renox), thyroid (p138(Tox)) and vascular smooth muscle cells (mox1), produce low levels of ROS upon stimulation (Ames et al., 1993; Bastian and Hibbs, 1994; Finkel, 1998; Suh et al., 1999; Dupuy et al., 1999; Geiszt et al., 2000). Overexpression of mox1, a homologue of the catalytic subunit (gp91phox) of the phagocytic NADPH oxidase, increases superoxide generation and cell growth, whereas antisense mox1 mRNA decreases superoxide generation and serum-stimulated growth (Suh et al., 1999).

Exogenous sources of ROS include a wide range of stress, toxic and infectious agents. Among them are UV-irradiation, chemotherapeutic agents, oxidative stress, pro-inflammatory cytokines, growth factors, and viral, bacterial and parasite infection (Sundaresan et al., 1995; Ohba et al., 1994; Lo and Cruz, 1995; Griendling et al., 1994; Thannickal et al., 1998; Kheradmand et al., 1998; Irani et al., 1997). These exogenous exposures generate ROS through altered expression and activity of diverse enzymes, including lipoxygenases, cyclooxygenases, NAPDH oxidases, and cytochrome p450s, as well as through direct action on the cellular pool of H_2O.

Constitutive generation of ROS results from chronic infection, inflammation or both, as well as from an altered balance of key regulatory proteins, including oncogenes and tumor suppressor genes.

3. ROS as second messengers in mitogenic signaling

The observations that ligand-stimulated H_2O_2 production is often linked to concomitant phosphorylation of downstream cellular targets, that a similar phosphorylation pattern could be also induced upon exogenous exposure to H_2O_2 and that either H_2O_2 or ligand-induced phosphorylation could be equally blocked by free radical scavengers collectively supported the notion that ROS may in fact serve as second messengers in mitogenic signaling (Finkel, 1998; Irani, 2000). One of the better-characterized examples of ROS as second messengers in mitogenic signaling comes from studies on vascular smooth muscle cells (VSMC) (Sundaresan et al., 1995; Griendling and Harrison, 1999). VSMC respond to growth factor stimulation with intracellular production of ROS. Such ligands include those acting via tyrosine kinase receptors such as platelet-derived growth factor (PDGF) and G protein-coupled receptors such as phenylephrine and thrombin (reviewed in Irani, 2000). Suppression of the PDGF-stimulated rise in H_2O_2 blocked this proliferative response. Increasing intracellular levels of catalase prevented elevation of ROS levels and blocked PDGF-induced tyrosine phosphorylation, thrombin-induced mitogenesis or phenylephrine-induced proliferation.

Similar to the effect of ligands such as PDGF, the expression of the small GTP-binding proteins Ras and Rac has reportedly led to production of ROS (Irani et al., 1997). Transformation of mouse fibroblasts with the constitutively active forms of Ras produced large amounts of the ROS superoxide. Superoxide production was suppressed by the expression of the dominant negative isoforms of Ras or Rac1, as well as by treatment with a farnesyltransferase or flavoprotein inhibitor or by the antioxidant NAC (Irani et al., 1997). Similarly, the activation of Rac1 is required for integrin-mediated reorganization of cell shape, which entails increased expression of collagenase-1 in rabbit synovial fibroblasts (Kheradmand et al., 1998). Increased collagenase-1 expression appears to be NF-κB-dependent, whereas NF-κB activation is supported by Rac1-generated ROS (Kheradmand et al., 1998). The dominant negative form of Rac1 efficiently suppressed ischemia/reperfusion-induced production of ROS and lipid peroxides and activation of NF-κB, and reduced acute liver necrosis (Ozaki et al., 2000).

4. Role of ROS in signal transduction

Almost every cellular signaling cascade is affected by ROS (Fig. 19.2) (reviewed in Adler et al., 1999b). Exposure of different tissue types to ROS-generating treatments, which in most cases were in the form of H_2O_2 or its derivatives, resulted in efficient activation of MAPK, ERK, JNK, p38, IKK, cAMP-dependent kinases, Fyn, JAKs, c-Abl, Src (Yoshizumi et al., 2000; Abe and Berk, 1999; Kharbanda et al., 1995), PI3K and PTK, among others (Deora et al., 1998; Adler et al., 1999b; Migliaccio et al., 1999). The mechanism(s) underlying the comprehensive activation of each of these different signaling pathways are not well understood and are likely to involve either one or a combination of the following three possibilities: (a) direct effect of ROS on signaling molecules; (b) effect of ROS on membrane composition and membrane-anchored receptor/regulatory proteins; and (c) effect on cysteine-rich molecules that have been implicated in the response to altered redox potential (Fig. 19.2). Current knowledge supports each of these possibilities, and it is likely that the primary mechanism or the combination of several pathways will depend on the type of cells, the nature of ROS (dose, duration) and the competency of the cellular antioxidant defense mechanisms.

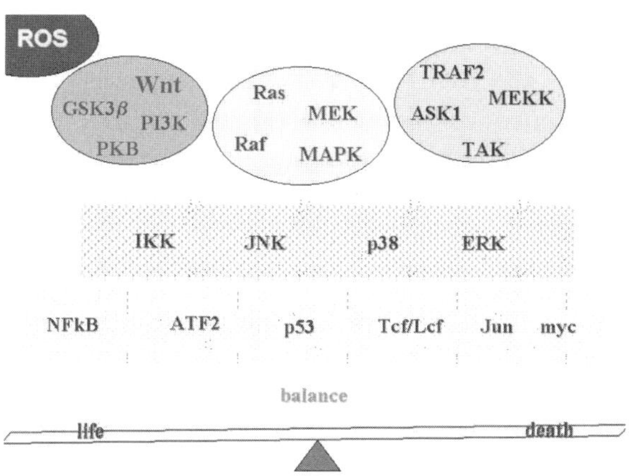

Fig. 19.2. ROS effect on cellular signaling. Shown are major signaling cascades that are affected by altered ROS, and consequently affect downstream effectors that play a key role in determining cell fate. It is the type of stress or damage, the dose administered and the tissue/cell affected which dictates the specific signaling that will be first activated, as well as the downstream kinases and transcription factors which will be phosphorylated to elicit altered expression of genes that would contribute to cell cycle arrest, DNA repair, proliferation, senescence or apoptosis.

Among membrane-anchored regulatory components which have been implicated in the response of signaling cascades to ROS is the dimerization or trimerization of growth factor receptors (e.g., EGFR, PDGFR; Huang et al., 1996; Rosette et al., 1996).

Central to the activation of the stress-signaling response are the MAPK pathways, a series of mitogen-activated protein kinases that are activated upon tyrosine phosphorylation, which is catalyzed by members of the MAPK extracellular-signal-regulated-kinase (ERK) MEK family (Cobb, 1999; Karin, 1998). MEK family members in turn are activated by Ser/Thr phosphorylation catalyzed by divergent array of protein kinases, which are referred to as MAPKKK (Ichijo, 1999).

The stress-activated protein kinases (SAPK, also known as Jun NH2 kinases: JNKs) and the p38s represent two MAPK subfamilies pivotal to the regulation of diverse immediate early-response genes implicated in the regulation of the cell cycle, DNA repair and apoptosis (Davis, 1999; Ono and Han, 2000). Thus far, 11 MAPKKK have been identified as upstream activators of the SAPKs and p38s, and yet little is known about the molecular mechanisms by which MAPKKK–MEK–MAPK core modules couple to processes at the cell membrane.

Some insight into the relationship between ROS production and the activity of key stress-signaling cascades comes from studies using three main sources for ROS: H_2O_2, UV-irradiation and tumor necrosis factor (TNF), each of which is likely to utilize an alternate cellular signaling pathway (Adler et al., 1999b; Liu et al., 1996).

Treatment of cells with TNF fosters the production of ROS and the concomitant change in the activation of stress-signaling molecules. Central to the cellular response to TNF are TRAF proteins, which consist of 6 family members. The most extensively characterized, TRAF2, was shown to shift TNF activation towards at least three signaling pathways, including MEKK1, ASK1, and NIK, resulting in the activation of AP1, ATF2 and NF-κB (Ichijo, 1999). ROS produced upon TNF treatment have been shown to affect the signaling molecule TRAF2. TRAF2 can be activated upon its dimerization or association with GCK, ASK1 or NIK (Ichijo, 1999). Over-expression of TRAF2 alone is sufficient to result in the production of ROS (Liu et al., 2000), which could be implicated in any of the many targets associated with this central regulatory molecule. It has been suggested that the ROS-dependent association of TRAF2 with ASK1 plays a role in the regulation of ASK1 activities. The use of free radical scavengers was sufficient to attenuate the TRAF2–ASK1 association and to prevent subsequent activation of ASK1. Interestingly, the regulation of ASK1's own activities as a kinase is ROS-dependent because of the association of ASK1 with the protein disulfide oxidoreductase thioredoxin (Trx) (Saitoh et al., 1998). Trx association with ASK1 inhibits ASK1's kinase activity in non-stressed cells. Treatment of cells with ROS-generating agents led to the dissociation of Trx from ASK1, which could be abolished upon treatment of cells with free radical scavengers. ASK1 has been implicated in the activation of MKK3/6, MKK4/7 and subsequently, the activation of p38 and JNK (Ichijo et al., 1999), which result in the phosphorylation of ATF2, c-jun and p53, which are key regulatory transcription factors implicated in the regulation of the cellular stress response.

In its regulation of JNK, glutathione S transferase pi (GSTp) has demonstrated a striking similarity to ASK1 regulation by Trx. In non-stressed cells, JNK exhibits a low level of basal activity which is attributed to the association with GSTp. GSTp inhibition of JNK activity requires its association with JNK as found in non-stressed cells. Upon treatment with H_2O_2 or UV, GSTp dissociates from JNK because of the formation of multimer forms of GSTp which are covalently linked. Free radical scavengers maintain the GSTp-JNK association and JNK inhibition by GSTp even in the presence of ROS-producing agents. Furthermore, the constitutively active form of MEKK1 efficiently phosphorylates both MKK4 and JNK, but not Jun, because of GSTp-mediated inhibition in cells over-expressing GSTp (Adler et al., 1999a). Along those lines, treatment with GSH or N-acetylcysteine (NAC) inhibits the induction of JNK by monofunctional alkylating agents, whereas depletion of GSH pools with L-buthionine S,R-sulfoximine super induces JNK activity (Wilhelm et al., 1997). Interestingly, increased expression of GSTp in non-stressed NIH3T3 cells also increases the phosphorylation of MKK4, p38, ERK and IKK, and reduces the phosphorylation of MKK7 and JNK (Yin et al., 2000). Whereas H_2O_2 treatment of cells induces JNK, p38 and IKK activities, in the presence of elevated GSTp expression H_2O_2 causes an additional increase in ERK, p38 and IKK activities and limited activity of JNK. These observations suggest that GST may serve as a coordinator of stress kinases under normal and ROS-generating conditions. The functional significance of such coordination is illustrated in cells subjected to H_2O_2, a potent inducer of cell death which can be prevented by GSTp. GSTp-mediated protection from H_2O_2-induced cell death is attenuated upon inhibition of p38, NF-κB or MAP kinase using the respective dominant negative or pharmacological inhibitors. Conversely, expression of a dominant negative JNK protects cells from H_2O_2-mediated death (Yin et al., 2000). Illustrating the pro-apoptotic role of JNK as opposed to the anti-apoptotic role of p38/ERKs, these findings provide an important insight regarding the regulation of stress kinases by ROS, highlighting the global role of GSTp as a general coordinator of these signaling cascades. How GSTp can affect such a diverse chain of stress-signaling molecules is yet to be determined. Among possible mediators are the upstream regulatory proteins such as the TRAFs, which are well positioned to affect each of the signal transduction cascades described above.

Since GSTp expression is often altered in human tumors, most of which exhibit over-expression whereas others exhibit under-expression, the possible changes implicated in tumor cells' ability to cope with stimuli that would normally mediate apoptosis may be attributed to altered stress signaling, a hypothesis that is currently under investigation.

5. Transcriptional regulation by ROS

One of the paradigms for understanding the mechanisms underlying ROS effects on gene expression comes from studies in *Escherichia coli*, which rely on the examples of OxyR and SoxR activation in response to H_2O_2 and superoxide-generating compounds. Activation of SoxR protein results from reversible one-electron oxidation of its iron-sulfur center (Hidalgo et al., 1997). Upon activation, SoxR induce SoxS transcription. The dramatic increase in SoxS level triggers transcription from 12–14 genes including Mn-SOD, the DNA repair enzyme endonuclease IV and glucose-6-phosphate dehydrogenase. OxyR is transiently activated by the formation of intramolecular disulfide bridges by H_2O_2 (Zhang et al., 1998). Activated OxyR rapidly induces the expression of several proteins, including hydrogen peroxidase I and a glutathione reductase, which protect against oxidative stress. Although the mammalian homologues of SoxR or OxyR are not known, redox-sensitive transcription factors that are direct targets of ROS have been identified, among which Ref1 is the best characterized.

Ref1 is a dual-function protein which is capable of eliciting repair of abasic sites that arise in DNA via its apurinic/apyrimidinic endonuclease activity and redox-responsive regulation of c-Jun/NFkB transcription. Whereas it has been suggested that phosphorylation by CKII plays an important role in its DNA repair activity (Yacoub et al., 1997), altered ROS determines the association and nature of Ref1's effect on the transcription factor's activity. For example, Trx activation of AP-1 is explained by its association with Ref1, which binds to AP-1 and increases its transcriptional activities. Mutations on specific cysteines on Trx which abolish its association with Ref-1 also result in diminished AP-1-transactivation.

Among the long list of transcription factors affected by altered ROS (reviewed in Adler et al., 1999b) AP-1 and NF-κB are best characterized. ROS's ability to alter transcriptional activities could be attributed to the changes in their respective upstream kinases by ROS or to the direct effect of ROS on transcription factors.

AP-1 consists of transcription factors of Fos and Jun and the ATF2 families, which form homodimers (Jun–Jun) or heterodimers (Fos–Jun, ATF2-Jun) and have a key regulatory role in the expression of a wide variety of genes (Lee et al., 1987; Sen and Packer, 1996; Pinkus et al., 1996). Activation of the AP-1 in response to altered ROS serves as an example of the immediate early genes that are induced as part of the complex stress response to oxidative damage. Mitogens, tumor promoters as well as H_2O_2, UV-C, UV-A, ionizing radiation, TNF-α, basic fibroblast growth factor, angiotensin II and dioxin represent the wide range of cellular stimuli which alter ROS and soon afterward result in the activation of AP-1 members (Puri et al.95; Lo and Cruz, 1995; Sen and Packer, 1996; Pinkus et al., 1996; Adler et al., 1999b). Central to the activities of AP-1 family members, as best illustrated for c-Jun, is its phosphorylation by upstream kinases. c-Jun phosphorylation by JNK on serines 63 and 73 suffices for its stabilization and transcriptional activation. More recent studies using Jun null cells made it possible to distinguish between c-Jun's effect on cellular proliferation and its effect on the stress response, using phosphomutant forms of c-Jun, thereby allowing separation of possible requirements for the Jun-mediated response to ROS vs. those required for normal cellular growth.

c-Jun activities are also modulated in response to Trx expression (Abate et al., 1990; Schenk et al.,

1994). Substitution of Cys154 in Fos or Cys-272 in Jun for serine resulted in loss of redox regulation and enhanced DNA binding (Abate et al., 1990), suggesting that in its reduced form AP-1 may exhibit greater DNA binding activity than in its oxidized form. Increase in the activity of AP-1 could be also attributed to altered cytosolic Ca2+ concentration in response to ROS, which function as an important second messenger for transcriptional induction of AP-1 (Hoyal et al., 1996).

6. ROS regulation of NF-κB

NF-κB plays a central role in the regulation of cellular defense mechanisms against cellular insult by pathogens and inflammation, and in response to cytokines and cell adhesion molecules (Schreck et al., 1991; Los et al., 1995; Sen and Packer, 1996; Mercurio and Manning, 1999; Li and Karin, 1999). NF-κB consists of a complex between members of the NF-κB (p50/p52) and the Rel families of proteins (p65). The NF-κB and the Rel families are present in the cytoplasm as inactive forms by virtue of their association with a class of inhibitory proteins, IκBs. Activation of NF-κB is achieved through the signal-induced proteolytic degradation of IκB, which is mediated by the 26S proteasome (Mercurio and Manning, 1999; Li and Karin, 1999). The critical event that initiates proteolytic degradation of IκB is the stimulus-dependent phosphorylation of IκB at specific N-terminal serine residues. Such phosphorylation enables recognition by the ROC1/βTrCP/HOS complex of E3 ligase, which efficiently ubiquitinates and degrades IκB (Fuchs et al., 1999; Tan et al., 1999a), leading to the release of NF-κB, which enters the nuclei to elicit its transcriptional output. Post translational modifications of the NF-κB complex occurring after IκB degradation can enhance transcriptional activation of the NF-κB dependent genes. Phosphorylation of the p65 subunit on serine 276 by PKA stimulates NF-κB transcriptional output (Zhong et al., 1997). The transcriptional co-activator CBP/p300 was found to associate with NF-κB RelA through two sites which require Rel A phosphorylation.

NF-κB, recognized as a redox-sensitive transcription factor, has been implicated in the cellular response to oxidative stress (Schreck et al., 1991; Los et al., 1995; Sen and Packer, 1996; Mercurio and Manning, 1999; Li and Karin, 1999). Inflammatory cytokines such as TNFα, IL-1, LPS and PMA, as well as DNA-damaging treatments such as ionizing radiation—all of which generate elevated levels of ROS in cells—efficiently induce NF-κB activation. Along those lines, whereas treatment of cells with H_2O_2 activates NF-κB (Schreck et al., 1991), a variety of antioxidants, including N-acetyl-L-cysteine (NAC), glutathione (GSH), thioredoxin (Trx) and also overexpression of antioxidant enzymes like superoxide dismutase or glutathione peroxidase, were reported to block ROS-induced NF-κB activity (Mercurio and Manning, 1999). It is important to note that H_2O_2-induced NF-κB activation is cell type specific, suggesting that H_2O_2 is not a general mediator of NF-κB activation (Li and Karin, 1999). Similarly, neither IKK (IκB kinase) activation nor the serine phosphorylation of IκBα was observed to occur after UV-C irradiation. Alternatively, treatment of cells with tyrosine phosphatase inhibitors (e.g., pervanadate) induces Tyr-42 phosphorylation of IκBα and results in its dissociation from NF-κB without its degradation (Imbert et al., 1996). Overexpression of manganese superoxide dismutase (Mn-SOD), a mitochondrial enzyme whose overexpression decreases O_2^- but not total ROS levels, efficiently suppresses TNF induced NF-κB and AP-1 activation (Manna et al., 1998; Moreno-Manzano et al., 2000). More recent studies revealed that the hydroxyl radical was primarily responsible for activation of NF-κB in Jurkat cells and macrophages. A hydroxyl radical-generating system that did not involve exogenous H_2O_2 was used to demonstrate that antioxidants, which scavenge the OH· radical or its precursor, inhibit NF-κB activation (Shi et al., 1999). In addition, metal chelators that render metal ions incapable of generating the hydroxyl radical from H_2O_2 were shown to block NF-κB activation, suggesting that the ability of the cell to convert H_2O_2 to OH· rather than to another metabolic by-product could influence the activation of NF-κB (Rouault and Klausner, 1996).

7. ROS in apoptosis

The redox state of a cell is an important factor in determining its susceptibility to different apoptotic stimuli (reviewed in Mignotte and Vayssiere, 1998). Both ROS and antioxidants play a critical role in regulating apoptosis following the application of an apoptotic stimulus, such as TNF, Fas signaling, glucocorticoids, thiol depletion, radiation, chemotherapeutic agents, ceramide and lymphocyte activation. Among the different cellular compartments, mitochondria are the predominant source of ROS produced in most apoptotic systems (Kroemer et al., 1998; Mignotte and Vayssiere, 1998; Chandel et al., 1998; Jacobson, 1996; Hampton et al., 1998; Melov, 2000). Mitochondrial homeostasis is critical in regulating apoptosis (Schulze-Osthoff et al., 1993; Green and Reed, 1998). Altered ATP/ADP ratio, thiol depletion and ROS induction are sufficient to induce mitochondrial membrane hyperpolarization and depolarization, matrix swelling and permeability of the outer membrane, resulting in the release of the proapoptotic proteins such as cytochrome c or apoptosis inducing factor. In many cases, such mitochondrial events are a prerequisite for the activation of cysteine containing proteases (caspases) known to be important mediators of apoptosis (Rathmell and Thompson., 1999). The anti-apoptotic family of Bcl-2 (Kane et al., 1993) proteins protects cells from hydrogen peroxide or thiol depletion-induced death and suppresses lipid peroxidation by at least two mechanisms. In the first, Bcl-2 maintains cells in a more reduced state by scavenging ROS, either directly or by up-regulating other ROS scavengers such as thiol compounds. Alternatively, or in addition, Bcl-2 may function to prevent the generation of ROS during apoptosis, probably by acting downstream of ROS formation (Hockenbery et al., 1993). Recent studies revealed that ROS scavengers do not interfere with the initial decrease in mitochondrial membrane potential ($\Delta\Psi m$) but significantly attenuate the subsequent mitochondrial collapse of $\Delta\Psi m$ and partially rescue cells from apoptosis, indicating that ROS contributes to mitochondrial membrane depolarization. Expression of Bcl-x_L blocked both the initial decrease in $\Delta\Psi m$ and subsequent increase in ROS levels following TNFα treatment (Gottlieb et al., 2000). Thus Bcl-x_L acts to promote mitochondrial recovery from oxidant-induced damage by partially inhibiting the decrease in mitochondrial respiration, thus maintaining mitochondrial physiology even under ROS-generating conditions, as shown for TNFα. Of interest is the observation that ROS-mediated T cell apoptosis, which was also attributed to rapid loss of mitochondrial transmembrane potential $\Delta\Psi m$), caspase-dependent DNA fragmentation and superoxide generation, was independent of either Fas or TNFα, which are major death signaling cascades (Hildeman et al., 1999).

Another source of ROS generation has been related to activation of the tumor suppressor gene protein product p53. For example, a high level of p53 expression revealed cell-type dependent sensitivity to apoptosis, which in certain cases is PI3K (p85) dependent (Yin et al., 1998). Cells sensitive to p53-mediated apoptosis produced ROS concomitantly with p53 over-expression, whereas cells resistant to p53 failed to produce ROS. In sensitive cells, both ROS production and apoptosis were inhibited by antioxidant treatment, suggesting that p53 in the regulation of the intracellular redox state and induces apoptosis by a pathway that is dependent on ROS production (Johnson et al., 1996; Li et al., 1999). The ability of p53 to affect levels of ROS is likely to be dependent on altered mitochondrial membrane potential ($\Delta\Psi m$) rather than cytochrome c release (Li et al., 1999). The recent discovery of several ROS-regulatory proteins among the products of p53-responsive genes provides an additional link to explain altered ROS levels in cells which are p53-dependent (Zhao et al., 2000).

References

Abate, C., Patel, L., Rauscher III, F.J. and Curran, T. (1990). Redox regulation of fos and jun DNA-binding activity *in vitro*. Science 249, 1157–1161.

Abe, J. and Berk, B.C. (1999). Fyn and JAK2 mediated Ras activation by reactive oxygen species. J. Biol. Chem. 274, 21003–21010.

Adler, V., Yin, Z., Fuchs, S.Y., Benezra, M., Rosario, L., Tew, K.D., Pincus, M.R., Sardana, M., Henderson, C.J., Wolf, C.R., Davis, R.J. and Ronai, Z. (1999a). Regulation of JNK signaling by GSTp. EMBO J. 18, 1321–1334.

Adler, V., Yin, Z., Tew, K.D. and Ronai, Z. (1999b). Role of redox potential and reactive oxygen species in stress signaling. Oncogene 18, 6104–6111.

Ames, B.N., Shigenaga, M.K. and Hagen, T.M. (1993). Oxidants, antioxidants, and the degenerative diseases of aging. Proc. Natl. Acad. Sci. USA. 99, 7915–7922.

Bastian, N. and Hibbs, Jr, J.B. (1994). Assembly and regulation of NADPH oxidase and nitric oxide synthase. Curr. Opin. Immunol. 6, 131–139.

Bogdan, C., Rollinghoff, M. and Diefenbach, A. (2000). Reactive oxygen and reactive nitrogen intermediates in innate and specific immunity. Curr. Opin. Immunol. 12, 64–76.

Chandel, N.S., Maltepe, E., Goldwasser, E., Mathieu, C.E., Simon, M.C. and Schumacker, P.T. (1998). Mitochondrial reactive oxygen species trigger hypoxia-induced transcription. Proc. Natl. Acad. Sci. USA 95, 11715–11720.

Cobb, M.H. (1999). MAP kinase pathways. Prog. Biophys. Mol. Biol. 71, 479–500.

Davis, R.J. (1999). Signal transduction by the c-Jun N-terminal kinase. Biochem. Soc. Symp. 64, 1–12.

Deora, A.A., Win, T., Vanhaesebroeck, B. and Lander, H.M. (1998). A redox-triggered ras-effector interaction. Recruitment of phosphatidylinositol 3′-kinase to Ras by redox stress. J. Biol. Chem. 273, 29923–29928.

Dupuy, C., Ohayon, R., Valent, A., Noel-Hudson, M.S., Deme, D. and Virion, A. (1999). Purification of a novel flavoprotein involved in the thyroid NADPH oxidase. Cloning of the porcine and human cDNAs. J. Biol. Chem. 274, 37265–37269.

Finkel, T. (1998). Oxygen radicals and signaling. Curr. Opin. Cell. Biol. 10, 248–253.

Fuchs, S.Y., Chen, A., Xiong, Y., Pan, Z.Q. and Ronai, Z. (1999). HOS, a human homolog of Slimb, forms an SCF complex with Skp1 and Cullin1 and targets the phosphorylation-dependent degradation of IκB and beta-catenin. Oncogene 18, 2039–2046.

Geiszt, M., Kopp, J.B., Varnai, P. and Leto, T.L. (2000). Identification of redox, an NAD(P)H oxidase in kidney. Proc. Natl. Acad. Sci. USA 97, 8010–8014.

Gottlieb, E., Vander Heiden, M.G. and Thompson, C.B. (2000). Bcl-xL prevents the initial decease in mitochondrial membrane potential and subsequent reactive oxygen species production during tumor necrosis factor alpha-induced apoptosis. Mol. Cell. Biol. 20, 5680–5689.

Green, D.R. and Reed, J.C. (1998). Mitochondria and apoptosis. Science. 281, 1309–1312.

Griendling, K.K. and Harrison, D.G. (1999). Dual role of reactive oxygen species in vascular growth. Circ. Res. 85, 562–563.

Griendling, K.K., Minieri, C.A., Ollerenshaw, J.D. and Alexander, R.W. (1994). Angiotensin II stimulates NADH and NADPH oxidase activity in cultured vascular smooth muscle cells. Circ. Res. 74, 1141–1148.

Hampton, M.B., Fadeel, B. and Orrenius, S. (1998). Redox regulation of the caspases during apoptosis. Biofactors 854, 328–335.

Hidalgo, E., Ding, H. and Demple, B. (1997). Redox signal transduction via iron–sulfur clusters in the SoxR transcription activator. Trends Biochem. Sci. 22, 207–220.

Hildeman, D.A., Mitchell, T., Teague, T.K., Henson, P., Day, B.J., Kappler, J. and Marrack, P.C. (1999). Reactive oxygen species regulate activation-induced T cell apoptosis. Immunity 10, 735–744.

Hockenbery, D.M., Oltvai, Z.N., Yin, X.M., Milliman, C.L. and Korsmeyer, S.J. (1993). Bcl-2 functions in an antioxidant pathway to prevent apoptosis. Cell 75, 241–251.

Hoyal, C.R., Thomas, A.P. and Forman, H.J. (1996). Hydroperoxide-induced increase in intracellular calcium due to annexin VI translocation and inactivation of plasma membrane Ca^{2+}-ATPase. J. Biol. Chem. 271, 29205–29210.

Huang, R.P., Wu, J.X., Fan, Y. and Adamson, E.D. (1996). UV activates growth factor receptors via reactive oxygen intermediates. J. Cell. Biol. 133, 211–220.

Ichijo, H. (1999). From receptors to stress-activated MAP kinase. Oncogene 18, 6087–6093.

Imbert, V., Ruper, R.A., Livolsi, A., Pahl, H.L., Traenckner, E.B., Mueller-Dieckmann, C., Farahifar, D., Rossi, B., Auberger, P., Baeuerle, P.A. and Peyron, J.F. (1996). Tyrosine phosphorylation of IκB-alpha activates NF-κB without proteolytic degradation of IκB-alpha. Cell 86, 787–798.

Irani, K. (2000). Oxidant signaling in vascular cell growth, death, and survival: A review of the roles of reactive oxygen species in smooth muscle and endothelial cell mitogenic and apoptotic signaling. Cir. Res. 87, 179–183.

Irani, K., Xia, Y., Zweier, J.L., Sollott, S.J., Der, C.J., Fearon, E.R., Sundaresan, M., Finkel, T. and Goldschmidt-Clermont, P.J. (1997). Mitogenic signaling mediated by oxidants in Ras-transformed fibroblasts. Science 275, 1649–1652.

Jacobson, M.D. (1996). Reactive oxygen species and programmed cell death. Trends Biochem. Sci. 21, 83–8

Johnson, T.M., Yu, Z.X., Ferrans, V.J., Lowenstein, R.A. and Finkel, T. (1996). Reactive oxygen species are downstream mediators of p53-dependent apoptosis. Proc. Natl. Acad. Sci. USA 93, 11848–11852.

Kane, D.J., Sarafian, T.A., Anton, R., Hahn, H., Gralla, E.B., Valentine, J.S., Ord, T. and Bredesen, D.E. (1993). Bcl-2 inhibition of neural death: decreased generation of reactive oxygen species. Science 262, 1274–1277.

Karin, M. (1998). Mitrogen-activated protein kinase cascades as regulators of stress responses. Ann. NY Acad. Sci. 30, 139–146.

Kharbanda, S., Ren, R., Pandy, P., Shafman, T.D., Feller, S.M., Weichselbaum, R.R. and Kufe, D.W. (1995). Activation of c-Abl tyrosine kinase in the stress response to DNA-damaging agents. Nature 376, 785–788.

Kheradmand, F., Werner, E., Tremble, P., Symons, M. and Werb, Z. (1998). Role of Rac1 and oxygen radicals in collagenase-1 expression induced by cell shape change. Science 280, 898–900.

Kroemer, G., Dallaporta, B. and Resche-Rigon, M. (1998). The mitochondrial death/life regulator in apoptosis and necrosis. Annu. Rev. Physiol. 60, 619–642.

Lander, H.M. (1997). An essential role for free radicals and derived species in signal transduction. FASEB J. 11, 118–124.

Lee, W., Mitchell, P. and Tjian, R. (1987). Purified transcription factor AP-1 interacts with TPA inducible elements. Cell 49, 741–752.

Li, N.X. and Karin, M. (1999). Is NF-kB the sensor of oxidative stress? FASEB J. 13, 1137–1143.

Li, P.F., Dietz, R. and Harsdorf, R. (1999). p53 regulates mitochondrial membrane potential through reactive oxygen species and induces cytochrome c-independent apoptosis blocked by Bcl-2. EMBO J. 18, 6027–6036.

Liu, H., Nishitoh, H., Ichijo, H. and Kyriakis, J.M. 2000. Activation of apoptosis signal-regulating kinase 1 (ASK1) by tumor necrosis factor receptor-associated factor 2 requires prior dissociation of the ASK1 inhibitor thioredoxin. Mol. Cell. Biol. 20, 2198–2208.

Liu, Z.G., Baskaran, R., Lea-Chou, E.T., Wood, L.D., Chen, Y., Karin, M. and Wang, J.Y.J. (1996). Three distinct signaling responses by murine fibroblasts to genotoxic stress. Nature 384, 273–276.

Lo, Y.Y. and Cruz, T.F. (1995). Involvement of reactive oxygen species in cytokine and growth factor induction of c-fos expression in chondrocytes. J. Biol. Chem. 270, 11727–11730.

Los, M., Schenk, H., Hexel, K., Baeuerle, P.A., Droge, W. and Schulze-Osthoff, K. (1995). IL-2 gene expression and NF-kB activation through CD28 requires oxygen production by 5-lipoxygenase. EMBO J. 14, 3731–3740.

Manna, S.K., Zhang, H.J., Yan, T., Oberley, L.W. and Aggarwal, B.B. (1998). Overexpression of manganese superoxide dismutase suppresses tumor necrosis factor-induced apoptosis and activation of nuclear transcription factor-κB and activated protein-1, J. Biol. Chem. 273, 13245–13254.

Melov, S. (2000). Mitochondrial oxidative stress. Physiologic consequences and potential for a role in aging. Ann. NY Acad. Sci. 908, 219–225.

Mercurio, F. and Manning, A.M. (1999). NF-kB as a primary regulator of the stress response. Oncogene 18, 6163–6171.

Migliaccio, E., Giorgio, M., Mele, S., Pelicci, G., Reboldi, P., Pandolfi, P.P., Lanfrancone, L. and Pelicci, P.G. (1999). The p66shc adaptor protein controls oxidative stress response and life span in mammals. Nature 402, 309–313.

Mignotte, B. and Vayssiere, J.L. (1998). Mitochondria and apoptosis. Eur. J. Biochem. 252, 1–15.

Moreno-Manzano, V., Ishikawa, Y., Lucio-Cazana, J. and Kitamura, M. (2000). Selective involvement of superoxide anion, but not downstream compounds hydrogen peroxide and peroxynitrite, in tumor necrosis factor-alpha-induced apoptosis of rat mesangial cells. J. Biol. Chem. 275, 12684–12691.

Ohba, M., Shibanuma, M., Kuroki, T. and Nose, K. (1994). Production of hydrogen peroxide by transforming growth factor-b1 and its involvement in induction of egr-1 in mouse osteoblastic cells. J. Cell. Biol. 126, 1079–1088.

Ono, K. and Han, J. (2000). The p38 signal transduction pathway: activation and function. Cell Signal. 12, 1–13.

Ozaki, M., Deshpande, S.S., Angkeow, P., Bellan, J., Lowenstein, C.J., Dinauer, M.C., Goldschmidt-Clermont, P.J. and Irani, K. (2000). Inhibition of the Rac1 GTPase protects against nonlethal ischemia/reperfusion-induced necrosis and apoptosis in vivo. FASEB J. 14, 418–429.

Pinkus, R., Weiner, L.M. and Daniel, V. (1996). Role of oxidants and antioxidants in the induction of AP-1, NF-kB, and glutathione S-transferase gene expression. J. Biol. Chem. 271, 13422–13429.

Powis, G., Briehl, M. and Oblong, J. (1995). Redox signaling and the control of cell growth and death. Pharmacol. Ther. 68, 149–173.

Puri, P.L., Avantaggiati, M.L., Burgio, V.L., Chirillo, P., Collepardo, D., Natoli, G., Balsano, C. and Levrero, M. (1995). Reactive oxygen intermediates mediate angiotensin II-induced c-Jun/c-Fos hererodimer DNA binding activity and proliferative hypertrophic responses in myogenic cells. J. Biol. Chem. 270, 22129–22134.

Rathmell, J.C. and Thompson, C.B. (1999). The central effectors of cell death in the immune system. Annu. Rev. Immunol. 17, 781–828.

Rosen, G.M., Pou, S., Ramos, C.L., Cohen, M.S. and Britigan, B.E. (1995). Free radicals and phagocytic cells. FASEB J. 9, 200–209.

Rosette, C. and Karin, M. (1996). Ultraviolet light and osmotic stress: activation of the JNK cascade through multiple growth factor and cytokine receptors. Science 274, 1194–1197.

Rouault, T.A. and Klausner, R.D. (1996). Iron–sulfur clusters as biosensors of oxidants and iron. Trends Biochem. Sci. 21, 174–177.

Saitoh,M., Nishitoh, H., Fijii, M., Takeda, K., Tobiume, K., Sawada, Y., Kawabata, M., Miyazono, K. and Ichijo, H. (1998). Mammalian thioredoxin is a direct inhibitor of apoptosis signal-regulating kinase (ASK)1. Mol. Cell. Biol. 17, 2596–2606.

Schenk, H., Klein, M., Erdbrugger, W., Droge, W. and Schulze-Osthoff, K. (1994). Distinct effect of thioredoxin and other antioxidants on the activation of NF-κB and AP-1. Proc. Natl. Acad. Sci. USA. 91, 1672–1676.

Schreck, R., Rieber, P. and Baeuerle, P.A. (1991). Reactive oxygen intermediates as apparently widely used messengers in the activation of the NF-κB transcription factor and HIV-1. EMBO J. 10, 2247–2252.

Schulze-Osthoff, K., Beyaert, R., Vandevoorde, Veronique, V., Haegeman, G. and Fiers, W. (1993). Depletion of the mitochondrial electron transport abrogates the cytotoxic and gene-induction effects if TNF. EMBO J. 12, 3095–3104.

Sen, C.K. and Packer, L. (1996). Antioxidant and redox regulation of gene transcription. FASEB J. 10, 709–720.

Shi, X., Dong, Z., Huang, C., Ma, W., Liu, K., Ye, J., Chen, F., Leonard, S.S., Ding, M., Castranova, V. and Vallyathan, V. (1999). The role of hydroxyl radical as a messenger in the activation of nuclear transcription factor NF-κB. Mol. Cell. Biochem. 194, 63–70.

Suh, Y.A., Arnold, R.S., Lassegue, B., Shi, J., Xu, X., Sorescu, D., Chung A.B., Griendling, K.K. and Lambeth, J.D. (1999). Cell transformation by the superoxide-generating oxidase Mox1. Nature 401, 79–82.

Sundaresan, M., Yu, Z.X., Ferrans, V.J., Irani,K. and Finkel, T. (1995). Requirement for generation of H_2O_2 for platelet-derived growth factor signal transduction. Science 270, 296–29.

Tan, P., Fuchs, S.Y., Chen, A., Wu, K., Gomez, C., Ronai, Z. and Pan, Z.Q. (1999a). Recruitment of a ROC1-CUL1 ubiquitin ligase by Skp1 and HOS to catalyze the ubiquitination of IκB alpha. Mol. Cell. 3, 527–533.

Tan, S., Sagata, Y., Liu, Y.b. Maher, P. and Schubert, D. (1998b). The regulation of reactive oxygen species production during programmed cell death. J. Cell. Biol. 141, 1423–1432.

Thannickal, V.J., Aldweib, K.D. and Fanburg, B.L. (1998). Tyrosine phosphorylation regulates H_2O_2 production in ling fibroblasts stimulated by transforming growth factor betal. J. Biol. Chem. 273, 23611–23615.

Wilhelm, D., Bender, K., Knebel, A. and Angel, P. (1997). The level of intracellular glutathione is a key regulator for the induction of stress-activated signal transduction pathway including Jun N-terminal protein kinases and p38 kinase by alkylating agents. Mol. Cell. Biol. 17, 4792–4800.

Winrow, V.R., Winyard, P.G., Morris, C.J. and Blake, D.R. (1993). Free radicals in inflammation: second messengers and mediators of tissue destruction. Br. Med. Bull. 49, 506–522.

Yacoub, A., Kelley, M.R. and Deutsch, W.A. (1997). The DNA repair activity of human redox/repair protein APE/Ref-1 is inactivated by phosphorylation. Cancer Res. 57, 5457–5459.

Yin, Y., Tarauchi, Y., Solomon, G.G., Aizawa, S., Rangarajan, P.N., Yazaki, Y., Kadowaki, T. and Barrett, J.C. (1998). Involvement of p85in p53-dependent apoptotic response to oxidative stress. Nature 391, 707–710.

Yin, Z., Ivanov, V.N., Habelhah, H., Tew, K. and Ronai, Z. (2000). Glutathione S-transferase p elicits protection against H_2O_2-induced cell dearth via coordinated regulation of stress kinase. Cancer Res. 60, 4053–4057.

Yoshizumi, M., Abe, J., Haendeler, J., Huang, Q. and Berk, B.C. (2000). Src and Cas mediate JNK activation but not ERK1/2 and p38 kinases by reactive oxygen species. J. Biol. Chem. 275, 11706–11712.

Zhang, M., Aslund, F. and Storz, G. (1998). Activation of the OxyR transcription factor by reversible disulfide bond formation. Science 279, 1718–1721.

Zhao, R., Gish, K., Murphy, M., Yin, Y., Notterman, D., Hoffman, W.H., Tom, E., Mack, D.H. and Levine, A.J. 2000. Analysis of p53-regulated gene expression patterns using oligonucleotides arrays. Genes Dev. 14, 981–193.

Zhong, H., SuYang, H., Erdjument-Bromage, H., Tempst, P. and Ghosh, S. (1997). The transcriptional activity of NF-κB is regulated by the IκB-associated PKAc subunit through a cyclic AMP-independent mechanism. Cell 89, 413–424.

Protein Adaptations and Signal Transduction. Edited by K.B. Storey and J.M. Storey

CHAPTER 20

Antioxidant defenses and animal adaptation to oxygen availability during environmental stress

Marcelo Hermes-Lima[1]*, Janet M. Storey[2] and Kenneth B. Storey[2]
[1]*Oxyradical Research Group, Departamento de Biologia Celular, Universidade de Brasília, Brasília, DF 70910-900, Brazil;* [2]*Institute of Biochemistry, College of Natural Sciences, Carleton University, Ottawa, Ontario, K1S 5B6, Canada*

1. Free radicals, antioxidant enzymes and oxidative stress

Oxygen is essential for most life forms. The full reduction of oxygen to H_2O by cytochrome oxidase is a key step in the mechanism of aerobic ATP formation. However, the partial reduction of oxygen results in the formation of various reactive oxygen species that can be damaging to cellular components. For example, about 1–4% of all oxygen consumed by cells is converted into superoxide radicals (O_2^-) by the "leaky" mitochondrial respiratory chain. Other relevant cellular sources of O_2^- include the activities of soluble oxidases (e.g. xanthine oxidase, aldehyde oxidase), NADPH oxidase at the plasma membrane of phagocytes, and the autoxidation of small molecules. Another reactive oxygen species is hydrogen peroxide (H_2O_2) which is formed by means of O_2^- dismutation (either spontaneous or catalyzed by superoxide dismutase; reaction 1) or by mixed function oxidase systems during the detoxification of xenobiotics. O_2^- and H_2O_2 have low oxidative toxicity themselves but they are readily converted into hydroxyl radicals (·OH) via the Haber–Weiss reaction which is catalyzed by iron or copper ions (reaction 2) (Cadenas, 1995; Fridovich, 1998). Furthermore, the reaction of O_2^- with nitric oxide (NO·) leads to the formation peroxynitrite (ONO_2^-), a non-radical oxidant species (Inoue et al., 1999) (reaction 3). Both ·OH and ONO_2^- are highly reactive species and cause serious damage to cellular macromolecules including lipid peroxidation, protein oxidation, and DNA damage (Cadenas, 1995) (Fig. 20.1).

$$2\,O_2^- + 2\,H^+ \rightarrow H_2O_2 + O_2 \quad (1)$$

$$O_2^- + H_2O_2 \rightarrow (Cu^{2+}/Fe^{3+}) \rightarrow O_2 + OH^- + \cdot OH \quad (2)$$

$$O_2^- + NO\cdot \rightarrow ONO_2^- \quad (3)$$

Enzymatic defenses against reactive oxygen species have evolved in all aerobic organisms and include superoxide dismutases (Mn-SOD, mitochondrial isoform; CuZn-SOD, cytosolic isoform; Fe-SOD, bacterial SOD), catalase and selenium-dependent glutathione peroxidase (Se-GPX). All SOD isoforms catalyze the dismutation of O_2^- into O_2 and H_2O_2. Catalase, a peroxisomal enzyme, has a major role in the decomposition of H_2O_2 (forming H_2O and O_2) as does Se-GPX, which similarly catalyzes the decomposition of H_2O_2 and organic hydroperoxides but uses glutathione (GSH) as its co-substrate (Ahmad, 1995; Fridovich, 1998; Hermes-Lima et al., 1998) (Fig. 20.1). Other peroxidases also have relevant roles for cellular H_2O_2 detoxification including ascorbate peroxidase and cytochrome C peroxidase (Ahmad, 1995; Campos et al., 1999). Glutathione S-transferases

*Corresponding author.

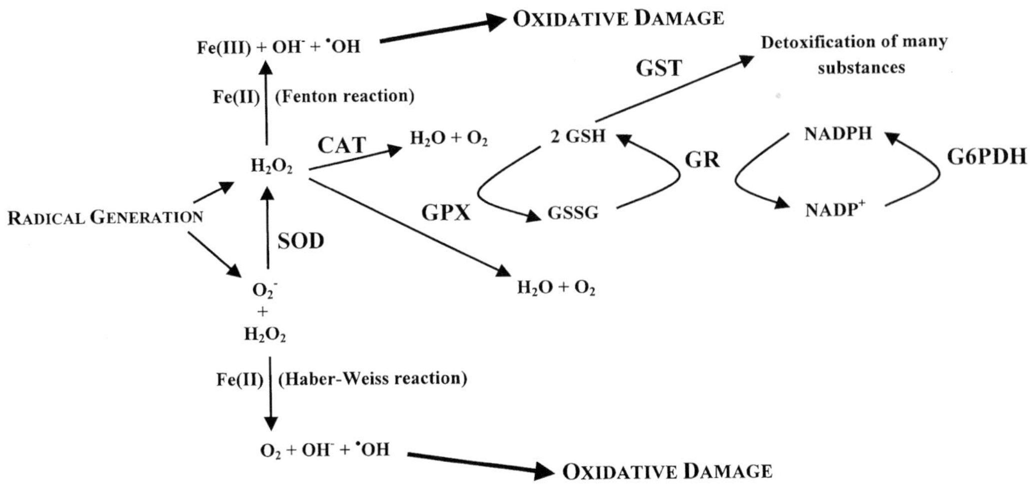

Fig. 20.1. Enzymatic antioxidant defense system of animal cells. Abbreviations are: CAT, catalase; SOD, superoxide dismutase; GPX, glutathione peroxidase; GST, glutathione-S-transferase; GR, glutathione reductase; G6PDH, glucose-6-phosphate dehydrogenase; GSH, reduced glutathione; GSSG, oxidized glutathione.

(GST) also play roles in antioxidant defense; these catalyze the conjugation of GSH to xenobiotics and also display selenium-independent GPX activity toward organic hydroperoxides (Habig and Jakoby, 1981; Joanisse and Storey, 1996).

Several auxiliary enzymes are involved in antioxidant defense. Glutathione reductase (GR) functions to recycle glutathione, converting the oxidized form of glutathione (GSSG) back into GSH using the reducing power of NADPH (Ahmad, 1995). Hexose monophosphate shunt enzymes including glucose-6-phosphate dehydrogenase (G6PDH) (Fig. 20.1) and 6-phosphogluconate dehydrogenase are major suppliers of the NAPDH required by GR. GSH synthetase is a key enzyme in the formation of GSH (Willmore and Storey, 1997b) and thiol-disulfide oxidoreductase enzymes (e.g. thioltransferase, thioredoxin) catalyze the removal of thiol compounds (usually GSH) from a sulfhydryl protein mixed disulfide. Mixed disulfides formed between GSH and enzymes accumulate in tissues that are subjected to oxidative stress. If a mixed disulfide forms on an enzyme, that enzyme can be inactivated or become more susceptible to proteolytic degradation (Starke et al., 1997; Willmore and Storey, 1997b).

Antioxidant enzyme systems are adaptable and changes in enzyme activities are well known to be triggered by oxidative stress. Specific responses of antioxidant enzymes do not follow set patterns but are stress-, tissue- and species-specific (Crawford et al., 2000). For example, activities (and mRNA transcript levels) of catalase, Se-GPX and Mn-SOD (but not CuZn-SOD) were elevated in tracheobronchial epithelia in response to H_2O_2 but only Mn-SOD mRNA increased in these cells following exposure to the oxyradical-generating enzyme, xanthine oxidase (Shull et al., 1991). Mice irradiated by X-rays (which induce free radical formation and hepatic lipid peroxidation), showed a compensatory increase in the activities of liver catalase and SOD (Mn- and CuZn-SOD) in radiation-resistant animals but not in radiation-sensitive ones (Hardmeier et al., 1997). However, the capacity of cells to modulate their antioxidant defenses has its limits and so situations can arise where oxyradical overproduction can inflict permanent damage or cell death.

Oxidative stress has been linked to varying extents with the natural aging process and with a number of diseases and disorders including ischemic heart disease, stroke, atherosclerosis, iron-overload diseases, diabetes, and several types of carcinogenic and neurodegenerative processes. In these cases, oxidative stress and oxyradical-induced damage results when oxyradical production exceeds the antioxidant buffering capacity of the tissue or when a stress-mediated reduction in the antioxidant capacity occurs (Beckman and Ames, 1998; Lipton, 1999; Crawford et al., 2000). In several disorders, the activities of antioxidant enzymes are increased in response to the stress but in many cases the adaptive response does not fully compensate for the increased stress. Examples of such partial compensatory responses are found in diabetic humans and rats (Kakkar et al., 1995; Crawford et al., 2000), during hypertrophy and heart failure in pigs (Dhalla and Singal, 1994) and in liver of rats during ethanol consumption (Crawford et al., 2000).

The present article considers oxidative stress and antioxidant defenses as they occur in situations where animals face high natural variation in oxygen availability including situations of environmental oxygen deprivation and natural episodes of hypoperfusion or ischemia/reperfusion. Both oxygen deprivation (as in ischemia) and its reintroduction (causing oxyradical overproduction; Ruuge et al., 1991) can cause serious metabolic injuries for most mammals, but a wide variety of lower vertebrate species can endure extended periods of hypoxia or anoxia ranging from hours to weeks and recover from these without injury. The adaptations of antioxidant defense systems that aid animal survival during natural cycles of oxygen deprivation and reintroduction are reviewed herein and provide examples of the design of antioxidant defenses for function under situations of widely varying oxygen availability.

2. Natural anoxia tolerance and adaptations to oxidative stress

The electron carriers of the mitochondrial respiratory chain become reduced during ischemic or hypoxic events. When oxygen is reintroduced an immediate reoxidation of these carriers takes place but with a transient overproduction of oxygen free radicals (Ruuge et al., 1991). This burst of oxyradical production can overwhelm existing cellular antioxidant defenses and cause damage to macromolecules including DNA and membrane lipids (Cordis et al., 1995, 1998). Moreover, post-ischemic peroxidation of endoplasmic reticulum causes an increase in cytoplasmic calcium concentration that can lead to uncontrolled activation of phospholipases and proteases. In mammalian systems undergoing reoxygenation or reperfusion, these oxyradical-induced events can lead to severe cell damage, apoptosis and organ failure (Bolli and Marban, 1999; Lipton, 1999). The use of exogenous antioxidants or the overexpression of antioxidant enzymes (by gene manipulation) has been shown to have beneficial effects in minimizing reperfusion injuries (Wang et al., 1998; Weisbrot-Lefkowitz et al., 1998).

Mammalian organ systems are basically designed to function under high oxygen conditions and every effort is made to maintain oxygen concentrations within a high and narrow range at all times. Situations of oxygen insufficiency (hypoxia) typically stimulate compensatory responses that increase oxygen carrying capacity (increased release or synthesis of red blood cells), stimulate capillary growth, or raise the glycolytic capacity of tissues (Bunn and Poyton, 1996). Mammals show limited tolerance for sustained endurance of low oxygen conditions and a very limited capacity for maintaining cellular homeostasis without oxygen. Furthermore, because every effort is made to minimize variation in tissue oxygenation, mammals also show a limited ability to deal with the oxyradical stress that accompanies the rapid reoxygenation of tissues after a hypoxic or ischemic episode (Halliwell et al., 1992; Lipton, 1999).

However, the mammalian (and avian) focus on sustaining consistently high rates of aerobic metabolism is not universal throughout the animal kingdom. Many invertebrates and cold-blooded vertebrates experience wide natural variations in oxygen availability to their tissues and many can live without oxygen for days, weeks or months at a time (Lutz and Storey, 1997). Numerous situations can limit oxygen availability. For example, many freshwater turtles hibernate underwater and go without breathing for 3–4 months at a time. Although some species use extrapulmonary mechanisms of oxygen uptake, others survive without oxygen for the entire time (Ultsch, 1989). Some fish also have well-developed anoxia tolerance; species such as carp and goldfish can survive the winter in small ice-locked ponds when oxygen levels fall to zero (van den Thillart, 1982; Lutz and Nilsson, 1997). Many marine molluscs and other invertebrates also have excellent anoxia tolerance that serves several situations: aerial exposure of gill-breathing species at low tide, burrowing in benthic sediments, and shell closure that protects the organisms from environmental insults (e.g. desiccation, toxins, predators, etc.) (Storey and Storey, 1990). Other organisms deal not just with anoxia but also with ischemia. This is illustrated best by freeze tolerant animals that endure the freezing of blood plasma and other extracellular body fluids during the winter. Hundreds of species of insects, many intertidal marine invertebrates and various amphibians and reptile species living in seasonally cold climates have developed this capacity (Storey and Storey, 1988). Most can survive for several weeks with up to 65% of their total body water frozen and with no breathing or circulation so that all cells must survive the freezing episode utilizing only their endogenous fuels and enduring long term oxygen deprivation (Storey et al., 1996). These anoxia- and ischemia-tolerant animals express a variety of biochemical adaptations that sustain survival during oxygen deprivation (Lutz and Storey, 1997; Brooks and Storey, 1997; Storey, 1996a, 1999).

Several years ago we hypothesized that anoxia tolerant species should also have good antioxidant defenses to deal with the oxidative stress that must occur with the reintroduction of oxygen into the tissues of these animals. After days, weeks or months without oxygen, its sudden reintroduction must create a danger of oxyradical overgeneration and oxidative damage to cells. We set out to determine whether biochemical adaptations also improved the ability of anoxia- or freeze-tolerant animals to deal with oxidative stress. There was little prior research on this topic, particularly with cold-blooded vertebrates. For example, in studying the South American freshwater turtle, *Phrynops hilarri*, which winters underwater, Reischl (1986, 1989) had proposed that sulfhydryl-rich hemoglobins could quench oxyradicals formed during reoxygenation stress. However, no systematic analysis of the antioxidant defenses of anoxia tolerant animals had been made. The present review summarizes our recent studies of the antioxidant defenses of a variety of vertebrate and invertebrate species that show good natural anoxia or freezing tolerance.

2.1. Antioxidants and garter snakes under anoxia exposure

The first study of the role of endogenous antioxidant defenses in the natural tolerance of oxygen deprivation was done with red-sided garter snakes, *Thamnophis sirtalis parietalis*, which are the most northerly distributed reptiles in North America (Pinder et al., 1992). This species from central Canada can endure anoxia exposure under nitrogen gas for up to 48 h with 100% recovery (Hermes-Lima and Storey, 1993a) and can also survive freezing at –2.5°C for several hours (an ischemic stress) (Churchill and Storey, 1992). The anoxia tolerance of this species might be useful for underwater survival, for example if hibernation dens became flooded during the spring melt. Indeed, for some garter snake populations, underwater hibernation may even be the norm (Costanzo, 1989). Not only does underwater hibernation avoid the threat of freezing but submergence initiates the diving reflex which sharply suppresses metabolic rate, thereby reducing the rate at which endogenous fuel stores are consumed during the nonfeeding season.

The effects of anoxia exposure (10 h under a nitrogen gas atmosphere at 5°C) on the activities of antioxidant enzymes and the levels of GSH in liver, lung and skeletal muscle of garter snakes were analyzed (Hermes-Lima and Storey, 1992, 1993a). Compared with control snakes held at 5°C, anoxia exposure caused an increase in SOD activity (Mn- plus CuZn-SOD) in skeletal muscle and liver by 59 and 118%, respectively (Table 20.1). Lung SOD was unaffected by anoxia as were the activities of catalase, GR, Se-GPX and GST in all three organs (Hermes-Lima and Storey, 1993a). The levels of GSH were unchanged in anoxic liver and lung but increased significantly in anoxic

Table 20.1. Effect of anoxia or freezing exposure on the activities of antioxidant enzymes and levels of GSH in garter snakes, *Thamnophis sirtalis*

	Control	Anoxic	Frozen
Liver			
Catalase (U/mg)	71.8 ± 5.5	91.7 ± 15.1	83.8 ± 6.7
SOD (U/mg)	8.3 ± 0.7	18.1 ± 2.9**a**	8.5 ± 1.2
Se-GPX (mU/mg)	155 ± 7	142 ± 5	130 ± 10**a**
GR (mU/mg)	10.9 ± 0.4	13.4 ± 2.2	11.7 ± 1.1
GST (mU/mg)	638 ± 77	720 ± 97	468 ± 18**a**
GSH (μmol/gww)	1.02 ± 0.09	0.82 ± 0.06	0.84 ± 0.05
Skeletal muscle			
Catalase (U/mg)	22.9 ± 2.8	16.6 ± 3.1	64.7 ± 12.1**a**
SOD (U/mg)	3.4 ± 0.2	5.3 ± 0.5**a**	3.7 ± 0.7
Se-GPX (mU/mg)	65.6 ± 10	50.8 ± 8.5	100 ± 4.7**a**
GR (mU/mg)	9.6 ± 1.0	9.1 ± 1.9	14.1 ± 3.0
GST (mU/mg)	59 ± 7	66 ± 10	47 ± 6
GSH (μmol/gww)	0.45 ± 0.04	0.71 ± 0.01**a**	0.38 ± 0.09

Data are means ± SEM, n = 3–5; enzyme activities are expressed per milligram protein and whereas reduced glutathione concentrations are per gram wet weight. **a**: Significantly different from the corresponding control values, $P < 0.05$. Results are from Hermes-Lima and Storey (1993a).

muscle (by 58%) although the ratio GSSG/GSH did not change significantly during anoxia in any organ. This increase in muscle GSH concentration could potentially stimulate muscle Se-GPX activity *in vivo* by ~50% due to the enzyme's low affinity for GSH (Km = 11 mM). Unfortunately, no analysis of antioxidant defenses were made during aerobic recovery after anoxia exposure in these garter snakes. However, the results suggest that anoxia-mediated increases in the defenses against O_2^- (by means of increased SOD activity) or peroxides (stimulation of Se-GPX activity) could prevent or minimize ·OH radical formation from Fenton reactions as well as oxidative damage during reoxygenation.

2.2. Antioxidants and leopard frogs under anoxia and reoxygenation

In a second study of oxidative stress, antioxidant defenses and anoxia tolerance we turned our attention to leopard frogs, *Rana pipiens*. Leopard frogs can readily survive extended periods of time in deoxygenated water at low temperatures (Pinder et al., 1992). This capacity undoubtedly aids their winter survival underwater. Although frogs can readily take up oxygen across their skin when submerged, the ice-locked ponds and lakes where they live frequently become oxygen-depleted as winter progresses (Ultsch, 1989).

To study the relationship between anoxia tolerance and antioxidant defenses, 5°C-acclimated leopard frogs were exposed to 10 or 30 h of anoxia followed by 1.5 or 40 h of aerobic recovery (Hermes-Lima and Storey, 1996). Thirty hours of anoxia exposure resulted in significant increases in the activities of skeletal muscle and heart catalase (by 53 and 47%), heart and brain Se-GPX (by 75 and 30%), and brain GST (by 66%). In most cases, enzyme activities had returned to control levels after 40 h recovery. Activities of SOD (Mn- plus CuZn-SOD) and GR were not affected in any organ, and anoxia/recovery had no effect on any of the enzymes in liver. Total glutathione levels (GSH-eq = GSH + 2 GSSG) remained constant in liver, skeletal muscle and heart during anoxia but decreased by 32% in anoxic brain. After 1.5 h reoxygenation, brain GSH-eq returned to control values and hepatic GSH-eq rose by 71%. We concluded from these results that organs of *R. pipiens* either increase or maintain their antioxidant capacity during anoxia exposure. The increase in catalase and Se-GPX activities are especially important for dealing with possible H_2O_2 overgeneration during reoxygenation.

The ratio GSSG/GSH-eq significantly increased in frog muscle and liver during anoxia exposure and returned to control levels after 90 min of recovery. This result could indicate a condition of oxidative stress during anoxia, where GSSG accumulates due to H_2O_2 overgeneration. Alternatively, the increased GSSG/GSH-eq ratio could also indicate a reduced capacity for GSH resynthesis caused by the hypometabolic condition, which might decrease the production of NAPDH, essential for the *in vivo* activity of GR (Hermes-Lima and Storey, 1996).

2.3. Antioxidants and goldfish under anoxia and reoxygenation

Tolerance to hypoxia/anoxia is a crucial survival strategy for many fish species, which can be exposed to transitory low oxygen availability in their aquatic environments. Some species have to deal with complete anoxia for extended periods of time in the waters where they live. For example, various lakes in the Amazon basin show seasonal and/or daily oscillations in oxygen concentration that can leave them severely hypoxic or fully anoxic for extended periods of time whereas in northern latitudes many lakes and ponds become oxygen depleted when ice-locked during the winter months. For example, studies with the Amazon cichlid, *Astronotus ocellatus*, show that these fish can survive 16 h of severe hypoxia and 4 h of anoxia at 28°C (Muusze et al., 1998). Goldfish, *Carassius auratus*, have a half-lethal time of 45 h under anoxia at 5°C and 22 h at 20°C. However, the champion of fish anoxia tolerance is the Crucian carp, *C. carassius*, that can survive 60–100 days under anoxia at 5°C. Strong metabolic depression (to about 20–30% of normal metabolic rates) during anoxia is a key determinant for survival of *Carassius* species, as well as the availability of very large tissue glycogen reserves and the ability to avoid lactic acidosis by further metabolizing lactate to produce ethanol and CO_2 that are excreted through the gills (Shoubridge and Hochachka, 1983; Lutz and Nilsson, 1997). The first suggestion that modulation of antioxidant defenses might be involved in fish anoxia tolerance came from Vig and Nemcsok (1989) who found a significant increase in liver, brain and gill SOD activity in carp, *Cyprinus carpio*, after several hours exposure to extreme hypoxia. Although these authors did not discuss the physiological significance of their findings, we can propose that elevated SOD activity is a relevant adaptive mechanism against post-hypoxic oxyradical insult.

The relationship between anoxia tolerance, reoxygenation and the modulation of antioxidant defenses has recently been investigated in four organs of goldfish (Lushchak et al., 2001). Exposure of goldfish to 8 h in N_2-bubbled water at 20°C induced an increase in the activities of liver catalase (by 38%) and brain G6PDH and Se-GPX (by 26 and 79%) (Table 20.2). However, kidney catalase activity was reduced by 17% during anoxia. After 14 h reoxygenation, liver catalase and brain Se-GPX activities remained higher than controls. Other tissue-specific changes also occurred during reoxygenation: white muscle and kidney GST rose by 31 and 91% and liver GR increased by 41% whereas white muscle G6PDH decreased by 47% (Luschack et al., 2001). The activity of SOD (Mn-plus CuZn-SOD) and the levels of GSH-eq were not affected by anoxia and reoxygenation in any tissue, except for a 14% decrease in kidney GSH-eq during anoxia. Levels of kidney GSH-eq returned to control values after 14 h reoxygenation. In addition to the up-regulation of liver catalase and two enzymes in brain of anoxic goldfish, these data also revealed relatively high constitutive activities of glutathione-dependent antioxidant enzymes (GR, Se-GPX and GST) in liver of anoxia-tolerant goldfish as compared with the activities in other lower vertebrate species (Table 20.3).

Table 20.2. Effect of 8 h anoxia exposure and reoxygenation (1 h or 14 h) on the levels of conjugated dienes and activities of selected antioxidant enzymes in tissues of goldfish, *Carassius auretus*

	Brain	Liver	Kidney	White muscle
Conjugated dienes (2nd derivative Abs/g wet wt)				
Control	96.0 ± 4.3	81.3 ± 2.5	152.5 ± 11.6	120.9 ± 7.3
Anoxia	111.6 ± 13,5	75.1 ± 5.5	167.2 ± 26.4	47.5 ± 3.6**a**
Recovery 1 h	99.0 ±13.7	173.6 ± 10.4**a**	162.2 ± 34.9	50.9 ± 3.6**a**
Recovery 14 h	167.8 ± 29.1**a**	162.2 ± 28.9**a**	120.7 ± 32.4	67.9 ± 14.4**a**
Catalase (U/mg protein)				
Control	3.8 ± 0.56	157.7 ± 16.4	19.15 ± 0.90	2.06 ± 0.27
Anoxia	4.00 ± 0.34	218.0 ± 14.5**a**	15.83 ± 0.83**a**	1.59 ± 0.12
Recovery 14 h	3.49 ± 0.32	247.0 ± 7.0**a**	14.08 ± 1.21**a**	1.59 ± 0.12
Se-GPX (mU/mg protein)				
Control	7.67 ± 1.15	554 ± 82	75.7 ± 3.9	17.9 ± 3.8
Anoxia	13.73 ± 1.42**a**	400 ± 94	63.5 ± 9.5	19.1 ± 1.4
Recovery 14 h	12.23 ± 0.79**a**	602 ± 111	75.7 ± 6.5	16.7 ± 1.4
G6PDH (mU/mg protein)				
Control	17.2 ± 1.1	269 ± 22	63.6 ± 6.5	1.41 ± 0.25
Anoxia	21.6 ± 0.3**a**	355 ± 31	66.2 ± 5.5	0.89 ± 0.12
Recovery 14 h	19.7 ± 0.9**a**	307 ± 25	45.9 ± 2.4	0.75 ± 0.12**a**

Data are means ± SEM, $n = 3$–6. **a**: Significantly different from the corresponding control values, $P < 0.05$. Data are from Lushchak et al. (2001).

From these data, it seems that the tissues of garter snakes, leopard frogs and goldfish respond to environmental anoxia in a manner that anticipates an overgeneration of oxyradicals at the termination of the anoxic insult. Thus, the observed increases in the activities of antioxidant enzymes in organs of these species suggest that antioxidant defenses are an important and necessary part of the adaptive machinery for natural anoxia tolerance. These antioxidant defenses are typically built up under anoxia, a time when oxyradical formation is not likely to occur, and hence they appear to be a preparatory event for the occurrence of oxidative stress following tissue reoxygenation. (Hermes-Lima and Storey, 1993a, 1996; Hermes-Lima et al., 1998; Storey, 1996b; Lushchak et al., 2001). Evolutionary pressure might have selected species with the ability to prepare for post-anoxic oxyradical formation.

2.4. Antioxidants and turtles under anoxia and reoxygenation

Willmore and Storey (1997a,b) studied the correlation between anoxia tolerance and oxidative stress in a freshwater turtle, the red-eared slider *Trachemys scripta elegans.* The organs of turtles are subjected to low oxygen tensions during diving, particularly during extended dives when circulatory adjustments can cause severe hypoxia in some organs due to the shunting of oxygen to vital organs (Storey and Storey, 1990; Storey, 1996a). More importantly, however, this reptile hibernates underwater and is one of the species with a very low capacity for oxygen uptake from the water by extrapulmonary means. Survival during winter submergence is therefore dependent on a very well-developed anaerobic capacity; indeed, animals have been shown to survive for at least 2–3

Table 20.3. Control values for activities of antioxidant enzymes and concentration of GSH-eq in vertebrate liver and snail hepatopancreas of several anoxia tolerant, freeze tolerant or estivating species in comparison with activities in rat liver*

	SOD (U/mg)	Catalase (U/mg)	GST (U/mg)	GR (mU/mg)	Se-GPX (mU/mg	GSH-eq (μmol/gww)
Vertebrate liver						
Rat[1]	75–85	340–380	0.4–0.5	25–35	600–700	7–9
Red-eared turtle[2]	43–53	220–240	1.9–2.3	30–35	280–320	3.0–3.5
Garter snake[3]	7–9	65–75	0.55–0.70	10–11	150–160	1.0–1.2
Wood frog[4]	35–40	200–250	0.58–0.62	15–20	120–150	1.2–1.5
Leopard frog[5]	15–20	500–600[6]	0.6–0.9	5–10	30–50	0.5–0.7
Spadefood toad[7]	40–60	1150–1400	1.5–1.6	8–12	60–80	1.5–1.8
Goldfish[8]	3.0–3.5[9,10]	140–170	0.5–0.6	24–28	470–630	2.5–3.0
Molluscs (hepatopancreas)						
O. lactea[11]	45–55	180–210	1.0–1.2	18–20	10–12	2.7–3
L. littorea[12]	20–30	17–20	0.3–0.4	12–16	13–15	0.3–0.35

*Table shows approximate range of enzyme activities (expressed per milligram protein) and GSH levels (per gram wet weight), calculated from published values for mean ± SEM.
[1]Data from Habig and Jakoby (1981) and Perez-Campo et al. (1993). [2]Willmore and Storey (1997a, 1997b). [3]Hermes-Lima and Storey (1993a).[4]Joanisse and Storey (1996). [5]Activities represent a range determined in two studies (Hermes-Lima and Storey 1996, 1998). [6]Catalase activity measured by Joanisse and Storey (1996) was about 100 U/mg protein. [7]Grundy and Storey (1998). [8]Lushchak et al. (2001). [9]SOD activity from trout liver was about 15 U/mg protein (Perez-Campo et al., 1993). [10]A more recent quantification of SOD activity in goldfish liver using an improved form of the assay yielded 10-fold higher values (M.V.R. Ferreira and M. Hermes-Lima, unpublished). [11]Hermes-Lima and Storey (1995a). [12]Pannunzio and Storey (1998).

months in fully deoxygenated water at 3°C (Ultsch, 1989).

Exposure of red-eared slider turtles to anoxic submergence (20 h in deoxygenated water at 5°C) brought about an interesting behavior by antioxidant enzyme activities. The activities of several antioxidant enzymes, including hydroperoxidase reductase (AHR), and GSH-synthetase (auxiliary players in the antioxidant defense system) were monitored in liver, brain, heart, kidney, red muscle and white muscle (Table 20.4) (Willmore and Storey, 1997a,b). Anoxia exposure led to selected decreases in enzyme activities which might be consistent with a reduced potential for oxidative damage while oxygen-limited. Heart showed the greatest reduction in antioxidant capacity during anoxic stress with 31–67% reductions in the activities of three enzymes (catalase, GR, GST) as well as in levels of GSH-eq. Reduced antioxidant capacity also occurred in liver and brain during anoxia; SOD activity decreased by 15–30%, brain catalase activity fell by 80% (Table 20.4) and liver GSH-eq concentration was reduced by 50%. Levels of GSH-eq and catalase also dropped by 41 and 68%, respectively, in kidney during anoxia exposure.

On the other hand, AHR activity increased during anoxia in heart and kidney, by 2- and 3.5-fold respectively (Table 20.4), and GR increased in anoxic liver (by 52%) and red muscle (by 80%) (Willmore and Storey, 1997b). GSH-synthetase activity also increased by 3-fold in anoxic white muscle (Willmore and Storey, 1997b). Most anoxia-induced changes were reversed after 24 h of aerobic recovery although turtle brain enzyme activities remained suppressed. In addition, other changes had occurred after 24 h reoxygenation: heart SOD and GR increased by 45 and 64%, red and white muscle AHR rose by 100 and 68%, respectively, and heart and brain GSH-synthetase activity doubled (all compared with controls) (Table 20.4). The activity of Se-GPX was

Table 20.4. Effect of anoxia exposure (20 h) and reoxygenation (24 h) on the activities of selected antioxidant enzymes in freshwater turtles, *Trachemys scripta elegans*

Tissue	Catalase (U/mg)	SOD (U/mg)	AHR (mU/mg)	GSH-synthetase (mU/mg)
Liver				
Control	229 ± 8.4	48.6 ± 5.8	12.6 ± 3.1	13.1 ± 2.6
Anoxia	203 ± 32.3	34.1 ± 3.5 **a**	15.7 ± 1.7	8.4 ± 1.1
Recovery	206 ± 19.2	40.7 ± 2.8	13.7 ± 1.6	6.7 ± 1.2
Heart				
Control	47.6 ± 3.0	29.4 ± 1.5	20.3 ± 2.1	18.3 ± 3.0
Anoxia	32.8 ± 4.0**a**	23.0 ± 1.8	42.5 ± 5.3**a**	26.0 ± 3.5
Recovery	35.0 ± 4.9	42.8 ± 2.9**a**	25.1 ± 2.7	40.8 ± 4.1**a**
Red muscle				
Control	46.9 ± 7.4	20.9 ± 2.7	12.0 ± 3.2	15.3 ± 0.9
Anoxia	73.5 ± 15.7	15.6 ± 2.0	12.7 ± 2.1	9.6 ± 1.4
Recovery	30.7 ± 13.3	13.9 ± 2.0	24.5 ± 2.5**a**	12.6 ± 5.8
White muscle				
Control	55.0 ± 5.6	34.1 ± 2.2	5.4 ± 1.0	3.0 ± 0.4
Anoxia	49.5 ± 9.8	26.9 ± 4.0	3.0 ± 0.5	9.3 ± 0.5**a**
Recovery	34.5 ± 9.6	33.0 ± 4.3	9.0 ± 1.6**a**	4.9 ± 1.2
Kidney				
Control	299 ± 12.4	49.4 ± 4.1	0.8 ± 0.1	12.3 ± 1.5
Anoxia	96.8 ± 18.9**a**	46.5 ± 6.7	2.7 ± 0.9**a**	10.0 ± 3.0
Recovery	169 ± 27.7**a**	51.3 ± 6.4	0.5 ± 0.1	18.0 ± 2.8
Brain				
Control	42.7 ± 13.3	18.4 ± 0.5	3.1 ± 0.4	15.5 ± 3.0
Anoxia	8.5 ± 0.8**a**	15.7 ± 0.8**a**	2.5 ± 0.7	17.8 ± 3.5
Recovery	6.0 ± 0.5**a**	13.6 ± 0.8**a**	1.1 ± 0.1**a**	33.8 ± 7.1**a**

Data are units or milliunits per milligram protein, means ± SEM, $n = 3–4$. **a**: Significantly different from the corresponding control values, $P < 0.05$. Results are from Willmore and Storey (1997a,b).

unchanged during anoxia/reoxygenation in all turtle organs (Willmore and Storey, 1997a).

Interestingly, turtle organs displayed high constitutive antioxidant enzyme activities and GSH-eq levels in control animals as compared with other cold-blooded animals (Table 20.3 for comparative liver data). Indeed, antioxidant defenses were frequently in the range of those found in endothermic mammals, despite the much lower aerobic metabolic rate of the ectothermic turtles (Table 20.3) (Storey, 1996a). Thus, even though some enzymatic activities and GSH-eq levels were reduced during anoxia in turtles, the remaining activities were still substantially higher than those commonly found in other cold-blooded vertebrates and would be sufficient to protect turtle tissues from damage caused by oxyradical generation during reoxygenation. Moreover, unlike the situation in garter snakes, leopard frogs and goldfish (see Sections 2.1 to 2.3 above), few relevant "anticipatory" adjustments were observed in the antioxidant defenses of turtles during anoxia exposure. Only some auxiliary antioxidant enzymes were up-regulated under anoxia. Turtles appear to rely mainly upon high constitutive activities of primary antioxidant enzymes to deal with any oxidative stress arising during tissue reoxygenation after prolonged diving or underwater hibernation.

2.5. Lipid peroxidation, xanthine oxidase, and post-anoxic reoxygenation in vertebrates

In order to determine whether oxidative stress is indeed taking place during post-anoxic reoxygenation, the levels of various lipid peroxidation products were measured. Lipid peroxidation has been reported to be a major contributor to the cellular damage caused by oxidative stress. For example, peroxidation of microsomal membranes can lead to calcium release and uncontrolled activation of Ca^{2+}-dependent proteases and phospholipases (Bolli and Marban, 1999; Lipton, 1999) and peroxidation of mitochondrial membranes can alter permeabilities and induce a disruption of cellular energetics (Bindoli, 1988; Hermes-Lima et al., 1995a). In brief, the general sequence of events in the peroxidation of polyunsaturated lipids is as follows. The first step is the removal of a hydrogen atom from a methylene group by ·OH or other radical species; typically this occurs at a methylene adjacent to an existing double bond in a fatty acid. This leads to a molecular rearrangement that produces a conjugated diene (alternating double–single–double bonds) which then reacts with oxygen to form a peroxy radical (LOO·). This radical abstracts a hydrogen atom from another lipid molecule (setting off a chain reaction) to become a lipid hydroperoxide (LOOH). Transition metal complexes (e.g. iron–citrate, iron–ATP; Castilho et al., 1999) can then catalyze the fission of the O–O bond to form an alkoxy radical (LO·) that causes beta-scission of the fatty acid chain to cleave off hydrocarbons and aldehydes of varying sizes, among them malondialdehyde (de Zwart et al., 1999; Cadenas, 1995).

Different assays quantify the extent of peroxidation at different stages in the process (de Zwart et al., 1999). Those used in our studies include the spectrophotometric measurement of conjugated dienes (Corongiu and Milia, 1983), measurement of lipid hydroperoxides via the ferrous oxidation/xylenol orange (FOX) assay (Hermes-Lima et al., 1995b; Storey, 1996b), and quantification of malondialdehyde and other aldehyde degradation products as thiobarbituric acid reactive substances (TBARS) (Bird and Draper, 1984). Lipid peroxidation products were evaluated during reoxygenation after anoxic exposure in leopard frogs, goldfish and red-eared turtles (Hermes-Lima and Storey, 1996; Willmore and Storey, 1997a; Lushchak et al., 2001).

In the case of goldfish, the levels of conjugated dienes were measured after 8 h anoxia exposure followed by reoxygenation. Conjugated diene levels had increased by 114% in liver at 60 min of aerobic recovery, and by 75% in brain after 14 h of recovery (Table 20.2). The increased activities of liver catalase and brain Se-GPX during anoxia exposure in the fish, as well as the relatively high constitutive activities of other antioxidant enzymes (see Section 2.3 above), probably contributed to keeping lipid peroxidation at a tolerable level during reoxygenation. Conjugated diene levels were unaffected in fish kidney (this organ also has a relatively high antioxidant capacity) and decreased (by 44–61%) during anoxia and reoxygenation in white muscle (Lushchak et al., 2001).

Lipid peroxidation was measured as TBARS in leopard frogs after 30 h anoxia or during reoxygenation (25 min, 90 min, or 40 h) (Hermes-Lima and Storey, 1996). TBARS concentration in control muscle and liver were 3 and 11 nmol/g wet wt, respectively, and did not change during either anoxia exposure or aerobic recovery. An anoxia-induced increase in catalase activity in frog muscle and the maintenance of relatively high antioxidant defenses in liver (see Section 2.2) might have contributed to preventing the accumulation of lipid peroxidation end products.

Conjugated dienes, FOX-reactive lipid hydroperoxides and TBARS were all quantified in organs of red-eared turtles after 20 h of submergence anoxia and 24 h aerobic recovery (Willmore and Storey, 1997a). Conjugated dienes were only detectable in liver and their levels were decreased by 38% during anoxia. The content of lipid hydroperoxides did not change in liver, white muscle and red muscle during anoxia/recovery (~5, 3 and 0.7 μmol/g wet weight, respectively), but kidney values decreased during recovery (from 6.5 to 4 μmol/gww). TBARS were similarly unaffected in liver, kidney and white muscle (~80, 25 and 4 nmol/gww, respectively) but decreased significantly in red muscle from 9.5 nmol/gww in controls to 6.6 nmol/gww during anoxia and to 3.8 nmol/gww after recovery. It appears, then, that the constitutive antioxidant defenses of turtle organs are sufficient to prevent oxidative damage to lipids during the reoxygenation of organs after anoxic submergence. As a caveat, however, it must be noted that peroxidative damage was assessed only after 24 h aerobic recovery whereas damage and damage repair might have taken place on a much shorter time scale.

Willmore and Storey (1997a) also investigated the effect of anoxia and recovery on activities of xanthine oxidase (XO) and xanthine dehydrogenase (XDH) in turtle organs. XO has been associated with ischemia/reperfusion stress in several mammalian systems (Terada et al., 1991; Greene and Paller, 1992; Lipton, 1999). Two factors are involved: (1) the breakdown of ATP to AMP in anoxia and the subsequent build-up of degradation products of AMP including hypoxanthine and xanthine that are substrates for XO (McCord, 1985), and (2) the rise in intracellular Ca^{2+} during hypoxia that stimulates Ca^{2+}-dependent proteases including those that convert XDH to XO (only XO produces O_2^-) (Roy and McCord, 1983). During reperfusion, then, the stage is set for increased O_2^- formation via XO.

In turtle organs, however, neither of these factors seems to be in play. During submergence anoxia in turtles there was no net reduction in organ adenylate levels or energy charge. A transient decrease occurred during the hypoxia transition period but ATP, ADP and AMP restabilized at control values within 5 h due to a coordinated suppression of the rates of ATP-utilizing reactions to a level that can be supported over the long term by anoxic ATP generation alone (Kelly and Storey, 1988; Storey and Storey, 1990). The proportion of total XO+XDH activity that was XO was very high in all turtle organs compared with mammals, ranging from 36 to 75%, and providing a high constitutive potential for oxyradical formation via XO (Willmore and Storey, 1997a). However, the %XO remained unchanged during 20 h anoxia and 24 h aerobic recovery in all six organs that were examined (XO activities in heart, liver and kidney were 120–160, 65–90 and 35–55 μU/mg protein; one unit XO produces 1 μmol isoxanthopterin/min). The only significant change observed was a rise in XO activity in brain during anoxia (from 7 to 22 μU/mg protein), although because total XO+XDH also rose, the %XO remained constant (Willmore and Storey, 1997a). Thus, in general, the data for most organs suggest that XO-mediated oxyradical formation would not contribute a significant oxidative stress during anoxia or recovery in turtles.

2.6. Oxidative stress and anoxia tolerance in a marine gastropod

Numerous gill-breathing marine invertebrates have well-developed biochemical adaptations that allow them to endure extended periods of oxygen deprivation. Many intertidal species deal with oxygen deprivation on a twice daily basis when the tide recedes and leaves them exposed to air for several hours at a time. Others may find themselves in a variety of situations where oxygen is depleted in the water (e.g. hot tide pools, benthic sediments) or where access to oxygen is interrupted due to shell valve closure or withdrawal into closed burrows as a means of avoiding inhospitable water conditions (e.g. high toxin levels, high silt conditions, wide changes in salinity) or predators. Biochemical adaptations supporting anaerobiosis in marine invertebrates have been extensively studied and include the maintenance of large reserves of fermentable fuels (e.g. glycogen, aspartate) in all tissues, the production of alternative end products of fermentative metabolism (e.g. succinate, alanine, propionate, acetate) that increase ATP yield compared with glycolysis alone, and strong metabolic rate depression that lowers ATP demands during anoxia by >90% (Storey, 1993; Brooks and Storey, 1997).

To determine whether natural anoxia tolerance by the marine gastropod mollusc, *Littorina littorea*, included an oxidative stress component and/or adaptations by antioxidant defenses, the effects of 6 days of anoxia exposure (under N_2 gas at 5°C) followed by aerobic recovery for 12 or 24 h were determined. Antioxidant enzyme activities, glutathione levels and lipid peroxidation damage were quantified in two tissues, foot muscle and hepatopancreas (Pannunzio and Storey, 1998).

In the case of most enzymatic antioxidant defenses (catalase, SOD, Se-independent GPX, GR and GST), anoxia exposure induced a 30–53% suppression of activities in hepatopancreas. However, Se-GPX activity was unchanged during anoxia. In foot muscle, enzyme activities were generally unaltered during anoxia except for a 44% decrease of SOD activity. On the other hand, anoxia induced an increase in GSH-eq levels by 2.8- and 1.6-fold in hepatopancreas and foot, respectively. Aerobic recovery after anoxia resulted in further modulation of glutathione pools with GSH-eq levels increasing in both organs by 2.4- and 3.5-fold in foot and hepatopancreas, respectively, after 24 h recovery (Pannunzio and Storey, 1998).

Oxidative damage to *L. littorea* tissues during anoxia (6 days) and recovery (0.5 to 12 h) was assessed using three methods for evaluating lipid peroxidation (Pannunzio and Storey, 1998). Hepatopancreas showed no change in either conjugated dienes or TBARS levels over anoxia or recovery, whereas the levels of FOX-reactive lipid hydroperoxides were suppressed by 62% during anoxia and remained low throughout recovery. Thus, the antioxidant defenses of hepatopancreas appear to be fully capable of handling any reoxygenation-induced oxidative stress. Foot muscle, however, showed an unexpected response. Levels of conjugated dienes and FOX-reactive lipid hydroperoxides had risen by 92 and 37%, respectively, after 6 days anoxia exposure but returned to control levels during reoxygenation (TBARS was unchanged over the experimental course) (Pannunzio and Storey, 1998). Further studies are need to understand the mechanism of anoxia-induced lipid peroxidation in *L. littorea* foot muscle.

3. Oxidative stress and natural freeze tolerance in vertebrates

The ability to endure the freezing of extracellular body fluids is an integral part of winter survival for a wide variety of invertebrates as well as a few species of cold-blooded vertebrates that hibernate on land. Several species of frogs have well developed freeze tolerance, the wood frog *Rana sylvatica* being the primary species that has been extensively studied, whereas selected species of turtles, garter snakes and lizards also have limited tolerances (Storey and Storey, 1988, 1992, 2001; Storey et al., 1996). Freeze tolerance requires adaptations to deal with several severe stresses including the potential for physical damage to tissues by ice, large reductions in cell volume and large increases in cellular ionic strength and osmolality due to the exit of a high percentage of cell water into extracellular ice, and prolonged ischemia due to plasma freezing (Storey and Storey, 2001). This latter means that for the duration of the freeze, all tissues are cut off from blood-borne supplies of oxygen and substrate. Indeed, over the course of a freezing episode tissues show the typical vertebrate response to oxygen limitation, a depletion of adenylates and an accumulation of the glycolytic end products, lactate and alanine (Storey, 1987; Storey and Storey, 1992). Hence, thawing is a reperfusion event and animals would experience the potential for oxidative stress associated with the reintroduction of oxygen to tissues once breathing resumes. The possible adaptations of antioxidant defenses that allow animals to undergo multiple cycles of freeze–thaw without oxidative injuries are of interest both for developing a better understanding of the mechanisms of natural freeze tolerance and for understanding the principles of ischemia/reperfusion endurance that could be applied to situations such as the cryopreservation of mammalian tissues and organs (Costanzo et al., 1995, Storey et al., 1996). Notably, empirical studies have shown that the inclusion of antioxidants in the perfusion medium before cryopreservation of mammalian cells or tissues improves viability after thawing so it is very possible that antioxidant defenses have a role to play in tissue protection during freeze/thaw (McAnulty and Huang, 1997; Bilzer et al., 1999).

3.1. Antioxidants and freeze tolerance in garter snakes

In the early 90s, we first examined the effects of survivable freezing on the activities of antioxidant enzymes in organs of the red-sided garter snake *T. s. parietalis* (Hermes-Lima and Storey, 1993a). Garter snakes from Manitoba, Canada, hibernate in communal underground dens for about 7 months of the year. This species can tolerate several hours of freezing at –2.5°C with 40–50% of their total body water frozen and 2 days frozen at –1°C with a lower ice content (34%) (Costanzo et al., 1988; Churchill and Storey, 1992). Freeze tolerance does not appear to be a mechanism for long term winter survival by snakes but may be important for dealing with overnight frosts during the autumn or spring when snakes are active above ground or offer limited protection if freezing penetrates into underground dens.

The effects of freezing exposure (5 h at –2.5°C) on tissue antioxidant defenses was assessed in autumn-collected garter snakes and compared with controls held at 5°C. Freezing stimulated an increase in the activities of catalase in skeletal muscle (by 183%) and lung (by 63%) and in muscle Se-GPX (by 52%) (Table 20.1) (Hermes-Lima and Storey, 1993a). Freezing exposure had no effect on these enzymes in liver and furthermore, SOD (Mn- plus CuZn-SOD) and GR activities as well as the levels of GSH-eq and GSSG were unaffected in the three organs. Hence, similar to the responses of garter snakes to anoxia stress (see Section 2.1), freezing elicited an increase in some antioxidant defenses and a maintenance of others. The effects of thawing on garter snake antioxidant defenses were not analyzed.

In another study, the *in vitro* oxidative inactivation of GST (by means of the Fenton reagents, H_2O_2 and Fe^{2+}, which induce ·OH formation) was examined in homogenates of snake muscle. The inactivation of GST was significantly lower in muscle homogenates from snakes that endured 5 h freeze exposure (at –2.5°C) than in controls. These data were explained by the presence of significantly elevated levels of catalase in muscle samples from freeze-exposed snakes (mean activities were 2.6 versus 1.5 μmol H_2O_2 decomposed/min/mL in muscle extracts from freeze-exposed versus control snakes, respectively), which may lead to faster decomposition of H_2O_2 in the frozen state and, therefore, diminish the oxidative inactivation of GST (Hermes-Lima and Storey, 1993b).

3.2. Oxidative stress and freeze tolerance in wood frogs

Another freeze tolerant species, the wood frog *Rana sylvatica*, also showed freeze-induced modulation of antioxidant defenses. Wood frogs can survive days or weeks at temperatures as low as –6 to –8°C with up to about 65% of total body water converted to extracellular ice and with no breathing, heart beat or muscle movement. Cellular metabolism during freezing is anaerobic (Storey, 1987) but upon thawing, heart beat and then breathing are restarted and the animal's organs are quickly reoxygenated.

Joanisse and Storey (1996) examined the effects of freezing exposure (24 h at –2.5°C) on the antioxidant systems of wood frog organs, comparing them to controls held at 5°C. Freezing lead to a 20–150% increase in total-GPX activity (Se-dependent plus Se-independent GPX) in five different tissues (Fig. 20.2). On thawing (24 h at 5°C) activity remained high in liver and brain but tended to decrease in the other tissues. Changes in Se-GPX activity paralleled total activity in four tissues, but in liver activity was unchanged. In general, activities of other antioxidant enzyme activities (catalase, SOD, GST, GR) as well as GSH levels were unaltered during freeze/thaw, except for a 23–57% decrease in the activity of SOD (Mn- plus CuZn-SOD) in muscle, kidney and heart

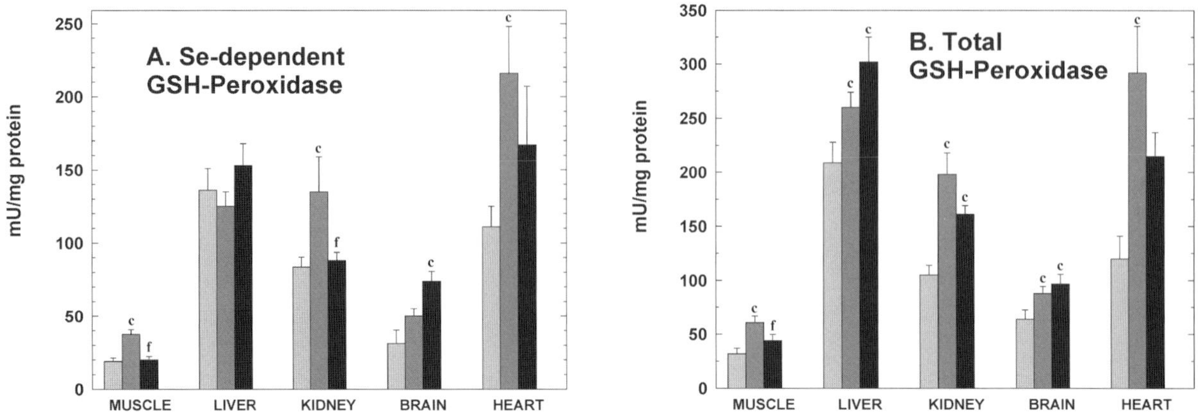

Fig. 20.2. Effect of freezing and thawing on the activities of (A) Se-dependent glutathione peroxidase and (B) total glutathione peroxidase in five organs of wood frogs, *Rana sylvatica*. Bar fills are: (light grey), control; (dark grey), 24 h freezing at –2.5°C; (black), 24 h thawing after 24 h freezing. Data are milliunits per mg soluble protein, means ± SEM, n = 4–6. c: Significantly different from the corresponding control value, $P < 0.05$; f: significantly different from the corresponding frozen value, $P < 0.05$. Data from Joanisse and Storey (1996).

during freezing (Joanisse and Storey, 1996). This reduction in SOD was reversed in heart and kidney during thawing. A comparison was also made with antioxidant defenses in leopard frogs (*Rana pipiens*) that are not freeze tolerant. In nearly every instance, the activities of antioxidant enzymes in wood frog tissues were significantly higher than those in leopard frog tissues as were the tissue concentrations of GSH and GSSG (see Table 20.3 for liver) (Joanisse and Storey, 1996). This adds support to the idea that high antioxidant defenses contribute significantly to natural freezing survival.

The increases in total GPX and Se-dependent GPX activities seen in most wood frog tissues during freezing were largely reversed after frogs had been thawed for 24 h at 5°C. This suggests that the stress necessitating GPX enhancement occurred while frozen (or possibly during the early stages of thawing) so that enhanced defenses needed to be in place before thawing began. That the antioxidant defenses of wood frogs were adequate to deal with any oxidative stress associated with the thaw-induced reintroduction of oxygen to tissues was confirmed by an analysis of the extent of oxidative damage to tissue lipids. Both TBARS and FOX-reactive lipid hydroperoxides were measured in wood frog tissues but neither increased significantly above controls during either freezing (24 h at –2.5°C) or thawing (0.5 to 4 h at 5°C). Moreover, the activity of XO was undetectable in muscle, and unchanged in liver and kidney (~50 and 110 μU/mg protein, respectively) and the percentage of total XO+XDH that was XO was small (10–15%) (Joanisse and Storey, 1996). This suggests that there would be minimal potential for XO-induced oxidative stress during thawing due to the catabolism of the products of adenylate degradation although, notably, total adenylate levels fall during freezing and take several days to recover after thawing (Storey and Storey, 1986). Thus, these results, taken as a whole, indicate that wood frogs have adequate antioxidant defenses to deal with the potential for oxyradical-mediated damage during cycles of winter freeze–thaw (Joanisse and Storey, 1996).

The biochemical adaptations to oxygen stress during freeze/thaw in garter snakes and wood frogs have similarities to those observed in anoxia tolerant vertebrates during anoxia/reoxygenation (see Sections 2.1 to 2.5 above). A mechanism of preparation for oxidative stress seems to be the main rule for dealing with potential oxidative stress during reperfusion (Hermes-Lima et al., 1998). Moreover, wood frogs and red-eared turtles also rely on constitutively high levels of antioxidant enzymes and metabolites so that they are well-prepared for dealing with oxidative stress arising after any bout of anoxia or freezing exposure. This makes sense because the energy-limited conditions of the anoxic and frozen states are not favourable for extensive protein synthesis. Hence, high constitutive antioxidant defenses are an appropriate biochemical strategy.

4. Oxidative stress and dehydration tolerance in leopard frogs

Loss of body water is a problem for most terrestrial organisms and animals have evolved numerous behavioral, physiological and biochemical strategies for avoiding extensive dehydration. Mammals, birds and reptiles rely heavily on a highly impermeable integument and, with the addition of various other adaptations, they are able to hold body water content (and hence cellular ionic strength and osmolality) within narrow limits. Amphibians, however, have a highly water permeable integument and, with a typical water loss of 6–9% of body weight per day for terrestrial and semi-aquatic amphibians (Hillman, 1980), it is understandable why most species are confined to damp habitats. As one consequence of the high rate of water loss across their skin, most amphibians have also developed a high tolerance for variation in their body water content and many are able to endure the loss of 25–40% of total body water, exhibiting no injuries after rehydration (Churchill and Storey, 1993). Desert anurans that estivate in underground burrows for 9–10 months of the year tolerate even higher body water losses (50–60%) (Pinder et al., 1992) as do freeze tolerant species whose cells undergo extreme dehydration as the result of the transfer of 50–65% of total body water into extracellular ice crystals (Churchill and Storey, 1993).

During cycles of dehydration/rehydration, animals must cope not only with wide variations in body fluid osmolality and ionic strength (Hillman, 1978) but also with wide variation in blood volume and viscosity. When dehydration is severe, the increase in blood viscosity and decrease in volume impairs the function of the cardiovascular system including a reduction in arterial pressure and pulse rate and reduced oxygen delivery to tissues (Hillman, 1987; Gatten, 1987). Internal organs become hypoxic during severe dehydration and products of anaerobic metabolism accumulate (Hillman, 1987; Churchill and Storey, 1993, 1995). During rehydration, a rapid uptake of water across the skin water restores blood volume and cardiovascular function and allows tissue perfusion with oxygenated blood to resume. Therefore, severe dehydration followed by rehydration has strong analogies with ischemia/reperfusion stress and we proposed that antioxidant defenses could play a role in the prevention of oxidative stress during recovery from dehydration.

The leopard frog (*R. pipiens*) can tolerate the loss of 50% of total body water at 5°C (Churchill and Storey, 1995). Using this species as a model animal, Hermes-Lima and Storey (1998) analyzed the responses by tissue antioxidant defenses to dehydration and rehydration. The activity of antioxidant enzymes (SOD, catalase, Se-GPX, GR and GST) and the levels of GSH-eq and TBARS were quantified (and expressed per mg soluble protein) in liver and skeletal muscle of leopard frogs over a cycle of 50% dehydration (lasting 92 h at 5°C) and full rehydration. During dehydration the activities of muscle catalase and liver Se-GPX increased by 52 and 74%, respectively. By contrast, muscle GR and SOD (Mn- plus CuZn-SOD) activities fell by 34–35%, whereas the other enzymatic activities as well as GSH-eq levels were unaffected. Liver GSH-eq was increased by 81% early in the rehydation process (30% recovery of total body water), but not muscle GSH-eq. Full rehydration restored the altered enzyme activities to control values. These minimal changes in antioxidant enzyme activities during stress/recovery also correlated with a lack of change in the levels of TBARS in liver and muscle during dehydration, 30% rehydration and full rehydration. This suggests that antioxidant defenses were able to protect frog liver and muscle from potential post-hypoxic oxidative stress during dehydration and recovery.

5. Estivation and oxidative stress in land snails and toads

5.1. Estivation in land snails and oxidative stress

Estivation is an aerobic dormancy where metabolic rate is typically lowered to 10–30% of normal resting rate. Estivation is typically associated with arid environmental conditions which also often include heat and lack of food availability (Storey and Storey, 1990; Storey, 2000). The response is common among desert animals and also in environments where water availability varies widely on a seasonal basis. Well-known examples of estivators include lungfish that surround themselves with mucous cocoons and enter torpor in riverbank burrows as the waters of the rainy season recede (Pinder, et al. 1992) and various frogs and toads that may spend 9–10 months underground each year and emerge with the first summer storms for a short season of feeding and breeding. Various terrestrial pulmonate snails are also active only under wet conditions and retreat into their shells (which they seal with a mucous epiphragm to minimize evaporative water loss) whenever environmental conditions dry out. Well studied species of estivating snails include *Otala lactea, Helix pomatia* and *H. aspersa* (Herreid, 1977; Barnhart, 1986; Storey and Storey, 1990; Ramos, 1999; Bishop and Brand, 2000). Estivators typically display a pattern of discontinuous or apnoic breathing that minimizes water loss across respiratory surfaces, but leads to wide variations in tissue oxygen levels. Oxygen is high just after a breath but then falls continuously to low levels; oppositely, pCO_2 rises until a threshold value is reached that triggers the next breath (Herreid, 1977; Barnhart, 1986). In snails, a rise in atmospheric humidity triggers arousal and snails emerge from their shells within just a few minutes. Due to the resumption of normal breathing patterns, pO_2 rises and stabilizes in tissues and oxygen consumption increases rapidly to a transient peak which in *O. lactea* is at least two-fold higher than control values and about six-fold higher than consumption in the dormant state (Herreid, 1977; Hermes-Lima et al., 1998).

Since it is known that the rate of production of O_2^- and H_2O_2 by the mitochondria is proportional to oxygen tension in many biological systems (Turrens et al., 1982; Cino and Del Maestro, 1989; Beckman and Ames, 1998), the rise in oxygen tension and consumption in snail organs during arousal could result in elevated production of oxyradicals. This must be dealt with rapidly by endogenous antioxidant defenses so that the snails do not sustain oxidative injury during these natural transitions from the hypometabolic estivating to the aroused active state. The idea of a natural oxidative stress, that is dealt with by specific adaptations, has also been proposed to occur during the arousal process in hibernating small mammals (Buzadzic et al., 1990), and also in instances of tolerable oxidative injuries linked to exercise (Barja de Quiroga, 1992) or to late gestational development in mammalian lung (Frank et al., 1996).

To determine whether adaptive changes in antioxidant defenses occurred in land snails to support transitions to and from the estivating state, Hermes-Lima and Storey (1995a) analyzed the endogenous antioxidant defenses and extent of lipid peroxidation in tissues of *O. lactea* under estivating versus aroused conditions. Snails were subjected to two cycles of 30 days of estivation with a 24 h period of arousal after each dormant period. Compared with 24 h aroused snails, foot muscle of estivating *O. lactea* showed significantly higher activities three antioxidant enzymes: 64% higher SOD (Mn- plus CuZn-SOD), 62% higher catalase, and 94% higher GST (Table 20.5). In hepatopancreas of estivating snails, SOD was also 68% higher as compared with active snails whereas Se-GPX was 117% higher. GR activity was not affected by estivation/arousal in either tissue. Within 40 min after arousal began, hepatopancreas Se-GPX activity had fallen again to control values, but SOD showed a further 70% rise in activity before returning to control levels by 80 min (Fig. 20.3). Estivation had no effect on GSH-eq concentration in tissues but GSSG content was about two-fold higher in both organs of 30-day dormant snails (Table 20.5). The increase in the GSSG/GSH-eq ratio that during estivation was

Table 20.5. Activities of antioxidant enzymes and levels of glutathione in tissues of land snails, *Otala lactea*, after 30 days of estivation followed by 24 h of arousal

Tissue	Estivating	Active
Hepatopancreas		
Catalase (U/mg)	174 ± 18	196 ± 15
SOD (U/mg)	84 ± 12	50 ± 6**a**
Se-GPX (mU/mg)	23 ± 4	10.6 ± 1.6**a**
GR (mU/mg)	16 ± 2	19 ± 2
GST (mU/mg)	1282 ± 215	1140 ± 83
GSH-eq (μmol/gww)	3.1 ± 0.3	2.8 ± 0.1
GSSG (μmol/gww)	0.46 ± 0.04	0.25 ± 0.02**a**
Foot muscle		
Catalase (U/mg)	5.5 ± 0.6	3.4 ± 0.1**a**
SOD (U/mg)	41 ± 6	25 ± 2**a**
Se-GPX (mU/mg)	4.4 ± 0.7	4.9 ± 0.4
GR (mU/mg)	6.2 ± 0.7	6.2 ± 1.2
GST (mU/mg)	223 ± 40	115 ± 20**a**
GSH-eq (μmol/gww)	0.99 ± 0.15	0.92 ± 0.05
GSSG (μmol/g ww)	0.18 ± 0.02	0.09 ± 0.01**a**

Snails were given two cycles of 30 days estivation followed by 24 h arousal with sampling after each experimental period (estivation, arousal, estivation, arousal). Values for both estivating and both active groups were virtually identical and so are averaged here. Data are presented as units or milliunits per milligram protein for enzymes or as μmoles per gram wet weight for metabolites, means ± SEM, n = 3–9. **a**: Significantly different from the corresponding values in estivating snails, $P < 0.05$. Data modified from Hermes-Lima and Storey (1995a).

attributed to a reduction in the rate of GSH recycling due to a possible decrease in NADPH supply in the hypometabolic state. A similar increase in GSSG/GSH-eq was also observed in organs of anoxia-exposed leopard frogs (see Section 2.2).

The extent of lipid peroxidation, as assessed by TBARS levels, was significantly enhanced by 25% in *O. lactea* hepatopancreas early in the arousal from estivation (within 20 min) (Fig. 20.3) and this suggested that oxidative stress and tissue damage was occurring at this time (Hermes-Lima and Storey, 1995a). After 40 minutes, however, TBARS levels had returned to control values, indicating an efficient mechanism for metabolizing the aldehydic products of peroxidation. Again, as proposed previously, it appears that the antioxidant defenses of tissues (hepatopancreas in this case) can be built up while an organism is in a hypometabolic state in order to be used to limit oxidative stress damage when oxygen levels and oxygen consumption abruptly return to normal.

Dykens and Shick (1988) proposed that XO would be major player in oxyradical formation in intertidal molluscs with poor tolerance to anoxia. Such species would be susceptible to oxyradical-mediated reperfusion injury (following aerial exposure at low tides) as the result of enhanced XO action during reimmersion at high tides. To assess the possibility of a similar occurrence during the estivation-arousal transition, we analyzed XO in *O. lactea* tissues. First of all, we found that XO

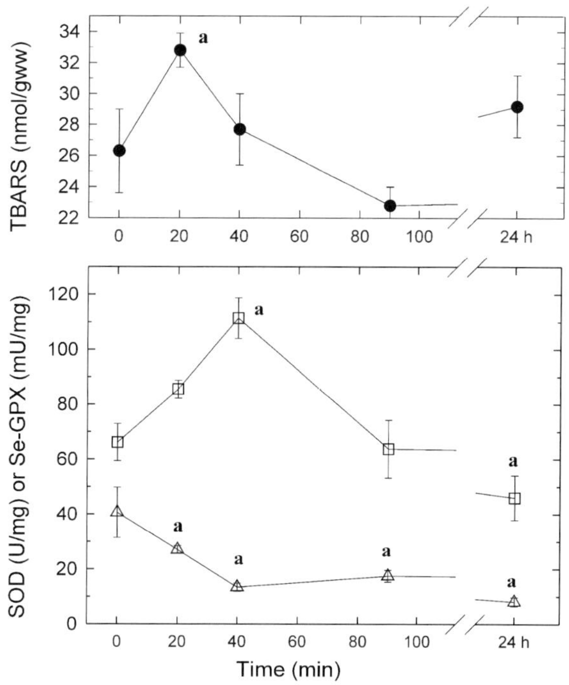

Fig. 20.3. Time course of changes in the levels of oxyradical-generated damage products and antioxidant enzyme activities in hepatopancreas of the land snail, *Otala lactea*, during arousal after 3 months of estivation. Upper panel: damage assayed as thiobarbituric acid reactive substances (TBARS, ●). Lower panel: activities of superoxide dismutase (SOD; ❑) and Se-dependent glutathione peroxidase (Se-GPX; Δ). Data are means ± SEM, n = 4–7. a: Significantly different from the corresponding value at 0 min. Modified from Hermes-Lima and Storey (1995a).

activity was absent from foot muscle of *O. lactea* and that only very low levels of XDH were present (Hermes-Lima and Storey, 1995b). XO activity was also very low in hepatopancreas, approximately ~10 μU/mg protein in 24 h awake animals or about 7% of the total XO+XDH activity. Although XO activity increased by three-fold during estivation, the expected rate of oxyradical formation via *O. lactea* XO would be too low to cause significant oxyradical production and the rate of oxyradical generation by XO could be easily handled by endogenous antioxidants, especially catalase (Hermes-Lima and Storey, 1995b).

Our results, taken as a whole, indicated that antioxidant defenses in *O. lactea* organs are increased while in the hypometabolic state (where oxyradical production is probably lower than in awake animals) as a preparation for oxidative stress during arousal. If the levels of endogenous antioxidants were not increased/maintained, TBARS could have reached toxic levels in arousing snails. Furthermore, another relevant adaptation of land snails is the maintenance of low levels of the oxyradical-generating enzyme XO in their tissues (Hermes-Lima and Storey, 1995a,b; Hermes-Lima et al., 1998).

5.2. Oxidative stress and estivation in a desert toad

Modulation of antioxidant capacity was also analyzed in vertebrate estivation using the spadefoot toad, *Scaphiopus couchii*, as the model animal. This native species of the American southwest spends 9–10 months of the year underground. Toads emerge when summer storms flood the desert, breed within the first 24 hours and then eat ravenously for the next few weeks to replenish their body fuel reserves before disappearing underground again as the desert dries out. While estivating, metabolic rate is typically suppressed to 20–30% of the resting rate in active toads and is fueled by the slow catabolism of lipid and protein reserves (Seymour, 1973). Toads enter dormancy with a very large reserve of water in their bladder which they resorb over time to replace tissue water that is lost across the skin as the soil dries out. When water stress becomes substantial toads switch from a near total dependence on lipid oxidation to a mixed metabolism that oxidizes an increasing proportion of protein reserves. The nitrogen released from protein catabolism is used to synthesize urea which rises to 200–300 mM in all fluid compartments and provides colligative resistance to the further loss of body water (Jones, 1980; Grundy and Storey, 1998). Nonetheless, as the soil dries out over time, net water loss can rise as high as 60% of total body water (47–50% of body mass) after several months (McClanahan, 1967).

During arousal from estivation, desert toads undergo a significant and transitory increase in oxygen uptake, not only reversing the metabolic suppression but expending large amounts of energy to dig out of the ground and immediately initiate breeding activities. Subsequently, metabolic rate remains high as the toads breed and feed over the next days/weeks. Since the rate of oxyradical formation is usually proportional to metabolic rate (see Section 5.1, above), it is possible that the rate of oxyradical formation would increase when toads awaken from estivation. This could require well-prepared antioxidant defenses to minimize oxyradical-mediated tissue damage. To analyze this, Grundy and Storey (1998) surveyed the activities of antioxidant enzymes and the levels of GSH and GSSG (measured per mg/soluble protein) in 2-month estivating toads (burrowed in soil at 21°C) and 10 day awakened toads. Catalase activity increased in heart and liver (by 42 and 73%) after awakening but decreased by 40% in kidney (Table 20.6). SOD activity rose in kidney and heart (by 47 and 110%, respectively) after arousal but decreased in liver and muscle (by 50 and 30%, respectively). The activity of total-GPX increased after arousal in liver, heart and lung (by 69, 73 and 134%, respectively) (Table 20.6) and Se-GPX more than doubled in liver and gut but decreased by 50% in kidney (Grundy and Storey, 1998). GR activity rose significantly after awakening in heart, kidney and liver (by 50, 64 and 200%, respectively). GST activity doubled in liver, heart and gut and increased by about 50% in lung and kidney after awakening.

Overall, of the six enzymatic activities tested by Grundy and Storey (1998), activities of five enzymes in liver and heart, three in kidney, and two in gut and lung had risen significantly when assayed 10 days of awakening. In total for the six enzymes in six tissues, 17 activities increased with the arousal from estivation, 14 were unchanged and only 5 decreased in aroused toads. In addition, metabolite antioxidants (total GSH-eq) rose in muscle, liver and lung after awakening by 20, 53 and 66%, respectively (Table 20.6). Hence, all organs showed evidence that antioxidant capacity was increased after arousal (and by implication was reduced during estivation) ranging from only an upward adjustment of GSH-eq in skeletal muscle to an extensive restructuring in liver (increases in five enzymes and GSH-eq).

The lower antioxidant capacity during estivation might also be the cause of the higher GSSG/GSH ratio (21–250% higher) in all organs, except leg muscle, of estivating toads. Increased GSSG/GSH suggests either a higher enzymatic use of GSH for peroxide detoxification during estivation or a decrease in the rate of GSSG recycling, perhaps due to a diminished rate of NADPH production while estivating (Grundy and Storey, 1998).

Moreover, in a second set of experiments where aroused toads (controls) were compared with toads that were subsequently allowed to estivate for two months, the levels of lipid peroxidation products generally increased during estivation. Levels of conjugated dienes (measured per mg lipid) increased significantly during estivation in all tissues (by 27–168%) except lung (40% reduction). FOX-reactive lipid hydroperoxide (measured per mg of protein) were increased during estivation in kidney and muscle (by 174 and 103%), reduced in heart and lung (by 70 and 93%) and unchanged in liver and gut. [Note: this is derived from a recalculation of the original data of Grundy and Storey (1998), converting values given per gram wet weight to re-express them per mg protein.] Since, levels of lipid peroxidation products should covary with oxygen consumption, the higher levels of these in tissues of estivating animals seems at odds with expectations. However, these results

Table 20.6. Activities of selected antioxidant enzymes and levels of GSH-eq of spadefoot toads, *Scaphiopus couchii*, after two months estivation or following 10 days arousal after estivation

Tissue	Estivating	Active
Liver		
Catalase	670 ± 40	1270 ± 140**a**
SOD	104 ± 10.3	49.6 ± 8.3**a**
Total-GPX	35.4 ± 3.3	59.8 ± 4.6**a**
GSH-eq	13.2 ± 0.8	21.9 ± 0.6**a**
Heart		
Catalase	51.5 ± 3.4	73.3 ± 11.6**a**
SOD	31.9 ± 6.3	67.7 ±9.2**a**
Total-GPX	30.7 ± 2.5	53.1 ± 7.5**a**
GSH-eq	13.2 ± 0.6	14.0 ± 0.9
Kidney		
Catalase	1910 ± 260	1160 ± 60**a**
SOD	33.8 ± 3.7	49.7 ± 3.6**a**
Total-GPX	34.1 ± 5.8	39.1 ± 12.4
GSH-eq	8.9 ± 1.2	9.6 ± 0.5
Leg muscle		
Catalase	38.3 ± 4.7	34.9 ± 3.3
SOD	44.9 ± 4.0	31.6 ± 3.0**a**
Total-GPX	12.1 ± 0.8	7.5 ± 1.8**a**
GSH-eq	18.0 ± 0.4	21.5 ± 0.8**a**
Lung		
Catalase	99.2 ± 7.1	115 ± 18
SOD	2.3 ± 0.1	3.6 ± 0.8
Total-GPX	26.5 ± 6.5	62.0 ± 7.5**a**
GSH-eq	18.7 ± 0.8	28.7 ± 1.5**a**
Gut		
Catalase	50.1 ± 5.3	58.2 ± 9.0
SOD	13.1 ± 1.4	16.4 ± 1.5
Total-GPX	42.6 ± 3.5	54.8 ± 7.7
GSH-eq	7.3 ± 0.5	8.1 ± 0.9

Data are expressed as units/mg protein (enzymes) or nmol/mg soluble protein (GSH-eq), means ± SEM, $n = 4–9$. **a**: Significantly different from the corresponding values in estivating toads, $P < 0.05$. Data from Grundy and Storey (1998).

might be explained by the reduced antioxidant defenses of the estivating animals, allowing lipid damage by oxyradicals formed at low rates. Notably, in lung, where antioxidant capacity was largely unchanged between estivating and aroused states, both conjugated dienes and FOX-reactive lipid hydroperoxides were significantly lower in estivating toads as compared to awake animals.

Thus, contrary to the situation observed for estivating land snails and anoxia or freeze tolerant vertebrates (see Sections 2, 3, and 5.1, above), estivating spadefoot toads do not show a mechanism of preparation for oxidative stress. Instead, they reduce antioxidant capacity during estivation, perhaps in response to reduced metabolic rates and tolerate increased levels of lipid peroxidation products in their tissues. It is possible that this species may use a different adaptive strategy for dealing with post-estivation oxidative stress; perhaps cellular mechanisms for the repair of oxidative damage to cellular components could be well developed in these animals.

6. Conclusions, speculations and perspectives

This chapter summarizes work on a range of vertebrate and invertebrate animals about the responses of antioxidant defenses to environmental stresses. Except for the case of desert spadefoot toads, all of the stress tolerant animals studied summarized here show a common response of maintaining or increasing in their antioxidant defenses (both antioxidant enzyme activities and GSH pools) under environmental stress situations when oxygen availability or delivery is low or cut off completely (Table 20.7). This includes the responses to anoxia exposure by marine molluscs, goldfish, leopard frogs, garter snakes and red-eared turtles, responses to freezing by wood frogs and garter snakes, responses to severe dehydration by leopard frogs, and responses to estivation by land snails. Peroxidative damage, rising above control levels, was not detected in any instance of anoxia/hypoxia stress, with the exception of the foot muscle of *L. littorea* (see Section 2.6). On the other hand, lipid

Table 20.7. Summary of the effects of environmental stress on the antioxidant defenses of stress-tolerant animals

Main antioxidant strategy during stress	Condition
SH-rich hemoglobins as putative antioxidant	
Freshwater turtle *Phrynops hilarri*	Long periods of winter diving[1]
Maintenance of antioxidant capacity	
Larvae of insect *Epiblema scudderiana*	Anoxia exposure[2]
Larvae of insect *Eurosta solidaginis*	Anoxia and freezing[2]
High constitutive levels of most antioxidant defenses	
Freshwater turtle *Trachemys scripta elegans*	Submergence anoxia[3]
Preparation for oxidative stress	
Carp *Cyprinus carpio*	Severe hypoxia[4]
Goldfish *Carassius auratus*	Anoxia exposure[5]
Leopard frog *Rana pipiens*	Anoxia[6] and severe dehydration[7]
Wood frog *Rana sylvatica*	Freezing exposure[8]
Garter snake *Thamnophis sirtalis parietalis*	Anoxia[9,10] and freezing[10,11]
Marine gastropod *Littorina littorea*	Anoxia exposure[12]
Land snails *Otala lactea* and *Helix aspersa*	Estivation[13,14]
Tolerance to oxidative stress	
Desert toad *Scaphiopus couchii*	Estivation (incl. dehydration of organs)[15]

[1]Reischl, (1986). [2]Joanisse and Storey (1998). [3]Willmore and Storey (1997a,b). [4]Data from Vig and Nemcsok (1989). [5]Lushchak et al. (2001). [6,7]Hermes-Lima and Storey (1996, 1998). [8]Joanisse and Storey (1996). [9]Hermes-Lima and Storey (1992). [10]Hermes-Lima and Storey (1993a). [11]Hermes-Lima and Storey (1993b). [12]Pannunzio and Storey (1998). [13]Hermes-Lima and Storey (1995a,b). [14]Increased Se-GPX activity in hepatopancreas of *H. aspersa* after 20 days of estivation at 25°C (Ramos, 1999). [15]Grundy and Storey (1998).

peroxidation was a significant event during reoxygenation in goldfish brain and liver and in hepatopancreas of awakening *O. lactea*. The modulation of antioxidant capacity under conditions of low oxygen availability or delivery seems to limit lipid peroxidation to tolerable levels during the recovery of full oxidative metabolism. Maintenance of the activity of antioxidant enzymes and levels of GSH were also seen during exposure to anoxia (15°C, 24 h; controls at 15°C) in two overwintering insect larvae, the freeze-tolerant fly *Eurosta solidaginis* and the freeze-intolerant moth *Epiblema scudderiana* and in response to freezing (–14°C, 24 h; controls at 3.5°C) in *E. solidaginis* (Joanisse and Storey, 1998). In addition, no peroxidative damage (TBARS and FOX-reactive lipid hydroperoxides) was observed in these insects in response to these stresses.

The lack of increased levels of lipid peroxidation in post-anoxic leopard frogs, red-eared turtles and insects (Hermes-Lima and Storey, 1996; Willmore and Storey, 1997a; Joanisse and Storey, 1998) as well as thawing or rehydrating frogs (Joanisse and Storey, 1996; Hermes-Lima and Storey, 1998) might suggest that other mechanisms could be also be functioning for the prevention of oxidative stress. For example, Barja et al. (1994) suggested that stringent control of the mitochondrial generation of oxygen radicals may limit oxidative stress.

The molecular basis of the modulation of antioxidant capacity during hypoxia (e.g. freezing, severe dehydration or estivation) or anoxia in stress-tolerant species is still a puzzle to be solved. Mitochondrial generation of oxyradicals under complete anoxia or hypoxia must be either shut down or much decreased due to the unavailability of oxygen. Moreover, anoxia/hypoxia shifts the redox state of the cell to the highly reduced side and, since antioxidant enzyme synthesis typically responds to the redox state of the cell, this should generally favor a decrease in the activity of antioxidant enzymes. This is what happens in mammalian organs under either acute ischemia or hypoxia (Shlafer et al., 1987; Kirshenbaum and Singal, 1992; Singh et al., 1993) or prolonged hypoxia exposure. For example, Costa (1990) observed decreased activities of hepatic antioxidant enzymes (catalase, SOD, Se-GPX) in rats subjected to severe hypoxia for 35 days. This reduction in antioxidant defenses is one of the reasons why mammalian organs are highly susceptible to oxidative stress when oxygen is reintroduced.

Another relevant point is that protein synthesis is a very costly process. For example, it accounts for 85% of the energy consumption in rainbow trout hepatocytes (Pannevis and Houlihan, 1992). Moreover, protein synthesis in Crucian carp is depressed by 95% in liver and 52–56% in red and white muscle during anoxia (Smith et al., 1996) whereas hepatocytes from anoxic-tolerant turtles suppress protein synthesis by 92% during anoxia (Land et al., 1993). Assuming that similar trends take place in other stress-tolerant species, the increase and/or maintenance of antioxidant enzyme activities (and GSH levels) could have high energy costs during the hypometabolic conditions of estivation and anoxia/hypoxia. Thus, only key proteins, such as certain antioxidant enzymes, would be continuously synthesized during anoxia.

It is possible that the molecular mechanisms involved in oxygen sensing and the associated transduction pathways, which regulate intermediary metabolism during anoxia/hypoxia (Hochachka et al., 1996, 1997), might also be involved in the activation and/or maintenance of antioxidant enzymes and GSH levels in response to anoxia/hypoxia signals in stress-tolerant species. These mechanisms might include erythropoietin-like heme proteins and transcription factors such as Fos, AP-1, NF-kappa-B and hypoxia-inducible factor-1 (HIF-1) (Hochachka et al., 1996, 1997; Walton et al., 1999; Zhu and Bunn, 1999; Semenza, 2000). Measurement of the concentrations and stability of mRNAs for certain antioxidant enzymes, and the role of transcription factors in estivation and anoxia/hypoxia tolerance is an essential step for future research.

Finally, activation and/or maintenance of antioxidant defenses in certain invertebrates and non-mammalian vertebrates is an integral part of the biochemical adaptive mechanism for tolerance to anoxia/hypoxia and reoxygenation. Further research is needed to elucidate the molecular mechanisms involved in oxygen sensing and regulation of oxyradical detoxification metabolism in these animal species. These studies may contribute to the development of biomedical strategies for the treatment and/or prevention of post-ischemic stress in humans.

Acknowledgements

This work was supported by grants from CNPq, PRONEX-97, FAP-DF (Brazil) and IFS (Sweden) to M. Hermes-Lima and the N.S.E.R.C. (Canada) to K.B. Storey. The authors thank G. Ramos-Vasconcelos (Universidade de Brasília, Brazil) for the artwork shown in Fig. 20.1. This manuscript is dedicated to Professor Etelvino J.H. Bechara (Biochemistry Department, Universidade de São Paulo, Brazil) for his tireless contribution on free radical research in Brazil and training of many "radical" graduate students (including M.H.-L.).

References

Ahmad, S. (1995). Antioxidant mechanisms of enzymes and proteins. In: Oxidative Stress and Antioxidant Defenses in Biology (Ahmad, S., Ed.), pp. 238–272. Chapman & Hall, New York.

Barja de Quiroga, G. (1992). Brown fat thermogenesis and exercise: two examples of physiological oxidative stress? Free Rad. Biol. Med. 13, 325–340.

Barja, G., Cadenas, S., Pérez-Campo, R. and López-Torres, M. (1994). Low mitochondrial free-radical production per unit O_2 consumption can explain the simultaneous presence of high longevity and high aerobic metabolic rate in birds. Free Radical Res. 21, 317–327.

Barnhart, M.C. (1986). Respiratory gas tensions and gas exchange in active and dormant land snails *Otala lactea*. Physiol. Zool 59, 733–745.

Beckman, K.B. and Ames, B.N. (1998). The free radical theory of aging matures. Physiol. Rev., 78, 547–581.

Bilzer, M., Paumgartner, G. and Gerbes, A.L. (1999). Glutathione protects the rat liver against reperfusion injury after hypothermic preservation. Gastroenterology 117, 200–210.

Bindoli, A. (1988). Lipid peroxidation in mitochondria. Free Rad. Biol. Med. 5, 247–261.

Bird, R.P. and Draper, H.H. (1984) Comparative studies on different methods of malonaldehyde determination. Meth. Enzymol. 105, 229–305.

Bishop, T. and Brand, M.D. (2000). Processes contributing to metabolic depression in hepatopancreas cells from the snail *Helix aspersa*. J. Exp. Biol. 203, 3603–3612.

Bolli, R. and Marban, E. (1999). Molecular and cellular mechanisms of myocardial stunning. Physiol. Rev. 79, 609–634.

Brooks, S.P.J. and Storey, K.B. (1997). Glycolytic controls in estivation and anoxia: a comparison of metabolic arrest in land and marine molluscs. Comp. Biochem. Physiol. A 118, 1103–1114.

Bunn, H.F. and Poyton, R.O. (1996). Oxygen sensing and molecular adaptation to hypoxia. Physiol. Rev. 76, 839–885.

Buzadzic, B., Spasic, M., Saicic, Z.S., Radojicic, R., Petrovic, V.M. and Halliwell, B. (1990). Antioxidant defenses in the ground squirrel *Citellus citellus*. 2. The effect of hibernation. Free Rad. Biol. Med. 9, 407–413.

Cadenas, E. (1995). Mechanism of oxygen activation and reactive oxygen species detoxification. In: Oxidative Stress and Antioxidant Defenses in Biology (Ahmad, S., Ed.), pp. 1–61. Chapman & Hall, New York.

Campos, E.G., Hermes-Lima, M., Smith, J.M. and Prichard, R.K. (1999). Characterization of *Fasciola hepatica* cytochrome c peroxidase as an enzyme with potential antioxidant activity *in vitro*. Int. J. Parasitol. 29, 655–662.

Castilho, R.F., Meinicke, A.R., Vercesi, A.E. and Hermes-Lima, M. (1999). Role of Fe(III) in Fe(II)citrate-mediated peroxidation of mitochondrial membrane lipids. Mol. Cell. Biochem. 196, 163–168.

Churchill, T.A. and Storey K.B. (1992). Freezing survival of the garter *snake Thamnophis sirtalis parietalis*. Can. J. Zool. 70, 99–105.

Churchill, T.A. and Storey K.B. (1993). Dehydration tolerance in wood frogs: a new perspective on development of amphibian freeze tolerance. Am. J. Physiol. 265, R1324–R1332.

Churchill, T.A. and Storey K.B. (1995). Metabolic effects of dehydration on an aquatic frog, *Rana pipiens*. J. Exp. Biol. 198, 147–154.

Cino, M. and Del Maestro, R.F. (1989). Generation of hydrogen peroxide by brain mitochondria: the effect of reoxygenation following postdecapitative ischemia. Arch. Biochem. Biophys. 269, 623–638.

Cordis, G.A., Maulik, G., Bagchi, D., Riedel, W. and Das, D.K. (1998) Detection of oxidative DNA damage to ischemic reperfused rat hearts by 8-hydroxydeoxyguanosine formation. J. Mol. Cell. Cardiol. 30, 1939–1944.

Cordis, G.A., Maulik, N. and Das, D.K. (1995). Detection of oxidative stress in heart by estimating the dinitrophenyl-hydrazine derivative of malonaldehyde. J. Mol. Cell. Cardiol. 27, 1645–1653.

Corongiu, F.P. and Milia, A.(1983) An improved method for determining diene conjugation in autoxidized polyunsaturated fatty acids. Chem. Biol. Interactions 44, 289–297.

Costa, L.E. (1990). Hepatic cytochrome P-450 in rats submitted to chronic hypobaric hypoxia. Am. J. Physiol. 259, C654–C659.

Costanzo, J.P. (1989) A physiological basis of prolonged submergence in hibernating garter snakes *Thamnophis sirtalis*: evidence for an energy-sparing adaptation. Physiol. Zool. 52, 580–592.

Costanzo, J.P., Claussen, D.L. and Lee, R.E. (1988) Natural freeze tolerance in a reptile. Cryo-Lett. 9, 380–385.

Costanzo, J.P., Lee, R.E., DeVries, A.L., Wang, T. and Layne, J.R. (1995). Survival mechanisms of vertebrate ectotherms at subfreezing temperatures: applications in cryomedicine. FASEB J. 9, 351–358.

Crawford, D.R., Suzuki, T. and Davies, K.J.A. (2000). Redox regulation of gene expression. In: Antioxidant and redox regulation of genes (Sen, C.K., Sies, H. and Baeuerle, P.A. Eds.), pp. 21–45. Academic Press, San Diego.

de Zwart, L.L., Meerman, J.H.N., Commandeur, J.N.M. and Vermeulen, N.E.P. (1999). Biomarkers of free radical damage. Applications in experimental animals and in humans. Free Radic. Biol. Med. 26, 202–226.

Dhalla, A.K. and Singal, P.K. (1994). Antioxidant changes in hypertrophied and failing guinea pig hearts. Am. J. Physiol. 266, H1280–H1285.

Dykens, J.A. and Shick, M. (1988). Relevance of purine catabolism to hypoxia and recovery in euryoxic and stenoxic marine invertebrates, particularly bivalve molluscs. Comp. Biochem. Physiol. 91C, 35–41.

Frank, L., Price, L.T. and Whitney, P.L. (1996). Possible mechanism for late gestational development of the antioxidant enzymes in the fetal rat lung. Biol. Neonate 70, 116–127.

Fridovich, I. (1998). Oxygen toxicity: a radical explanation. J. Exp. Biol. 201, 1203–1209.

Gatten Jr, R.E. (1987). Activity of anuran amphibians: tolerance to dehydration. Physiol. Zool. 60, 576–585.

Greene, E.L. and Paller, M.S. (1992) Xanthine oxidase produces O_2^- in posthypoxic injury of renal epithelial cells. Am. J. Physiol. 263, F251–F255.

Grundy, J.E. and Storey, K.B. (1998). Antioxidant defenses and lipid peroxidation damage in estivating toads, *Scaphiopus couchii.* J. Comp. Physiol. B 169, 132–142.

Habig, W.H. and Jakoby, W.B. (1981). Glutathione S-transferases (rat and human). Meth. Enzymol. 77, 218–231.

Halliwell, B., Gutteridge, J.M.C. and Cross, C.E. (1992). Free radicals, antioxidants, and human disease: Where are we now? J. Lab. Clin. Med. 119, 598–620.

Hardmeier, R., Hoeger, H., Fang-Kircher, S., Khoschsorur, A. and Lubec, G. (1997). Transcription and activity of antioxidant enzymes after ionizing irradiation in radiation-resistant and radiation-sensitive mice. Proc. Natl. Acad. Sci. USA 94, 7572–7576.

Hermes-Lima, M. and Storey, K.B. (1992). Antioxidants as part of the machinery for the natural tolerance to cold anoxia in the garter snake (29th Annual Meeting of the Society for Cryobiology, Ithaca, New York, abstract 160). Cryobiology 29, 760.

Hermes-Lima, M. and Storey, K.B. (1993a). Role of antioxidants in the tolerance of freezing and anoxia by garter snakes. Am. J. Physiol. 265, R646–R652.

Hermes-Lima, M. and Storey, K.B. (1993b). In vitro oxidative inactivation of glutathione S-transferase from a freeze tolerant reptile. Mol. Cell. Biochem 124, 149–158.

Hermes-Lima, M. and Storey, K.B. (1995a). Antioxidant defenses and metabolic depression in a pulmonate land snail. Am. J. Physiol. 268, R1386–R1393.

Hermes-Lima, M. and Storey, K.B. (1995b). Xanthine oxidase and xanthine dehydrogenase from an estivating land snail. Z. Naturforschung. 50 C, 685–694.

Hermes-Lima, M. and Storey, K.B. (1996). Relationship between anoxia exposure and antioxidant status of the frog *Rana pipiens*. Am. J. Physiol. 271, R918–R925.

Hermes-Lima, M. and Storey, K.B. (1998). Role of antioxidant defenses in the tolerance of severe dehydration by anurans. The case of the leopard frog *Rana pipiens*. Mol. Cell. Biochem. 189, 79–89.

Hermes-Lima, M., Castilho, R.F., Meinicke, A.R. and Vercesi, A.E. (1995a). Characteristics of Fe(II)ATP complex-induced damage to the rat liver mitochondrial membrane. Mol. Cell. Biochem. 145, 53–60.

Hermes-Lima, M., Willmore, W.G. and Storey, K.B. (1995b). Quantification of lipid peroxidation in tissues based on the Fe(III)xylenol orange complex formation. Free Radic. Biol. Med. 19, 271–280.

Hermes-Lima, M., Storey, J.M. and Storey, K.B. (1998). Antioxidant defenses and metabolic depression. The hypothesis of preparation for oxidative stress in land snails. Comp. Biochem. Physiol. B 120, 437–448.

Herreid, C.F. (1977). Metabolism of land snails (*Otala lactea*) during dormancy, arousal and activity. Comp. Biochem. Physiol. A 56, 211–215.

Hillman, S. (1978). The roles of oxygen delivery and electrolyte levels in the dehydrational death of *Xenopus laevis*. J. Comp. Physiol. 128, 169–175.

Hillman, S. (1980). Physiological correlates of differential dehydration tolerance in anuran amphibians. Copeia 1980, 125–129.

Hillman, S (1987). Dehydrational effects on cardiovascular and metabolic capacity in two amphibians. Physiol. Zool. 60, 608–613.

Hochachka, P.W., Buck, L.T., Doll, C.J. and Land, S.C. (1996). Unifying theory of hypoxia tolerance: molecular/metabolic defense and rescue mechanisms for surviving oxygen lack. Proc. Natl. Acad. Sci. USA 93, 9493–9498.

Hochachka, P.W., Land, S.C. and Buck, L.T. (1997). Oxygen sensing and signal transduction in metabolic defense against hypoxia: lessons from vertebrate facultative anaerobes. Comp. Biochem. Physiol. A 118, 23–29.

Inoue, M., Nishikawa, M., Sato, E.F., Ah-Mee, P., Kashiba, M., Takehara, Y. and Utsumi, K. (1999). Cross-talk of NO, superoxide and molecular oxygen, a majesty of aerobic life. Free Radic. Res. 31, 251–260.

Joanisse, D.R. and Storey K.B. (1996). Oxidative damage and antioxidants in *Rana sylvatica*, the freeze tolerant wood frog. Am. J. Physiol. 271, R545–R553.

Joanisse, D.R. and Storey, K.B. (1998). Oxidative stress and antioxidants in stress and recovery of cold-hardy insects. Insect Biochem. Mol. Biol. 28, 23–30.

Jones, R.M. (1980). Metabolic consequences of accelerated urea synthesis during seasonal dormancy of spadefoot toads, *Scaphiopus couchii* and *Scaphiopus multiplicatus*. J. Exp. Zool. 212, 255–267.

Kakkar, R., Kalra, J., Mantha, S.V. and Prasad, K. (1995). Lipid peroxidation and activity of antioxidant enzymes in diabetic rats. Mol. Cell. Biochem. 151, 113–119.

Kelly, D.A. and Storey, K.B. (1988). Organ-specific control of glycolysis in anoxic turtles. Am. J. Physiol. 255, R774–R779.

Kirshenbaum, L.A. and Singal, P.K (1992). Changes in antioxidant enzymes in isolated cardiac myocytes subjected to hypoxia-reoxygenation. Lab. Invest. 67, 796–803.

Land, S.C., Buck, L.T. and Hochachka, P.W. (1993). Response of protein synthesis to anoxia and recovery in anoxia-tolerant hepatocytes. Am. J. Physiol. 256, R41–R48.

Lipton, P. (1999). Ischemic cell death in brain neurons. Physiol. Rev. 79, 1431–1568.

Lushchak, V.I., Lushchak, L.P., Mota, A.A. and Hermes-Lima, M. (2001). Oxidative stress and antioxidant defenses in goldfish *Carassius auratus* during anoxia and reoxygenation. Am. J. Physiol. 280, R100–107.

Lutz, P.L. and Nilsson, G.E. (1997). Contrasting strategies for anoxic brain survival. Glycolysis up or down. J. Exp. Biol. 200, 411–419.

Lutz, P.L. and Storey, K.B. (1997). Adaptations to variations in oxygen tension by vertebrates and invertebrates. In: Handbook of Physiology, Section 13: Comparative Physiology (Dantzler, W.H., Ed.), Vol. 2, pp. 1479–1522. Oxford University Press, Oxford.

McAnulty, J.F. and Huang, X.Q. (1997). The efficacy of antioxidants administered during low temperature storage of warm ischemic kidney tissue slices. Cryobiology 34, 406–415.

McClanahan, L. (1967). Adaptations of the spadefoot toad, *Scaphiopus couchii*, to desert environments. Comp. Biochem. Physiol. 20, 73–79.

McCord, J.M. (1985). Oxygen-derived free radicals in postischemic tissue injury. N. Engl. J. Med.312, 159–163.

Muusze, B., Marcon, J., van den Thillart, G. and Almeida-Val, V. (1998). Hypoxia tolerance of Amazon fish respirometry and energy metabolism of the cichlid *Astronotus ocellatus*. Comp. Biochem. Physiol. A. 120, 151–156.

Pannevis, M.C. and Houlihan, D.F. (1992). The energetic cost of protein synthesis in isolated hepatocytes of rainbow trout (*Oncorhyncus mykiss*). J. Comp. Physiol. B 162, 393–400.

Pannunzio, T.M. and Storey, K.B. (1998). Antioxidant defenses and lipid peroxidation during anoxia stress and aerobic recovery in the marine gastropod *Littorina littorea*. J. Exp. Marine Biol. Ecol. 221, 277–292.

Perez-Campo, R., Lopez-Torres, M., Rojas, C., Cadenas, S. and Barga, G. (1993). A comparative study of free radicals in vertebrates. I. Antioxidant enzymes. Comp. Biochem. Physiol. B 105, 749–755.

Pinder, A.W., Storey, K.B. and Ultsch, G.R. (1992). Estivation and hibernation. In: Environmental Biology of the Amphibia. (Feder, M.E. and Burggren, W.W., Eds.), pp. 250–274. University of Chicago Press, Chicago.

Ramos, G.R. (1999). Estresse oxidativo e hipometabolismo em gastrópodes *Helix aspersa*. M.Sc. Thesis, Universidade de Brasilia, Brazil.

Reischl, E. (1986). High sulphydryl content in turtle erythrocytes: Is there a relation with resistance to hypoxia? Comp. Biochem. Physiol. 85 B, 723–726.

Reischl, E. (1989). Sulfhydryl-rich hemoglobins in reptiles: a defence against reactive oxygen species? In: Non-mammalian Animal Models for Biomedical Research. (Woodhead, A.V., Ed.), pp 309–318. CRC Press, Boca Raton.

Roy, R.S. and McCord, J.M. (1983). Superoxide and ischemia: conversion of xanthine dehydrogenase to xanthine oxidase. In: Oxy Radicals and their Scavenger Systems (Greenwald, R.A. and Cohen, G. Eds.), Vol 2, pp 145–153. Elsevier, New York.

Ruuge, E.K., Ledenev, A.N., Lakomkin, V.L., Konstantinov, A.A. and Ksenzenko, M.Y. (1991). Free radical metabolites in myocardium during ischemia and reperfusion. Am. J. Physiol. Suppl (Oct.) 261, 81–86.

Semenza, G.L. (2000) HIF-1: mediator of physiological and pathophysiological responses to hypoxia. J. Appl. Physiol. 88, 1474–1480.

Seymour, R.S. (1973). Energy metabolism of dormant spadefoot toads (*Scaphiopus*). Copeia 1973, 435–445.

Shlafer, M., Myer, C.L. and Adkins, S. (1987). Mitochondrial hydrogen peroxide generation and activities of glutathione peroxide and superoxide dismutase following global ischemia. J. Mol. Cell. Cardiol. 19, 1195–1206.

Shoubridge, E.A. and Hochachka, P.W. (1983). The integration and control of metabolism in the anoxic goldfish. Mol. Physiol. 4, 165–195.

Shull, S., Heintz, N.H., Periasamy, M., Manohar, M., Janssen, Y.M., Marsh, J.P. and Mossman, B.T. (1991). Differential regulation of antioxidant enzymes in response to oxidants. J. Biol. Chem. 266, 24398–24403.

Singh, I., Gulati, S., Orak, J.K. and Singh, A.K. (1993). Expression of antioxidant enzymes in rat kidney during ischemia-reperfusion injury. Mol. Cell. Biochem. 125, 97–104.

Smith, R.W., Houlihan, D.F., Nilsson, G.E. and Brechin, J.G. (1996). Tissue-specific changes in protein synthesis rates *in vivo* during anoxia in crucian carp. Am. J. Physiol. 271, R897–R904.

Starke, D.W., Chen, Y., Bapna, C.P., Lesnefsky, E.J. and Mieyal, J.J. (1997). Sensitivity of protein sulfhydryl repair enzymes to oxidative stress. Free Rad. Biol. Med. 23, 373–384.

Storey, K.B. (1987). Organ-specific metabolism during freezing and thawing in a freeze-tolerant frog. Am. J. Physiol. 253, R292–R297.

Storey, K.B. (1993). Molecular mechanisms of metabolic arrest in mollusks. In: Surviving Hypoxia: Mechanisms of Control and Adaptation (Hochachka, P.W., Lutz, P.L., Sick, T.J., Rosenthal, M. and van den Thillart, G., Eds.), pp. 253–269. CRC Press, Boca Raton.

Storey, K.B. (1996a). Metabolic adaptations supporting anoxia tolerance in reptiles: recent advances. Comp. Biochem. Physiol. B 113, 23–35.

Storey, K.B. (1996b). Oxidative stress: animal adaptations in nature. Braz. J. Med. Biol. Res. 29, 1715–1733.

Storey, K.B. (1999). Stress-induced gene expression in freeze tolerant and anoxia tolerant vertebrates. In: Environmental Stress and Gene Regulation (Storey, K.B, Ed.), pp. 1–23. BIOS Scientific Publishers, Oxford.

Storey, K.B. (2001). Turning down the fires of life: metabolic regulation of hibernation and estivation. In: Molecular Mechanisms of Metabolic Arrest (Storey, K.B., Ed.), pp. 1–21, BIOS Scientific Publishers, Oxford.

Storey, K.B. and Storey, J.M. (1986) Freeze tolerant frogs: cryoprotectants and tissue metabolism during freeze/thaw cycles. Can. J. Zool. 64, 49–56.

Storey, K.B. and Storey, J.M. (1988). Freeze tolerance in animals. Physiol. Rev. 68, 27–84.

Storey, K.B. and Storey, J.M. (1990). Facultative metabolic rate depression: molecular regulation and biochemical adaptations in anaerobiosis, hibernation, and estivation. Quart. Rev. Biol. 65, 145–174.

Storey, K.B. and Storey, J.M. (1992). Natural freeze tolerance in ectothermic vertebrates. Ann. Rev. Physiol. 54, 619–637.

Storey, K.B. and Storey, J.M. (2001). Signal transduction and gene expression in the regulation of natural freezing survival. In: Cell and Molecular Responses to Stress. (Storey, K.B. and Storey, J.M., Eds.), Vol. 2, pp. 1–19. Elsevier, Amsterdam.

Storey, K.B., Mosser, D.D., Douglas, D.N., Grundy, J.E. and Storey, J.M. (1996). Biochemistry below 0°C: Nature's frozen vertebrates. Braz. J. Med. Biol. Res. 29, 283–307.

Terada, L.S., Rubinstein, J.D., Lesnefsky, E.J., Horwitz, L.D., Leff, J.A. and Repine, J.E. (1991). Existence and participation of xanthine oxidase in reperfusion injury of ischemic rabbit myocardium. Am. J. Physiol. 260, H805–H810.

Turrens J.F., Freeman, B.A., Levitt, J.G. and Crapo J.D. (1982). The effect of hyperoxia on superoxide production by lung submitochondrial particles. Arch. Biochem. Biophys. 217, 401–410.

Ultsch, G.R. (1989). Ecology and physiology of hibernation and overwintering among freshwater fishes, turtles, and snakes. Biol. Rev. 64, 435–516.

van den Thillart, G. (1982). Adaptations of fish energy metabolism to hypoxia and anoxia. Mol. Physiol. 8, 393–409.

Vig, E. and Nemcsok, J. (1989). The effect of hypoxia and paraquat on the superoxide dismutase activity in different organs of carp, *Cyprinus carpio* L. J. Fish Biol. 35, 23–25.

Walton, M., Connor, B., Lawlor, P., Young, D., Sirimanne, E., Gluckman, P., Cole, G. and Dragunow, M. (1999). Neuronal death and survival in two models of hypoxic-ischemic brain damage. Brain Res. Brain. Res. Rev. 29, 137–168.

Wang, P., Chen, H., Qin, H., Sankarapandi, S., Becher, M.W., Wong, P.C. and Zweier, J.L. (1998). Overexpression of human copper,zinc-superoxide dismutase (SOD1) prevents postischemic injury. Proc. Natl. Acad. Sci. USA 95, 4556–4560.

Weisbrot-Lefkowitz, M., Reuhl, K., Perry, B., Cahn, P.H., Inouye, M. and Mirochnitchenko, O. (1998). Overexpression of human glutathione peroxidase protects transgenic mice against focal cerebral ischemia/reperfusion damage. Molec. Brain Res. 53, 333–338.

Willmore, W.G. and Storey, K.B. (1997a). Antioxidant systems and anoxia tolerance in a freshwater turtle, *Trachemys scripta elegans*. Mol. Cell. Biochem. 170, 177–185.

Willmore, W.G. and Storey, K.B. (1997b). Glutathione systems and anoxia tolerance in turtles. Am. J. Physiol. 273, R219–R225.

Zhu, H. and Bunn, H.F. (1999). Oxygen sensing and signaling: impact on the regulation of physiologically important genes. Respir. Physiol. 115, 239–247.

Index